CONSTRUCTION WATERPROOFING HANDBOOK

CONSTRUCTION WATERPROOFING HANDBOOK

Michael T. Kubal

McGraw-Hill

New York San Francisco Washington, D.C. Auckland Bogotá
Caracas Lisbon London Madrid Mexico City Milan
Montreal New Delhi Dan Juan Singapore
Sydney Tokyo Toronto

Library of Congress Cataloging-in-Publication Data

Kubal, Michael T.
Construction waterproofing handbook / Michael T. Kubal
p. cm.
ISBN 0-07-135162-0
1. Buildings—Protection. 2. Waterproofing. 3. Dampness in buildings. I. Title.
TH9031.K78 1999
693.8'92—dc21 99-0544043
CIP

McGraw-Hill

*A Division of The **McGraw-Hill** Companies*

7 8 9 BKM BKM 0 9 8 7 6

ISBN 0-07-135162-0

The sponsoring editor of this book was Larry Hager. The editing supervisor was Sally Glover, and the production supervisor was Pamela Pelton. It was set in the HB1 design in Times Roman. It was composed by Paul Scozzari of McGraw-Hill's Professional Publishing Book Group in Hightstown, N.J.

CONTENTS

PREFACE

Since the printing several years ago of the original edition of this book, *Waterproofing the Building Envelope*, the term building envelope has become commonly used in the architectural, engineering, and construction professions. In fact, a whole new profession, "building envelope consultants," has been created.

This new edition, *Construction Waterproofing Handbook*, continues the pursuit for standards of excellence within the design, construction, and specifically the waterproofing profession. While this text moves beyond discussing only waterproofing related to the building envelope, the importance of this topic cannot be understated. Only when the industry recognizes that all exterior building skin components must be made watertight into a single cohesive unit will the continual problems regarding leakage into interior occupied spaces go away.

The waterproofing contractor is *not* the only contractor responsible for ensuring watertight effectiveness of the building facade. The mechanical equipment, electric signage, carpentry work, and each and every item on the exterior skin of a structure must be waterproof and adequately transitioned into adjacent components.

The Preface of the original edition of this book called upon the industry to acknowledge these performance standards in order to resolve water-intrusion problems at the job site rather than in the courtroom. The industry has since made strides towards making these improvements, but—recognizing that waterproofing remains a common frequent cause of construction complaints and claims—much work lies ahead of us.

The industry is beginning to recognize that successful waterproofing of any structure goes well beyond Division 7000, and this book remains dedicated to achieving that goal.

CHAPTER 1
WATERPROOFING PRINCIPLES—THE BUILDING ENVELOPE

INTRODUCTION

Since our beginnings, we have sought shelter as protection from the elements. Yet, even today, after centuries of technological advances in materials and construction techniques, we are still confronted by nature's elements contaminating our constructed shelters. This is not due to a lack of effective waterproofing systems and products. Waterproofing problems continue to plague us due to the increasing complexity of shelter construction, a disregard for the most basic waterproofing principles, and an inability to coordinate interfacing between the multitude of construction systems involved in a single building.

Adequately controlling groundwater, rainwater, and surface water will prevent damage and avoid unnecessary repairs to building envelopes. In fact, water is the most destructive weathering element of concrete, masonry, and natural stone structures. Water continues to damage or completely destroy more buildings and structures than war or natural disasters.

Waterproofing techniques preserve a structure's integrity and usefulness through an understanding of natural forces and their effect during life-cycling. Waterproofing also involves choosing proper designs and materials to counter the detrimental effects of these natural forces.

Site construction requires combining numerous building trades and systems into a building skin to prevent water infiltration. Our inability to tie together these various components effectively causes the majority of water and weather intrusion problems. Actual experience has shown that the majority of water intrusion problems occur within a relatively, minute portion of a building's total exposed surface area. An inability to control installation and details, linking various building facade components that form the building's exterior skin, create the multitude of problems confronting the design and construction industry.

While individual waterproofing materials and systems continue to improve, no one provides attention to improving the necessary and often critical detailing that is required to transition from one building facade component to the next. Furthermore, we seem to move further away from the superior results achieved by applying basic waterproofing principles, such as maximizing roof slopes, to achieve desired aesthetics values instead. There is no reason that aesthetics cannot be fully integrated with sound waterproofing guidelines.

THE BUILDING ENVELOPE

The building envelope is equivalent to the skin of a building. Essentially a structure must be *enveloped* from top to bottom to prevent intrusion from nature's elements into interior spaces and protect the structural components from weathering and deterioration. Envelopes complete numerous functions in a building's life cycle, including:

- Preventing water infiltration
- Controlling water vapor transmission
- Controlling heat and air flow, into and out of interior spaces
- Providing a shield against ultraviolet rays and excessive sunlight
- Limiting noise infiltration
- Providing structural integrity for the façade components
- Providing necessary aesthetics

While the main purpose of any building envelope is to provide protection from all elements, including wind, cold, heat, and rain, this book concentrates on the controlling of water and leakage for all construction activities including the building envelope. Making a building envelope waterproof also provides protection against vapor transmission, and serves to prevent the unnecessary passage of wind and air into or out of a building, assisting in the controlling of heating and cooling requirements. Before considering each specific type of waterproofing system (e.g., below-grade), some basic concepts of waterproofing and how they affect the performance of a building envelope are important to understand.

INTRODUCTION TO WATERPROOFING AND ENVELOPE DESIGN

Waterproofing is the combination of materials or systems that prevents water intrusion into structural elements of a building or its finished spaces. Basic waterproofing and envelope design incorporates three steps to ensure a watertight and environmentally sound interior:

1. Understanding water sources likely to be encountered.
2. Designing systems to prevent leakage from these sources.
3. Finalizing the design by properly detailing each individual envelope component into adjacent components.

Water Sources

Water likely to penetrate building envelopes is most commonly from rainwater on above-grade components and groundwater intrusion below-grade. Other sources should also be considered as appropriate such as melting snow, overspray from cooling towers, landscaping sprinklers, and redirected water from such sources as downspouts and gutters.

The presence of any of these water sources alone, though, will not cause leakage; for leakage to occur, three conditions must be present. First, water in any of its forms must be present;

second, the water must be moved along by some type of force including wind and gravity for above-grade envelope components and hydrostatic pressure or capillary action for below-grade components. Finally and most importantly, there must be a breach (hole, break, or some type of opening) in the envelope to facilitate the entry of water into the protected spaces.

Available water is moved into the interior of a structure by numerous forces that include

- Natural gravity
- Surface tension
- Wind/air currents
- Capillary action
- Hydrostatic pressure

The first three are typically encountered on above-grade portions of the envelope, while the last two are recognized at-grade or below-grade areas of buildings or structures. For above-grade envelope components, horizontal areas are very prone to gravitational forces, and should never be designed completely flat. Water must be drained away from the structure as quickly as possible and this includes walkways, balconies, and other necessary "flat" areas. In building components such as these, a minimum 1/4 inch per foot slope should be incorporated rather than the 1/8 inch that is often used as a standard. The faster the water is directed off the envelope the less chance for leakage.

Consider the teepee, built from materials that are hardly waterproof in themselves; the interior areas remain dry simply due to the design that sheds water off instantaneously. The same is true for canvass tents; the material keeps the occupants dry as long as the water is diverted off the canvass immediately, but use the same material in a horizontal or minimally sloped area and the water will violate the canvas material. Figure 1.1 emphasizes the importance of slope to prevent unnecessary infiltration.

In fact, incorporating adequate slope into the design could prevent many of the common leakage problems that exist today. Simply compare residential roofs that incorporate a slope as high as 45 degrees to commercial roofs that are designed with a minimum 1/8-inch slope. Although the materials used in the commercial application are more costly and typically have superior performance capabilities than asphalt shingles used on residential projects, the commercial roofs continue to have leakage problems at a far greater incident rate than residential roofing.

Surface tension is the momentum that occurs when water being moved by gravity approaches a change in building plane (e.g., face brick to lintel) and clings to the underside of the horizontal surface, continuing with momentum into the building by adhering to the surface through this tension. This situation frequently occurs at mortar joints, where water is draw into a structure by this tension force as shown in Fig. 1.2.

This is the reason for drip edges and flashings to become a standard part of any successfully building envelope. Drip edges and flashings break the surface tension and prevent water from being attracted to the inside of a building by this force. Some common drip edge and flashing details to prevent water tension infiltration are shown in Fig. 1.3.

When wind is present in a rainstorm, envelopes become increasingly suspect to water infiltration. Besides the water being directly driven into envelopes by the wind currents themselves, wind can create sufficient air pressure that creates hydrostatic pressure on the

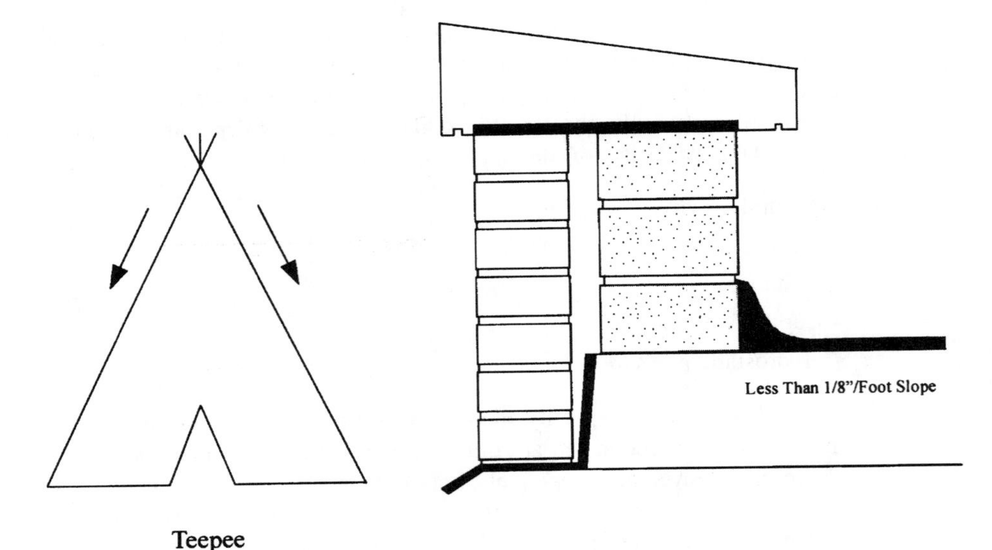

FIGURE 1.1 Sloping of envelope components maximizes drainage of water away from envelope. The flat roof design shown is often the cause for leakage problems simply due to the water stands or "ponds" on the envelope surface.

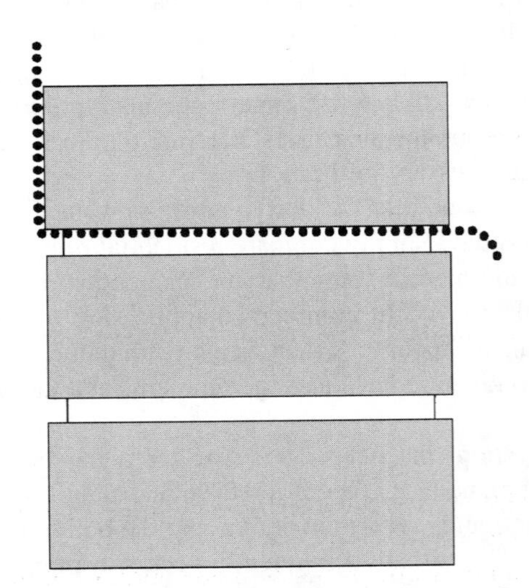

FIGURE 1.2 Surface tension accelerates water infiltration.

facade that can force water upward and over envelope components. Again, flashing is frequently used to prevent this phenomenon from causing water penetration into a structure. This typical detailing is shown in Fig. 1.4.

Capillary action occurs in situations where water is absorbed into an envelope substrate by a wicking action. This situation is most likely to occur with masonry or concrete portions of the envelope at or below-grade levels. These materials have a natural high degree of minute void space within their composition, making them susceptible to capillary water intrusion. These minute voids actually create a capillary suction force that draws water into the substrate when standing water is present. This is similar to the action of a sponge that is laid in water and begins absorbing water. Materials that have large voids or are very porous are not as susceptible to capillary action and in some cases are actually used to prevent this reaction on a building. For example, sand is often used as a fill material below concrete slabs placed directly on grade to prevent

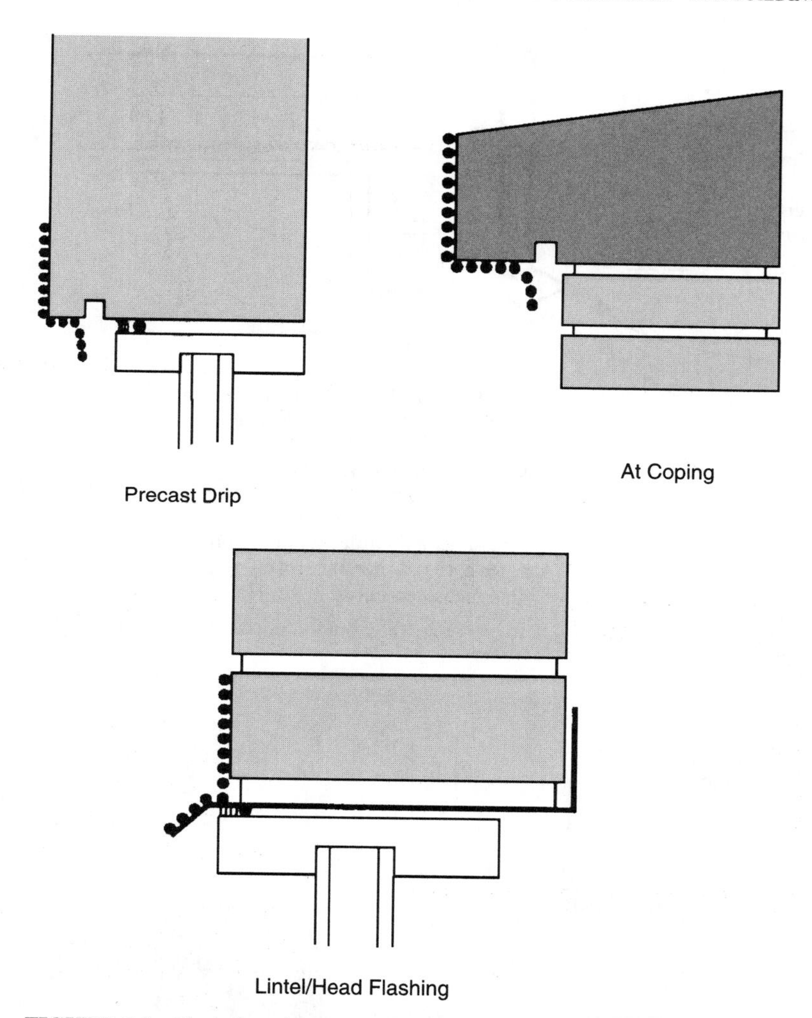

FIGURE 1.3 Typical and common uses of "drip edges" to prevent tension infiltration.

the concrete from drawing water from the soil through a capillary action. Typical methods to prevent capillary action in envelopes are shown in Fig. 1.5.

Hydrostatic pressure most commonly affects below-grade portions of the envelope that are subject to groundwater. Hydrostatic pressure on an envelope is created by the weight of water above that point (e.g., the height of water due to its weight creates pressure on lower areas referred to as *hydrostatic pressure*). This pressure can be significant, particularly in below-grade areas, where the water table is near the surface or rises near the surface during heavy rainfalls. Water under this significant pressure will seek out any failures in the envelope, especially the areas of weakness—the terminations and transitions

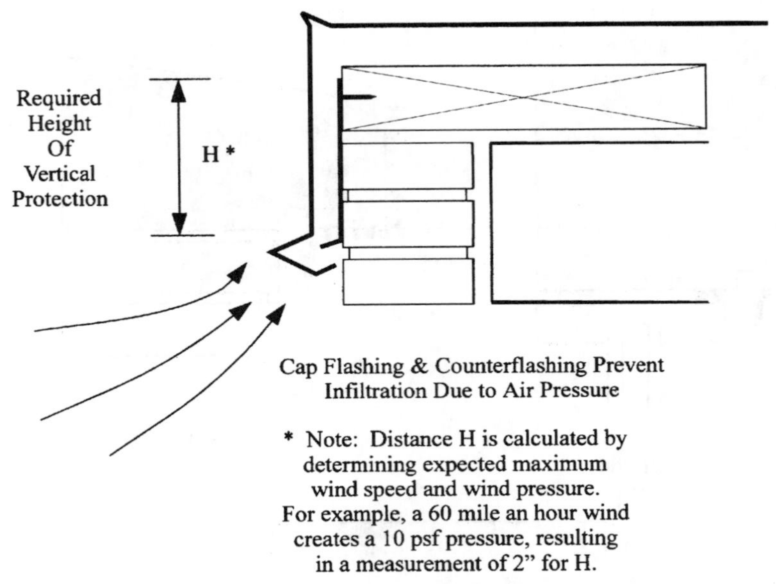

FIGURE 1.4 Flashing used to prevent water under pressure from entering envelope.

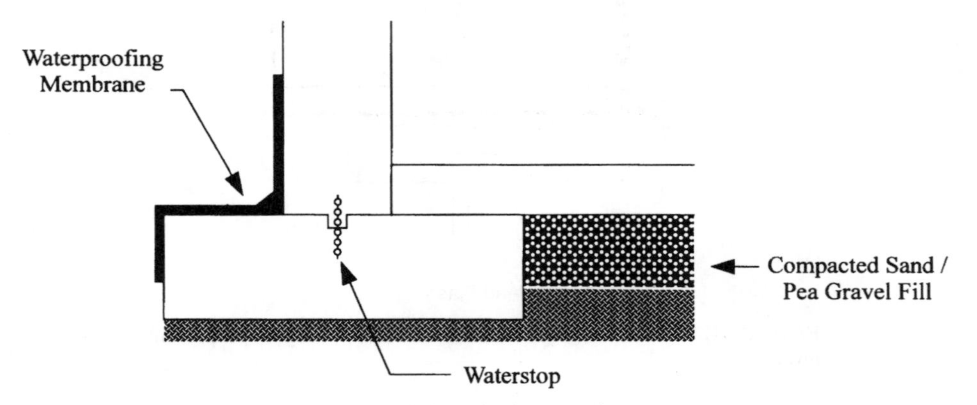

FIGURE 1.5 Preventing capillary water infiltration into envelopes.

between the envelope components. This is why certain envelope substrates used below-grade have to be better protected against water infiltration than those above grade are. For example, concrete above-grade is often only protected with a water repellent, while below-grade the same concrete must be protected with a waterproofing material to prohibit leakage into the structure.

Designing to Prevent Leakage

Once a complete understanding of the potential sources of water and forces that can move this water into an envelope is gained for a particular structure, the design must incorporate effective systems to prevent this intrusion. Expected conditions for a particular geographic area that will affect the above-grade envelope are available from the national weather service at www.nws.noaa.gov. Below-grade water tables are determined by testing actual site conditions.

It should also be understood that substrate water penetration and absorption do not necessarily cause leakage to interior spaces. Water absorption occurs regularly in masonry facades, but the masonry is either large enough to absorb the penetrating water without passing it on to interior finishes or this water is collected and redirected back to the exterior by the use of dampproofing systems. Water penetration also occurs at the microscopic and larger voids in the masonry mortar joints, but again the masonry absorbs it or the water is redirected back out through the dampproofing system.

For definition purposes, water infiltration and leakage are used interchangeably in this book, since each is not an expected outcome in envelope design. All envelope components are designed to prevent leakage or infiltration by one of three systems:

1. Barrier
2. Drainage
3. Diversions

Barrier systems are, as their name implies, effective and complete barriers to water infiltration. They include actual waterproofing systems such as below-grade urethane membranes, and other envelope components such as glass. They completely repel water under all expected conditions including gravity and hydrostatic pressure. Refer to Fig. 1.6.

Drainage systems are envelope components that might permit water absorption and some infiltration through the substrate, but provide a means to collect this water and divert it back out to the exterior before it causes leakage. Examples include masonry walls with

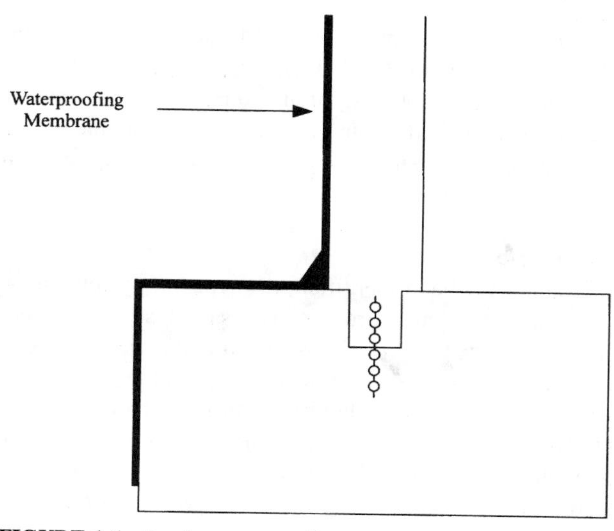

FIGURE 1.6 Barrier waterproofing system.

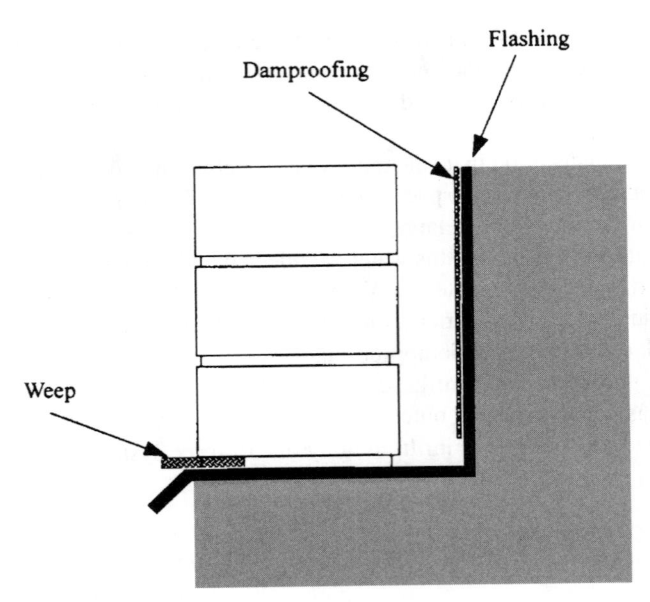

FIGURE 1.7 Drainage waterproofing system.

cavity wall dampproofing and flashing to divert penetrating water and water vapor back to the exterior. Refer to Fig. 1.7.

Diversions actually redirect water being forced against envelope components and divert it elsewhere before it infiltrates or absorbs into the substrate. These might include sloping of roof decks and balconies, vertical drainage mats applied to below-grade walls, gutters and downspouts, flashings, and wind screens. Refer to Fig. 1.8 for typical examples of diversion systems.

Building facades usually contain combinations of these systems, each preventing water infiltration at their location on the envelope. However, regardless of how well the individual systems function, if they are not properly transitioned into other envelope components or terminated sufficiently, leakage will occur. These situations become the major issues preventing effective building envelope and waterproofing functioning, and are the cause of most leakage that occurs in all structures.

Completing the Envelope

Once the sources of water have been identified, the types of systems to prevent leakage identified, and the materials selected to provide necessary aesthetics to the finished product, the envelope design must be carefully constructed and reviewed to ensure successful performance of the completed product. To prevent all possible water intrusion causes, a building must be enveloped from top to bottom with barrier or drainage systems, with divertor components added where appropriate to increase performance of the envelope. These systems must then interact integrally to prevent water infiltration. Should any one of these systems fail or not act integrally with all other envelope components, leakage will occur.

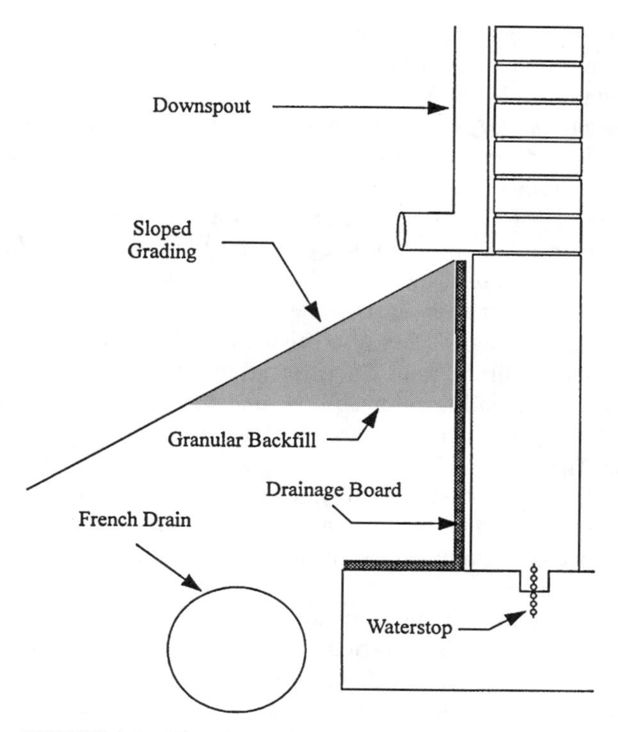

FIGURE 1.8 Diversion systems.

Even with continual technological advances in building materials, water continues to create unnecessary problems in completed construction products. This is most often due to an envelope's inability to act as an integrated system preventing water and pollutant infiltration. All too often several systems are designed into a building, chosen independently and acting independently rather than cohesively.

Detailing transitions from one component to another, or terminations into structural components, are often overlooked. Product substitutions that do not act integrally with other specified systems create problems and leakage. Inadequate attention to movement characteristic of a structure can cause stress to in-place systems that they are not able to withstand. All these situations acting separately, or in combination, will eventually cause water intrusion.

BASIC ENVELOPE DESIGN

To understand the complete enveloping of a structure, several definitions as well as their relationship to one another must be made clear:

Roofing. That portion of a building that prevents water (usually from gravitational forces) in horizontal or slightly inclined elevations. Although typically applied to the surface and exposed to the elements, roofing systems can also be internal, or sandwiched, between other building components.

Below-grade waterproofing. Materials that prevent water under hydrostatic pressure from entering into a structure or its components. These systems are not exposed or subjected to weathering such as by ultraviolet rays.

Above-grade waterproofing. A combination of materials or systems that prevent water intrusion into exposed structure elements. These materials can be subject to hydrostatic pressure from wind conditions, and are exposed to weathering and pollutant attack.

Dampproofing. Materials resistant to water vapor or minor amounts of moisture that act as backup systems to barrier systems or an integral part of drainage systems.

Flashing. Materials or systems installed to redirect water entering through the building skin back to the exterior. Flashings are installed as integral components of waterproofing, roofing, and dampproofing systems. They also can act as divertor systems.

Diversions. Diversions redirect water being forced against envelope components and divert it elsewhere before it infiltrates or absorbs into the substrate. Examples include flashings, downspouts, sloped concrete decks, and drainage mats.

Building envelope. The combination of roofing, waterproofing, dampproofing, flashing, and divertor systems in combination with all exterior facade elements, acting cohesively as a complete barrier to natural forces and elements, particularly water and weather intrusion. These systems envelop a building or structure from top to bottom, from below-grade to the roof.

The entire exterior building skin must be enveloped to prevent water infiltration. It is important to recognize that every component used in the envelope or building skin must be waterproof. This would include many features that most people do not recognize as having to be waterproof to maintain the integrity of the envelope, including exterior lighting fixtures, mechanical equipment, signs and all other types of decorative features.

Each item used or attached to the building envelope should be made waterproof and then appropriately connected to other envelope components to ensure that there are no breaches in the envelope's integrity. All envelopes contain combinations of several systems, such as the building's main facade material (e.g., brick), glass curtain walls, or punch windows, and decorative features such as concrete eyebrows.

These main facade elements are typically barrier waterproofing systems (e.g., glass is actually a barrier system), or drainage systems as in the case of brick. Installing divertors where necessary or appropriate for additional protection against water infiltration then completes the envelope.

Each individual system must then act integrally with all others as a total system for complete effectiveness as a weather-tight building envelope. Figure 1.9 illustrates the interrelationships of the various components of a simplified building envelope.

In Fig. 1.9, the horizontal roofing membrane terminates in a vertical parapet at the metal counterflashing that also transitions the parapet waterproofing into the membrane roofing. In this specific case, the flashing acts as a transition component between the roofing and parapet materials, ensures the watertightness of the envelope at this transition enabling these two separate components to act cohesively.

A similar detail occurs at the coping cap. This flashing detail provides transitioning between the brick facade, water repellent on the brick, cavity wall dampproofing, wood blocking beneath the coping, and the parapet waterproofing. Note also that sealant in this case was also added to protect against any hydrostatic water pressure or

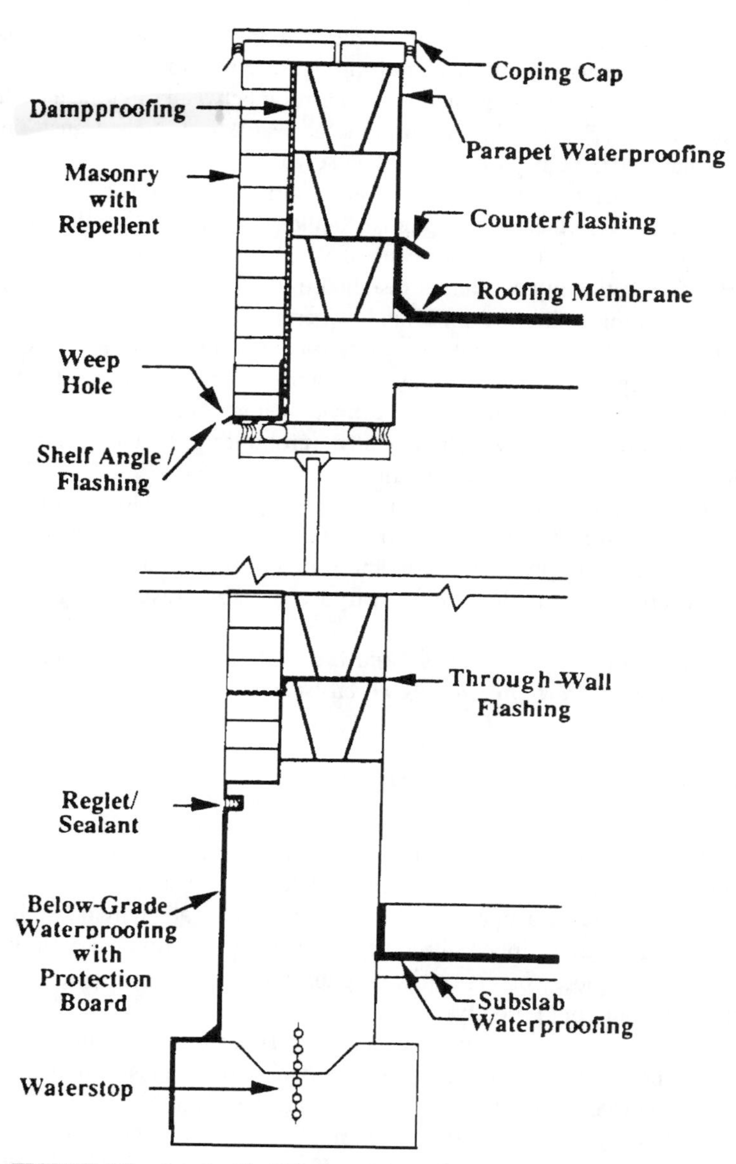

FIGURE 1.9 A typical building envelope.

wind-driven rain from forcing water up under the flashing. Without this, transitioning and termination detailing the various independent systems involved could not function cohesively to provide building envelope watertightness.

On the vertical facade, vertical and horizontal control joints (not shown) finished with sealant allow for adequate space for the masonry to move during thermal expansion and contraction while maintaining a watertight facade. Note that the brick also has been detailed with through-wall flashing, diverting intruding water vapor and moisture that was collected by the dampproofing, back out through the provided weep holes. Additionally, sealant at the window perimeters acts as a transition between brick facade shelf angle and the window frame. The window frame then acts as a watertight transition between the frame and glass, both being waterproof themselves.

To transition between the barrier waterproofing system used below-grade to the drainage system (brick facade), a reglet is installed. This reglet provides the detailing necessary to transition between the two systems while maintaining the watertight integrity of the envelope. Additionally, sealant is installed in the reglet to allow the systems to move independently at this point but still remain waterproof.

Even the waterstop shown in the concrete foundation provides a very important transitioning and waterproofing detailing that is often overlooked. In this wall section, the waterstop effectively ties together the vertical waterproofing to the horizontal slab waterproofing, providing a watertight seal by prohibiting the lateral movement of water along the concrete wall to foundation joint.

The Fig. 1.9 wall section also details divertor systems by sloping of the adjacent soil or landscaping and installing a French drain system. Each system, while not in itself necessary for the waterproofing of the building envelope, quickly removes water away from the structure, eliminating unnecessary hydrostatic pressure against the foundation walls.

As illustrated in Fig. 1.9, *each separate waterproofing material effectively joins together to form a watertight building envelope.*

THE MOST IMPORTANT WATERPROOFING PRINCIPLE

Each separate envelope trade contractor's work, regardless of its being thought of as a waterproofing system or not (e.g., exterior mechanical apparatus) must become part of a totally watertight building envelope. Equally important, all individual envelope systems must be adequately transitioned into other components or provided with watertight terminations. Often the tradesworkers completing this work are not aware, trained, or supervised in enveloping a building properly.

The resulting improper attention to details is responsible for countless problems in construction. Properly detailing a building's envelope presents an enormous task. From inception to installation, numerous obstacles occur. Highlighting this interrelationship of various envelope systems is the most important principle of waterproofing:

The 90%/1% principle: As much as 90 percent of all water intrusion problems occur within 1 percent of the total building or structure exterior surface area.

This 1 percent of a building's exterior skin area contains the termination and transition detailing as previously discussed with Fig. 1.9. This 1 percent area all too frequently leads to breaches and complete failure of the effectiveness of the building envelope.

Not until industry members including contractors, designers, and manufacturers recognize the importance of the 90%/1% principle will the building industry see a marked decline in the all-too numerous claims and lawsuits over envelope failures involving watertightness. Architects must recognize the importance of these termination and transition detailing, manufacturers provide the appropriate details with their specifications, and general contractors provide the coordination and oversight of the numerous subcontractors involved in a single envelope for the completed product to perform as expected.

The 90%/1% principle is the reason that despite continuing technological advances, waterproofing continues to be one of the major causes of legal claims in the design and construction profession. It is not the actual manufactured waterproofing systems or envelope components that leak but the field construction details involving terminations and transitions.

THE SECOND MOST IMPORTANT PRINCIPLE OF WATERPROOFING

The inattention to details is often exacerbated by overall poor workmanship that presents the next most important principle of waterproofing:

The 99%/principle: Approximately 99 percent of waterproofing leaks are attributable to causes other than material or system failures.

When considering the millions of square feet of waterproofing systems installed, both barrier and drainage systems, and miles of sealant involved in building envelopes, it can be estimated that only 1 percent of envelope failure and resulting leakage is actually attributable to materials or systems actually failing. The reasons typically involved in failures include human installation errors, the wrong system being specified for in-place service requirements (e.g., thermal movement encountered exceeds the material's capability), the wrong or no primer being used, inadequate preparatory work, incompatible materials being transitioned together, and insufficient—or in certain cases such as sealants, too much—material being applied.

Today, with quality controls and testing being instituted at the manufacturing stage, it is very infrequent that actual material failures occur. For example, it is rare to have an outright material failure of a below-grade liquid applied membrane, as presented in Chapter 2. More often than not, the leakage would be attributable to improper application including insufficient mileage, improper substrate preparation, or applying over uncured concrete, among numerous other possible installation errors. Furthermore, it is likely that the leakage is also attributable to the 90%/1% principle, with inattention to proper detailing of terminations and transitions with the below-grade membrane occurring.

These two important principles of waterproofing work in unison to represent the overall majority of problems encountered in the waterproofing industry. *By considering these two principles together, it can be expected that 1 percent of a building's exterior area will typically involve actual and direct leakage and that the cause will have a 99 percent chance of being anything but material failure.*

PREVENTING WATER INFILTRATION

Considering that these two simple principles cover the vast majority of leakage problems, it would seem that preventing water infiltration problems would be easy. Certainly, prevention of envelope failures must be a proactive process implemented before actual field construction activities commence. One of the first steps to implement this quality control process is to encourage preconstruction envelope meetings that include all subcontractors involved in the building envelope and cover the following topics:

- Review of the building facade components
- Review of the proposed waterproofing and roofing systems related to the building envelope
- Following the envelope barrier/drainage systems front line, to ensure complete continuity (covered in Chapter 9)
- Reviewing all transitions between envelope components to insure effectiveness and compatibility
- Reviewing all termination details for waterproofing adequateness
- Instructing all attendees on the necessity of meeting the 90%/1% and 99% principles
- Assigning the responsibility for each termination and transition detail

The last issue is often the root of the 90%/1% principle, the fact that many leaks are directly attributable to transition details that are never installed because the general contractor overlooks assigning responsibility for this details in their subcontracts. For example, refer again to Fig. 1.9; whose responsibility would it be to install the reglet detail provided for the below-grade waterproofing to dampproofing transition? The general contractor might easily neglect assigning the completion of this detail to one of the involved subcontractors.

Since the waterproofing membrane would be installed first in most cases, it would be more appropriate for the dampproofing applicator to finish this detail. Although the masonry contractor as part of their contract often applies dampproofing, few masonry contractors would understand the importance of this detail. What if the dampproofing used is a coal-tar-based product that is incompatible with the urethane waterproofing membrane? Further complicating the situation, an acrylic sealant might be used to finish the detail that is not compatible with either the membrane or dampproofing.

Such situations continually occur during field construction activities and result in the facilitating the 90%/1% principle failures.

Unfortunately, all too often waterproofing is considered an isolated subcontracting requirement and few architects, engineers, general contractors, and subcontractors understand the importance of knowing the requirements of successfully designing and constructing a watertight building envelope. It must be clearly recognized that all components of a building exterior facade, from the backfill soil selected to the mechanical rooftop equipment, are integral parts of the building envelope, and all are equally affected by the 99%/1% and 99% principles.

BEYOND ENVELOPE WATERPROOFING

Besides preventing water infiltration, waterproofing systems prevent structural damage to building components. In northern climates, watertightness prevents spalling of concrete, masonry, or stone due to freeze-thaw cycles. Watertightness also prevents rusting and deterioration of structural or reinforcing steel encased in exterior concrete or behind facade materials.

Waterproofing also prevents the passage of pollutants that cause steel deterioration and concrete spalling, such as chloride ions (salts, including road salts used for deicing) into structural components. This is especially true in horizontal exposed areas such as balcony decks and parking garages. Prevention of acid rain contamination (sulfites mixed with water to form sulfuric acid) and carbon acids (vehicle exhaust—carbon dioxide that forms carbonic acid when mixed with water) is also an important consideration when choosing proper waterproofing applications.

Building envelopes also provide energy savings and environmental control by acting as weather barriers against wind, cold, and heat. Additionally, envelopes must be resistant to wind loading and wind infiltration. These forces, in combination with water, can multiply the magnitude of damage to a structure and its interior contents. Direct wind load pressure can force water deeper into a structure through cracks or crevices where water might not normally penetrate. It also creates vertical upward movement of water (hydrostatic pressure) over windowsills and through vents and louvers. Air pressure differentials due to wind conditions may cause water that is present to be sucked into a structure because of the negative pressure in interior areas.

This situation occurs when outside air pressure is greater than interior air pressure. It also occurs through a churning effect, where cool air is pulled into lower portions of a building, replacing warmer air that rises and escapes through higher areas. To prevent this forced water infiltration and associated energy loss, a building envelope must be resistant and weather-tight against wind as well.

SUCCESSFUL ENVELOPE CONSTRUCTION

For envelopes to function as intended requires proper attention to:

- Selection and design of compatible materials and systems
- Proper detailing of material junctions and terminations
- Installation and inspection of these details during construction
- Ability of composite envelope systems to function during weathering cycles
- Maintenance of the completed envelope by building owners

From the multitude of systems available to a designer, specific products that can function together and be properly transitioned must be chosen carefully. Once products are chosen and specified, proposed substitutions by contractors must be thoroughly reviewed. Similar products may not function nor be compatible with previously chosen components. Substitutions of specified components with multiple, different systems only further complicate the successful installation of a building envelope.

Improper attention to specified details of terminations, junctures, and changes in materials during installation can cause water infiltration. Once construction begins, installation procedures must be continually monitored to meet specified design and performance criteria and

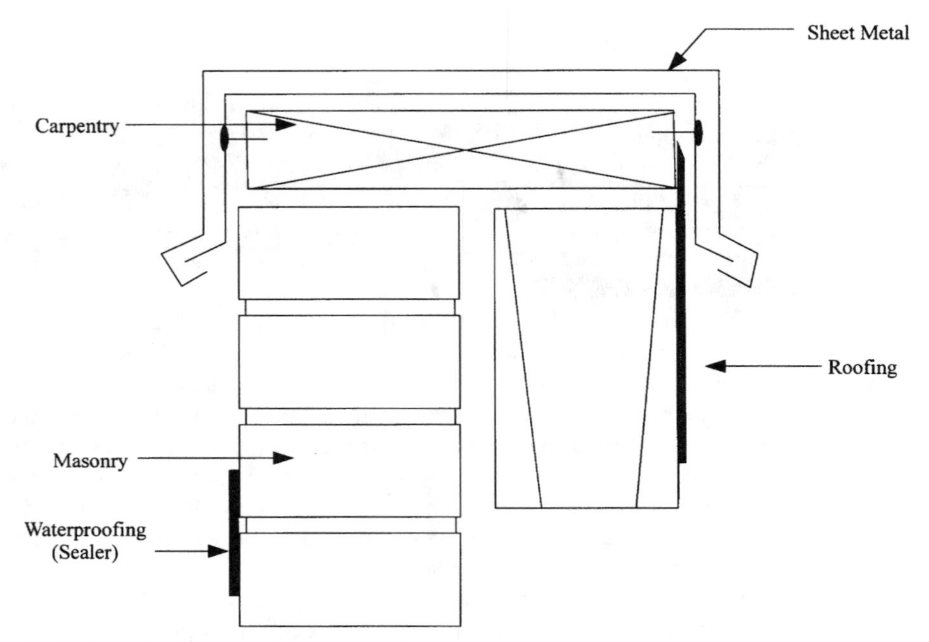

FIGURE 1.10 Typical coping cap detailing and the subcontractors involved.

manufacturers' recommendations. Detailing problems compound by using several different crafts and subcontractors in a single detail. For instance, a typical coping cap detail, Fig. 1.10, involves roofing, carpentry, masonry, waterproofing, and sheet metal contractors. One weak or improperly installed material in this detail will create problems for the entire envelope.

Finally, products chosen and installed as part of a building envelope must function together during life-cycling and weathering of a structure. For example, an installed precast panel might move over $1/2$ inch during normal thermal cycling, but the sealant installed in the expansion joint might be capable of withstanding only $1/4$-inch movement.

Proper maintenance after system installation is imperative for proper life-cycling. Will shelf angles be adequate to support parapet walls during wind or snow loading? Will oxidation of counterflashing allow water infiltration into a roof system, causing further deterioration?

From these processes of design, construction, and maintenance, 99 percent of a building envelope will typically function properly. The remaining 1 percent creates the magnitude of problems. This 1 percent requires much more attention and time by owners, architects, engineers, contractors, and subcontractors to ensure an effective building envelope.

The most frequent problems of this 1 percent occur because of inadequate detailing by architects, improper installation by contractors and subcontractors, and improper maintenance by building owners. Typical frequent envelope errors include:

- *Architects and engineers.* Improper detail specifications (90%/1% principle); no allowance for structural or thermal movement; improper selection of materials; use of substitutes that do not integrate with other components of the envelope.

- *Component manufacturers.* Insufficient standard details provided for terminations and transitions; inadequate training for installers of materials; insufficient testing for compatibility with other envelope components.
- *General contractors and subcontractors.* Improper installations (99% principle); inattention to details; no coordination between the various envelope subcontractors; use of untrained mechanics to complete the work.
- *Building owners and managers.* No scheduled maintenance programs; use of untrained personnel to make repairs; no scheduled inspection programs; postponement of repairs until further damage is caused to the envelope and structural components.

Manufacturers are now concentrating on making technological improvements in the materials themselves rather than technological advances, specifically making their products "idiot-proof." They realize that meeting industry standards does not correlate to success in field applications. In reality, products are subjected to everything that can possible go wrong, from environmental conditions during installation to untrained mechanics installing the product. Never are products installed in the pristine conditions of a laboratory. Making their products with "belt and suspenders" protection increases the likelihood of success, at least for the individual system—for example, products that no longer require primers, no mixing of two component materials but now one-part materials, pressure rinse versus pressure wash preparation, and 300% elongation rather than 100% to add additional protection against excess movement or in-place service requirements.

Similar quality advances at the job-site level by contractors to adequately apply the precautions necessary to protect against the 90%/1% and 99% principles will eliminate the vast majority of waterproofing problems that now plague the industry.

USING THIS MANUAL

Each chapter in this manual supplements a specific area of the building envelope. Every component involved in waterproofing envelopes is reviewed in detail, including below-grade waterproofing systems, above-grade systems, sealants, expansion joints, and admixtures.

Chapter 3 also provides coverage for nonenvelope waterproofing, including interior applications such as shower stalls and civil/infrastructure waterproofing. The preventative systems in chapters 2 through 6 are then supplemented by a presentation on remedial waterproofing covering restoration systems available to repair failed envelopes or existing envelope problems.

Chapter 8 presents a detailed discussion on terminations and transitions, the actual putting together of a successful envelope. Chapter 9 furthers this discussion by covering life cycles, quality and maintenance issues of envelopes to prevent unnecessary 99% and 90%/1% principle problems. Chapter 10 presents detailed coverage of testing envelope components before construction. Chapter 11 presents methods to determine and pinpoint the cause of water leakage. Chapter 12 discusses safety issues involving waterproofing materials including VOC requirements

The manual then includes two important resource chapter. First, Chapter 13 presents a series of guide specifications for commonly used waterproofing methods used in new construction. Then Chapter 14 presents an in-depth resource guide to available waterproofing material and systems manufacturers, building associations and other resources to further assist the reader in gathering sufficient information to successfully complete any water-proofing installation or repair that might be encountered.

SUMMARY

While the manual is structured to discuss the various waterproofing systems individually, they should always be regarded as one component in a series of envelope elements. Remember, no matter how good a material is used, a building envelope will succeed only if all components act as a cohesive unit. One weak detail on an envelope can cause the entire facade to deteriorate and allow water penetration to interior spaces.

CHAPTER 2
BELOW-GRADE WATERPROOFING

INTRODUCTION

Water in the form of vapor, liquid, and solids presents below-grade construction with many unique problems. Water causes damage by vapor transmission through porous surfaces, by direct leakage in a liquid state, and by spalling of concrete floors in a frozen or solid form. Water conditions below-grade make interior spaces uninhabitable not only by leakage but also by damage to structural components as exhibited by reinforcing steel corrosion, concrete spalling, settlement cracks, and structural cracking.

Below-grade waterproofing materials are subject to water conditions that are typically more severe than above-grade envelope areas. Structure elements below-grade are often exposed to hydrostatic pressure from ground water tables that can rise significantly during periods of heavy rainfall. At the same time, below-grade materials are not subject to the harsh environmental conditions of exposed envelope components, including wind-driven rain, ultraviolet weathering, and acid rain.

Manufacturers of below-grade waterproofing systems can then concentrate on the properties to ensure effective barriers to water penetration without having to contend with the elements encountered above-grade. For example, membranes used below-grade can have substantial elongation properties since the manufacturer does not have to supplement the product with ultraviolet resistant properties that tend to limit elongation capabilities.

Below-grade systems are all barrier systems; there are no appropriate new construction drainage systems designed for adequate protection under hydrostatic pressure. Diversion systems are frequently included in the design of below-grade waterproofing, and in fact are highly recommended for use in conjunction with any below-grade system, with the possible exception of hydrous clay materials that require the presence of adequate water supply to maintain their hydration and waterproofing properties.

Proper below-grade design begins with adequate control of water conditions. There is no reason to subject any below-grade envelope components to unnecessary amounts of water that could otherwise be diverted away form the structure for supplementary protection. Both surface and groundwater should be diverted immediately away from the structure at all times.

SURFACE WATER CONTROL

Water present at below-grade surfaces is available from two sources—surface water and groundwater. Beyond selection and installation of proper waterproofing materials, all waterproof installations must include methods for control and drainage of both surface and groundwater.

Surface water from sources including rain, sprinklers, and melting snow should be directed immediately away from a structure. This prevents percolation of water directly adjacent to perimeter walls or water migration into a structure. Directing water is completed by one or a combination of steps. Soil adjacent to a building should be graded and sloped away from the structure. Slopes should be a minimum of 1/2 in/ft for natural areas, paved areas, and sidewalks sloped positively to drain water away from the building.

Automatic sprinklers directed against building walls can saturate above-grade walls causing leakage into below-grade areas. Downspouts or roof drains, as well as trench drains installed to direct large amounts of water into drains, direct water away from a building. Recommended controls for proper water control are summarized in Fig. 2.1.

GROUNDWATER CONTROL

Besides protection from normal groundwater levels, allowance is made for temporary rises in groundwater levels to protect interior areas. Groundwater levels rise due to rain

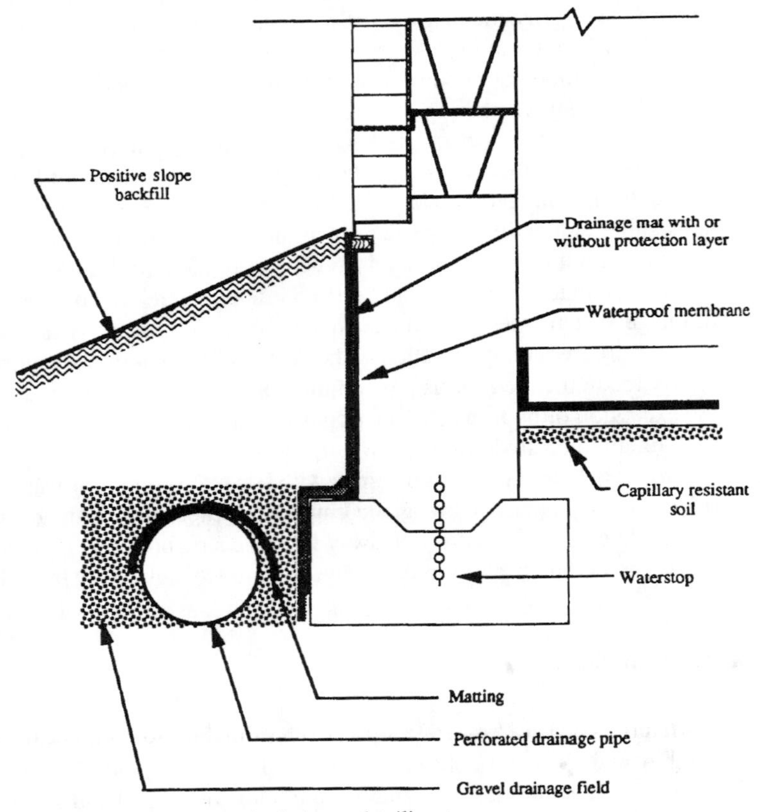

FIGURE 2.1 Below-grade drainage detailing.

accumulations and natural capillary action of soils. Waterproof materials must be applied in heights sufficient to prevent infiltration during temporarily raised groundwater levels.

With every below-grade installation, a system for collecting, draining, and discharging water away from envelopes is recommended. Foundation drains are effective means for proper collection and discharge. They consist of a perforated pipe installed with perforations set downward in a bed of gravel that allows water drainage. Perforated drain piping is usually polyvinyl chloride (PVC). Vitreous clay piping is sometimes used, but it is more susceptible to breakage. Drain piping is installed next to and slightly above the foundation bottom, not below foundation level to prevent the washing away of soil under the foundation that can cause structural settlement.

Coarse gravel is installed around and over drainage piping for percolation and collection of water. Additionally, mesh or mats are installed over the gravel to prevent soil buildup, which can seal drainage piping perforations and prevent water drainage. Collected water must be drained by natural sloping of pipe to drain fields or pumped into sump pits.

PREFABRICATED FOUNDATION AND SOIL DRAINAGE SYSTEMS

These field-constructed foundation drainage systems are obviously very difficult to build properly and often perform poorly over time due to infiltration into the drainage piping by silt, sand, and soil that will eventually clog the entire system. Manufacturers have responded by developing "idiot-proof" systems to replace these now-antiquated field constructed systems. These prefabricated systems are relatively inexpensive and make them completely reasonable for use as additional water control for practically any construction project including residential, multifamily, commercial and civil structures. These systems add superior protection for minor costs to any project. For example, concrete slabs without reinforcing can withstand hydrostatic pressure equal to approximately 2.5 times the slab's thickness. In practically every structural design, it becomes much more economical to add under-slab drainage than to increase the thickness of the slab.

Prefabricated plastic soil drainage systems are available from a number of manufacturers (refer to Chapter 14). These products are manufactured in a variety of plastic composite formulations including polypropylene, polystyrene, and polyethylene. Figure 2.2 pictures a typical manufactured drainage product. The systems combine specially designed drainage cores covered with geotextile fabric in prepackaged form that eliminates all field construction activities except trenching and backfilling operations.

The systems are idiot-proof in that the product is merely laid into the area designated for a drainage field. Only appropriate sloping of the trench to collection points is required. Figure 2.3 presents a simplified isometric detail of a drainage system installation. The product is puncture-resistive to protect its performance during backfill. Manufacturers also provide ample accessories (including termination and transition detailing) to complete the installation. Figures 2.4 and 2.5 show available accessories including a tee connection to join one branch of a drainage to another, and an outlet connection for collection of water that terminates at a drain box or culvert.

Materials are available in a variety of widths (up to 36 inches) and lengths provided in rolls of up to 500 feet long. The product should be puncture-resistant with some elongation capability for movement after installation, and be resistant to the natural or human-made elements to be found within the intended service area.

FIGURE 2.2 Typical foundation premanufactured drainage system with geotextile attached. (*Courtesy of American Wick Drain Corporation*)

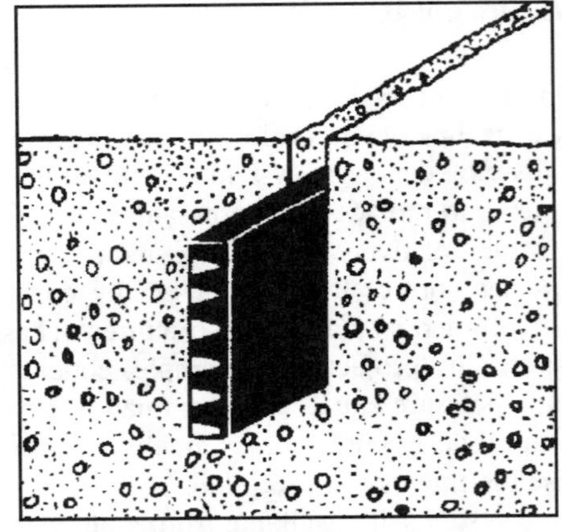

FIGURE 2.3 Isometric detail of drainage system. (*Courtesy of American Wick Drain Corporation*)

Prefabricated Drainage System Installation

The product can be laid into preexisting trenches available from foundation construction or trenches constructed specifically for the drainage field. The width of the trench is typically 2 to 6 inches wide. The depth of the trench is determined upon the actual site conditions and soil permeability. Figure 2.6 represents a typical drainage detail.

The prefabricated plastic drains usually permit the excavated soil to be used as backfill, eliminating the requirement for special backfill material. The backfill must be mechanically compacted in layers. Geotextile covering is selected based on the soil conditions. Here are the basic geotextiles required for typical soil conditions:

- High clay content—nonwoven needle-punched geotextile
- Sandy soils—woven materials with high permeability
- High silt content—small-opening geotextiles

Soils of any combinations of the above types generally require testing to be performed and specific recommendation by the drainage system manufacturer.

Manufacturer-provided tees, splicing connectors and outlet connectors should be used as designed. The system is designed to collect and drain water in a variety of ways that meet specific site requirements. Drainage can be as simple as outflow to bare soil away from the structure as surface drainage, or it can be designed to outflow into municipal storm drains.

MANUFACTURED DRAINAGE SYSTEMS

In addition to the premanufactured foundation and soil drainage systems, there are also available drainage systems used in conjunction with both vertical and horizontal below-grade waterproofing systems. These drainage systems provide additional protection against water infiltration and effectively reduce hydrostatic pressure against below-grade envelope components.

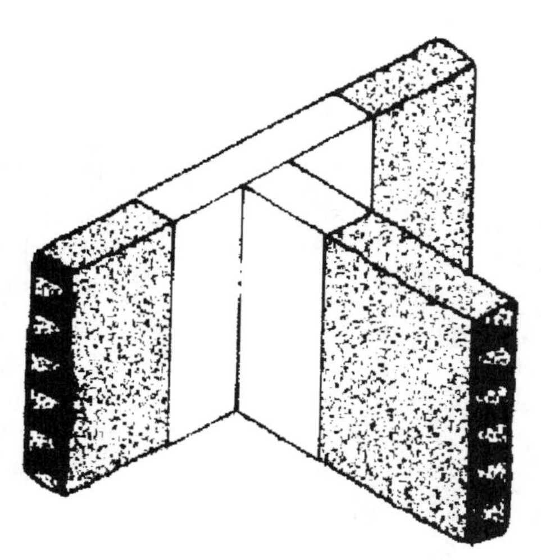

FIGURE 2.4 "T" connector for drainage system. (*Courtesy of American Wick Drain Corporation*)

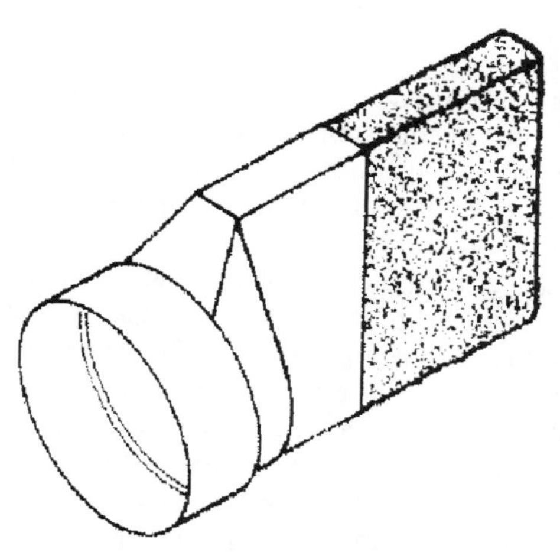

FIGURE 2.5 Outlet connection for drainage system. (*Courtesy of American Wick Drain Corporation*)

The products aid in the drainage of groundwater by collecting and conveying the water to appropriate collection points for drainage away from the structure. A simplified typical design is shown in Figure 2.7. The products provide low-cost insurance against water infiltration and should be used with every below-grade waterproofing application (with the possible exception of hydrous-clay materials). More often than not, the drainage systems can be used in lieu of protection board for most membrane applications, effectively negating any additional costs for the system's superb protection.

Besides the additional drainage protection for occupied spaces, the systems are also used alone for protecting various civil structures such as landfills and retaining walls or abutments. Among the many uses for manufactured drainage systems:

- Below-grade walls and slabs
- Retaining walls and abutments
- Tunnels and culverts
- Lagging
- Embankments
- Landfills
- French and trench drains (described in the previous section)
- Drainage fields for golf courses and other park and play field structures
- Specialized drainage requirements
- Above-grade plaza decks and similar installations (covered in Chapter 3)

The system is similar to the prefabricated soil drainage systems only available in larger sheets and drainage cores to facilitate drainage. The material consists of a formed plastic three-dimensional core that acts as the collector and drainage transporter of the water, as shown in Figure 2.8. The plastic drainage product is also covered with a geotextile fabric to prevent the silt, soil, clay, and sand from clogging the drainage system. The systems usually have some type of plastic sheeting adhered to one side to protect from indenting waterproofing membranes as well as acting as an initial waterproofing system.

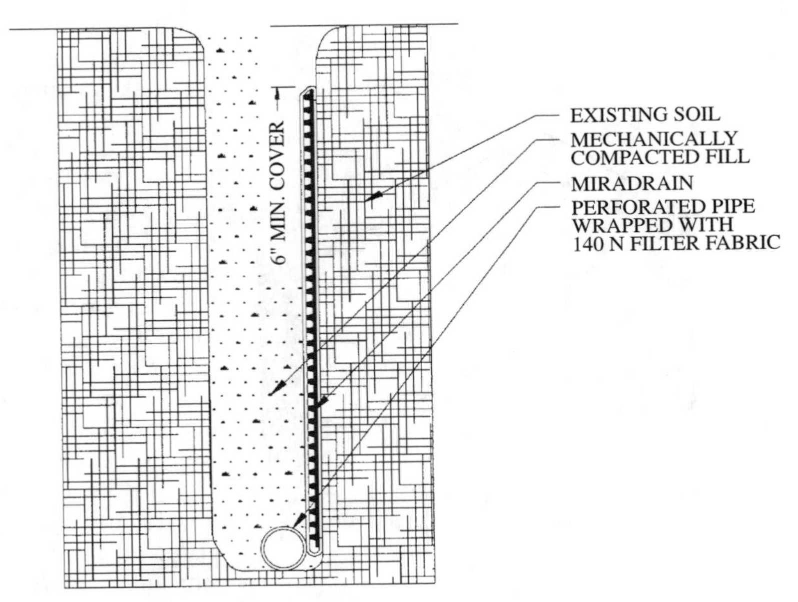

FIGURE 2.6 Typical detailing for foundation drainage system. (*Courtesy of TC Mira DRI*)

The systems not only eliminate the need for protection board, but also eliminate the requirement for special backfill material consisting of sand or gravel materials to promote drainage. Typically, the existing soil is used as backfill material, reducing the overall costs of new construction. A typical below-grade wall detailing, using the drainage as protection for the waterproofing membrane, is detailed in Figure 2.9.

The systems also provide a drainage flow rate (depending on the size of plastic core structure, which varies from 1/4 inch to 1/2 inch) 3 to 5 times the capacity of commonly used drainage backfill materials such as sand or small aggregate fill. The material is obviously lightweight, with one person capable of carrying the average roll of material that covers as much as 200 square feet of substrate, the equivalent of a small dump truck of aggregate backfill.

Materials selected should have a high compressive strength to protect waterproofing applications (a minimum of 10,000 psf). Also, the system should be resistant to any chemicals it might be exposed to, such as hydrocarbon materials at airports.

Among the many advantages of manufactured drainage systems over conventional aggregate backfill:

- Cost effectiveness.
- Attached filter fabric or geotextile eliminates the usual clogging of traditional systems.
- High-strength material can be used in lieu of protection board for membranes.
- Provides belt and suspender protection for below-grade spaces by quickly channeling ground and surface water away from the structure.
- Permits backfilling with the excavated soils.
- Lightweight and idiot-proof installations.

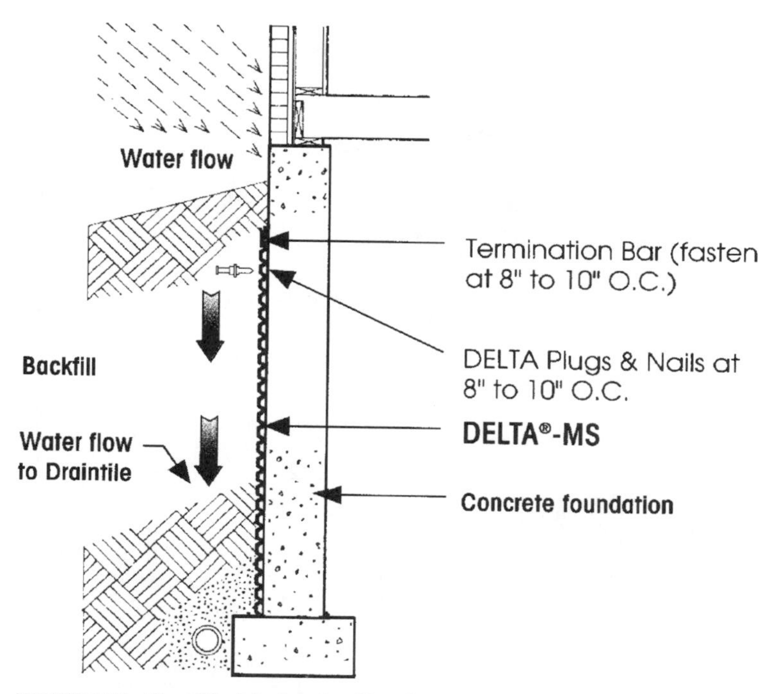

FIGURE 2.7 Simplified design detailing for premanufactured drainage system. (*Courtesy of Cosell-Dorken*)

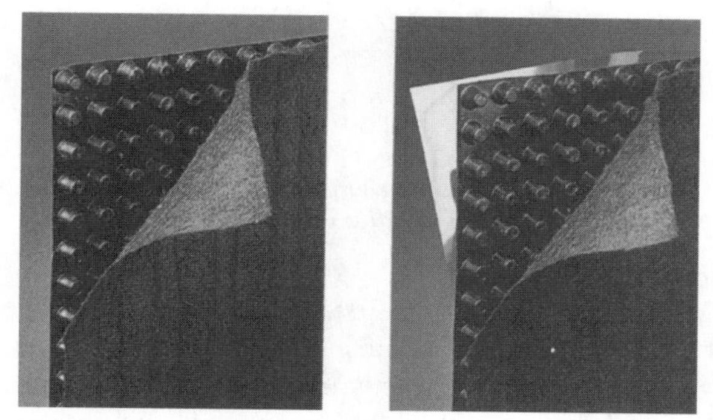

FIGURE 2.8 Typical manufactured drainage systems. (*Courtesy of American Hydrotech*)

Manufactured Drainage System Installation

Material is generally supplied in rolls that is simply applied to the waterproofed walls by using double-sided masking tape, sealant, or other adhesives recommended by the waterproofing membrane manufacturer; see installation photograph, Figure 2.10. The material is installed like roofing shingles, overlapping in the direction of water flow,

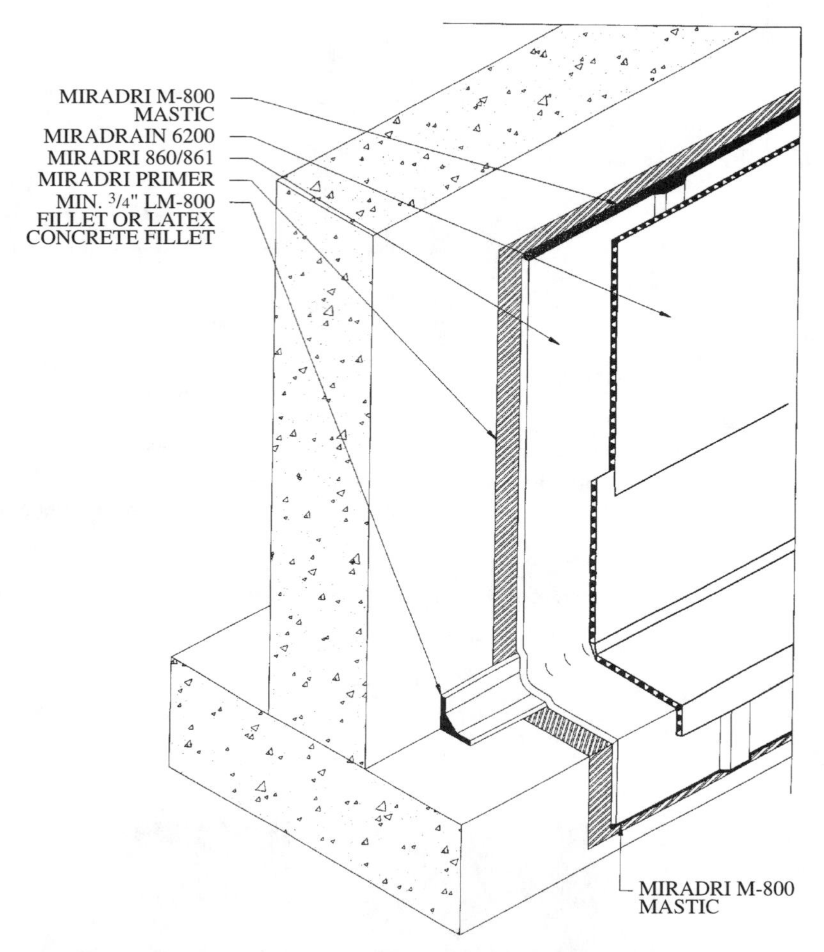

FIGURE 2.9 Below-grade waterproofing application with drainage board used as protection. (*Courtesy of TC Mira DRI*)

starting with the lower portion first, lapping higher elevation goods over the already installed piece to match the manufacturer-supplied flange edges (Figure 2.11). Drainage systems can also be applied directly to lagging prior to concrete placement (Figure 2.12)

The filter fabric material is always applied facing out, and manufacturers provide additional fabric at ends to overlap all seams. The terminated ends of the material are covered with the fabric by tucking it behind the plastic core sheet. Side edges of the sheet are typically attached together by overlapping and applying an adhesive. Figure 2.13 shows a partially completed drainage system installed with appropriate drain field gravel backfill.

Figure 2.14 details the use of drainage systems for under-slab drainage. Figure 2.15 details the use of these systems for horizontal transitioning to vertical drainage at a below-grade tunnel installation.

FIGURE 2.10 Application of drainage system system using termination bar directly over terminating edge of waterproofing membrane. (*Courtesy of Coseall-Dorken*)

FIGURE 2.11 Application of drainage system. (*Courtesy of Webtec, Inc.*)

FIGURE 2.12 Drainage system being applied directly to foundation lagging. (*Courtesy of TC Mira DRI*)

FIGURE 2.13 Installation of drainage field adjacent to foundation for completion of prefabricated drainage system. (*Courtesy of Webtech, Inc.*)

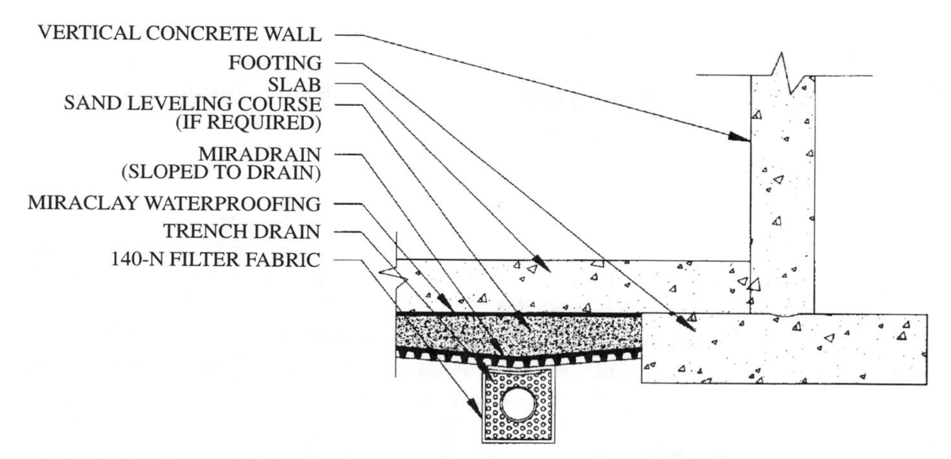

FIGURE 2.14 Manufactured drainage system used for below-slab drainage. (*Courtesy of TC Mira DRI*)

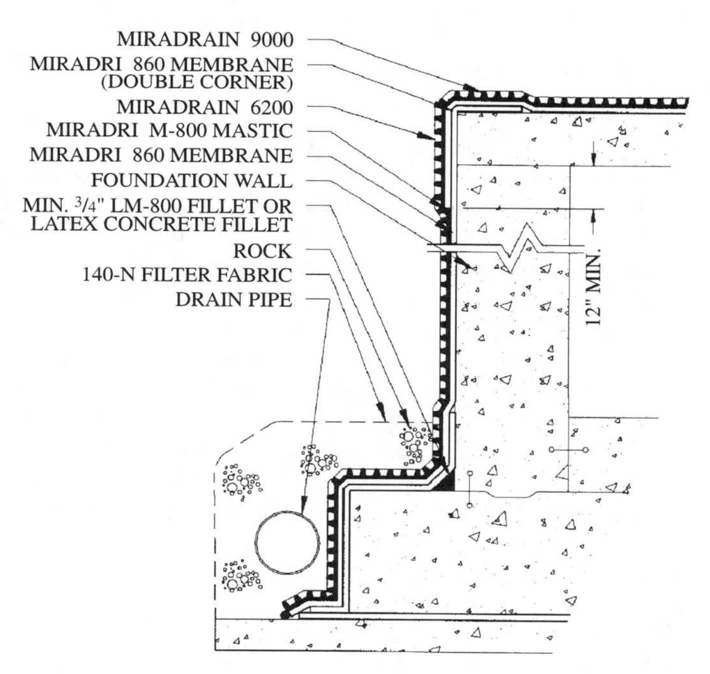

FIGURE 2.15 Below-grade tunnel waterproofing using both horizontal and vertical drainage application. (*Courtesy of TC Mira DRI*)

Backfilling should take place as soon as possible after installation; using the available site soil is acceptable. Backfill should be compacted as required by specifications using plate vibratory compactors. Caution should be taken during compaction not to damage the fabric material.

WATERSTOPS

Whenever a construction joint occurs in a below-grade concrete structure, a waterstop should be installed in the joint to prevent the transmission of water through the joint. Construction joints, also referred to as "cold-joints," occur when one section of concrete is placed and cured or partially cured before the adjacent concrete placement occurs. This occurs frequently in concrete structures at locations including

- Transitions between horizontal and vertical components
- When formwork is insufficient to finish the structure in one placement, such as long lengths of wall area
- Where design elements require a change in form design
- When concrete placement is stopped, for schedule reasons or end of workday

In most of these cases a joint is not actually formed; the cold or construction joint reference refers to the area of concrete structures where two different concrete placements have occurred (properties of concrete preventing it from forming an excellent bond to itself and the previously placed concrete). In addition, control joints are added to a poured-in-place concrete structure to control cracking that occurs from shrinkage in large placements. Control joints are typically recommended for installation at no more than 30 feet apart. The joints are typically the weakest points of the concrete structural components, but not subject to movement other than structural settlement.

Below-grade conditions present conditions that make it very likely that water, which is present under hydrostatic pressure, will infiltrate through these construction joints. To prevent this from occurring, waterstops are commonly specified for installation at every construction joint on concrete work below-grade. The capability of waterstops to prevent infiltration at these weak points in the structure is critical to successful waterproofing of below-grade structures, so their importance should never be underestimated.

Waterstops are used for waterproofing protection on a variety of below-grade concrete structures including

- Water treatment facilities
- Sewage treatment structures
- Water reservoirs
- Locks and dams
- Basement wall and floors
- Parking structures
- Tunnels
- Marine structures

Waterstops are premanufactured joint fillers of numerous types, sizes, and shapes. Waterstops are available in a variety of compositions including

- Polyvinyl chloride (PVC)
- Neoprene rubber
- Thermoplastic Rubber
- Hydrophilic (modified chlorophene)
- Bentonite clay
- Asphalt plastic

The first three, PVC and rubber types, are manufactured exclusively for use in poured-in-placed concrete structural elements. The remaining three, while mainly used for concrete installations, can be used with other building materials such as concrete block and are also excellent where installations involve metal protrusions in or adjacent to the construction joint. Manufacturers also make waterstops that are resistant to chemicals and adverse groundwater conditions. A summary of the properties of the various type waterstop is shown in Table 2.1. As with many products, manufacturers have begun making systems that approach "idiot-proof" installations.

PVC waterstops have long been the standard within the construction industry. They are provided in a variety of shapes and sizes for every situation to be encountered, as shown in Figure 2.16.

PVC waterstops with the dumbbell shape in the middle are used for installation where actual movement is expected in the substrate, typically not thermal movement but structural movement. Figure 2.17 shows an expansion joint installation with the bulb portion of the waterstop left exposed to permit movement. However, waterproofing applications require the joint to be filled with a properly designed sealant joint to permit a waterproofing below-grade membrane to run continuously over the joint.

The problem with PVC waterstops is their susceptibility to improper installation (99% principle) or damage during the concrete placement. The waterstop must be held in place properly during the first half of the concrete placement. This is accomplished by a variety of methods as shown in Figures 2.18 and 2.19. This situation is not idiot-proof and must be carefully monitored for quality control to ensure that the waterstop remains positioned during both halves of the concrete placement activities. Far too often, the PVC waterstop ends up folded over, preventing it from functioning properly. In addition, workers installing the reinforcing bars will often burn, puncture, or cut the waterstop.

TABLE 2.1 Comparison of Various Waterstop Types

Waterstop Type	Advantages	Disadvantages
PVC Neoprene Rubber Thermoplastic Rubber	• Rugged and durable material • Numerous manufactured shapes	• Installation can be difficult • Tendency to fold over during concrete placement
Hydrophilic Bentonite	• Ease of installation • No requirement for first half of concrete placement	• Subject to damage by rain or other wetting • No expansion joint installation
Asphalt Plastic	• Ease of installation • Not subject to swelling by rainfall	• No expansion joint installation • Substrate prep required

Ribbed Flat

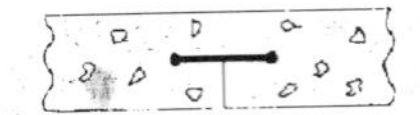

Construction or contraction joints where little or no movement is expected. Ribbed shapes provide a better seal than dumbbell shapes.

Dumbbell

Construction or contraction joints where little or no movement is expected.

Base Seal

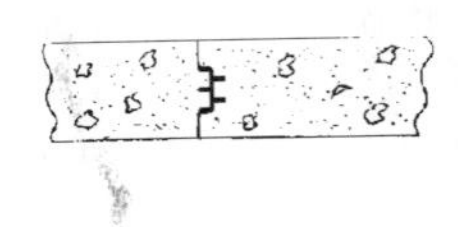

Ideal for slab-on-grade joints or walls which will be back-filled. Easy to form.

Labyrinth

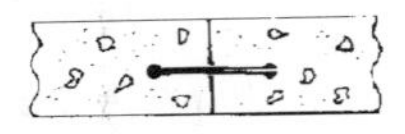

Primarily used in vertical joints where little or no differential movement is expected. Does not require split forming and adds a key to the joint. Difficult to use in horizontal joints.

Split Waterstop

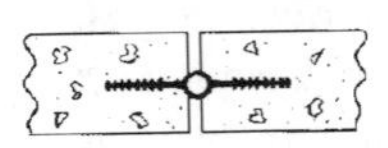

Available in ribbed with center-bulb and dumbbell shapes. Eliminates the need for split form bulkhead.

Ribbed with Centerbulb

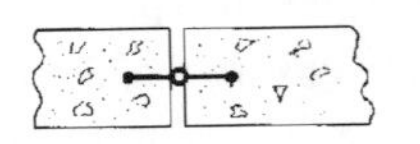

The most versatile type of waterstop available. The centerbulb accommodates lateral, transverse and shear movement. Larger centerbulbs accommodate larger movements.

Dumbbell with Centerbulb

Accommodates lateral, transverse and shear movements. **Ribbed shapes provide better sealing characteristics.**

Tear Web

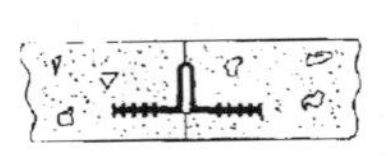

Accommodates large movements. Upon joint movement the tear web ruptures allowing the U-bulb to deform without putting the material in tension.

FIGURE 2.16 Typical PVC waterstops and their properties. (*Courtesy of Greenstreak*)

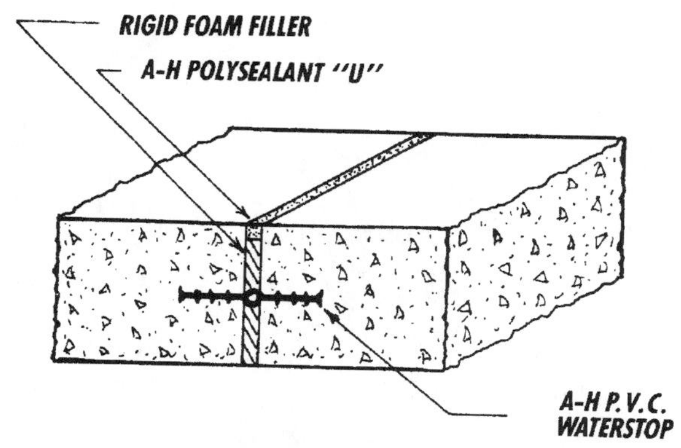

FIGURE 2.17 Use of PVC waterstop in expansion joint. (*Courtesy of AntiHydro*)

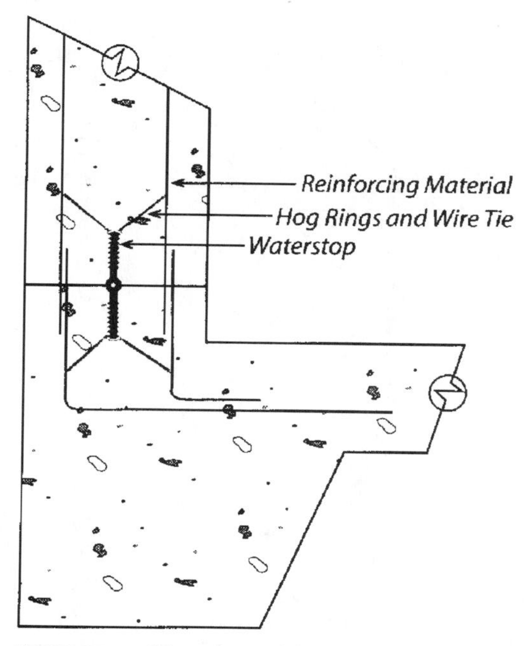

FIGURE 2.18 Placement and securing of waterstops at construction joints. (*Courtesy of J.P. Specialties, Inc.*)

In striving to make waterstops idiot-proof, manufacturers have created several alternatives to the PVC standard including many hydrophilic derivatives. These systems, along with the bentonite and asphalt plastic, are used mainly for control joints and not provided for expansion joints. These systems are simple to install, and do not have to be installed in both sections of concrete placements. The material is adhered directly to the edge of the first concrete placement in preparation for the second placement of concrete. Note this detailing in Figure 2.20 and in the photograph of the installed product, Figure 2.21.

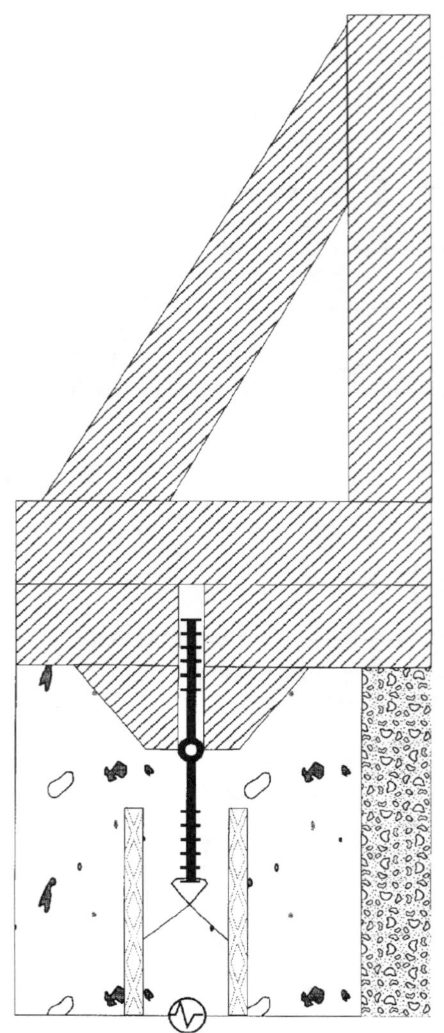

FIGURE 2.19 Placement of waterstop for first half of concrete placement. (*Courtesy of J.P. Specialties, Inc.*)

The materials generally expand after being wetted by the water contained in the concrete mixture. This swelling action enables the materials to fill the voids within the joint to form a watertight construction joint. Since these products expand in the presence of water, they must not be wetted prematurely. This requires that the second concrete placement take place almost immediately after the waterstop placement, otherwise the joint might expand if exposed to rain or dew. The asphalt plastic is not susceptible to moisture like bentonite or hydrophilic materials, but their limited elastomeric capabilities might prevent the complete sealing of the joint if some areas are not bonded properly.

The materials are easily installed in a variety of positions for properly detailing watertight joints below-grade as shown in Figure 2.22. None is meant for exposure to the elements and must be completely covered by the concrete placement. As such, they present limited expansion capabilities for the substrate. When an expansion waterstop material is required, the PVC or rubber types are required.

Waterstop size is determined by the expected head of water pressure to be encountered at the joint. Table 2.2 summaries the recommended waterstop and minimum depth of embedment into the concrete substrate for various head pressures. Actual site conditions vary, and these measurements should be used only as approximations. Waterstop manufacturers will recommend actual joint design when actual job conditions are submitted for review.

WATERSTOP INSTALLATION

PVC/Neoprene Rubber/Thermoplastic Rubber

Waterstops of this type are placed in the concrete formwork and tied or secured to firmly position the material during concrete placement. It is imperative that the waterstop is never allowed to fold over during concrete placement. Figure 2.23 shows some typical methods to secure the waterstop prior to the first concrete placement. Figure 2.24 details the method for installing waterstop using a keyed joint.

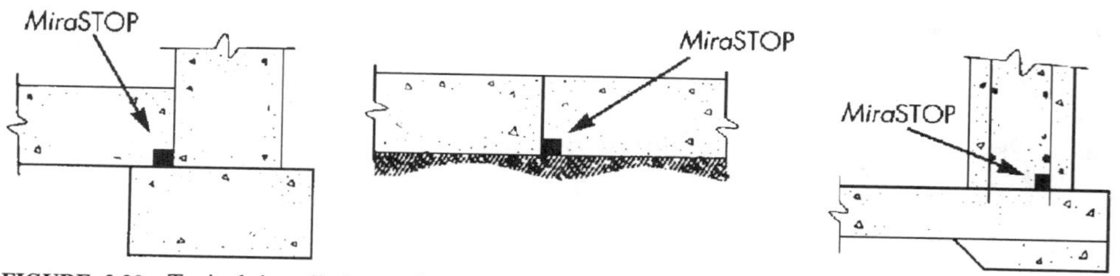

FIGURE 2.20 Typical installations of hydrophilic or similar waterstop materials (*Courtesy of TC MiraDRI*)

FIGURE 2.21 Installed asphaltic waterstop. (*Courtesy of Vinylex Corp.*)

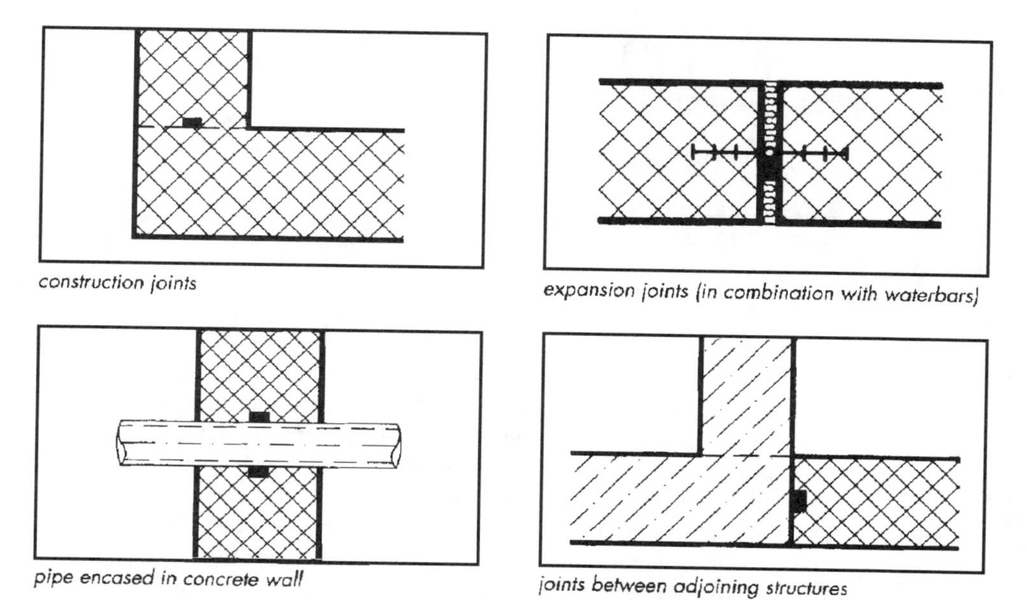

FIGURE 2.22 Several recommended uses of hydrophilic waterstop. (*Courtesy of Vandex*)

TABLE 2.2 Suggested Waterstop Sizing for General Conditions (*Courtesy of Vinylex Corporation*)

Size (flange /thickness), inches	Maximum allowable head of water (per Army Corps Engineering Manual EM 1110-22101) feet	Minimum embedment of flange into concrete inches	Minimum distance to edge of slab/wall, inches
4 × 3/16	50	1.250	2.000
9 × 3/16	100	2.875	2.875
9 × 3/8	150	2.875	2.875
12 × 1/2	250	4.000	4.000

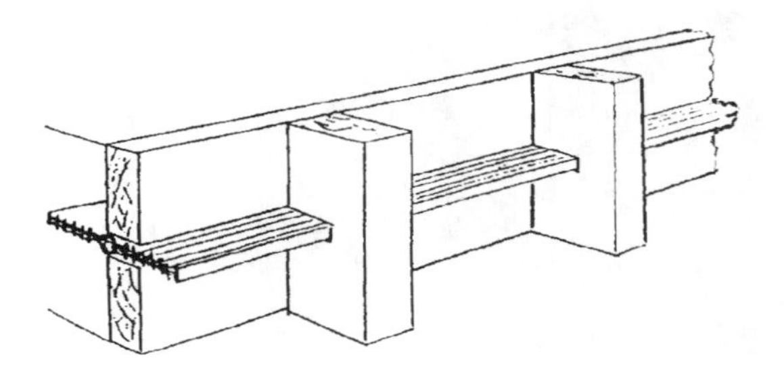

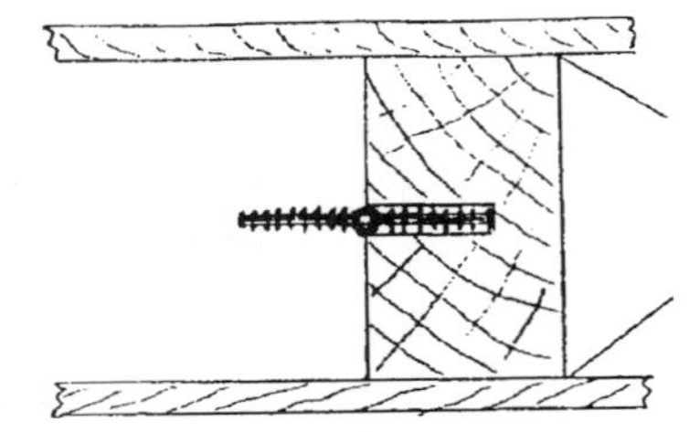

FIGURE 2.23 Securing of PVC waterstop for first concrete placement. (*Courtesy of Tamms Industries*)

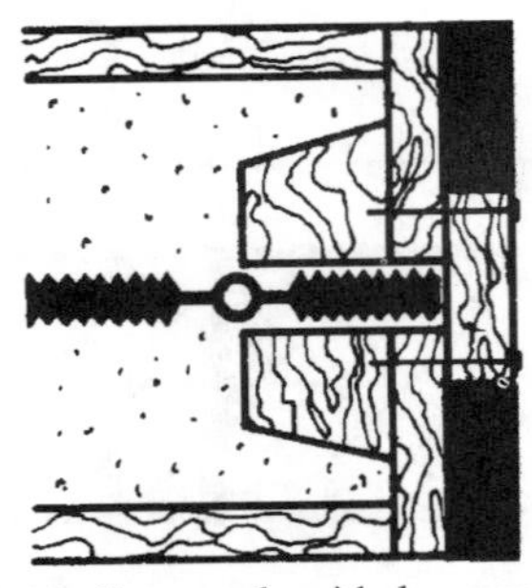

FIGURE 2.24 Formwork with keyway joint system. Note the bulb is centered directly in midpoint of the joint to ensure proper functioning as an expansion joint. (*Courtesy of Tamms Industries*)

To secure the flange in place for both concrete placements, the waterstop is generally secured using wires tied to the reinforcing steel every 12 inches. The wire should be tied through the first or second ribs of the waterstop flange, never going beyond the second flange as shown in Figure 2.25. Note that in each of these details the center bulb is directly in the midpoint of the joint. This is to ensure that the waterstop acts properly as an expansion joint during any structural movement.

The bulb must never be placed completely in one side of the placement or it will lose all its capability to act as an expansion material. Nails or any other construction debris should not be allowed to puncture the waterstop bulb or any part of the flange near the bulb.

When using waterstops at construction joints, material with bulb ends makes securing to reinforcing steel easier by using wire rings that pass through the bulb but not the flange. Figure 2.26 details the steps using this system for both halves of the concrete placement.

Placing PVC and rubber waterstops in the field usually requires some welding to joint ends of rolls or making necessary changes in plane. Waterstop should never be installed by merely lapping the ends together. The material must be heat-welded to fuse the ends together by using manufacturer-supplied splicing irons that melt the ends that are then held together until they cool, forming one continuous piece. Refer to Figure 2.27 for a field weld application.

Testing of failed joints usually reveals that failures were either the cause of improperly positioned material, Figure 2.28, (folded over during concrete placement) or where directional changes occurred in structure that the waterstop did not conform to. Whenever heat welding is used, the material is adversely affected at this point and its properties are not equal to the original material. Therefore it is recommended that whenever major directional changes are designed into a structure, the contractor should secure prefabricated fittings. Waterstop manufacturers will usually provide a variety of premolded splice pieces for directional changes, as shown in Figure 2.29. Also, most manufacturers will offer to custom-make the required splices to ensure the successful applications with their material.

At all penetrations in below-grade slabs or walls, waterstop should also be installed continuously around the penetration to protect against water penetration. Figure 2.30 details the use of waterstop installed continuously around a structural steel column that penetrates the concrete slab over the foundation.

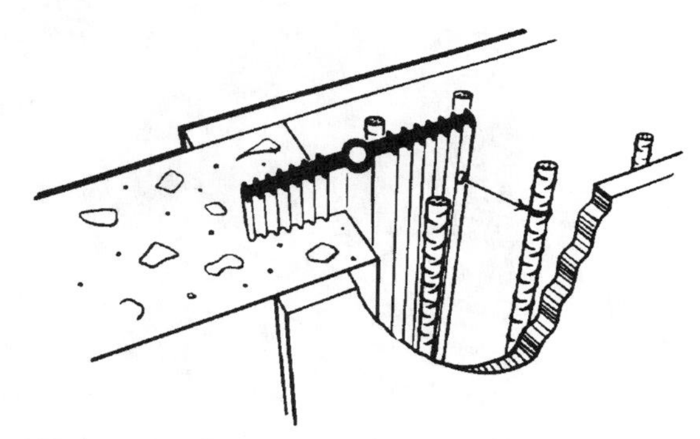

FIGURE 2.25 Securing PVC waterstop for second concrete placement. (*Courtesy of Vinylex Corp.*)

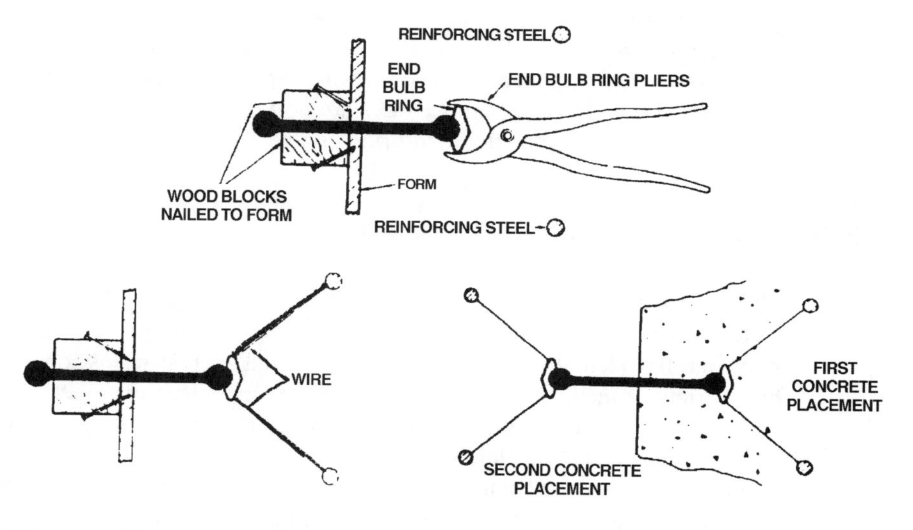

FIGURE 2.26 Placement procedures for end bulb waterstop. (*Courtesy of Tamms Industries*)

FIGURE 2.27 Field welding operation of PVC waterstop. (*Courtesy of J.P. Specialties*)

Photographs in Figures 2.31 and 2.32 show how complicated waterstop installations can become. Such detailing necessitates the use of premanufactured weld splices to ensure watertight applications. These photographs also emphasize how important proactive job-site quality-control procedures are, to verify that the PVC waterstop is installed and maintains proper positioning during concrete placement.

HYDROPHILIC/BENTONITE/ASPHALT RUBBER

These systems are all installed after the first concrete placement has occurred, with the materials attached directly to the first half by a variety of methods. The waterstop is supplied in rolls in lengths of several hundred feet and the material is adhered to the substrates by a variety of methods as recommended by the manufacturer.

Typically, the concrete does not need to be cured completely, as this would interfere with the placement schedule of the concrete. Substrate preparation is usually minimal, ensuring that there are no form release agents, fins, or other protrusions that can damage or puncture the waterstop during installation. Attachment is completed by a variety of methods, some as simple as nailing the strip to the concrete to hold it in place temporarily until the second half of concrete placement occurs.

FIGURE 2.28 Improperly positioned, placed, and secured waterstop. (*Courtesy of Coastal Construction Products*)

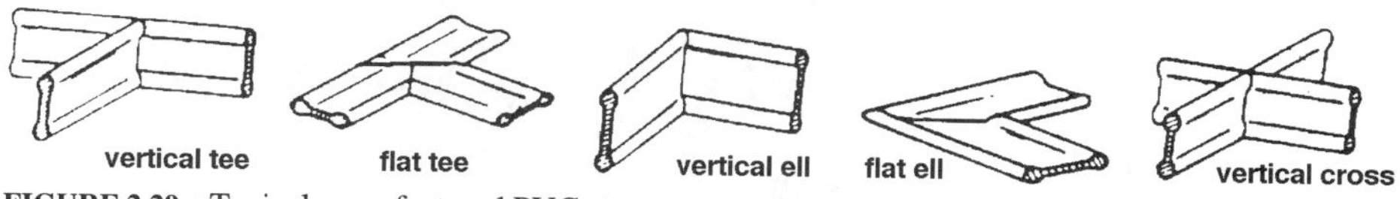

FIGURE 2.29 Typical manufactured PVC waterstop splices and transition pieces. (*Courtesy of Tamms Industries*)

It is imperative that the hydrophilic and bentonite materials are not left exposed to rainfall before concrete is placed. If this occurs, the material will swell and lose all its capability to seal the joint after concrete placement. The photographs in Figure 2.33 show a typical waterstop installation using a swelling material that is adhered to the substrate with sealant.

CAPILLARY ACTION

Construction details must be included to prevent natural capillary action of soils beneath foundations or below-grade floors. Capillary action is upward movement of water and vapor through voids in soil from wet lower areas to drier high areas. Capillary action is dependent upon the soil type present. Clay soils promote the most capillary action, allowing more than 10 ft. of vertical capillary action. Loose coarse gravel prevents capillary action, with this type of soil promoting virtually no upward movement.

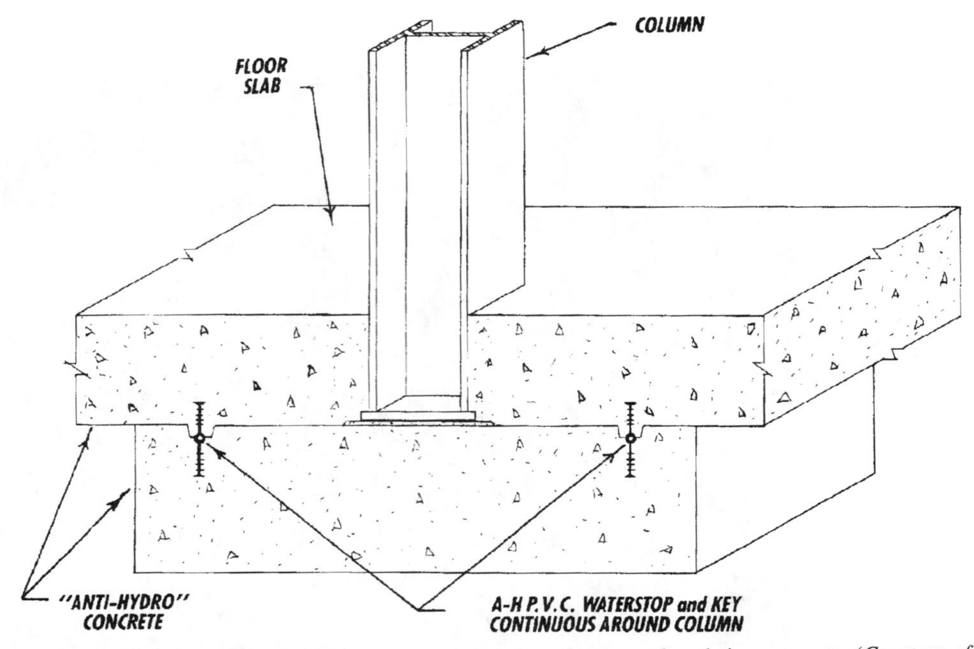

FIGURE 2.30 Waterstop application around structural steel column foundation supports. (*Courtesy of AntiHydro*)

FIGURE 2.31 Placement of spliced PVC joint in formwork. (*Courtesy of J.P. Specialties*)

FIGURE 2.32 Field quality-control procedures ensure successful installation of waterstop. (*Courtesy of J.P. Specialties*)

Capillary action begins by liquid water saturating lower areas adjacent to the water source. This transgresses to a mixture of liquid and vapor above the saturation layer. Finally, only vapor exists in upper soil areas. This vapor is as damaging as water to interior building areas. Soil capillary action can add as much as 12 gallons of water per day per 1,000 sf of slab-on-grade area if insufficient waterproofing protection is not provided.

Microscopic capillaries and pores that naturally occur in concrete substrates create the ability for the concrete to allow water and moisture to move readily through below-grade walls and floors. This process is particularly sustainable when the interior space of the structure has lower humidity than the 100% humidity of the adjacent water-saturated soil and when the occupied space is warmer than the soil. These conditions present ideal circumstances for water to be actually drawn into the occupied space if not protected with waterproofing materials or at minimum vapor barriers where appropriate.

Water vapor penetrates pores of concrete floors, condensing into water once it reaches adjacent air-conditioned space. This condensation causes delamination of finished floor

FIGURE 2.33 Installation of swell-type waterstop. (*Courtesy of Vandex*)

surfaces, mildew, and staining. Therefore, it is necessary to prevent or limit capillary action, even when using waterproof membranes beneath slabs. Excavating sufficiently below finished floor elevation and installing a bed of capillary-resistant soil provides drainage of water beneath slabs on grade.

This combination of foundation drainage and soil composition directs water away from a structure and is necessary for any waterproofing and envelope installation. Refer again to Figure 2.1 for recommended controls for proper surface and groundwater.

POSITIVE AND NEGATIVE SYSTEMS

In new and remedial installations, there are both *negative side* and *positive side* below-grade systems. Positive-side waterproofing applies to sides with direct exposure to water or a hydrostatic head of water. Negative-side waterproofing applies to the opposite or interior side from which water occurs. Examples are shown in Fig. 2.34.

Although both systems have distinct characteristics, as summarized in Table 2.3, the majority of available products are positive-type systems. Negative systems are limited to cementitious-based materials, which are frequently used for remedial applications. Some materials apply to negative sides of a structure for remedial applications but function as positive-side waterproofing. These materials include chemical grouts, epoxy grouts, and pressure grouts. Admixtures (material added or mixed into mortars, plaster, stucco, and concrete) have both positive and negative features but are not as effective as surface-applied systems.

The principal advantage of a negative system is also its principal disadvantage. It allows water to enter a concrete substrate, promoting both active curing and the corrosion and deterioration of reinforcing steel if chlorides are present. Positive-side waterproofing produces an opposite result—no curing of concrete, but protection of reinforcing steel and of the substrate itself.

Positive and negative below-grade systems include

- Cementitious systems
- Fluid-applied membranes
- Sheet-membrane systems
- Hydros clay
- Vapor barriers

CEMENTITIOUS SYSTEMS

Cementitious waterproofing systems contain a base of Portland cement, with or without sand, and an active waterproofing agent. There are four types of cementitious systems: metallic, capillary system, chemical additive systems, and acrylic modified systems. Cementitious systems are effective in both positive and negative applications, as well as in remedial applications. These systems are brushed or troweled to concrete or masonry surfaces and become an integral part of a substrate.

Cementitious systems are excellent materials for use with civil and infrastructure projects, both above and below-grade, using both positive and negative applications. These projects generally consist of large concrete components that make the same generic composition cementitious systems relatively easy to specify and install without compatibility problems. Among the types of structures cementitious systems are used for:

- Tunnels
- Underground vaults
- Water reservoirs
- Water and sewage treatment facilities
- Elevator and escalator pits
- Below-grade concrete structures

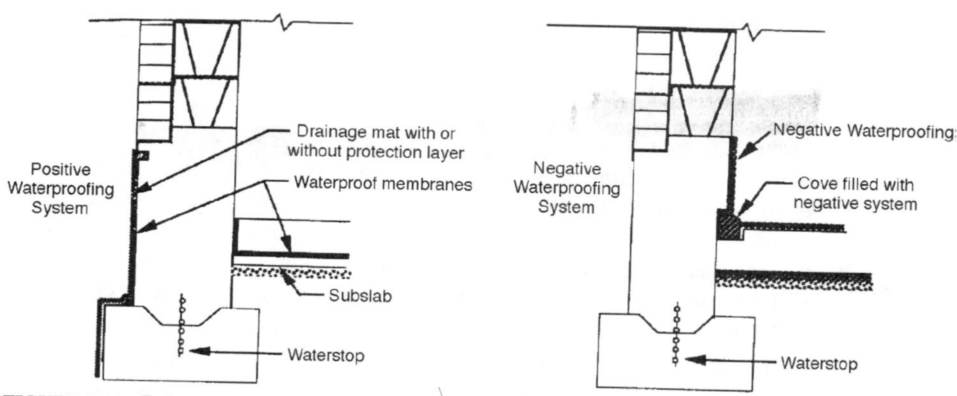

FIGURE 2.34 Below-grade positive and negative waterproofing details.

TABLE 2.3 Comparison of Positive and Negative Waterproofing Systems*

Positive systems	Negative systems
Advantages	
Water is prevented from entering substrate surface Substrate protected from freeze-thaw cycles Substrate is protected from corrosive chemicals in groundwater	Accessible after installation Concrete substrate is allowed to moist cure Eliminates need for subslabs and well pointing for foundation waterproofing
Disadvantages	
Concrete may not cure properly System inaccessible for repairs after installation Subslabs and well pointing necessary for foundation waterproofing	Limited to application of cementitious systems No protection from freeze-thaw cycles No protection of substrate or reinforcing steel from ground water and chemicals

*With all positive systems, concrete should cure properly (up to 21 days) before application of any waterproofing materials.

- Swimming pools
- Cooling tower basins

In new construction, where costs and scheduling are critical, these systems are particularly effective. They do not require a completely dry substrate, and concrete does not need to be fully cured before application. This eliminates well pointing and the need for water control during construction. These systems apply to both walls and floors at one time, thereby eliminating staging of waterproofing operations. No subslabs are required for horizontal applications in new construction preventative waterproofing installations. Finally, in cases such as elevator pits, the waterproofing is completed almost any time during construction as best fits scheduling.

All cementitious systems are similar in application and performance but repel water differently by the proprietary additives of a manufacturer's formulations. Cementitious systems have several mutual advantages, including seamless application after which no protection board installation is necessary.

All cementitious materials lack crack bridging or elastomeric properties but are successfully applied to below-grade areas that do not experience thermal movement. However, below-grade areas are subject to freeze-thaw cycling and structural settlement. If these cause movement or cracking, a cementitious system will crack, allowing water infiltration.

Metallic systems

Metallic materials contain a mixture of sand and cement with finely graded iron aggregate or filings. When mixed with water to form a slurry for application, the water acts as an agent permitting the iron filings to oxidize. These materials expand due to this oxidizing, which then effectively seals a substrate and prohibits further transmission of water through the material. This system is one of the oldest methods used for waterproofing (first patented in 1906) and remains today an effective waterproofing system. (See Fig. 2.35.)

Metallic systems are applied in two or three coats, with the final coat a sand and cement mixture providing protection over base coat waterproofing where exposed. This final coat seals the metallic coats and prevents leaching or oxidization through paints or

FIGURE 2.35 Negative application of cementitious waterproofing. (*Courtesy of Vandex*)

finishes applied over waterproofed areas. To prevent excessive wear, concrete toppings are installed over horizontal exposed surfaces subject to pedestrian or vehicular traffic.

If drywall or paneling is installed over the waterproofing, furring strips are first applied by gluing them directly to the cementitious system. This eliminates nailing the boards through the cementitious membrane. Carpet perimeter tracks should be applied in the same manner to prevent damage.

Capillary/crystalline systems

Capillary/crystalline systems are mixtures of cement and sand in combination with proprietary chemical derivatives in dry or liquid form. The systems are applied in trowel, brush, or spray applications. Unlike other cementitious systems however, capillary have the additional advantage of an application using only the dry mix product that is broadcast directly over concrete that has not yet reached final set and cure. This is referred to as the "dry-shake" method, commonly used on slab components as a vapor barrier, as additional protection with below-grade slab waterproofing systems, or as a stand-alone waterproofing system. A typical dry-shake application is shown in Fig. 2.36.

A capillary/crystalline system not only waterproofs, as a system itself; the chemical additives are able to penetrate into the concrete wall or slab and react with the calcium hydroxide and available capillary water present to form crystalline structures within the concrete itself. These crystalline structures block transmission of water through the substrate, adding additional water repellency to the envelope components.

The chemical process begins immediately upon application of the waterproofing system but can take as many as thirty days to fully reach maximum repellency. Once fully cured, capillary/crystalline systems have been tested to withstand hydrostatic pressures as great as 400 feet of water head. These systems have other advantages compared to other cementitious systems, including the following:

FIGURE 2.36 Dry-shake application of crystalline cementitious waterproofing. (*Courtesy of Vandex*)

- No need for a protection layer.
- Some products have stated capability to seal hairline cracks that occur after installation.
- Most are not harmed in the presence of chemicals and acids, making their application ideal for storage tanks, sewage treatment facilities, and similar structures.
- Penetrate and react with the concrete substrate to form additional "belt and suspenders" protection.

Curing installed systems is critical for adequate crystalline growth. The curing should continue 24–48 hours after installation. Concrete or masonry substrates must be wet to apply these systems, which may be installed over uncured concrete.

In exposed interior applications, coating installation should be protected by plastic, drywall, or paneling applied over furring strips. Floor surfaces are protected by concrete overlays, carpet, or tile finishes.

Chemical additive systems

Chemical cementitious systems are a mixture of sand, cement, and proprietary chemicals (inorganic or organic), which when applied to masonry or concrete substrates provide a watertight substrate by chemical action. Proprietary chemicals are unique to each manufacturer, but typically include silicate and siloxane derivatives in combination with other chemicals. While the chemicals do not penetrate the substrate like the other cementitious systems, chemical systems also effectively become an integral part of the substrate after application.

Chemical cementitious systems, approximately $^1/_{16}$ in. thick, are thinner applications than other cementitious products. As with all cementitious systems, concrete or masonry substrates need not be dry for application. Chemical systems do not require curing, but capillary systems do.

Acrylic modified systems

Acrylic modified cementitious systems add acrylic emulsions to a basic cement-and-sand mixture. These acrylics add waterproofing characteristics and properties to in-place materials. Acrylic systems are applied in two trowel applications, with a reinforcing mesh added into the first layer immediately upon application. This mesh adds some crack-bridging capabilities to acrylic installations. However, since the systems bond tenaciously to concrete or masonry substrates, movement capability is limited.

Acrylic cementitious systems are applicable with both positive and negative installations. Concrete substrates can be damp, but must be cured for acrylic materials to bond properly. Alkaline substrates can deter performance of acrylic-modified cementitious systems.

Acrylic-modified materials are applied in a total thickness of approximately 1/8 in. Reinforcing mesh eliminates the need for protective covering of the systems on floor areas in minimal or light-traffic interior areas.

The properties of all types of cementitious systems are summarized in Table 2.4.

CEMENTITIOUS SYSTEM APPLICATION

Before applying cementitious systems, substrates must be free of dirt, laitance, form release agents, and other foreign materials. Manufacturers typically require concrete surfaces to be acid-etched, lightly sand-blasted, or bush-hammered to a depth of cut of approximately 1/16 in. This ensures adequate bonding to a substrate.

All tie holes, honeycomb, and cracks must be filled by packing them with an initial application of the cementitious system. Refer to Fig. 2.37. Construction joints, wall-to-floor joints, wall-to-wall intersections, and other changes in plane should be formed or grooved with a 1-in by 1-in cutout to form a cove. This cove is then packed with cementitious material before initial application. This is a critical detail for cementitious systems, as they do not allow for structural or thermal movement. This cove prevents water infiltration at weak points in a structure where cracks typically develop. At minimum, if a cove is not formed, place a cant of material at the intersections, using a dry mix of cementitious material.

Cementitious systems do not require priming of a substrate before application. However, wetting of the concrete with water is necessary.

Cementitious systems are available in a wide range of packaging. They may be premixed with sand and cement in pails, or chemicals and iron may be provided in separate

TABLE 2.4 Properties of Cementitious Waterproofing Systems

Advantages	Disadvantages
Positive or negative applications	No movement capability
Remedial applications	Job-site mixing required
No subslabs or well pointing required	Not for high traffic areas

FIGURE 2.37 Patching of concrete substrate prior to waterproofing application. (*Courtesy of Vandex*)

containers and added to the sand-and-cement mixture. Products are mixed in accordance with manufacturers' recommendations, adding only clean water.

Typically, cementitious systems are applied in two coats after the initial preparatory work is complete. First coats may be proprietary materials only. Second coats are usually the chemical or metallic materials within a cement-and-sand mixture. Third coats are applied if additional protection is necessary. They consist only of sand and cement for protecting exposed portions or adding texture. Acrylic systems often require a reinforcing mesh to be embedded into the first-coat application.

Thickness of a system depends upon the sand and cement content of the coatings. The systems are applied by trowel, brush, or spray. Refer to Fig. 2.38. Certain systems are dry-broadcast over just-placed concrete floors to form a waterproofing surface integral with the concrete.

This method is referred to as the *dry-shake application* method. Broadcasting powder onto green concrete is followed by power troweling to finish the concrete and distribute the chemicals that are activated by the concrete slurry. This method should not be used for critical areas of a structure subject to water head, as it is difficult to monitor and control. Refer again to Fig. 2.36.

To protect exposed floor applications, a 2-in. concrete topping, carpet, tile, or other finish is applied over the membrane. Walls can be finished with a plaster coating or furred out with adhesively applied drywall or other finish systems.

These systems require proper curing of the cementitious waterproof coating, usually a wet cure of 24–48 hours. Some systems may have a chemical additive to promote proper curing.

These systems do not withstand thermal or structural substrate movement. Therefore they require special detailing at areas that are experiencing movement, such as wall-floor intersec-

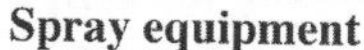

Spray equipment

A trowel

A brush

FIGURE 2.38 Spray, trowel, and brush applications of cementitious waterproofing. (*Courtesy of Vandex*)

tions. It is advantageous to install negative cementitious systems after a structure is completely built. This allows structural movement such as settling to occur before application.

A typical installation for all cementitious systems is elevator and escalator pits similar to Fig. 2.39, which details the installation for this type application. Note that the system calls for two coats with no protection course and the typical cove detailing at the wall-floor intersection. This detailing would be improved by the installation of waterstop at these intersections. Some cementitious manufacturers will permit the use of their product to supplement protection of these intersections when waterstop is not used, as shown in Fig. 2.40. In this detail, the cementitious product is installed continuously on the floor-foundation slab under the wall area intersection. Further protection could be added by installing premanufactured drainage systems on the walls and below-slab locations.

While not often recommended for below-grade applications subject to hydrostatic pressure, concrete block walls are sometimes used as the wall component. The mortar joints are the weak points in this design, and cementitious systems often are not able to protect against the settlement cracking that occurs. Typically, fluid-applied membranes or sheet-good systems would be preferred; however, some cementitious manufacturers do provide detailing for this type of installation, as shown in Fig. 2.41. Note that the cementitious system is applied as a positive system in this situation, with two heavy slurry coats applied to afford the necessary protection required. Also note that a cove cant is added to the exterior side also at the wall-floor intersection in addition to the cove installation on the floor-wall negative application. Since waterstop is not applicable for the concrete block, the manufacturer prescribes a detail coat of the material on the foundation before block is laid. A drainage

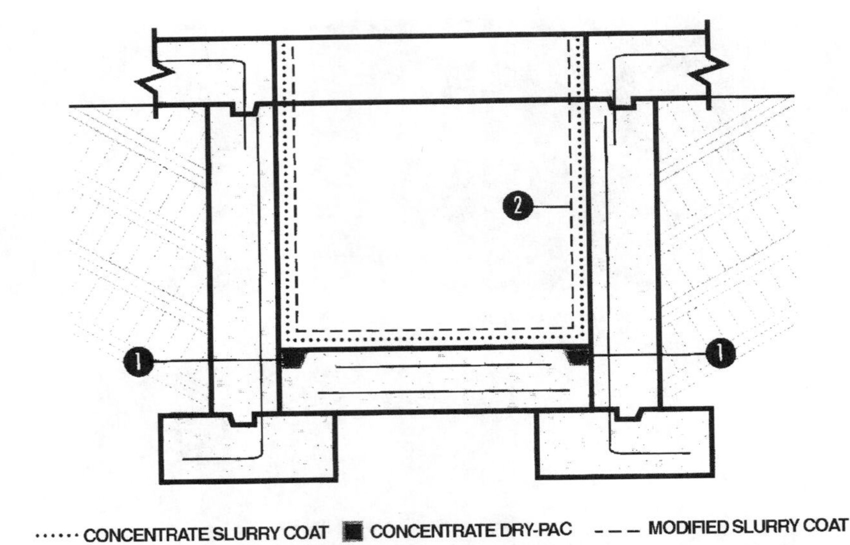

FIGURE 2.39 Typical cementitious cetailing for elevator pit waterproofing. (*Courtesy of Xypex*)

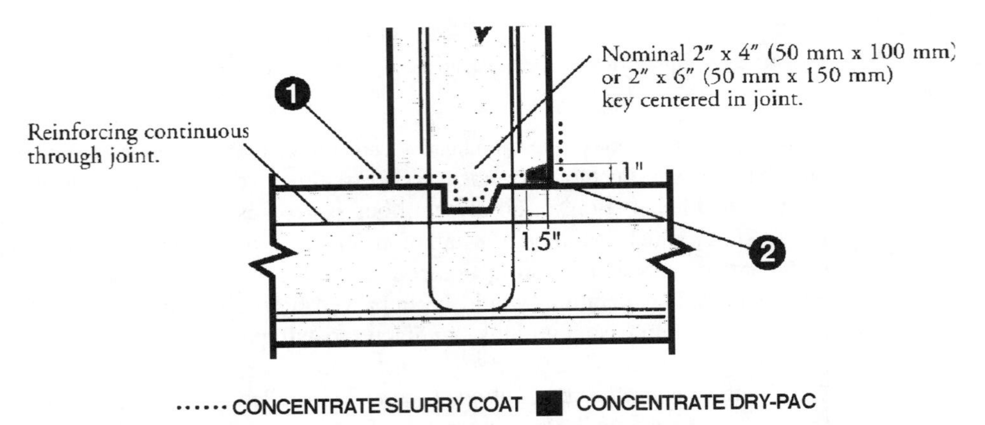

FIGURE 2.40 Application details for critical wall-to-floor juncture, with product run continuously under the wall structure. (*Courtesy of Xypex*)

system is installed for additional protection, as should a vertical drainage system. Again, such a detail should be used with caution because of the mortar joint weakness. Figure 2.42 details a manufacturer's alternative installation suggestion using the cementitious system in both interior and exterior waterproofing applications to ensure watertightness under a "belt and suspenders" system.

Cementitious systems are ideal for many below-grade civil structures that are typically concrete structures. Figure 2.43 provides a manufacturer's cementitious system for an underground vault. Note in this detail how the cementitious system is used as a negative system on the slab-

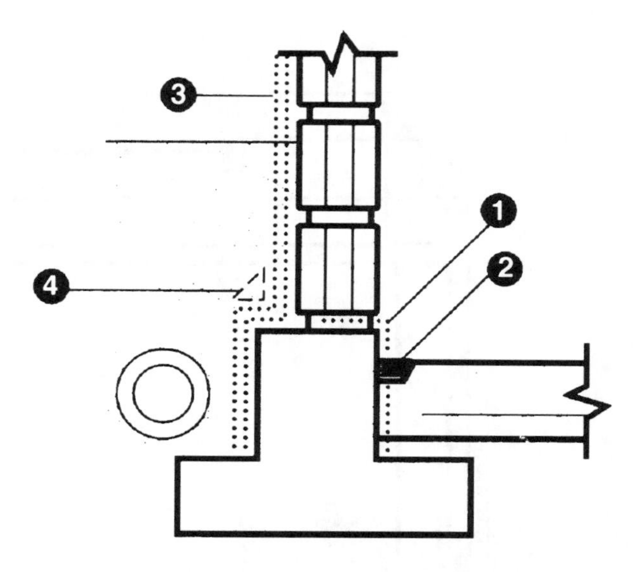

FIGURE 2.41 Application of detail for cementitious waterproofing over concrete block. (*Courtesy of Xypex*)

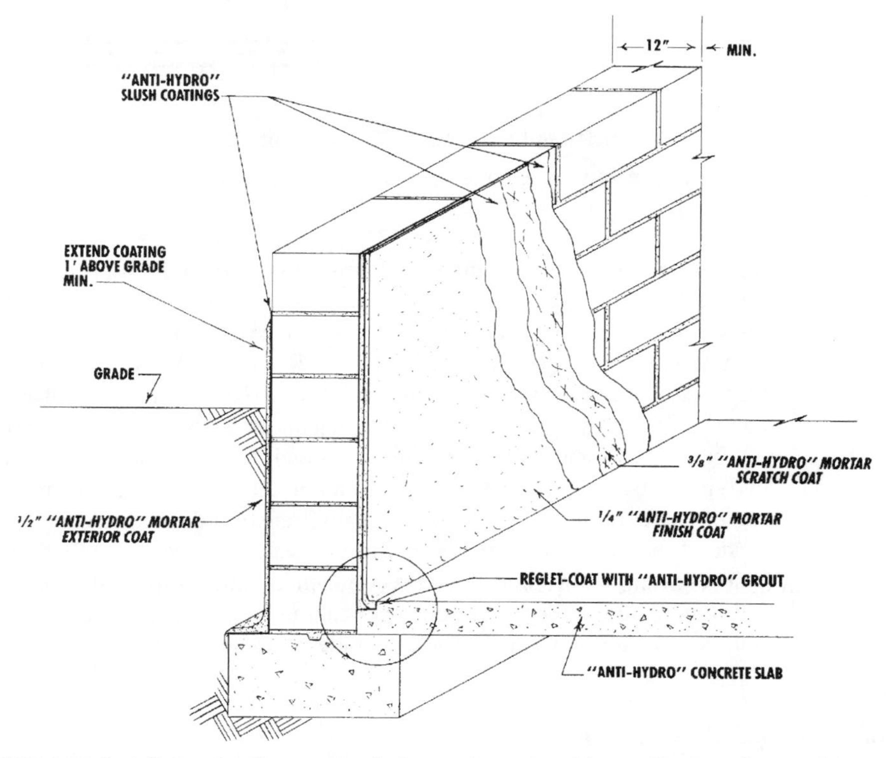

FIGURE 2.42 Installation detail suggesting both negative and positive application of cementitious waterproofing. (*Courtesy of AntiHydro*)

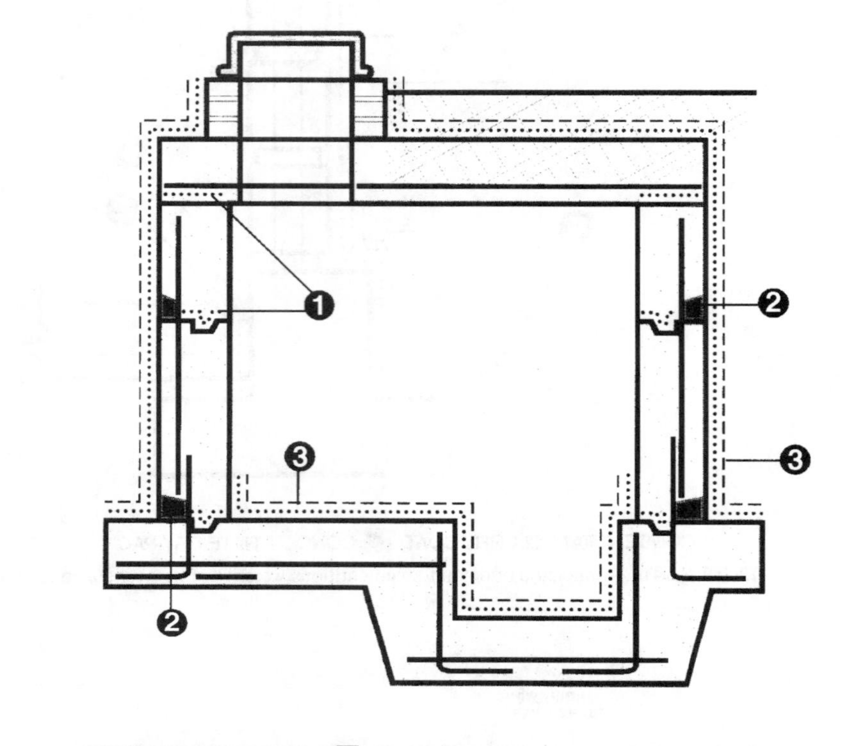

FIGURE 2.43 Civil structure cementitious waterproofing detailing. (*Courtesy of Xypex*)

on-grade portion, then transitions to a positive system on the walls and ceiling structure. This transfer from negative to positive is accomplished by applying the cementitious system continuously along the slab including the area beneath the exterior walls. Cove details packed with the manufacturer's materials occur at the floor-wall juncture. A cove is also used for additional protection at the construction joint in the wall. It would be recommended that waterstop be used at the wall floor juncture and the construction joint for the most complete envelope protection.

Figures 2.44 and 2.45 present typical detailing for cementitious systems on two other civil projects, a sewage treatment digester and a swimming pool structure. Similar detailing of cementitious systems can be easily transferred to other concrete structures below-grade. Again, combining the proper use of waterstop, cove installations at structure weak points (note the cove installation recommendations at pipe and other similar penetrations in Figure 2.46), and cementitious applications in accordance with manufacturer's instructions will result in watertight below-grade concrete structures.

FLUID-APPLIED SYSTEMS

Fluid-applied waterproof materials are solvent-based mixtures containing a base of urethanes, rubbers, plastics, vinyls, polymeric asphalts, or combinations thereof. Fluid membranes are

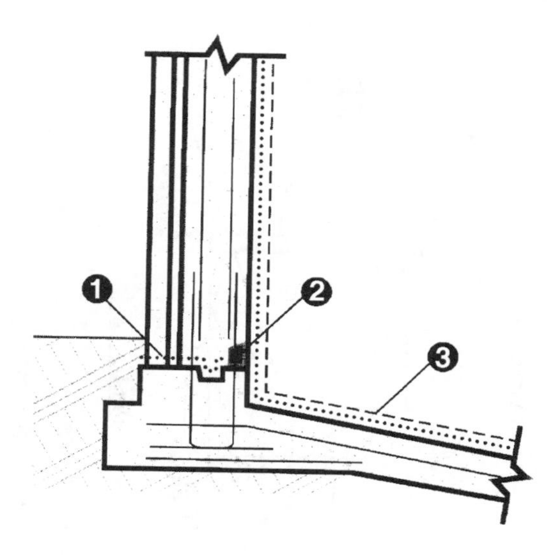

······ **CONCENTRATE SLURRY COAT** ■ **CONCENTRATE DRY-PAC** – – – **MODIFIED SLURRY COAT**

FIGURE 2.44 Civil structure cementitious waterproofing detailing. (*Courtesy of Xypex*)

applied as a liquid and cure to form a seamless sheet. Since they are fluid applied, controlling thickness is critical during field application (see Fig. 2.47).

Therefore, field measurements must be made (wet or dry film) for millage control. The percentages of solids in uncured material vary. Those with 75 percent solids or less can shrink, causing splits, pinholes, or insufficient millage to waterproof adequately.

Fluid systems are positive waterproofing side applications and require a protection layer before backfilling. Fluid-applied systems are frequently used because of their ease of application, seamless curing, and adaptability to difficult detailing, such as penetrations and changes in plane. These systems allow both above- and below-grade applications, including planters and split-slab construction. Fluid systems are not resistant to ultraviolet weathering and cannot withstand foot traffic and, therefore, are not applied at exposed areas.

Several important installation procedures must be followed to ensure performance of these materials. These include proper concrete curing (minimum 7 days, 21–28 days preferred), dry and clean substrate, and proper millage. Should concrete substrates be wet, damp, or uncured, fluid membranes will not adhere and blisters will occur. Proper thickness and uniform application are important for a system to function as a waterproofing material.

Materials can be applied to both vertical and horizontal surfaces, but with horizontal applications, a subslab must be in place so that the membrane can be applied to it. A topping, including tile, concrete slabs, or other hard finishes, is then applied over the membrane. Fluid materials are applicable over concrete, masonry, metal, and wood substrates. Note the application to below-grade concrete block wall in Fig. 2.48.

Fluid-applied systems have elastomeric properties with tested elongation over 500 percent, with recognized testing such as ASTM C-836. This enables fluid-applied systems to bridge substrate cracking up to $^{1}/_{16}$ in. wide.

An advantage with fluid systems is their self-flashing installation capability. This application enables material to be applied seamless at substrate protrusions, changes in planes,

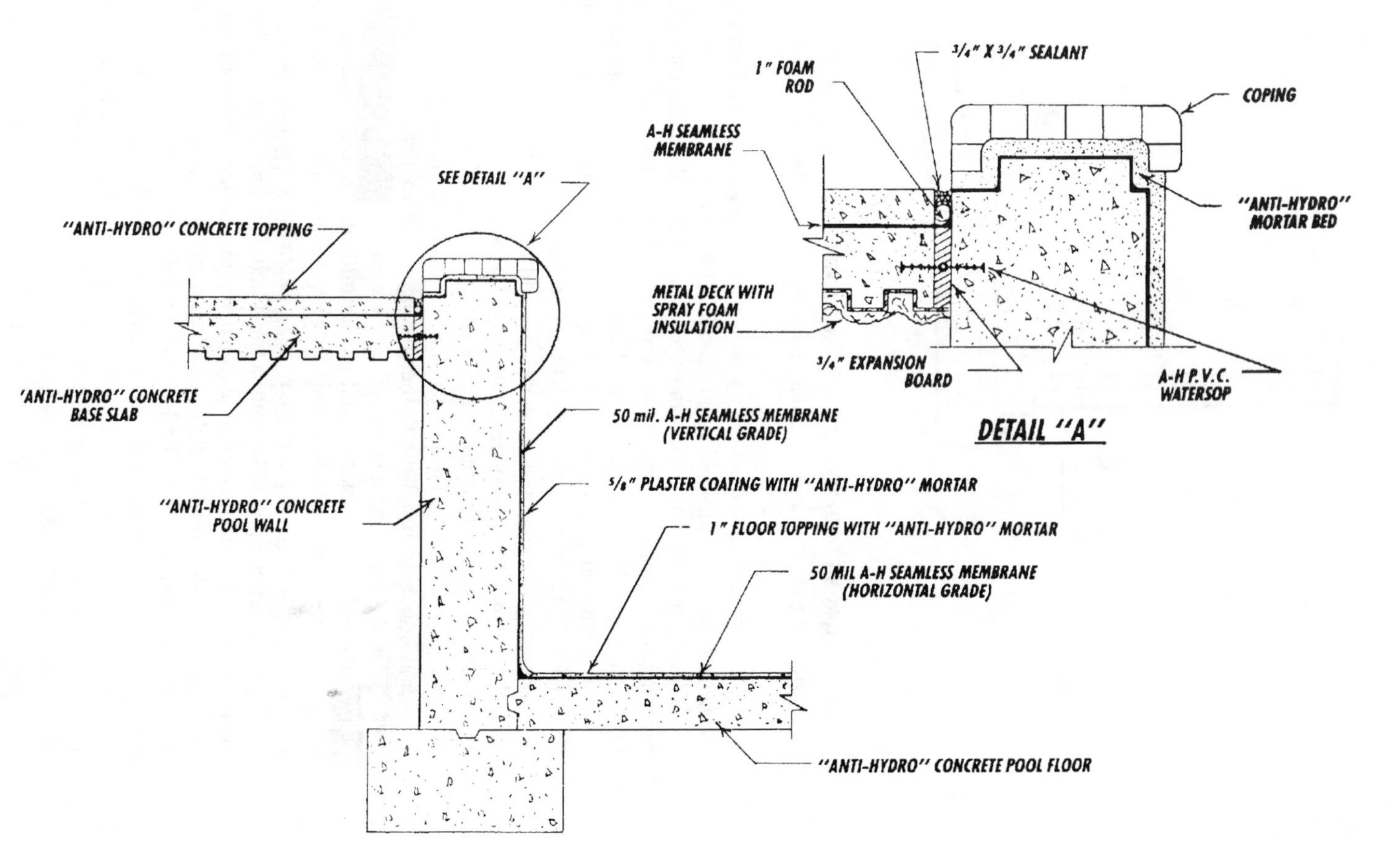

FIGURE 2.45 Civil structure cementitious waterproofing detailing. (*Courtesy of AntiHydro*)

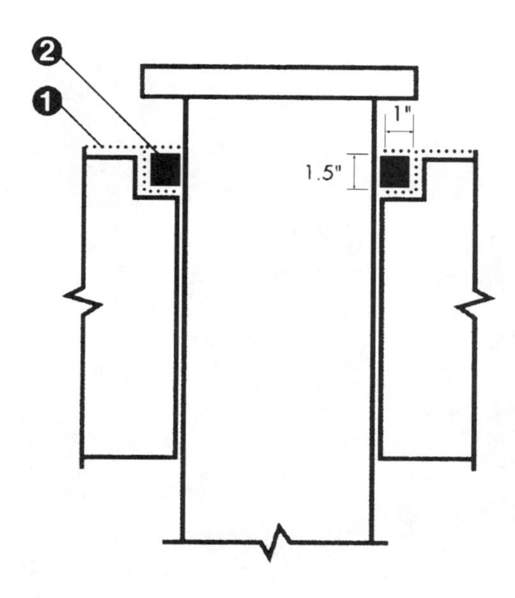

······ CONCENTRATE SLURRY COAT ■ CONCENTRATE DRY-PAC

FIGURE 2.46 Penetration detailing using cementitious waterproofing. (*Courtesy of Xypex*)

FIGURE 2.47 Spray application of fluid-applied membrane. (*Courtesy of LBI Technologies*)

FIGURE 2.48 Fluid-applied membrane application to below-grade block wall. (*Courtesy of Rubber Polymer Corporation*)

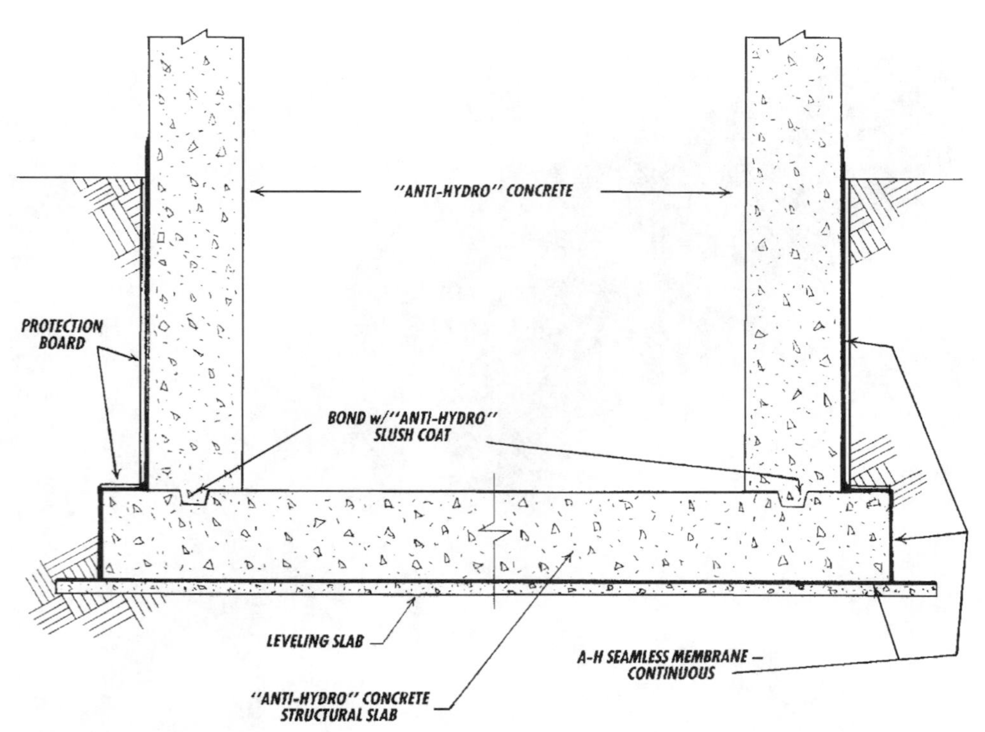

FIGURE 2.49 Typical below-grade application detailing for fluid-applied membranes. (*Courtesy of AntiHydro*)

and floor-wall junctions. Figure 2.49 details a typical below-grade application using fluid membranes. Fluid materials are self-flashing, with no other accessories required for transitions into other building envelope components. However, a uniform 50–60 mil is difficult to control in field applications, and presents a distinct disadvantage with fluid systems.

These systems contain toxic and hazardous chemicals that require safety protection during installation and disposal of materials. Refer to Chapter 12 and the discussion on V.O.C. materials.

Fluid-applied systems are available in the following derivatives: urethane (single or two-component systems), rubber derivatives (butyl, neoprene, or hypalons), polymeric asphalt, coal tar, or asphalt modified urethane, PVC, and hot applied systems (asphalt).

Urethane

Urethane systems are available in one- or two-component materials. Black coloring is added only to make those people who believe waterproofing is still "black mastic" comfortable with the product. Urethanes are solvent-based, requiring substrates to be completely dry to avoid membrane blistering.

These systems have the highest elastomeric capabilities of fluid-applied membranes, averaging 500–750 percent by standardized testing. Urethanes have good resistance to all chemicals likely to be encountered in below-grade conditions, as well as resistance against alkaline conditions of masonry substrates.

Rubber derivatives

Rubber derivative systems are compounds of butyls, neoprenes, or hypalons in a solvent base. Solvents make these materials flammable and toxic. They have excellent elastomeric capability, but less than that of urethane membranes.

Rubber systems are resistant to environmental chemicals likely to be encountered below grade. As with most fluid membranes, toxicity requires safety training of mechanics in their use and disposal.

Polymeric asphalt

A chemical polymerization of asphalts improves the generic asphalt material qualities sufficiently to allow their use as a below-grade waterproofing material. Asphalt compounds do not require drying and curing of a masonry substrate, and some manufacturers allow installation of their asphalt membranes over uncured concrete.

However, asphalt materials are not resistant to chemical attack as are other fluid systems. These membranes have limited life-cycling and are used less frequently than other available systems.

Coal tar or asphalt-modified urethane

Coal tar and asphalt-modified urethane systems lessen the cost of the material while still performing effectively. Extenders of asphalt or coal tar limit the elastomeric capabilities and chemical resistance of these membranes.

Coal tar derivatives are especially toxic, and present difficulties in installing in confined spaces such as small planters. Coal tar can cause burns and irritations to exposed skin areas. Field mechanics should take necessary precautions to protect themselves from the material's hazards.

Polyvinyl chloride

Solvent-based PVC or plastics are not extensively used in liquid-applied waterproofing applications. These derivatives are more often used as sheet membranes for roofing. Their elastomeric capabilities are less than other fluid systems and have higher material costs. They do offer high resistance to chemical attack for below-grade applications.

Hot-applied fluid systems

Hot-applied systems are improvements over their predecessors of coal tar pitch and felt materials. These systems add rubber derivatives to an asphalt base for improved performance, including crack-bridging capabilities and chemical resistance.

Hot systems are heated to approximately 400°F in specialized equipment and applied in thickness up to 180 mil, versus urethane millage of 60 mil (see Fig. 2.50). Asphalt extenders keep costs competitive even at this higher millage. These materials have a considerably extended shelf life compared to solvent-based products, which lose their usefulness in 6 months to 1 year.

Since these materials are hot-applied, they can be applied in colder temperatures than solvent-based systems, which cannot be applied in weather under 40°F. Manufacturers often market their products as self-healing membranes, but in below-grade conditions this is a questionable characteristic. Properties of typical fluid-applied systems are summarized in Table 2.5.

FLUID SYSTEM APPLICATION

Substrate preparation is critical for proper installation of fluid-applied systems. See Fig. 2.51 for typical fluid system application detail. Horizontal concrete surfaces should have a light broom finish for proper bonding. Excessively smooth concrete requires acid etching or sandblasting to roughen the surface for adhesion. Vertical concrete surfaces with plywood form finish are satisfactory, but honeycomb, tie holes, and voids must be patched, with fins and protrusions removed (Fig. 2.52).

Wood surfaces must be free of knotholes, or patched before fluid application. Butt joints in plywood decks should be sealed with a compatible sealant followed by a detail coat of membrane. On steel or metal surfaces, including plumbing penetrations metal must be cleaned and free of corrosion. PVC piping surfaces are roughened by sanding before membrane application.

Curing of concrete surfaces requires a minimum of 7 days, preferably 28 days. On subslabs, shorter cure times are acceptable if concrete passes a mat dryness test. Mat testing is accomplished by tapping visquene to a substrate area. If condensation occurs within 4 hours, concrete is not sufficiently cured or is too wet for applying material.

Blistering will occur if materials are applied to wet substrates, since they are nonbreathable coatings. Water curing is the recommended method of curing, but some manufacturers allow sodium silicate curing compounds. Most manufacturers do not require primers over concrete or masonry surfaces; however, metal substrates should be primed and concrete if required (Fig. 2.53).

All cold joints, cracks, and changes in plane should be sealed with sealant followed by a 50–60-mil membrane application, 4-in wide. Figure 2.54 details typical locations where

FIGURE 2.50 Application process for hot-applied membrane. (*Courtesy of American Hydrotech*)

TABLE 2.5 Properties of Fluid-Applied Systems

Advantages	Disadvantages
Excellent elastomeric properties	Application thickness controlled in field
Ease of application	Not applicable over damp or uncured surfaces
Seamless application	Toxic chemical additives

additional layers of membrane application are required for reinforcement.

Cracks over 1/16 in. should be sawn out, sealed, then coated. Refer to Fig. 2.55 for typical detailing examples.

At wall-floor intersections, a sealant cant approximately 1/2–1 in. high at 45° should be applied, followed with a 50-mil detail coat. All projections through a substrate should be similarly detailed. Refer again to Fig. 2.56 for typical installation detailing. At expansion joints and other high-movement details, a fiberglass mesh or sheet flashing is embedded in the coating material. This allows greater movement capability.

Figure 2.57 provides a perspective view of a typical below-grade fluid-applied membrane application using a sheet material to reinforce the horizontal-to-vertical transition. The detail coat applied at this point provides additional protection at the same transition. This detail emphasizes the 90%/1% principle, assuming that the weak point in this structure (wall to floor juncture) is a likely candidate for water infiltration. Recognizing this, the manufacturer has tried to idiot-proof the detail by adding several layers of protection, including the waterstop and drainage board that properly completes the waterproof installation.

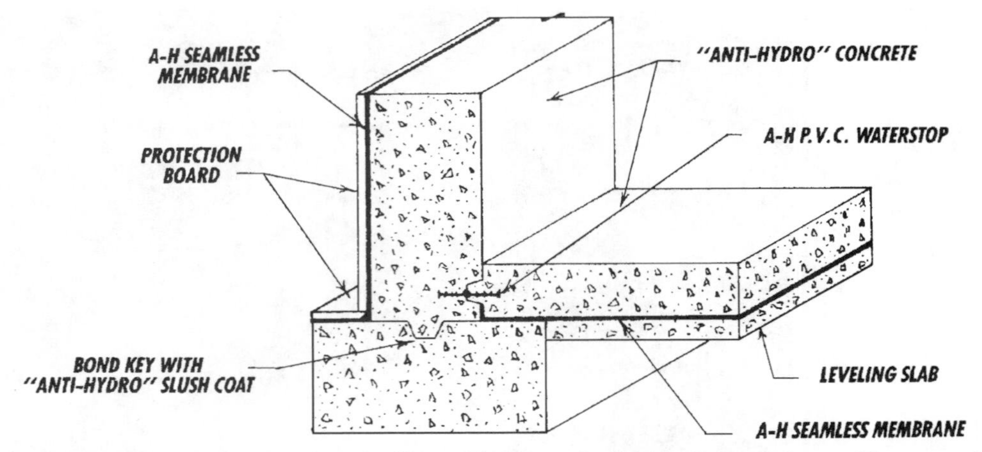

FIGURE 2.51 Typical application detailing of below-grade fluid-applied membrane. (*Courtesy of AntiHydro*)

FIGURE 2.52 Preparation of block wall prior to membrane application. (*Courtesy of Rubber Polymer Corporation*)

FIGURE 2.53 Roller application of fluid-applied membrane. (*Courtesy of American Hydrotech*)

The detailing provided in Fig. 2.58 shows a fluid membrane application that runs continuously on the horizontal surface, including beneath the wall structure. Many engineers will not permit such an application due to the membrane acting as a bond break between the wall and floor components that might present structural engineering problems.

In Fig. 2.59, the manufacturer has detailed the use of a liquid membrane over foundation lagging using a fluid-applied membrane before the concrete is placed. In this detail, the membrane is applied to a sheet-good fabric that acts as the substrate. This is applied over a premanufactured drainage mat to facilitate water drainage and hydrostatic pressure. This would be a difficult application, and not as idiot-proof as using a clay system in a similar installation as outlined later in this chapter.

All penetrations occurring through a membrane application must be carefully detailed to prevent facilitating water infiltration at this "90%/1% principle" envelope area. Figure 2.60 shows a recommend installation at a pipe penetration. Note that the concrete has been

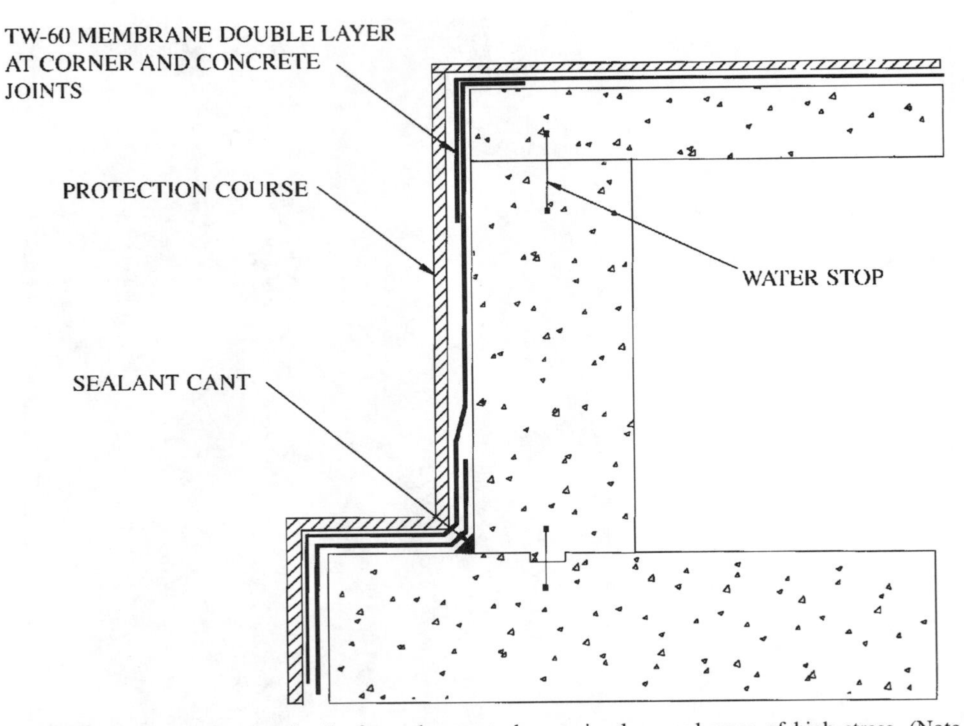

FIGURE 2.54 Reinforcement detail of membrane at changes-in-plane and areas of high stress. (Note sealant cant added at floor-wall juncture, and membrane layers at changes-in-plane.) (*Courtesy of Tamko Waterproofing*)

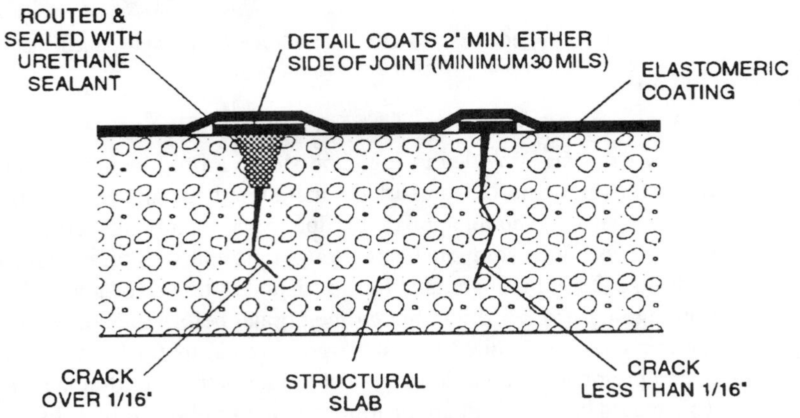

FIGURE 2.55 Substitute crack detailing and preparation for membrane appliation. (*Courtesy of Neogard*)

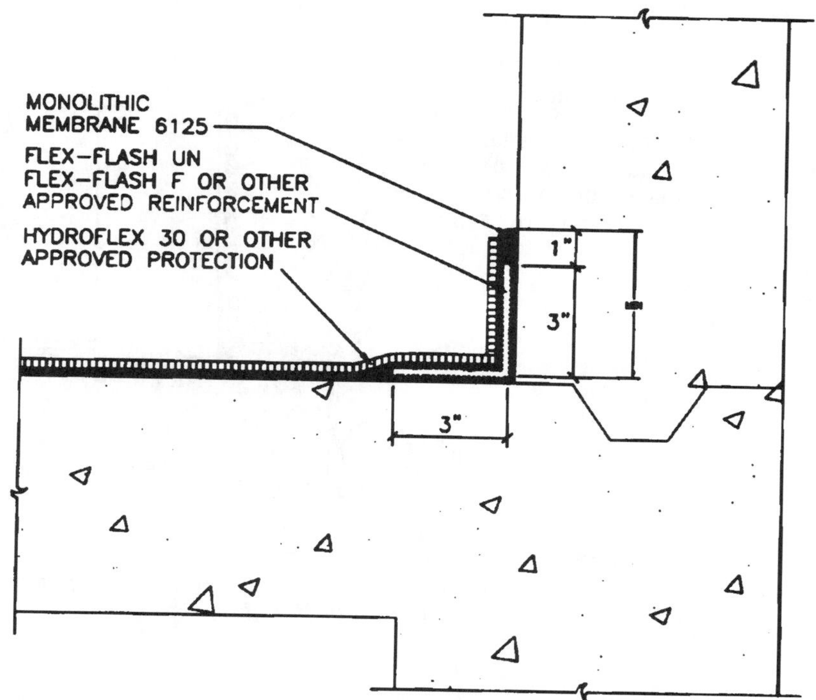

FIGURE 2.56 Transition detailing for membrane applications. (*Courtesy of American Hydrotech*)

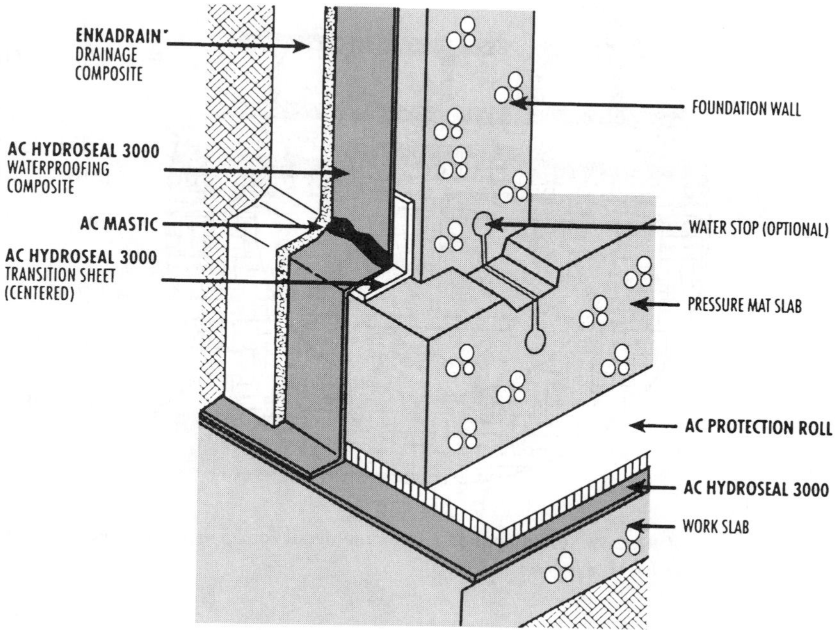

FIGURE 2.57 Perspective detail emphasizing the reinforcing of the wall-to-floor transition. (*Courtesy of of NEI Advanced Composite Technology*)

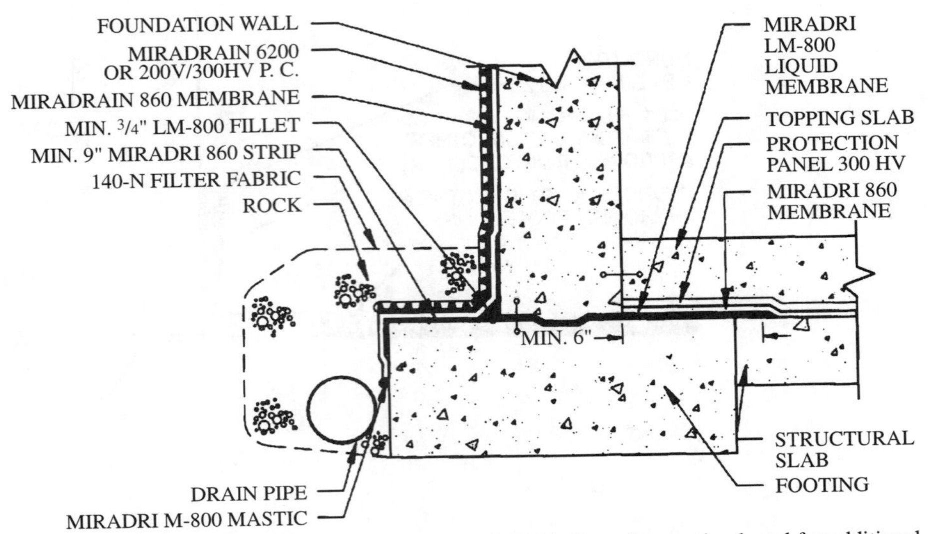

FIGURE 2.58 Application detailing using drainage board in lieu of protection board for additional waterproofing protection. (*Courtesy of TC MiraDRI*)

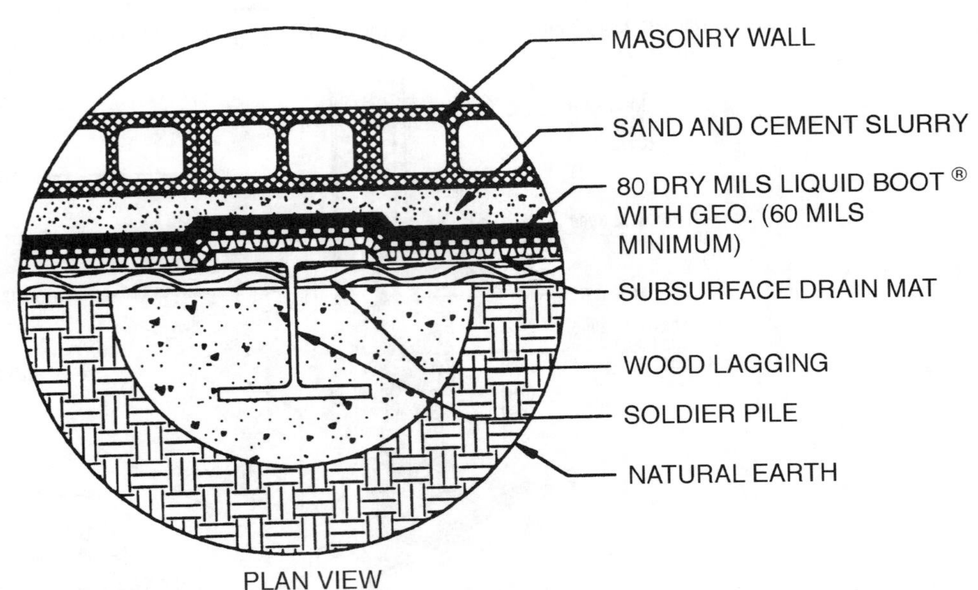

FIGURE 2.59 Fluid-applied membrane detail for application directly to foundation lagging. (*Courtesy of LBI Technologies*)

notched to install sealant along the perimeter of the pipe. The waterproof membrane is then detail-coated around the pipe, followed by the regular application.

Fluid-applied membrane applications all require that the termination of the membrane be carefully completed to prevent disbonding at the edge and resulting water infiltration. Figure 2.61 shows the membrane terminating with a sealant of manufacturer-supplied mastic. Figure 2.62 details the use of a reglet to terminate and seal the membrane, which could also simultaneously be used to terminate above-grade waterproofing.

Control coating thickness by using notched squeegees or trowels. If spray equipment is used, take wet millage tests at regular intervals during installation. Application by roller is not recommended. Pinholes in materials occur if a substrate is excessively chalky or dusty, material cures too fast, or material shrinks owing to improper millage application.

Fluid membranes are supplied in 5- or 55-gal containers. Their toxicity requires proper disposal methods of containers after use. Since these materials rapidly cure when exposed to atmospheric conditions, unopened sealed containers are a necessity.

These materials are not designed for exposed finishes. They will not withstand traffic or ultraviolet weathering. Apply protection surfaces to both horizontal and vertical applications. On vertical surfaces, a 1/2-in. polystyrene material or other lightweight protection system is used. For horizontal installations a 1/8-in., asphalt-impregnated board is necessary. On curved surfaces, such as tunnel work, 90-lb. roll roofing is usually acceptable protection. For better protection and detailing, use premanufactured drainage board in lieu of these protection systems (Fig. 2.63).

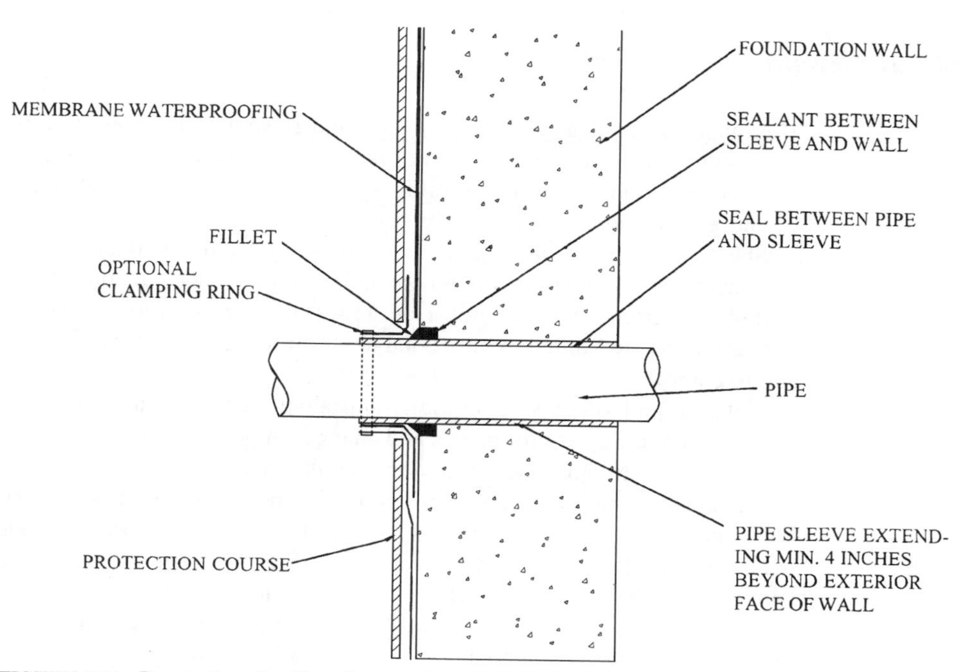

FIGURE 2.60 Penetration detailing for membrane waterproofing applications. (*Courtesy of Tamko Waterproofing*)

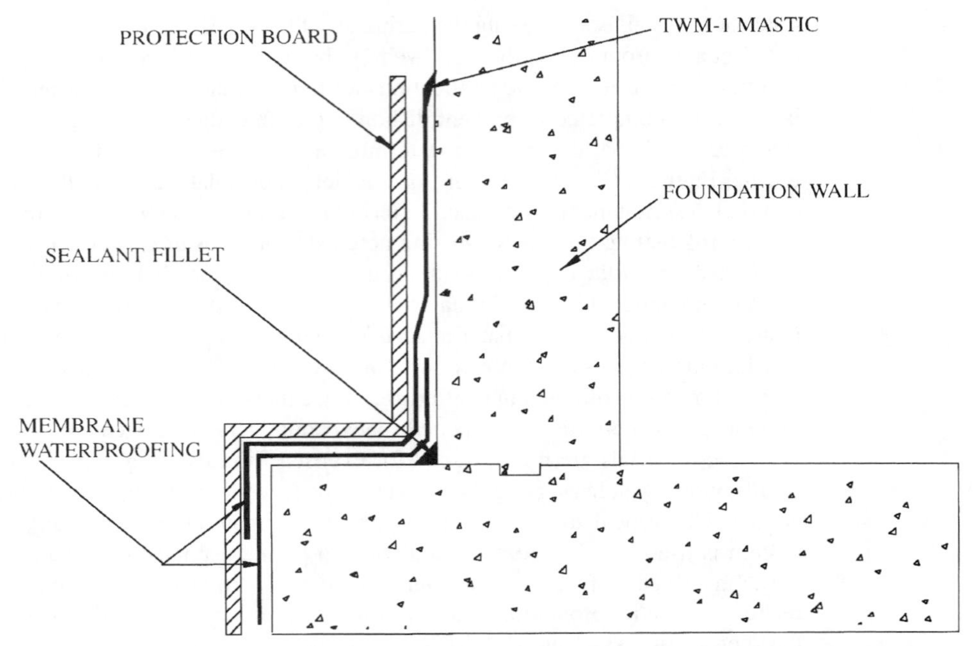

FIGURE 2.61 Termination detailing for membrane waterproofing. (*Courtesy of Tamko Waterproofing*)

SHEET MEMBRANE SYSTEMS

Thermoplastics, vulcanized rubbers, and rubberized asphalts used in waterproofing applications are also used in single-ply roofing applications. Although all systems are similar as a generic grouping of waterproofing systems, consider their individual characteristics whenever you choose systems for particular installations.

Sheet membranes have thickness controlled by facto manufacturing. This ensures uniform application thickness throughout an installation. Sheet manufactured systems range in thickness from 20 to 120 mil. Roll goods of materials vary in width from 3 to 10 ft. Larger widths are limited to horizontal applications, because they are too heavy and difficult to control for vertical applications.

Unlike liquid systems, sheet system installations involve multiple seams and laps and are not self-flashing at protrusions and changes in plane. This is also true for terminations or transitions into other members of the building envelope.

Applications below grade require protection board during backfill operations and concrete and steel placements. Fins and sharp protrusions in substrates should be removed before application, or they will puncture during installation. Materials used in vertical applications should not be left exposed for any length of time before backfilling. Weathering will cause blistering and disbonding if backfill operations must begin immediately after membrane application.

Vertical single-ply applications are more difficult than fluid applications, due to the difficulty of handling and seaming materials. Seams are lapped and sealed for complete

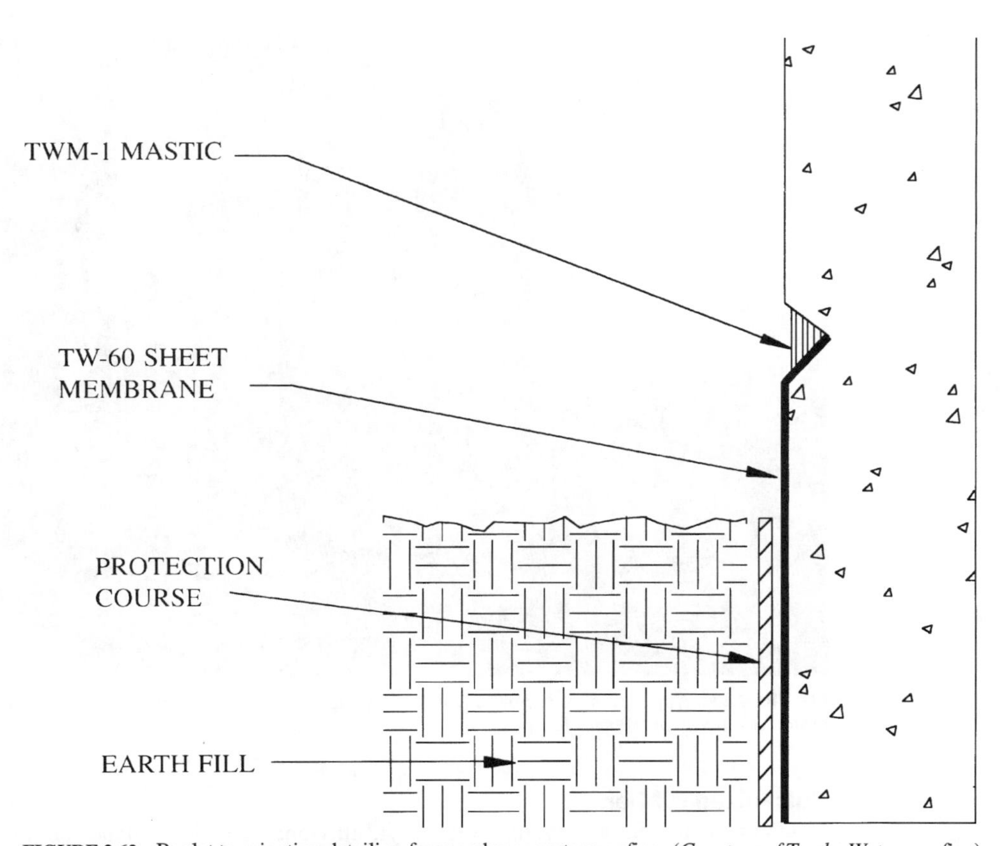

FIGURE 2.62 Reglet termination detailing for membrane waterproofing. (*Courtesy of Tamko Waterproofing*)

waterproofing. In small, confined areas such as planter work, vertical installation and transitions to horizontal areas become difficult and extra care must be taken.

Thermoplastics

Thermoplastic sheet-good systems are available in three compositions: PVC, chlorinated polyurethane (CPE), and chlorosulfonated polyethylene (CSPE), which is referred to as hypalon. Materials are manufactured in rolls of varying widths, but difficulty with vertical applications makes smaller widths more manageable.

On horizontal applications, wider roll widths require fewer seams; therefore, it is advantageous to use the widest workable widths. All three systems adhere by solvent-based adhesives or heat welding at seams.

PVC membranes are available in thicknesses of 30–60 mil. CPE systems vary by as much as 20–120 mil, and hypalon materials (CSPE) are 30–35 mil. All derivatives have excellent hydrostatic and chemical resistance to below-grade application conditions. PVC membranes are generically brittle materials requiring plasticizers for better elastomeric properties, but elongation of all systems is acceptable for below-grade conditions.

FIGURE 2.63 Application of premanufactured drainage board in lieu of protection board to protect membrane. (*Courtesy of Webtec, Inc.*)

Vulcanized rubbers

Vulcanized rubbers are available in butyl, ethylene propylene diene monomer (EPDM), and neoprene rubber. These materials are vulcanized by the addition of sulfur and heat to achieve better elasticity and durability properties. Membrane thickness for all rubber systems ranges from 30–60 mil. These materials are nonbreathable, and will disbond or blister if negative vapor drive is present.

As with thermoplastic materials, vulcanized rubbers are available in rolls of varying widths. Seam sealing is by a solvent-based adhesive, as heat welding is not applicable. A separate adhesive application to vertical areas is necessary before applying membranes. Vulcanized rubber systems incorporate loosely laid applications for horizontal installations.

Although other derivatives of these materials, such as visquene, are used beneath slabs as dampproofing membranes or vapor barriers, they are not effective if hydrostatic pressure exists. Material installations under slabs on grade, by loose laying over compacted fill and sealing joints with adhesive or heat welding, are useful in limited waterproofing applications.

This is a difficult installation procedure and usually not specified or recommended. Loosely laid applications do, however, increase the elastomeric capability of the membrane, versus fully adhered systems that restrict membrane movement.

Rubberized asphalts

Rubberized asphalt sheet systems originally evolved for use in pipeline protection applications. Sheet goods of rubberized asphalt are available in self-adhering rolls with a polyethylene film attached. Self-adhering membranes adhere to themselves, eliminating the need for a seam adhesive. Sheets are manufactured in varying widths of 3–4 ft. and typically 50-ft. lengths.

Also available are rubberized asphalt sheets reinforced with glass cloth weave that require compatible asphalt adhesives for adhering to a substrate. Rubber asphalt products require a protection layer, to prevent damage during backfill or concrete placement operations.

Self-adhering asphalt membranes include a polyethylene film that acts as an additional layer of protection against water infiltration and weathering. The self-adhering portion is protected with a release paper, which is removed to expose the adhesive for placement. Being virtually self-contained, except for primers, this system is the simplest of all sheet materials to install. Figure 2.64 details a typical below-grade installation.

Self-adhering membranes are supplied in 60-mil thick rolls, and accessories include compatible liquid membranes for detailing around protrusions or terminations. Rubberized asphalt systems have excellent elastomeric properties but are not used in above-grade exposed conditions. However, membrane use in sandwich or split-slab construction for above-grade installations is acceptable.

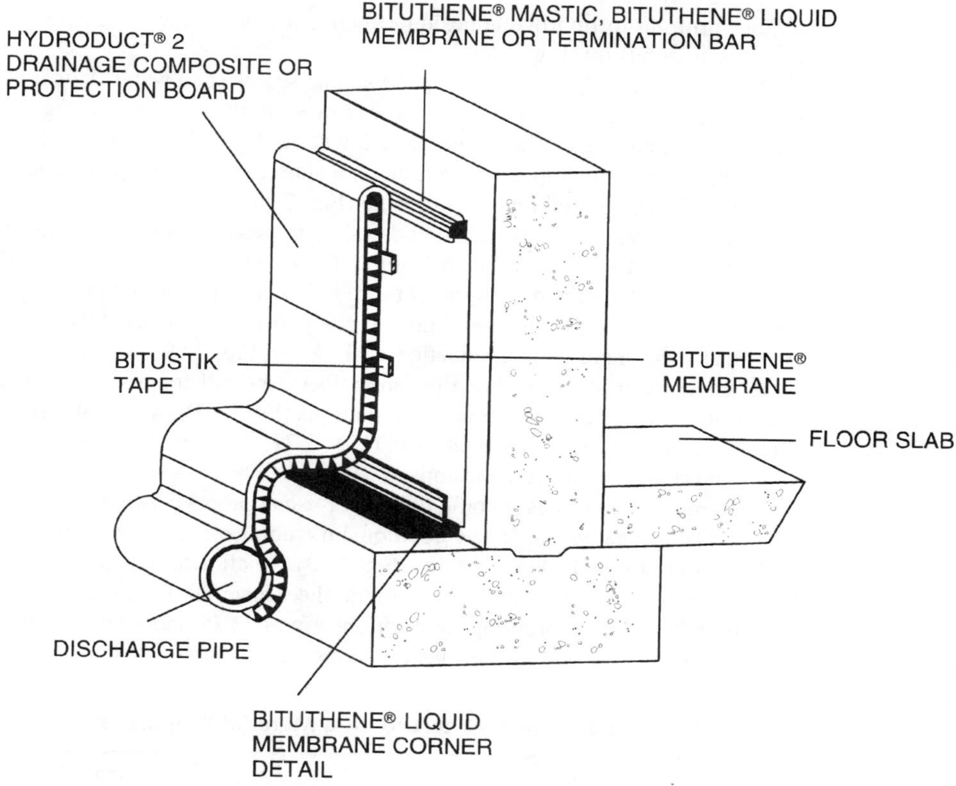

FIGURE 2.64 Below-grade sheet waterproofing system detailing. (*Courtesy of Grace Construction Products*)

Glass cloth–reinforced rubber asphalt sheets, unlike self-adhering systems, require no concrete curing time. Separate adhesive and seam sealers are available. Glass cloth rubber sheets are typically 50 mil thick and require a protection layer for both vertical and horizontal applications. Typical properties of sheet materials are summarized in Table 2.6.

SHEET SYSTEM APPLICATION

Unlike liquid-applied systems, broom-finished concrete is not acceptable, as coarse finishes will puncture sheet membranes during application. Concrete must be smoothly finished with no voids, honeycombs, fins, or protrusions. Concrete curing compounds should not contain wax, oils, or pigments. Concrete surfaces must be dried sufficiently to pass a mat test before application.

Wood surfaces must be free of knotholes, gouges, and other irregularities. Butt joints in wood should be sealed with a 4-in-wide membrane detail strip, then installed. Masonry substrates should have all mortar joints struck flush. If masonry is rough, a large coat of cement and sand is required to smooth surfaces.

Metal penetrations should be cleaned, free of corrosion, and primed. Most systems require priming to improve adhesion effectiveness and prevent concrete dust from interfering with adhesion (Fig. 2.65).

All sheet materials should be applied so that seams shed water. This is accomplished by starting at low points and working upward toward higher elevations (Fig. 2.66). With adhesive systems, adhesives should not be allowed to dry before membrane application. Self-adhering systems are applied by removing a starter piece of release paper or polyethylene backing, adhering membrane to substrate (Fig. 2.67).

With all systems, chalk lines should be laid for seam alignment. Seam lap requirements vary from 2 to 4 in. (Fig. 2.68). Misaligned strips should be removed and reapplied, with material cut and restarted if alignments are off after initial application. Attempts to correct alignment by pulling on the membrane to compensate may cause "fish mouths" or blisters. A typical sheet membrane application is shown in Fig. 2.69.

At changes in plane or direction, manufacturers call for a seam sealant to be applied over seam end laps and membrane terminations (Fig. 2.70). Materials are back-rolled at all seams for additional bonding at laps (Fig. 2.71). Any patched areas in the membrane should be rolled to ensure adhesion.

Each manufacturer has specific details for use at protrusions, joints, and change in plane (Fig. 2.72). Typically, one or two additional membrane layers are applied in these areas and sealed with seam sealant or adhesive (Fig. 2.73). Small detailing is sealed with liquid membranes that are compatible and adhere to the sheet material. Figure 2.74 details a typical column foundation waterproofing application. Figure 2.75 shows the proper treatment of a control or expansion joint using sheet systems.

TABLE 2.6 Sheet Waterproofing Material Properties

Advantages	Disadvantages
Manufacturer-controlled thickness	Vertical applications difficult
Wide rolls for horizontal applications	Seams
Good chemical resistance	Detailing around protrusions difficult

FIGURE 2.65 Applying primer to concrete substrate in preparation for sheet system. (*Courtesy of TC Mira DRI*)

FIGURE 2.66 Application of sheet membrane. (*Courtesy of TC MiraDRI*)

FIGURE 2.67 Removing release paper backing from self-adhering sheet membrane. (*Courtesy of TC MiraDRI*)

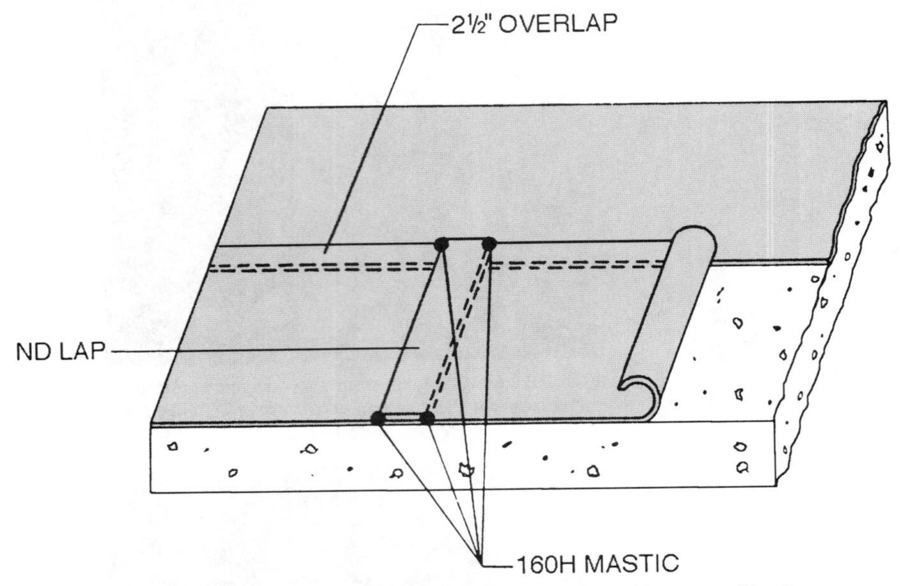

FIGURE 2.68 Seam lap detailing for sheet membranes. (*Courtesy of Protecto Wrap*)

Protection systems are installed over membranes before backfilling, placement of reinforcing steel, and concrete placement. Hardboard, $^1/_8$–$^1/_4$ in thick, made of asphalt-impregnated material is used for horizontal applications. Vertical surfaces use polystyrene board, $^1/_2$ in thick, which is lightweight and applied with adhesives to keep it in place during backfill. Sheet systems cannot be left exposed, and backfill should occur immediately after installation.

Protrusions through the membrane must be carefully detailed as shown in Fig. 2.76. Manufacturers require an additional layer of the sheet membrane around the penetration that is turned on or into the protrusion as appropriate. A bead of sealant or mastic is applied along the edges of the protrusion. For expansion joints in below-grade walls or floors, the installation should include appropriate waterstop and the required additional layers of membrane (Fig. 2.77). Sheet systems must be terminated appropriately as recommended by the manufacturer. Termination details prohibit water from infiltrating behind the sheet and into the structure. Termination bars are often used as shown in Fig. 2.78. Reglets can be used (Fig. 2.79); these also permit the termination of above-grade waterproofing in the same reglet that then becomes a transition detail.

HOT-APPLIED SHEET SYSTEMS

Hot-applied systems are effectively below-grade roofing systems. They use either coal tar pitch or asphalts, with 30-lb roofing felts applied in three to five plies. Waterproofing technology has provided better-performance materials and simpler applications, limiting hot systems usage to waterproofing applications.

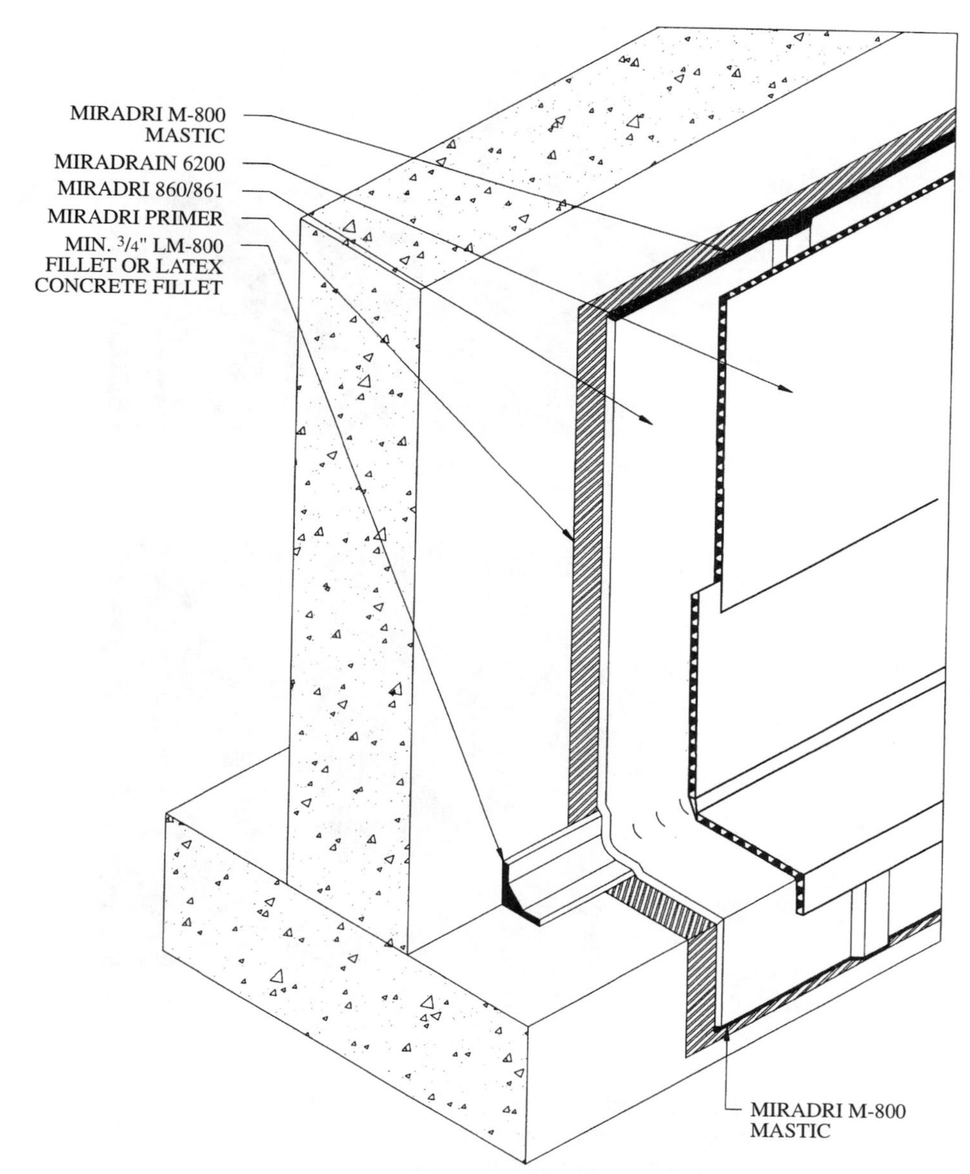

FIGURE 2.69 Typical sheet membrane application detailing. (*Courtesy of TC MiraDRI*)

Hot systems are extremely difficult to apply to vertical surfaces due to the weight of felts. Also, roofing asphalts and coal tars are self-leveling in their molten state, which causes material to flow down walls, during application. Safety concerns are multiplied during their use as waterproofing, because of difficulties in working with the confined areas encountered at below-grade details.

FIGURE 2.70 Applying mastic termination detailing. (*Courtesy of TC MiraDRI*)

FIGURE 2.71 Back-rolling membrane at seams to ensure bonding. (*Courtesy of TC MiraDRI*)

Hot-applied sheet systems have installation and performance characteristics similar to those of roofing applications. These systems are brittle and maintain very poor elastic properties. Extensive equipment and labor costs offset inexpensive material costs. Below-grade areas must be accessible to equipment used for heating materials. If materials are carried over a distance, they begin to cool and cure, providing unacceptable installations. Properties of typical hot-applied sheet systems are summarized in Table 2.7.

CLAY SYSTEMS

Natural clay systems, commonly referred to as *bentonite*, are composed primarily of montmorillonite clay. This natural material is used commercially in a wide range of prod-

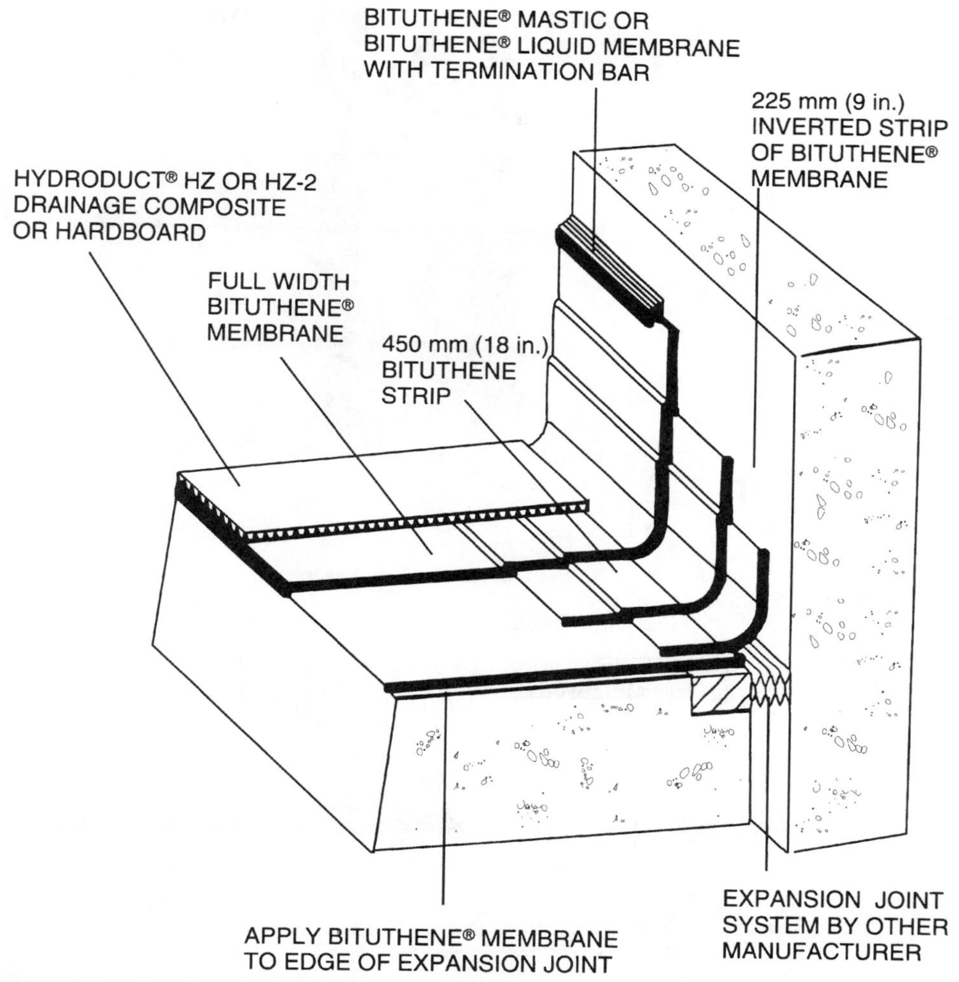

FIGURE 2.72 Transition detailing for sheet membranes. (*Courtesy of Grace Construction Products*)

ucts including toothpaste. Typically, bentonite waterproofing systems contain 85–90 percent of montmorillonite clay and a maximum of 15 percent natural sediments such as volcanic ash.

After being installed in a dry state, clay, when subjected to water, swells and becomes impervious to water. This natural swelling is caused by its molecular structural form of

TABLE 2.7 Material Properties of Hot-Applied Sheet Systems

Advantages	Disadvantages
Material costs	Safety
Similar to builtup roofing	Difficult vertical installations
Some installations are self-healing	Poor elastomeric properties

FIGURE 2.73 Applying reinforcement strips at transition details. (*Courtesy of TC MiraDRI*)

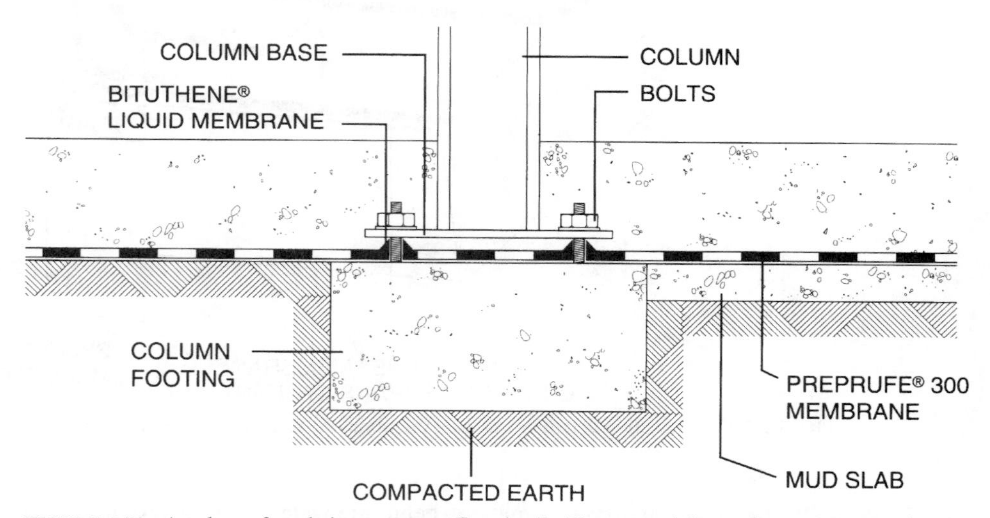

FIGURE 2.74 A column foundation waterproofing detail. (*Courtesy of Grace Construction Products*)

expansive sheets that can expand massively. The amount of swelling and the ability to resist water is directly dependent on grading and clay composition. Clay swells 10–15 percent of its dry volume under maximum wetting. Therefore, it is important to select a system high in montmorillonites and low in other natural sediments.

Bentonite clay is an excellent waterproofing material, but it must be hydrated properly for successful applications. Clay hydration must occur just after installation and backfilling, since the material must be fully hydrated and swelled to become watertight. This hydration and swelling must occur within a confined area after backfill for the water-

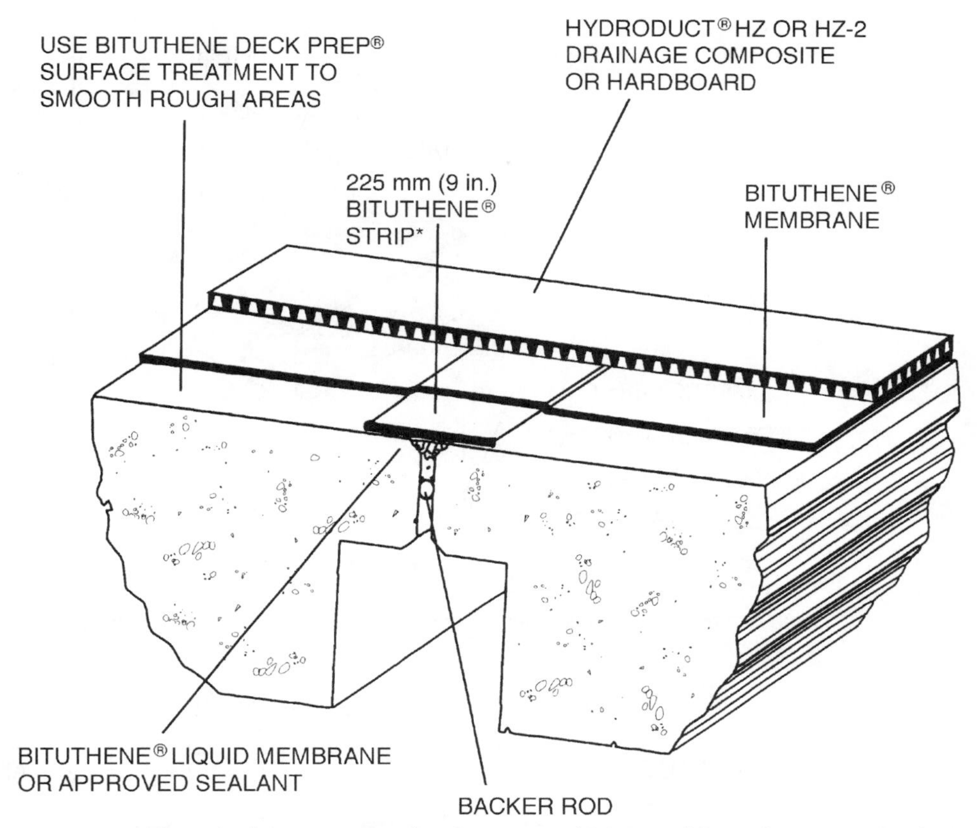

FIGURE 2.75 Expansion joint treatment using sheet system. (*Courtesy of Grace Construction Product*)

proofing properties to be effective. Precaution must be taken to ensure the confined space is adequate for clay to swell. If insufficient, materials can raise floor slabs or cause concrete cracking due to the swelling action.

Clay systems have the major advantage of being installed in various stages during construction to facilitate the shortening of the overall building schedule or reducing any impact the waterproofing system installation might have. Clay systems can be installed before concrete placement by adhering the waterproofing product to the excavation lagging system as shown in Fig. 2.80, or against slurry walls or similar excavation and foundation support systems as detailed in Fig. 2.81 and 2.82.

Clay systems can also be applied to the inside face of concrete formwork that is intended to be left in place due to site access constrictions; a similar installation photograph in shown in Fig. 2.83. These application methods permit the contractor to provide an effective waterproofing installation without having to delay the schedule awaiting the concrete placement and curing time necessary for other types of below-grade products.

This also holds true for the typical waterproofing of elevator pits shown in Fig. 2.84. Here the clay panels are laid directly on the compacted soils before concrete placement,

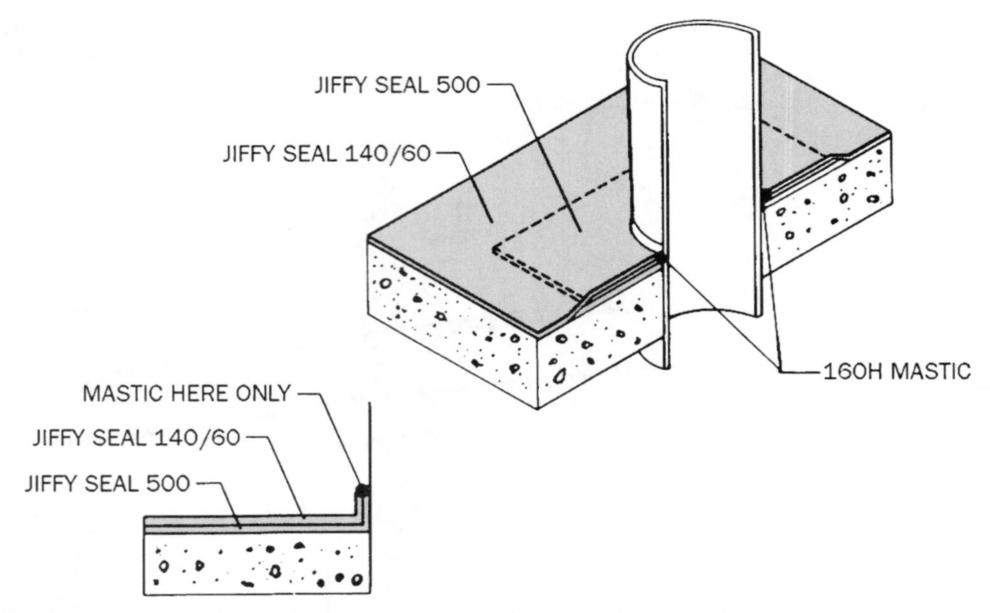

FIGURE 2.76 Protrusion detailing for sheet systems. (*Courtesy of Protecto Wrap*)

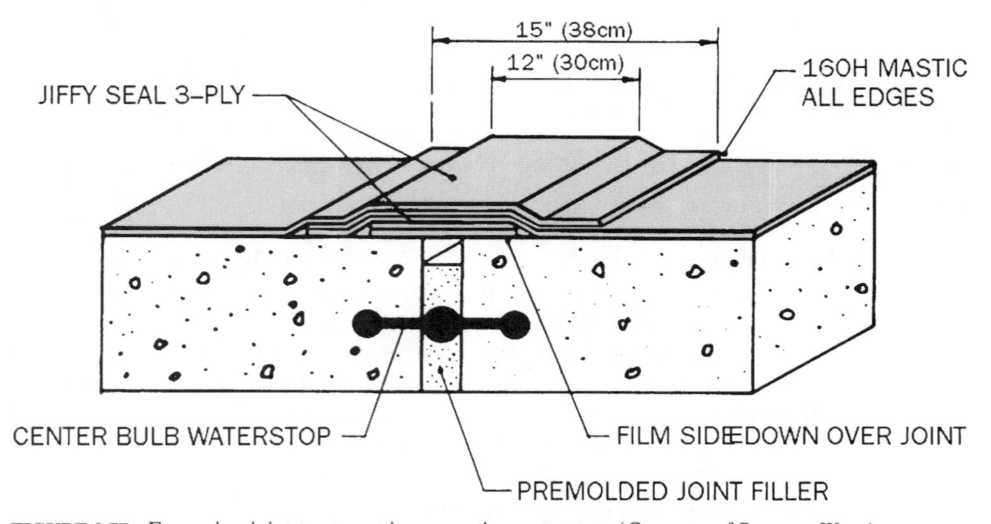

FIGURE 2.77 Expansion joint treatment incorporating waterstop. (*Courtesy of Protecto Wrap*)

without a working or mud slab required for the waterproofing installation. Again, this can save not only construction time but associated costs as well.

There is no concrete cure time necessary, and minimal substrate preparation is necessary. Of all waterproofing systems, these are the least toxic and harmful to the environment. Clay systems are self-healing, unless materials have worked away from a

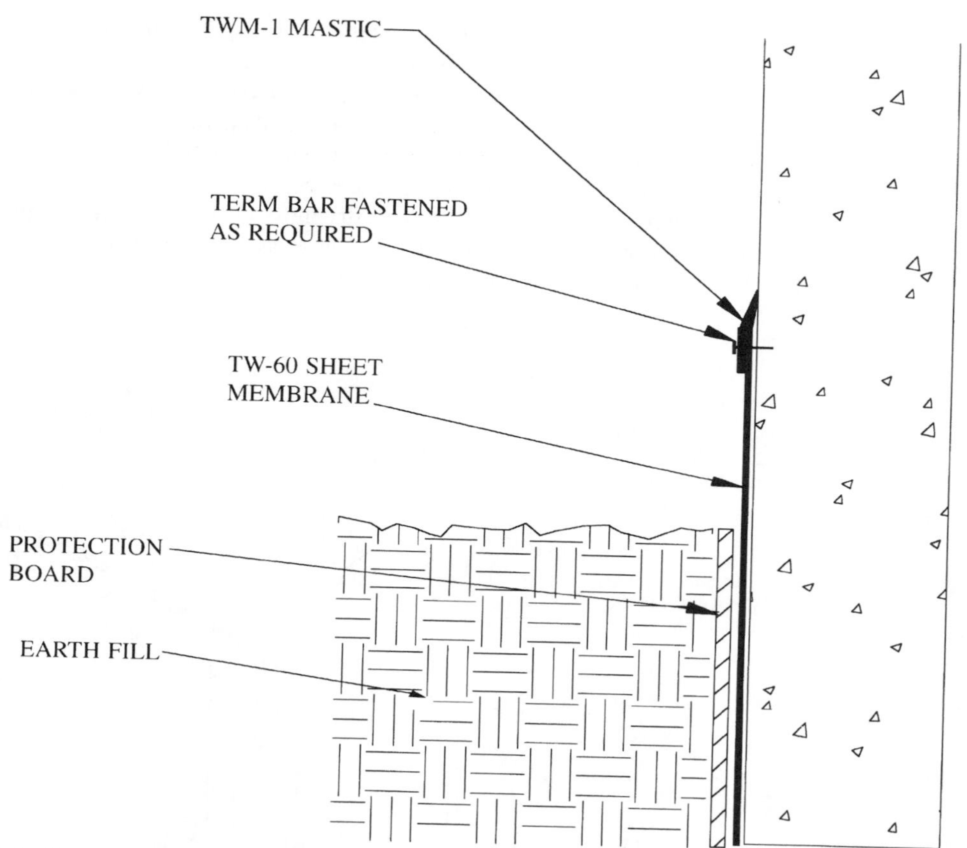

FIGURE 2.78 Termination of sheet membrane using termination bar. (*Courtesy of Tamko Waterproofing*)

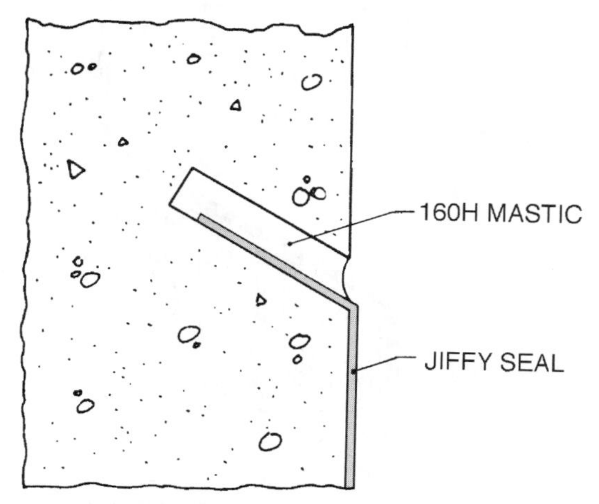

FIGURE 2.79 Termination of sheet membrane using reglet. (*Courtesy of Protecto Wrap*)

substrate. Installations are relatively simple, but clay is extremely sensitive to weather during installation. If rain occurs or groundwater levels rise and material is wetted before backfilling, hydration will occur prematurely and waterproofing capability will be lost, since hydration occurred in an unconfined space.

Immediate protection of applications is required, including uses of polyethylene covering to keep materials from water sources before backfill. If installed in below-grade conditions where constant wetting and drying occurs, clay will eventually deteriorate and lose its waterproofing capabilities. These systems should not be installed where free-flowing groundwater occurs, as clay will be washed away from the substrate.

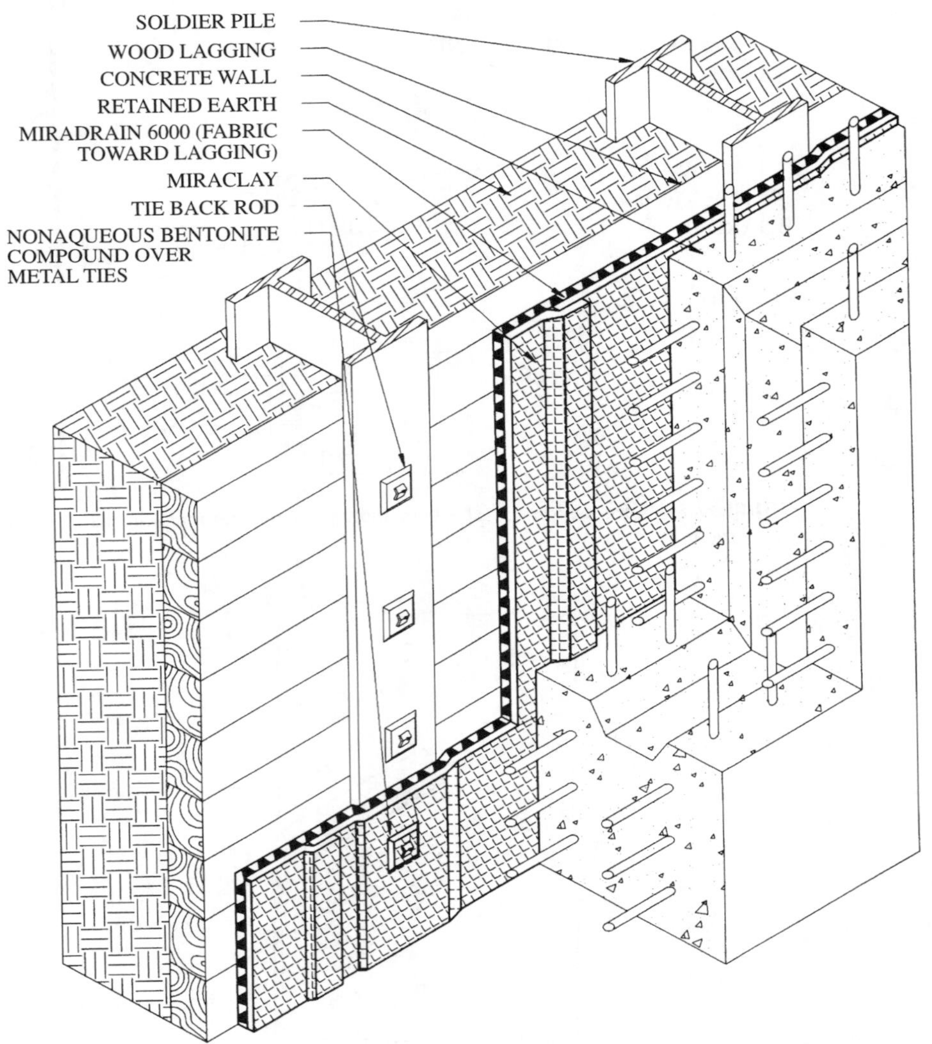

FIGURE 2.80 Clay system applied directly to foundation lagging. (*Courtesy of TC MiraDRI*)

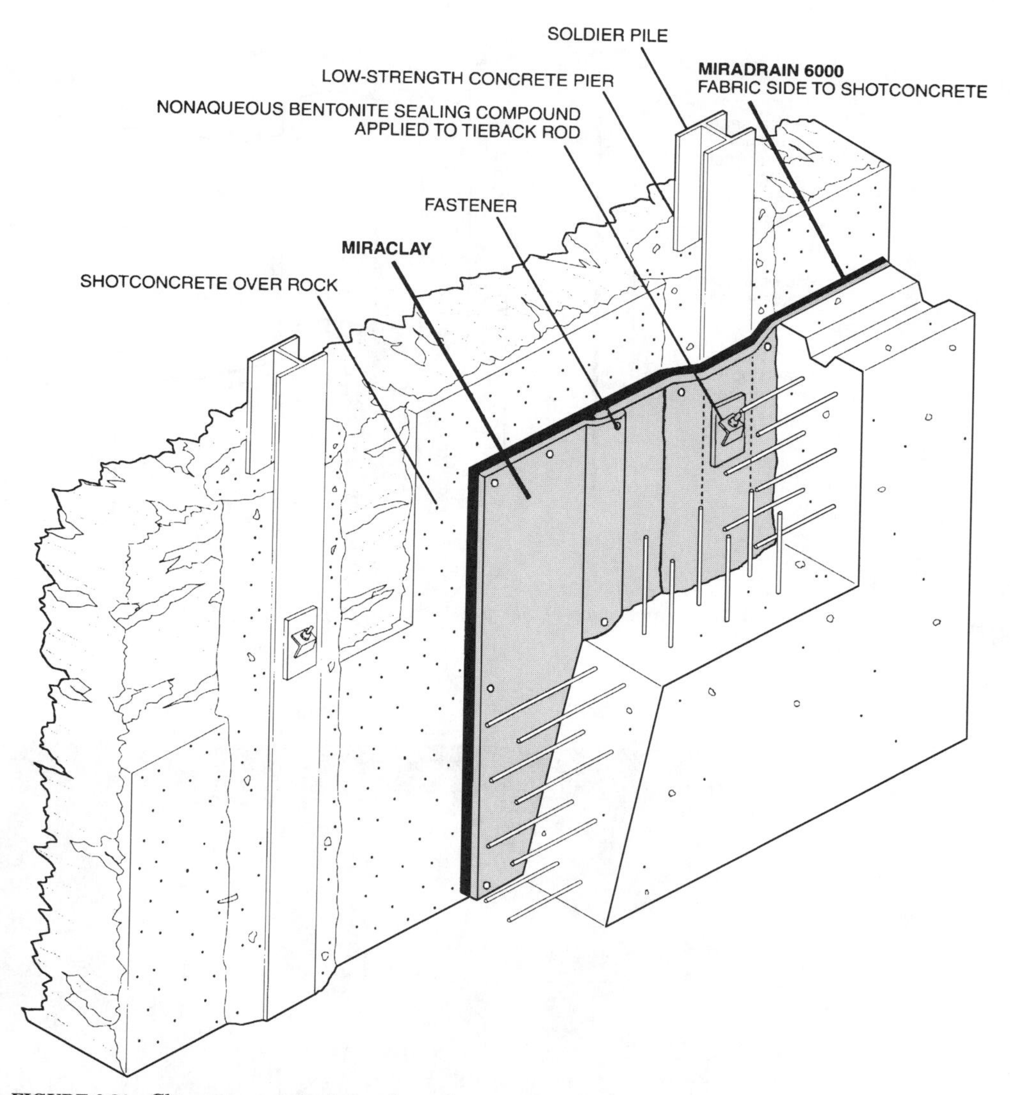

FIGURE 2.81 Clay system applied directly to shotconcrete foundation wall. (*Courtesy of TC MiraDRI*)

Bentonite clays are not particularly resistant to chemicals present in groundwater such as brines, acids, or alkalines.

Bentonite material derivatives are now being added to other waterproofing systems such as thermoplastic sheets and rubberized asphalts. These systems were developed because bulk bentonite spray applications cause problems, including thickness control and substrate adhesion. Bentonite systems are currently available in the following forms:

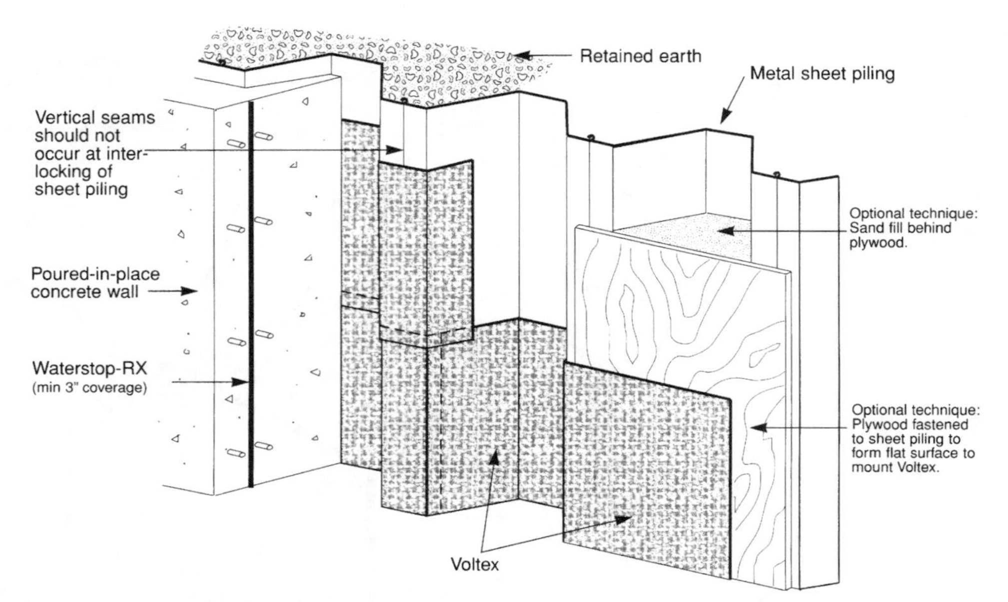

FIGURE 2.82 Application of clay panels directly to foundation sheet piling. (*Courtesy of Cetco Building Materials Group*)

FIGURE 2.83 Clay membrane applied to inside of concrete formwork. (*Courtesy of TC MiraDRI*)

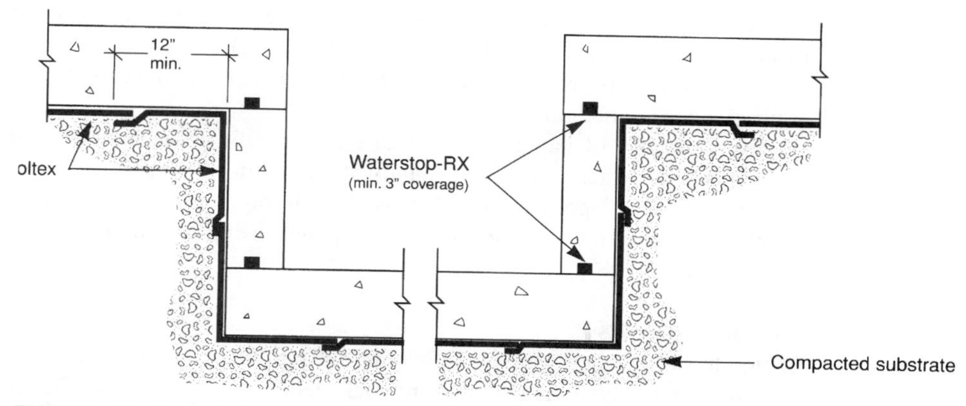

FIGURE 2.84 Typical clay system detailing for elevator pit with no mud slab required. (*Courtesy of Cetco Building Materials Group*)

- Bulk
- Fabricated paper panels
- Sheet goods
- Bentonite and rubber combination sheets
- Textile mats

Bulk bentonite

Bulk bentonite is supplied in bulk form and spray-applied with an integral adhesive to seal it to a substrate. Applications include direct installations to formwork or lagging before foundation completion in lieu of applications directly to substrates. Materials are applied at quantities of 1–2 lb/ft^2.

Bulk bentonite spray applications provide seamless installations. Controls must be provided during application to check that sufficient material is being applied uniformly. Materials should be protected by covering them with polyethylene after installation. Due to possibilities of insufficient thickness during application, manufacturers have developed several clay systems controlling thickness by factory manufacturing, including boards, sheets, and mat systems.

Panel systems

Bentonite clays are packaged in cardboard panels usually 4 ft. square, containing 1 lb/ft^2 of bentonite material. Panels are fastened to substrates by nails or adhesives. Upon backfilling, panels deteriorate by anaerobic action, allowing groundwater to cause clay swelling for waterproofing properties. On horizontal applications the panels are simply laid on the prepared substrates and lapped (Fig. 2.85).

These systems require time for degradation of cardboard panels before swelling and watertightness occurs. This can allow water to penetrate a structure before swelling occurs. As such, manufacturers have developed systems with polyethylene or butyl backing to provide temporary waterproofing until hydration occurs.

Panel clay systems require the most extensive surface penetration of clay systems. Honeycomb and voids should be filled with clay gels before panel application. Special prepackaged clay is provided for application to changes in plane, and gel material is used at protrusions for detailing.

Several grades of panels are available for specific project installation needs. These include special panels for brine groundwater conditions (Fig. 2.86), and reinforced panels for horizontal applications where steel reinforcement work is placed over panels. Panels are lapped onto all sides of adjacent panels using premarked panels that show necessary laps.

Bentonite sheets

Bentonite sheet systems are manufactured by applying bentonite clay at 1 lb/ft^2 to a layer of chlorinated polyethylene. They are packaged in rolls 4 ft. wide. The addition of polyethylene adds temporary waterproofing protection during clay hydration. This polyethylene also protects clay material from prematurely hydrating if rain occurs before backfilling and adds chemical-resistant properties to these systems.

Some manufacturers have developed sheet systems for use in above-grade split or sandwich slab construction. However, constant wetting and drying of this system can alter the clay's natural properties, and waterproofing then depends entirely upon the polyethylene sheet.

Bentonite and rubber sheet membranes

Bentonite and rubber sheet membrane systems add clay to a layer of polyethylene, but also compound the bentonite in a butyl rubber com position. Materials are packaged in rolls 3 ft wide that are self-adhering using a release paper backing. They are similar to rubberized asphalt membranes in application and performance characteristics.

These combination sheet systems are used for horizontal applications, typically split-slab construction in parking or plaza deck construction. As with rubberized asphalt systems, accessories must be used around protrusions, terminations, and changes in plane. The polyethylene, butyl rubber, and bentonite each act in combination with the others, providing substantial waterproofing properties.

Unlike other clay systems, concrete substrates must be dry and cured before application. Care must be taken in design and construction to allow for adequate space for clay swelling.

Bentonite mats

Bentonite mat systems apply clays at 1 lb/ft^2 to a textile fabric similar to a carpet backing. This combination creates a carpet of bentonite material. The coarseness of the fabric allows immediate hydration of clay after backfilling, versus a delayed reaction with cardboard panels.

The textile material is not self-adhering, and adhesives or nailing to vertical substrates is necessary. Protection with a polyethylene sheet after installation is used to prevent premature hydration. This system is particularly effective in horizontal applications where the large rolls eliminate unnecessary seams. This lowers installation costs as well as prevents errors in seaming operations.

Properties of typical clay systems are summarized in Table 2.8.

FIGURE 2.85 Clay sheets installed under horizontal concrete slab; note the waterstop installed in the cold joint. (*Courtesy of Coastal Construction Products*)

TABLE 2.8 Material Properties of Clay Systems

Advantages	Disadvantages
Self-healing characteristics	Clay subject to hydration before backfilling
Ease of application	Not resistant to chemical in soil
Range of systems and packaging	Must be applied in confined conditions for proper swelling conditions

CLAY SYSTEM APPLICATION

Natural clay waterproofing materials require the least preparatory work of all below-grade systems. Concrete substrates are not required to be cured except for rubberized asphalt combination systems. Concrete can be damp during installation, but not wet enough to begin clay hydration.

Large voids and honeycombs should be patched before application. Minor irregularities are sealed with clay gels. Most concrete curing agents are acceptable with clay systems. Masonry surfaces should have joints stricken flush. Note the standard application details in Figs. 2.87, 2.88, and 2.89.

Bentonite materials combined with butyl rubber require further preparation than other clay systems, including a dry surface, no oil or wax curing compounds, and no contaminants, fins, or other protrusions that will puncture materials.

FIGURE 2.86 Saltwater panel application. (*Courtesy of Coastal Construction Products*)

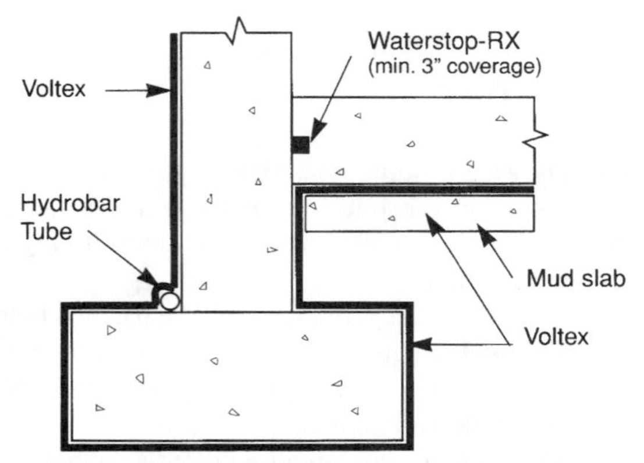

FIGURE 2.87 Clay system detail for foundation waterproofing using mub slab. (*Courtesy of Cetco Building Materials Group*)

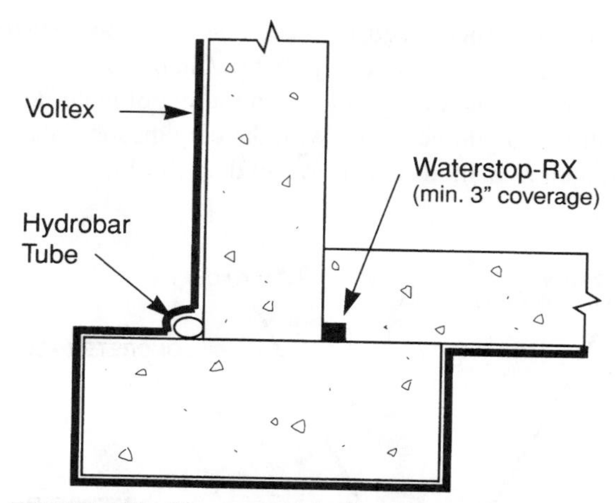

FIGURE 2.88 Clay system detail for foundation waterproofing without, with horizontal membrane applied directly to grade. (*Courtesy of Cetco Building Materials Group*)

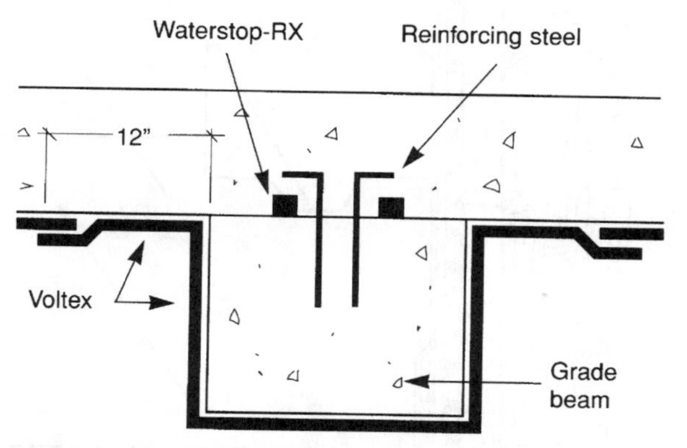

FIGURE 2.89 Grade beam detailing for clay system. (*Courtesy of Cetco Building Materials Group*)

The variety of bentonite systems available means that applications will vary considerably and have procedures similar to the waterproofing systems they resemble in packaging type (e.g., sheet goods). Bulk clay is applied like fluid membranes. Panels and sheets as sheet-good systems, and butyl compound-polyethylene systems are applied virtually identically to rubberized asphalt systems.

With bulk systems, proper material thickness application is critical as it is with fluid-applied systems. Bulk systems are sprayed or troweled, applied at 1–2 lb/ft^2 of substrate.

Panel and mat systems are applied to vertical substrates by nailing. Horizontal applications require lapping only. These systems require material to be lapped 2 in. on all sides. Cants of bentonite material are installed at changes in plane, much the same way as cementitious or sheet-applied systems. Bentonite sheet materials are applied with seams shedding water by starting applications at low points.

Outside corners or turns receive an additional strip of material usually 1 ft wide for additional reinforcement (Fig. 2.90). Chalk lines should be used to keep vertical applications straight and to prevent fish mouthing of materials. All end laps, protrusions, and terminations should be sealed with the clay mastic, as shown in Figs. 2.91 and 2.92. Proper termination methods are shown in detail in Figs. 2.93 and 2.94.

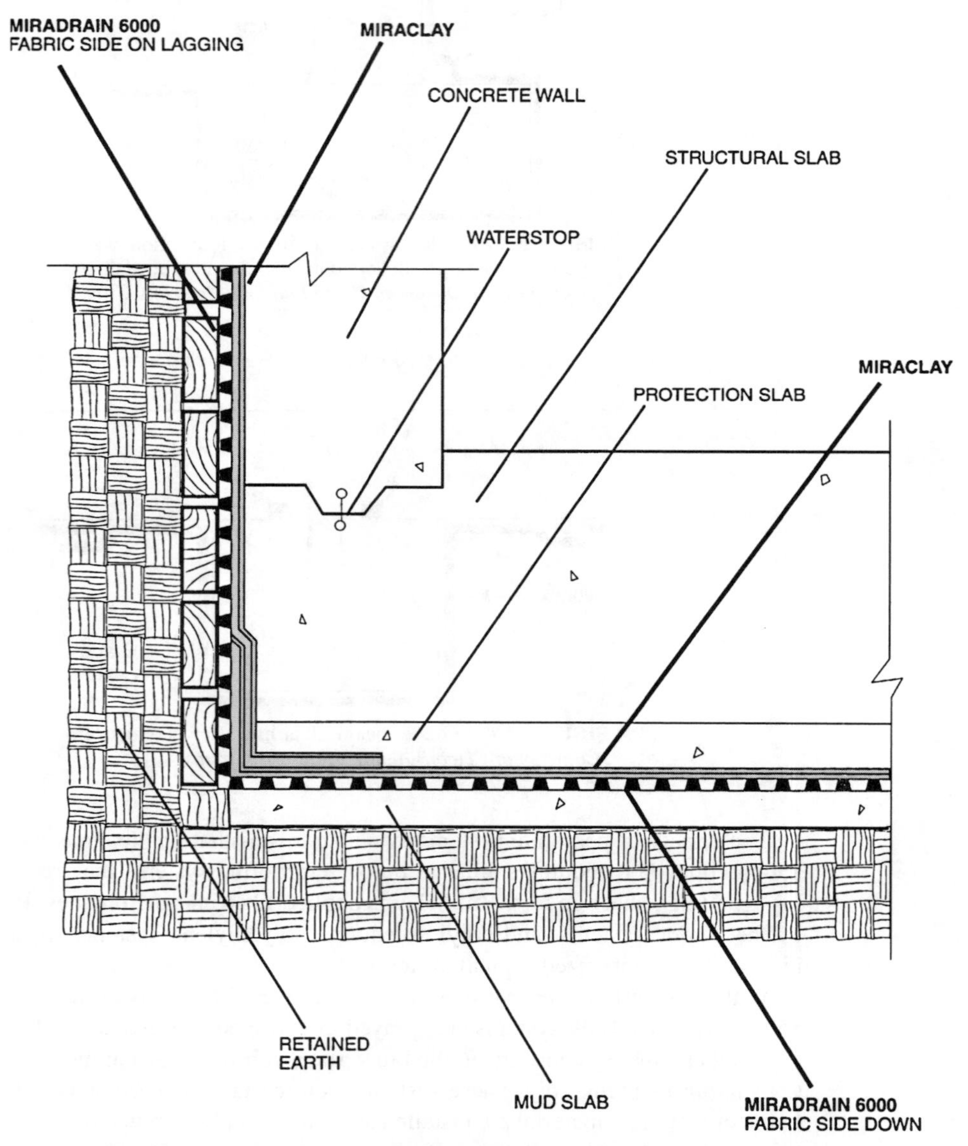

FIGURE 2.90 Clay system applied to lagging detailing. Note reinforcement at corner. (*Courtesy of TC MiraDRI*)

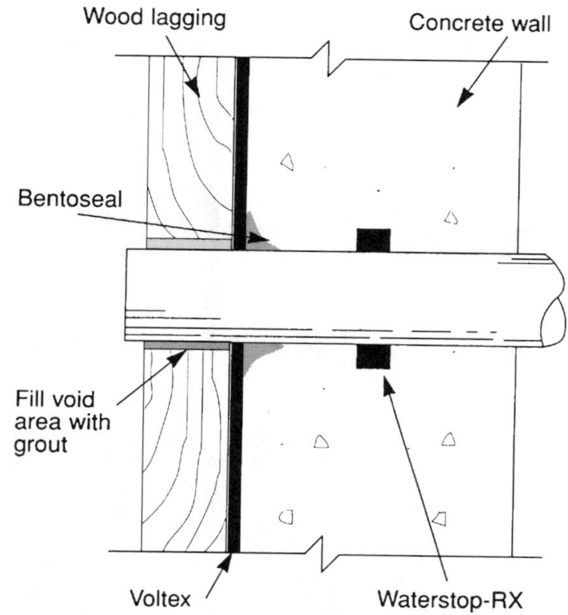

FIGURE 2.91 Typical penetration detailing for clay system. (*Courtesy of Cetco Building Materials Group*)

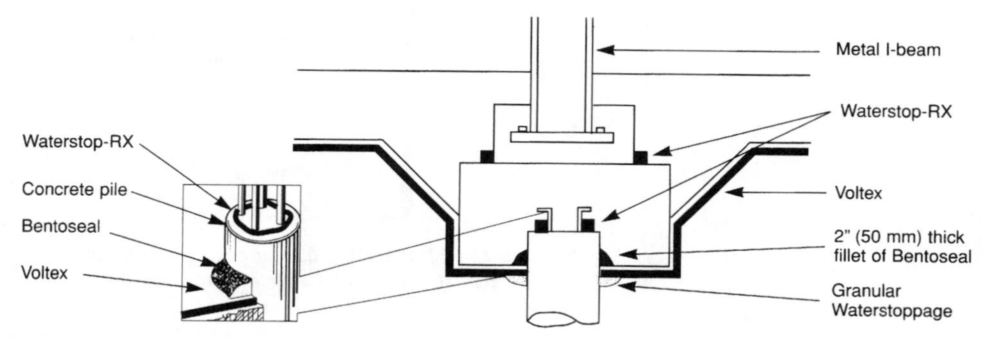

FIGURE 2.92 Pile cap detailing for clay system. (*Courtesy of Cetco Building Materials Group*)

VAPOR BARRIERS

Vapor barriers are not suitable for waterproofing applications. As their name implies, they prevent transmission of water vapor through a substrate in contact with the soil. Typically used at slabs-on-grade conditions, they also are used in limited vertical applications.

Vapor barriers are sometimes used in conjunction with other waterproofing systems, where select areas of the building envelope are not subject to actual water penetration. Vapor barriers are discussed only to present their differences and unsuitability for envelope waterproofing.

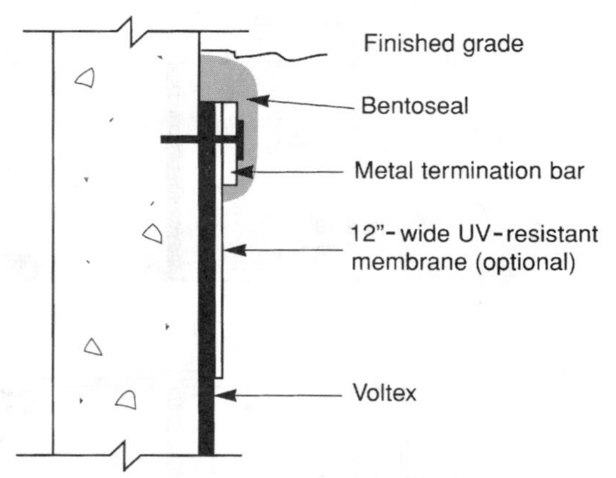

FIGURE 2.93 Termination detailing for clay system. (*Courtesy of Cetco Building Materials Group*)

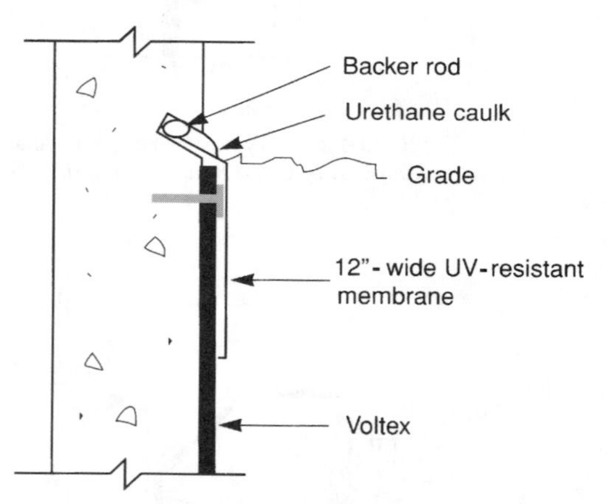

FIGURE 2.94 Termination detailing for clay system using reglet. (*Courtesy of Cetco Building Materials Group*)

As previously discussed, soils have characteristic capillary action that allows the upward movement or migration of water vapor through the soil. Beginning as water and saturating the soil immediately adjacent to the water source, the capillary action ends as water vapor in the upper capillary capability limits of the soil.

Vapor barriers prevent upward capillary migration of vapor through soils from penetrating pores of concrete slabs. Without such protection, delamination of flooring materials, damage to structural components, paint peeling, mildew formation, and increased humidity in finished areas will occur. Vapor barriers can also prevent infiltration by alkaline salts into the concrete slab and flooring finish.

Vapor barriers are produced in PVC, combinations of reinforced waterproof paper with a polyvinyl coating, or polyethylene sheets (commonly referred to as *visquene*).

Polyethylene sheets are available in both clear and black colors in thicknesses ranging from 5 to 10 mil. PVC materials are available in thickness ranging from 10 to 60 mil. Typical properties of vapor barriers are summarized in Table 2.9.

Vapor barriers are rolled or spread out over prepared and compacted soil, with joints lapped 6 in. Vapor barriers can be carried under, up, and over foundations to tie horizontal floor applications into vertical applications over walls. This is necessary to maintain the integrity of a building envelope.

Mastics are typically available from manufacturers for adhering materials to vertical substrates. In clay soil, where capillary action is excessive, laps should be sealed with a mastic for additional protection. Proper foundation drainage systems should be installed, as with all waterproofing systems.

Vapor barriers are installed directly over soil, which is not possible with most waterproofing systems. Protection layers or boards are not used to protect the barrier during reinforcement application or concrete placement.

SUMMARY

Systems available for below-grade waterproofing are numerous and present sufficient choices for ensuring the integrity of below-grade envelopes. Project conditions to review when choosing an appropriate below-grade waterproofing system include

- Soil conditions; rock or clay soils can harm waterproofing systems during backfill.
- Chemical contamination, especially salts, acids, and alkalines.
- Freeze–thaw cycling and envelope portions below frost line.
- Expected movement, including settlement and differential.
- Concrete cold joints to see if they are treatable for the system selected.
- Positive or negative system to see which is better for job site conditions.
- Large vertical applications, which are difficult with certain systems.
- Difficult termination and transition detailing, which prevents use of many systems.
- Length of exposure of installed system due to project conditions before backfilling.
- Safety concerns at project.
- Site/foundation access limitations.
- Dewatering requirements during construction.
- Concrete cure time available before backfill or other construction must commence.
- Adjacent envelope systems that the waterproofing system must be comparable with or not damage.
- Scheduling requirements.
- Access for repairs after construction is complete.

Although not a substitute for referring to specific manufacturer information on a specific material, Table 2.10 is a summary and comparison of major below-grade waterproofing systems. One system or material may not be sufficient for all situations encountered on a particular project. Once below-grade materials are chosen, they must be detailed into above-grade envelope materials and systems. This detailing is critical to the success of the entire building envelope and is discussed further in Chap. 8.

TABLE 2.9 Material Properties of Vapor Barriers

Advantages	Disadvantages
Ease of horizontal applications	Noneffective waterproofing materials
Prevent moisture transmission	Seams
No subslab required	Difficult vertical installations

TABLE 2.10 Summary Properties of Below-Grade Materials

Property	Cementitious	Fluid applied	Sheet goods	Clay system
Elongation	None	Excellent	Good	Fair to good
Chemical and weathering resistance	Good	Fair to good	Good	Fair to good
Difficulty of installation	Moderate	Simple	Difficult	Simple
Thickness 1/8–7/16 in	60 mil	average	20–60 mil	1/4–1/2 in.
Horizontal subslab required	No	Yes	Yes	No
Positive or negative system	Both	Positive	Positive	Positive
Areas requiring inspections	Coves and cants at changes in plane; control joint detailing	Millage, especially at turnups; detailing and priming at penetrations	Laps and seams; penetration detailing; transition	Laps, penetration detailing, changes in plane
Repairs	Simple	Simple	Moderate to difficult	Moderate
Protection required	*No	*Yes	*Yes	*No

*Note—Manufactured drainage systems should be used in lieu of protection whenever possible and preferably installed with all positive side applications (Clay system may prohibit use of drainage board.)

CHAPTER 3
ABOVE-GRADE WATERPROOFING

INTRODUCTION

Waterproofing of surfaces above grade is the prevention of water intrusion into exposed elements of a structure or its components. Above-grade materials are not subject to hydrostatic pressure but are exposed to detrimental weathering effects such as ultraviolet light.

Water that penetrates above-grade envelopes does so in five distinct methods:

- Natural gravity forces
- Capillary action
- Surface tension
- Air pressure differential
- Wind loads

The force of water entering by gravity is greatest on horizontal or slightly inclined envelope portions. Those areas subject to ponding or standing water must be adequately sloped to provide drainage away from envelope surfaces.

Capillary action is the natural upward wicking motion that can draw water from ground sources up into above-grade envelope areas. Likewise, walls resting on exposed horizontal portions of an envelope (e.g., balcony decks) can be affected by capillary action of any ponding or standing water on these decks.

The molecular surface tension of water allows it to adhere to and travel along the underside of envelope portions such as joints. This water can be drawn into the building by gravity or unequal air pressures.

If air pressures are lower inside a structure than on exterior areas, water can be literally sucked into a building. Wind loading during heavy rainstorms can force water into interior areas if an envelope is not structurally resistant to this loading. For example, curtain walls and glass can actually bend and flex away from gaskets and sealant joints, causing direct access for water.

The above-grade envelope must be resistant to all these natural water forces to be watertight. Waterproofing the building envelope can be accomplished by the facade material itself (brick, glass, curtain wall) or by applying waterproof materials to these substrates. Channeling water that passes through substrates back out to the exterior using flashing, weeps, and dampproofing is another method. Most envelopes include combinations of all these methods.

Older construction techniques often included masonry construction with exterior load bearing walls up to 3 feet thick. This type of envelope required virtually no attention to waterproofing or weathering due to the shear impregnability of the masonry wall.

Today, however, it is not uncommon for high-rise structures to have an envelope skin thickness of 1/8 in. Such newer construction techniques have developed from the need for lighter-weight systems to allow for simpler structural requirements and lower building costs. These systems, in turn, create problems in maintaining an effective weatherproof envelope.

Waterproof building surfaces are required at vertical portions as well as horizontal applications such as balconies and pedestrian plaza areas. Roofing is only a part of necessary above-grade waterproofing systems, one that must be carefully tied into other building envelope components.

Today roofing systems take many different forms of design and detailing. Plaza decks or balcony areas covering enclosed spaces and parking garage floors covering an occupied space all constitute individual parts of a total roofing system. Buildings can have exposed roofs as well as unexposed membranes acting as roofing and waterproofing systems for preventing water infiltration into occupied areas.

DIFFERENCES FROM BELOW-GRADE SYSTEMS

Most above-grade materials are breathable in that they allow for negative vapor transmission. This is similar to human skin; it is waterproof, allowing you to swim and bathe but also to perspire, which is negative moisture transmission. Most below-grade materials will not allow negative transmission and, if present, it will cause the material to blister or become unbonded.

Breathable coatings are necessary on all above-grade wall surfaces to allow moisture condensation from interior surfaces to pass through wall structures to the exterior. The sun causes this natural effect by drawing vapors to the exterior. Pressure differentials that might exist between exterior and interior areas create this same condition.

Vapor barrier (nonbreathable) products installed above grade cause spalling during freeze–thaw cycles. Vapor pressure buildup behind a nonbreathable coating will also cause the coating to disbond from substrates. This effect is similar to window or glass areas that are vapor barriers and cause formation of condensation on one side that cannot pass to exterior areas.

Similarly, condensation passes through porous wall areas back out to the exterior when a breathable coating is used, but condenses on the back of nonbreathable coatings. This buildup of moisture, if not allowed to escape, will deteriorate structural reinforcing steel and other internal wall components.

Below-grade products are neither ultraviolet-resistant nor capable of withstanding thermal movement experienced in above-grade structures. Whereas below-grade materials are not subject to wear, above-grade materials can be exposed to wear such as foot traffic. Below-grade products withstand hydrostatic pressure, whereas above-grade materials do not. Waterproofing systems properties are summarized in Table 3.1.

Since many waterproofing materials are not aesthetically acceptable to architects or engineers, some trade-off of complete watertightness versus aesthetics is used or specified. For instance, masonry structures using common face brick are not completely waterproof due to water infiltration at mortar joints. Rather than change the aesthetics of brick by applying a waterproof coating, the designer chooses a dampproofing and flashing system. This dampproofing system diverts water that enters through the brick wall back out to the exterior.

Table 3.1 Waterproofing Systems Differences

Below-grade systems:	Above-grade systems
Hydrostatic pressure resistant	Ultraviolet-resistant
Structural movement capability	Thermal and structural movement capability
Most inaccessible after installation	Breathable
Both positive and negative applications	Traffic wear and weathering exposure
Mostly barrier systems	Aesthetically pleasing
Drainage enhancement a must	Freeze–thaw cycle resistant

Application of a clear water repellent will also reduce water penetration through the brick and mortar joints. Such sealers also protect brick from freeze–thaw and other weathering cycles.

Thus, waterproofing exposed vertical and horizontal building components can include a combination of installations and methods that together compose a building envelope. This is especially true of buildings that use a variety of composite finishes for exterior surfacing such as brick, precast, and curtain wall systems. With such designs, a combination of several waterproofing methods must be used. Although each might act independently, as a whole they must act cohesively to prevent water from entering a structure. Sealants, wall flashings, weeps, dampproofing, wall coatings, deck coatings, and the natural weathertightness of architectural finishes themselves must act together to prevent water intrusion (Figure 3.1).

This chapter will cover vertical waterproofing materials, including clear water repellents, elastomeric coatings, cementitious coatings, and related patching materials. It will also review horizontal waterproofing materials including deck coatings, sandwich slab membranes, and roofing.

VERTICAL APPLICATIONS

Several systems are available for weatherproofing vertical wall envelope applications. Clear sealers are useful when substrate aesthetics are important. These sealers are typically applied over precast architectural concrete, exposed aggregate, natural stone, brick, or masonry.

It is important to note that clear sealers are not completely waterproof; they merely slow down the rate of water absorption into a substrate, in some situations as much as 98 percent. However, wind-driven rain and excessive amounts of water will cause eventual leakage through any clear sealer system. This requires flashings, dampproofing, sealants, and other systems to be used in conjunction with sealers, to ensure drainage of water entering through primary envelope barriers.

This situation is similar to wearing a canvas-type raincoat. During light rain, water runs off; but should the canvas become saturated, water passes directly through the coat. Clear sealers as such are defined as water repellents, in that they shed water flow but are not impervious to water saturation or a head of water pressure.

Elastomeric coatings are high-solid-content paints that produce high-millage coatings when applied to substrates. These coatings are waterproof within normal limitations of movement and proper application. Elastomeric coatings completely cover and eliminate any natural substrate aesthetics. They can, however, add a texture of their own to an envelope system, depending on the amount of sand, if any, in the coating.

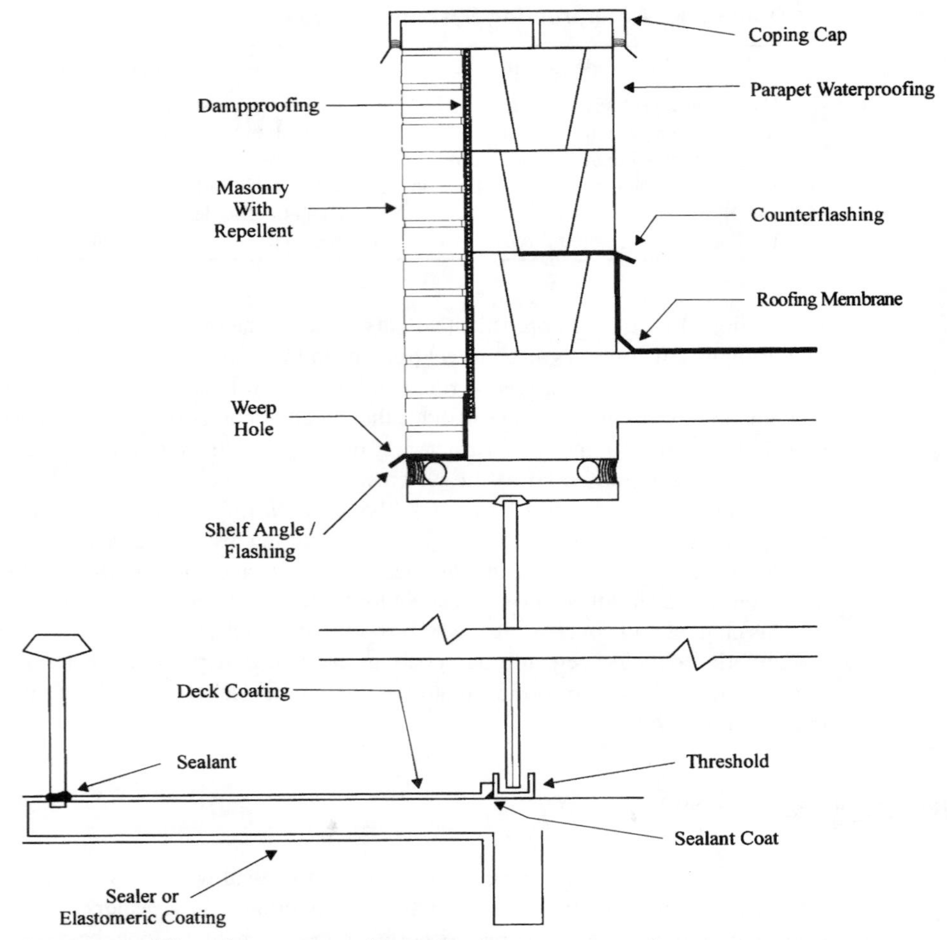

FIGURE 3.1 All envelope waterproofing applications must act together to prevent water intrusion.

To waterproof adequately with an elastomeric coating, details must be addressed, including patching cracks or spalls in substrates, allowing for thermal movement, and installation of flashings where necessary.

Cementitious coatings are available for application to vertical masonry substrates, which also cover substrates completely. The major limitation of cementitious above-grade product use is similar to its below-grade limitation. The products do not allow for any substrate movement or they will crack and allow water infiltration. Therefore, proper attention to details is imperative when using cementitious materials. Installing sealant joints for movement and crack preparation must be completed before cementitious coating application.

With all vertical applications, there are patching materials used to ensure water tightness of the coating applied. These products range from brushable-grade sealants for small cracks, to high-strength, quick-set cementitious patching compounds for repairing spalled substrate areas.

HORIZONTAL APPLICATIONS

Several types of systems and products are available for horizontal above-grade applications, such as parking garages and plaza decks. Surface coatings, which apply directly to exposed surfaces of horizontal substrates, are available in clear siloxane types or solid coatings of urethane or epoxy. Clear horizontal sealers, as with vertical applications, do not change existing substrate aesthetics to which they are applied. They are, however, not in themselves completely waterproof but only water-resistant.

Clear coatings are often specified for applications, to prevent chloride ion penetration into concrete substrates from such materials as road salts. These pollutants attack reinforcing steel in concrete substrates and cause spalling and structural deterioration.

Urethane, epoxy, or acrylic coatings change the aesthetics of a substrate but have elastomeric properties that allow bridging of minor cracking or substrate movement. Typically, these coatings have a "wearing coat" that contains silicon sand or carbide, which allows vehicle or foot traffic while protecting the waterproof base coat.

Subjecting coatings to foot or vehicular wear requires maintenance at regular frequency and completion of necessary repairs. The frequency and repairs are dependent on the type and quantity of traffic occurring over the envelope coating.

As with vertical materials, attention to detailing is necessary to ensure watertightness. Expansion or control joints must be properly sealed, cracks or spalls in the concrete must be repaired before application, and allowances for drainage must be created.

Several types of waterproof membranes are available for covered decks such as sandwich slab construction or tile-topped decks. These membranes are similar to those used in below-grade applications, including liquid-applied and sheet-good membranes. Such applications are also used as modified roofing systems.

ABOVE-GRADE EXPOSURE PROBLEMS

All above-grade waterproof systems are vulnerable to a host of detrimental conditions due to their exposure to weathering elements and substrate performance under these conditions. Exposure of the entire above-grade building envelope requires resistance from many severe effects, including the following:

- Ultraviolet weathering
- Wind loading
- Structural loading due to snow or water
- Freeze–thaw cycles
- Thermal movement

- Differential movement
- Mildew and algae attack
- Chemical and pollution attack from chloride ions, sulfates, nitrates, and carbon dioxide

Chemical and pollution attack is becoming ever more frequent and difficult to contend with. Chloride ions (salts) are extremely corrosive to the reinforcing steel present in all structures, whether it is structural steel, reinforcing steel, or building components such as shelf angles.

Even if steel is protected by encasement in concrete or is covered with a brick facade, water that penetrates these substrates carries chloride ions that attack the steel. Once steel begins to corrode it increases greatly in size, causing spalling of adjacent materials and structural cracking of substrates.

All geographic areas are subject to chloride ion exposure. In coastal areas, salt spray is concentrated and spread by wind conditions; in northern climates, road salts are used during winter months. Both increase chloride quantities available for corrosive effects on envelope components.

Acid rain now affects all regions of the world. When sulfates and nitrates present in the atmosphere are mixed with water, they create sulfuric and nitric acids (acid rain), which affect all building envelope components. Acids attack the calcium compounds of concrete and masonry surfaces, causing substrate deterioration. They also affect exposed metals on a structure such as flashing, shelf angles, and lintel beams.

Within masonry or concrete substrates, a process of destructive weathering called *carbonation* occurs to unprotected, unwaterproofed surfaces. Carbonation is the deterioration of cementitious compounds found in masonry substrates when exposed to the atmospheric pollutant carbon dioxide (automobile exhaust).

Carbon dioxide mixes with water to form carbonic acid, which then penetrates a masonry or concrete substrate. This acid begins deteriorating cementitious compounds that form part of a substrate.

Carbonic acid also causes corrosion of embedded reinforcing steel such as shelf angles by changing the substrate alkalinity that surrounds this steel. Reinforcing steel, which is normally protected by the high alkalinity of concrete, begins to corrode when carbonic acid change lowers alkalinity while also deteriorating the cementitious materials.

Roofing systems will deteriorate because of algae attack. Waterproof coatings become brittle and fail due to ultraviolet weathering. Thermal movement will split or cause cracks in a building envelope. This requires that any waterproof material or component of the building envelope be resistant to all these elements, thus ensuring their effectiveness and, in turn, protecting a building during its life-cycling.

Finally, an envelope is also subject to building movement, both during and after construction. Building envelope components must withstand this movement; otherwise, designs must include allowances for movement or cracking within the waterproofing material.

Cracking of waterproofing systems occur because of structural settlement, structural loading, vibration, shrinkage of materials, thermal movement, and differential movement. To ensure successful life-cycling of a building envelope, allowances for movement must be made, including expansion and control joints, or materials must be chosen that can withstand expected movement.

All these exposure problems must be considered when choosing a system for waterproofing above-grade envelope portions. Above-grade waterproofing systems include the following horizontal and vertical applications:

- Vertical
  - Clear repellents
  - Cementitious coatings
  - Elastomeric coatings
- Horizontal
  - Deck coatings
  - Clear deck sealers
  - Protected membranes

CLEAR REPELLENTS

Although clear sealers do not fit the definition of true waterproofing systems, they do add water repellency to substrates where solid coatings as an architectural finish are not acceptable (see Figure 3.2.) Clear sealers are applied on masonry or concrete finishes when a repellent that does not change substrate aesthetics is required. Clear sealers are also specified for use on natural stone substrates such as limestone. Water repellents prevent chloride ion penetration into a substrate and prevent damage from the freeze–thaw cycles.

There is some disagreement over the use of sealers in historic restoration. Some prefer stone and masonry envelope components to be left natural, repelling or absorbing water and aging naturally. This is more practical in older structures that have massive exterior wall substrates than in modern buildings. Today exterior envelopes are as thin as 1/8 inch, requiring additional protection such as clear sealers.

The problem with clear sealers is not in deciding when they are necessary but in choosing a proper material for specific conditions. Clear repellents are available in a multitude of compositions, including penetrates and film-forming materials. They vary in percentage of solids content and are available in tint or stain bases to add uniformity to the substrate color.

The multitude of materials available requires careful consideration of all available products to select the material appropriate for a particular situation. Repellents are available in the compositions and combinations shown in Table 3.2. Sealers are further classified into penetrating and film-forming sealers.

Clear sealers will not bridge cracks in the substrate, and this presents a major disadvantage in using these materials as envelope components. Should cracks be properly prepared in a substrate before application, effective water repellency is achievable. However, should further cracking occur, due to continued movement, a substrate will lose its watertightness. Properly designed and installed crack-control procedures, such as control joints and expansion joints, alleviate cracking problems.

Figure 3.3 shows a precast cladding after rainfall with no sealer applied. Water infiltrating the precast can enter the envelope and bypass sealant joints into interior areas. Figure 3.4 demonstrates just how effective sealers can be in repelling water.

FIGURE 3.2 Repellency of sealer application. (*Courtesy of Saver Systems*)

TABLE 3.2 Repellent Types and Compositions

Penetrating sealers	Film-forming sealers
Siloxanes	Acrylics
Silanes	Silicones
Silicone Rubber	Aliphatic urethane
Siliconates	Aromatic urethane
Epoxy-modified siloxane	Silicone resin
Silane–siloxane combination	Methyl methacrylate
Siloxane–acrylic combination	Modified stearate

Film-forming sealers

Film-forming, or surface, sealers have a viscosity sufficient to remain primarily on top of a substrate surface. Penetrating sealers have sufficiently low viscosity of the vehicle (binder and solvent) to penetrate into masonry substrate pores. The resin molecule sizes of a sealer determine the average depth of penetration into a substrate.

Effectiveness of film-forming and penetrating sealers is based upon the percentage of solids in the material. High-solid acrylics will form better films on substrates by filling open pores and fissures and repelling a greater percentage of water. Higher-solids-content materials are necessary when used with very porous substrates; however, these materials may darken or impart a glossy, high sheen appearance to a substrate.

FIGURE 3.3 Precast concrete building with no sealer permits water absorption. (*Courtesy of Coastal Construction Products*)

FIGURE 3.4 Effectiveness of sealer application is evident after a rainfall. (*Courtesy of Coastal Construction Products*)

Painting or staining over penetrating sealers is not recommended, as it defeats the purpose of the material. With film-forming materials, if more than a stain is required, it may be desirable to use an elastomeric coating to achieve the desired watertightness and color.

Most film-forming materials and penetrates are available in semitransparent or opaque formulations. If it is desired to add color or a uniform coloring to a substrate that may contain color irregularities (such as tilt-up or poured-in-placed concrete), these sealers offer effective solutions. (See Table 3.3)

Penetrating sealers

Penetrating sealers are used on absorptive substrates such as masonry block, brick, concrete, and porous stone. Some penetrating sealers are manufactured to react chemically with these substrates, forming a chemical bond that repels water. Penetrating sealers are not used over substrates such as wood, glazed terra cotta, previously painted surfaces, and exposed aggregate finishes.

On these substrates, film-forming clear sealers are recommended (which are also used on masonry and concrete substrates). These materials form a film on the surface that acts as a water-repellent barrier. This makes a film material more susceptible to erosion due to ultraviolet weathering and abrasive wear such as foot traffic.

Penetrating sealers are breathable coatings, in that they allow water vapor trapped in a substrate to escape through the coating to the exterior. Film-forming sealers' vapor transmission (perm rating) characteristics are dependent on their solids content. Vapor transmission or perm ratings are available from manufacturers. Permeability is an especially important characteristic for masonry installed at grade line. Should an impermeable coating be applied here, moisture absorbed into masonry by capillary action from ground sources will damage the substrates, including surface spalling.

Many sealers fail due to a lack of resistance to alkaline conditions found in concrete and masonry building materials. Most building substrates are high in alkalinity, which causes a high degree of failure with poor alkaline-resistant sealers.

Penetrating materials usually have lower coverage rates and higher per-gallon costs than film materials. Penetrating sealers, however, require only a one-coat application versus two for film-forming materials, reducing labor costs.

Penetrating and film-forming materials are recognized as effective means of preventing substrate deterioration due to acid rain effects. They prevent deterioration from air and water pollutants and from dirt and other contaminants by not allowing these pollutants to be absorbed into a substrate. (See Table 3.4.)

TABLE 3.3 Film-Forming Sealer Properties

Advantages	Disadvantages
High solids content; able to fill minor cracks in substrate	Not effective in weathering
Opaque stains available to cover repair work in substrate	Not resistant to abrasive wear
Applicable to exposed aggregate finishes and wood substrates	Film adhesion dependent on substrate cleanliness and preparation

TABLE 3.4 Penetrating Sealer Properties

Advantages	Disadvantages
Resistant to ultraviolet weathering	Can damage adjacent substrates, especially glass and aluminum
Effective for abrasive wear areas	Causes damage to plants and shrubs
Excellent permeability ratings	Not effective on wood or hard finish materials such as glazed tile

CHOOSING THE APPROPRIATE REPELLENT

Without any doubt, choosing the correct water repellent for a specific installation can be a difficult task. Sealer manufacturers offer you little assistance as you try to find your way through the maze of products available, reported to be as many as 500 individual systems. Even though there is a finite number of families of sealers, as outlined in the following sections, within each family manufacturers will try to differentiate themselves from all others, even though most are very similar systems.

There are numerous chemical formulations created using the basic silicon molecule that forms the basis for most of the penetrating sealers. These formulations result in the basic family groups of: Silicones, Silicates, Silanes, and Siloxanes. There is often confusion as to the basic families of sealers; for example, some will classify Siliconates as a family even though it begins as a derivative of a Silane. From these basic groups, manufacturers formulate numerous minor changes that offer little if any improvements and only tend to confuse the purchaser into thinking they are buying something totally unique.

Derivatives include Alkylalkoxysiloxane (siloxane), Isobutyltrialkoxysilane (silane), Alkylalkoxysilane (silane), methylsiloxanes, and many blends of the family groups such as a silane/siloxane combination. These formulations or chemical combinations should not confuse a prospective purchaser. With a few basic guidelines, the best selection for each individual installation can be made easily.

First, any water repellent used should have the basic characteristics necessary for all types of installations: sufficient water repellency, and long life-cycling under alkaline conditions. The latter, performance in alkaline conditions, usually controls how well the product will perform as a repellent over extended life cycling. For the penetrating sealers listed above, no matter how well the product repels water during laboratory testing, the product will virtually become useless after installation if it cannot withstand the normal alkaline conditions of concrete or masonry substrates. Concrete in particular has very high alkaline conditions that can alter the chemical stability of penetrating sealers, resulting in a complete loss of repellency capability.

Therefore when reviewing manufacturer's guide specifications, the high initial repellency rates should not be depended upon solely; rather emphasize the test results of accelerated weathering, especially when application is used on concrete or precast concrete substrates. Verify that the accelerated weathering is tested on a similar substrate, as masonry or most natural stones will not have alkaline conditions as high as concrete.

In addition, when the proposed application is over concrete substrates with substantial reinforcing steel embedded, the resistance of the repellent to chloride ion infiltration should be highlighted. Chlorides attack the reinforcing steel and can cause structural damage after extended weathering. Many sealers have very poor chloride resistance.

Since water penetration begins on the surface, depth of penetration is not a particularly important consideration. While all penetrating sealers must penetrate sufficiently to react chemically with the substrate, many penetration depth claims are made on the solvent carrier rather than the chemical solids that form the repellency. The effective repellency must be at the surface of the substrate to repel water. Water should only penetrate the surface if there is cracking in the substrate, and if this is the case, no repellent can bridge cracking or penetrate sufficiently to repel water in the crack crevice (see Fig. 3.5).

Penetrating capability is a better guide for a sealer's protection against UV degradation. Having the active compounds deeper into the substrate surface protects the molecules from the sun's ultraviolet rays that can destroy a sealers repellency capability.

When comparing the capability of sealers to penetrate into a substrate be sure to review what is referred to as the *uniform gradient permeation (UPG)*, which measures the penetration of the active ingredient rather than the solvent carrier. Most alcohol carriers will penetrate with the active ingredient deeper than those using a petroleum-based carrier will.

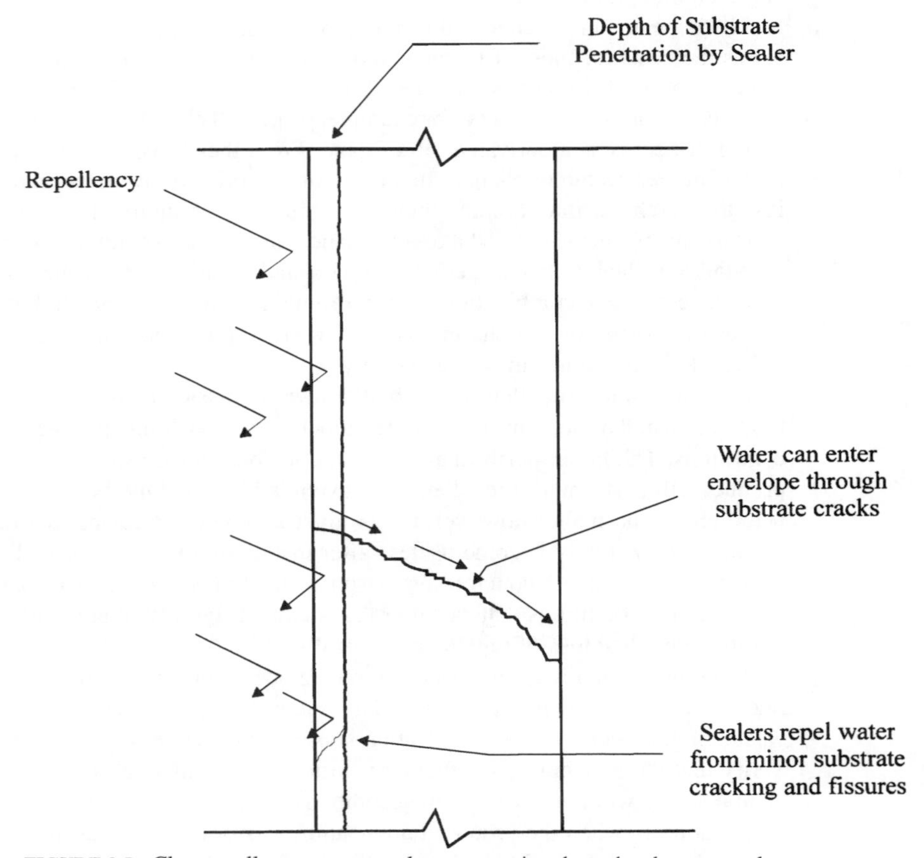

FIGURE 3.5 Clear repellents cannot repel water entering through substrate cracks.

Some manufacturers will make claims as to the size of their active molecules being so small that they penetrate better than other compounds using larger molecules. While this may be the case, compounds with larger molecules usually repel water better than those using smaller molecules.

The amount of solids or active ingredient is always a much-trumpeted point of comparison. Certainly, there is a minimum amount of solids or active agent to produce the required repellency, but once this amount is exceeded there is no logic as to what a greater concentration will do. For the majority of penetrating sealers, 10 percent active compounds seems to be the minimum to provide sufficient water repellency, with 20 percent moving towards the maximum return for the amount of active agent necessary. While manufacturers will often exceed this to increase a product's sales potential, the value of its in-place service capability is often no more than those with a smaller percentage of active compounds.

When considering film-forming repellents, a greater percentage of solids is important since these solids are deposited directly on the surface of the substrate and left to repel water directly and without the assistance of the substrate environment. With film-forming repellents, the closer to 100 percent solids, the more likely the repellent will be capable of repelling water.

When trying to compare products through the maze of contradictory and confusing information available, it is best to review the results of completed standard and uniform tests that are most appropriate for the substrate and service requirements required. The next section expands on the most frequently used testing to compare products, a much better guide than reading sales literature about percent solids, size of molecule, and chemical formulations. In most cases it is not appropriate to make comparison without the use of standard testing, and no product should be considered without this critical information being provided. Recognize however, that these tests are conducted in the pristine conditions of a laboratory that are never duplicated under actual field conditions. This requires that a sufficient margin of error or safety factor be used for actual expectations of performance results in actual installations.

Sealer testing

Several specific tests should be considered in choosing clear sealers. Testing most often referred to is the National Cooperative Highway Research Program (NCHRP). This is the most appropriate test for concrete substrates including bridges and other civil construction projects. Although often used for testing horizontal applications, it remains an effective test for vertical sealers as well. NCHRP test 244, Series II, measures the weight gain of a substrate by measuring water absorption into a test cube submerged after treatment with a selected water repellent. To be useful, a sealer should limit weight gain to less than 15 percent of original weight and preferably less than 10 percent. Test results are also referred to as "a reduction in water absorption from the control [untreated] cube." These limits should be an 85 to 100 percent reduction, preferably above 90 percent.

Testing by ASTM includes ASTM D-514, Water Permeability of Masonry, ASTM C-67, Water Repellents Test, and ASTM C-642, Water Absorption Test. Also, federal testing by test SS-W-110C includes water absorption testing.

Any material chosen for use as a clear sealer should be tested by one of these methods to determine water absorption or repellency. Effective water repellency should be above 85 percent, and water absorption should be less than 20 percent, preferably 15 to 10 percent.

Weathering characteristics are important measures of any repellent, due to the alkaline conditions of most masonry and concrete substrates that will deter or destroy the water repellency capabilities of penetrating sealers. In addition, UV degradation affects the life-cycle repellency capabilities for both film-forming and penetrating sealers. Accelerated weathering testing, ASTM 793-75, is an appropriate test to determine the capabilities of a sealer to perform over an extended period. Be sure that the testing is used on a similar substrate, however, as the alkaline conditions of concrete are more severe that masonry products.

Of course, it is always appropriate to test for the compatibility of the sealer with other envelope components and on the exact substrate on which it will be applied. This testing will ensure that there will be no staining of the substrate, that the sealer can penetrate sufficiently, and that the sealer does not damage adjacent envelope components such as glass or aluminum curtain wall etching and sealants, as well as surrounding landscaping. Appropriate field testing methods are reviewed in Chapter 10.

Acrylics

Acrylics and their derivatives, including methyl methacrylates, are film-forming repellents. Acrylics are formulated from copolymers of acrylic or methocrylic acids. Their penetration into substrates is minimal, and they are therefore considered film-forming sealers. Acrylic derivatives differ by manufacturer, each having its own proprietary formulations.

Acrylics are available in both water- and solvent-based derivatives. They are frequently used when penetrating sealers are not acceptable for substrates such as exposed aggregate panels, wood, and dense tile. They are also specified for extremely porous surfaces where a film buildup is desirable for water repellency.

Acrylics do not react chemically with a substrate, and form a barrier by filming over surfaces as does paint. Solids content of acrylics varies from 5 to 48 percent. The higher a solid's content, the greater the amount of sheen imparted to a substrate. High-solids materials are sometimes used or specified to add a high gloss or glazed appearance to cementitious finish materials such as plaster. Methyl methacrylates are available in 5 to 25 percent solids content.

Most manufacturers require two-coat applications of acrylic materials for proper coverage and uniformity. Coverage rates vary depending on the substrate and its porosity, with first coats applied at 100–250 ft^2/gal. Second coats are applied 150–350 ft^2/gal. Acrylics should not be applied over wet substrates, as solvent-based materials may turn white if applied under these conditions. They also cannot be applied in freezing temperatures or over a frozen substrate.

Higher-solids-content acrylics have the capability of being applied in sufficient millage to fill minor cracks or fissures in a substrate. However, no acrylic is capable of withstanding movement from thermal or structural conditions. Acrylic sealers have excellent adhesion when applied to properly prepared and cleaned substrates. Their application resists the formation of mildew, dirt buildup, and salt and atmospheric pollutants.

Acrylics are available in transparent and opaque stains. This coloring enables hiding or blending of repairs to substrates with compatible products such as acrylic sealants and patching compounds. Stain products maintain existing substrate textures and do not oxidize or peel as paint might.

Acrylics are compatible with all masonry substrates including limestone, wood, aggregate panels, and stucco that has not previously been sealed or painted. Acrylic sealers are not effective on very porous surfaces such as lightweight concrete block. The surface of this block contains thousands of tiny gaps or holes filled with trapped air. The acrylic coatings cannot displace this trapped air and are ineffective sealers over such substrates. (See Table 3.5.)

Silicones

Silicone-based water repellents are manufactured by mixing silicone solids (resins) into a solvent carrier. Most manufacturers base their formulations on a 5 percent solids mixture, in conformance with the requirements of federal specification SS-W-110C.

Although most silicone water repellents are advertised as penetrating, they function as film-forming sealers. Being a solvent base allows the solid resin silicone to penetrate the surface of a substrate, but not to depths that siloxanes or quartz carbide sealers penetrate. The silicone solids are deposited onto the capillary pores of a substrate, effectively forming a film of solids that repels water.

All silicone water repellents are produced from the same basic raw material, silane. Manufacturers are able to produce a wide range of repellents by combining or reacting different compounds with this base silane material. These combinations result in a host of silicone-based repellents, including generic types of siliconates, silicone resins, silicones, and siloxanes. The major difference in each of these derivatives is its molecular size.

Regardless of derivative type, molecular size, or compound structure, all silicone-based repellents repel water in the same way. By penetrating substrates, they react chemically with atmospheric moisture, by evaporation of solvents, or by reaction with atmospheric carbon dioxide to form silicone resins that repel water.

Only molecular sizes of the final silicone resin are different. Silicone-based products require that silica be present in a substrate for the proper chemical actions to take place. Therefore, these products do not work on substrates such as wood, metal, or natural stone.

A major disadvantage of silicone water repellents is their poor weathering resistance. Ultraviolet-intense climates can quickly deteriorate these materials and cause a loss of their water repellency. Silicone repellents are not designed for horizontal applications, as they do not resist abrasive wearing.

Silicone repellents are inappropriate for marble or limestone substrates, which discolor if these sealer materials are applied. Discoloring can also occur on other substrates such as precast concrete panels. Therefore, any substrate should be checked for staining by a test application with the proposed silicone repellent.

Lower-solid-concentration materials of 1 to 3 percent solids are available to treat substrates subject to staining with silicone. These formulations should be used on dense sur-

TABLE 3.5 Acrylic Water Repellent Properties

Advantages	Disadvantages
High solids materials can fill minor substrate cracks and fissures	Poor weathering resistance
Stain colors available; compatible with patching materials	Can pick up dirt particles during cure stage
Breathable coating, allows vapor transmission	Poor crack-bridging capabilities

face materials such as granite to allow proper silicone penetration. Special mixes are manufactured for use on limestone but also should be tested before actual application. Silicones can yellow after application, aging, or weathering.

As with most sealers, substrates will turn white or discolor if applied during wet conditions. Silicones do not have the capabilities to span or bridge cracking in a substrate. Very porous materials, such as lightweight or split-face concrete blocks, are not acceptable substrates for silicone sealer application. Adjacent surfaces such as windows and vegetation should be protected from overspray during application. (See Table 3.6.)

Urethanes

Urethane repellents, aliphatic or aromatic, are derivatives of carbonic acid, a colorless crystalline compound. Clear urethane sealers are typically used for horizontal applications but are also used on vertical surfaces. With a high solids content averaging 40 percent, they have some ability to fill and span nonmoving cracks and fissures up to 1/16 inch wide. High-solids materials such as urethane sealers have low perm ratings and cause coating blistering if any moisture or vapor drive occurs in the substrate.

Urethane sealers are film-forming materials that impart a high gloss to substrates, and they are nonyellowing materials. They are applicable to most substrates including wood and metal, but adhesive tests should be made before each application. Concrete curing agents can create adhesion failures if the surface is not prepared by sandblasting or acid etching.

Urethane sealers can also be applied over other compatible coatings, such as urethane paints, for additional weather protection. They are resistant to many chemicals, acids, and solvents and are used on stadium structures for both horizontal and vertical seating sections. The cost of urethane materials has limited their use as sealers. (See Table 3.7.)

TABLE 3.6 Silicone Water-Repellent Properties

Advantages	Disadvantages
Breathable coating, allows vapor transmission	Poor ultraviolet resistance
Easy application	Can stain or yellow a substrate such as limestone
Cost	Contamination of substrate prohibits other materials' application over silicone

TABLE 3.7 Urethane Water-Repellent Properties

Advantages	Disadvantages
Applications over wood and metal substrates	Poor vapor transmission
Horizontal applications also	Blisters that occur if applied over wet substrates
Chemical-, acid-, and solvent-resistant applications	Higher material cost

Silanes

Silanes contain the smallest molecular structures of all silicone-based materials. The small molecular structure of the silane allows the deepest penetration into substrates. Silanes, like siloxanes, must have silica present in substrates for the chemical action to take place that provides water repellency. These materials cannot be used on substrates such as wood, metal, or limestone that have no silica present for chemical reaction.

Of all the silicone-based materials, silanes require the most difficult application procedures. Substrates must have sufficient alkalinity in addition to the presence of moisture to produce the required chemical reaction to form silicone resins. Silanes have high volatility that causes much of the silane material to evaporate before the chemical reaction forms the silicon resins. This evaporation causes a high silane concentration, as much as 40 percent, to be lost through evaporation.

Should a substrate become wet too quickly after application, the silane is washed out from the substrate-prohibiting proper water-repellency capabilities. If used during extremely dry weather, after application substrates are wetted to promote the chemical reaction necessary. The wetting must be done before all the silane evaporates.

As with other silicone-based products, silanes applied properly form a chemical bond with a substrate. Silanes have a high repellency rating when tested in accordance with ASTM C-67, with some products achieving repellency over 99 percent. As with urethane sealers, their high cost limits their usage. (See Table 3.8.)

Siloxanes

Siloxanes are produced from the CL-silane material, as are other silicone masonry water repellents. Siloxanes are used more frequently than other clear silicones, especially for horizontal applications. Siloxanes are manufactured in two types, oligomerous (short chain of molecular structure) and polymeric (longer chain of molecular structure) alkylalkoxysiloxanes.

Most siloxanes produced now are oligomerous. Polymeric products tend to remain wet or tacky on the surface, attracting dirt and pollutants. Also, polymeric siloxanes have poor alkali resistance, and alkalis are common in masonry products for which they are intended. Oligomerous siloxanes are highly resistant to alkaline attack, and therefore can be used successfully on high alkaline substrates such as cement-rich mortar.

Siloxanes react with moisture, as do silanes, to form the silicone resin that acts as the water-repellent substance. Upon penetration of a siloxane into a substrate it forms

TABLE 3.8 Silane Water-Repellent Properties

Advantages	Disadvantages
Deepest penetration capabilities of all silicone-based products	High evaporation rate during application
Forms chemical bond with substrate with good permeability rating	Dry substrates must be wetted to ensure chemical reactions before evaporation
Good weathering characteristics	High cost of material

a chemical bond with the substrate. The advantage of siloxanes over silanes is that their chemical structure does not promote a high evaporation rate.

The percentage of siloxane solids used is substantially less (usually less than 10 percent for vertical applications), thereby reducing costs. Chemical reaction time is achieved faster with siloxanes, which eliminates a need for wetting after installation. Repellency is usually achieved within 5 hours with a siloxane.

Siloxane formulations are now available that form silicone resins without the catalyst—alkalinity—required. Chemical reactions with siloxanes take place even with a neutral substrate as long as moisture, in the form of humidity, is present.

These materials are suitable for application to damp masonry surfaces without the masonry turning white, which might occur with other materials. Testing of all substrates should be completed before full application, to ensure compatibility and effectiveness of the sealer.

Siloxanes do not change the porosity or permeability characteristics of a substrate. This allows moisture to escape without damaging building materials or the repellent. Since siloxanes are not subject to high evaporation rates, they can be applied successfully by high-pressure sprays for increased labor productivity.

Siloxanes, as other silicone-based products, may not be used with certain natural stones such as limestone. They also are not applicable to gypsum products or plaster. Siloxanes should not be applied over painted surfaces, and if surfaces are to be painted after treatment they should first be tested for compatibility. (See Table 3.9.)

Silicone rubber

These systems are a hybrid of the basic silicone film-forming and the silicone derivatives penetrating sealers. The product is basically a silicone solid dissolved in a solvent carrier that penetrates into the substrate, carrying the solids to form a solid film that is integral with the substrate. Unlike the penetrating derivatives, silicone rubbers do not react with the substrate to form the repellency capability.

The percentage solids, as high as 100 percent, carried into the substrate supposedly create a thickness of product millage internally in the substrate to a film thick enough to bridge minute hairline cracking in the substrate. This elongation factor, expressed as high as 400 percent by some manufacturers, does not produce substantial capacity to bridge cracks, since the millage of the film that creates movement capability is minimal with clear repellents. Only existing cracks less than 1/32 inch are within the capability of these materials to seal, and new cracks that develop will not be bridged since the material is integral with the substrate and cannot move as film-forming membranes are allowed to do.

Through chemical formulations and the fact that they penetrate into the substrate, the silicone rubber products have been UV-retardant, unlike basic silicone film-forming sealers. At the same time they retain sufficient permeability ratings to permit applications to typical clear repellent substrates. These systems are also applicable to wood, canvas, and terra cotta substrates that other penetrating sealers are not applicable, since the rubber systems do not have to react with the substrate to form their repellency.

Silicone rubber systems are applicable in both horizontal and vertical installations and make excellent sealers for civil project sealing including bridges, overpasses, and parking garages. Like the generic silicone compounds, silicone rubber does not permit any other material to bond to it directly. Therefore, projects sealed with these materials can not be painted over in the future without having to remove the sealer with caustic chemicals such as solvent paint removers. This can create problems on projects where some applications are required over the substrate once sealed, such as parking-stall painted stripes in a parking garage. Manufacturers of the silicone rubber sealers should be contacted directly for recommendations in such cases.

These materials generally have excellent repellency rates in addition to acceptable permeability rates. Overspray precautions should be taken whenever using the product near glass or aluminum envelope components, since the material is difficult if not almost impossible to remove from such substrates. (See Table 3.10).

Sodium silicates

Sodium silicate materials should not be confused with water repellents. They are concrete densifiers or hardeners. Sodium silicates react with the free salts in concrete such as calcium or free lime, making the concrete surface more dense. Usually these materials are sold as floor hardeners, which when compared to a true, clear deck coating have repellency insufficient to be considered with materials of this section.

TABLE 3.9 Siloxanes Water-Repellent Properties

Advantages	Disadvantages
Not susceptible to alkali degradation	Not applicable on natural stone substrates
Bonds chemically with substrates with high permeability rating	Can damage adjacent substrates and vegetation
High repellency rating and excellent penetration depth	Cost

Table 3.10 Silicone Rubber Water-Repellent Properties

Advantages	Disadvantages
Application to a wider range of substrates including canvas and wood	Cannot be painted over
Bonds integrally with substrate	Can damage other envelope components such as glass or aluminum
Can fill minor cracks and fissures	

WATER-REPELLENT APPLICATION

General surface preparations for all clear water-repellent applications require that the substrate be clean and dry. (Siloxane applications can be applied to slightly damp surfaces, but it is advisable to try a test application.) All release agents, oil, tar, and asphalt stains, as well as efflorescence, mildew, salt spray, and other surface contaminants, must be removed.

Application over wet substrates will cause either substrate discoloring, usually a white film formation, or water-repellent failure. When in doubt of moisture content in a substrate, do a moisture test using a moisture meter or a mat test using visquene taped to a wall, to check for condensation. Note that some silicone-based systems, such as silanes, must have moisture present, usually in the form of humidity, to complete the chemical reaction.

Substrate cracks are repaired before sealer application. Small cracks are filled with nonshrink grout or a sand–cement mixture. Large cracks or structural cracking should be epoxy-injected. If a crack is expected to continue to move, it should be sawn out to a minimum width of 1/4 inch and sealed with a compatible sealant.

Note that joint sealers should be installed first, as repellents contaminate joints, causing sealant-bonding failure. Concrete surfaces, including large crack patching, should be cured a minimum of 28 days before sealer application.

All adjacent substrates not being treated, including window frames, glass, and shrubberies, should be protected from overspray. Natural stone surfaces, such as limestone, are susceptible to staining by many clear sealers. Special formulations are available from manufacturers for these substrates. If any questions exist regarding an acceptable substrate for application, a test area should first be completed.

All sealers should be used directly from purchased containers. Sealers should never be thinned, diluted, or altered. Most sealers are recommended for application by low-pressure spray (20 lb/in^2), using a Hudson or garden-type sprayer. Brushes or rollers are also acceptable, but they reduce coverage rates. High-pressure spraying should be used only if approved by the manufacturer.

Applicators should be required to wear protective clothing and proper respirators, usually the cartridge type. Important cautionary measures should be followed in any occupied structure. Due to the solvents used in most clear sealers, application areas must be well ventilated. All intake ventilation areas must be protected or shut off, to prevent the contamination of interior areas from sealer fumes. Otherwise, evacuation by building occupants is necessary.

Most manufacturers require a flood coating of material, with coverage rates dependent upon the substrate porosity. Materials should be applied from the bottom of a building, working upward (Figure 3.6). Sealers are applied to produce a rundown or saturation of about 6 in of material below the application point for sufficient application. If a second coat is required, it should be applied in the same manner. Coverage rates for second coats increase, as fewer materials will be required to saturate a substrate surface.

Testing should be completed to ensure that saturation of surfaces will not cause darkening or add sheen to substrate finishes. Dense concrete finishes may absorb insufficient repellent if they contain admixtures such as integral waterproofing or form-release agents. In these situations, acid etching or pressure cleaning is necessary to allow sufficient sealer absorption. Approximate coverage rates of sealers over various substrates are summarized in Table 3.11.

Priming is not required with any type of clear sealer. However, some manufacturers recommend that two saturation coats be applied instead of one coat. Some systems may require a mist coat to break surface tension before application of the saturation coat.

CEMENTITIOUS COATINGS

Cementitious-based coatings are among the oldest products used for above-grade waterproofing applications. Their successful use continues today, even with the numerous clear

FIGURE 3.6 Spray application of clear repellent. (*Courtesy of Saver Systems*)

TABLE 3.11 Coverage Rates for Water Repellents*

Surface	Coverage (ft/gal)
Steel-troweled concrete	150–300
Precast concrete	100–250
Textured concrete	100–200
Exposed aggregate concrete	100–200
Brick, dense	100–300
Brick, coarse	75–200
Concrete block, dense	75–150
Concrete block, lightweight	50–100
Natural stones	100–300
Stucco, smooth	125–200
Stucco, coarse	100–150

*Manufacturer's suggested rates should be referred to for specific installations. If a second coat is required, coverage will be higher for second application.

and elastomeric sealers available. However, cementitious systems have several disadvantages, including an inability to bridge cracks that develop in substrates after application. This can be nullified by installation of control or expansion joints to allow for movement. In remedial applications where all settlement cracks and shrinkage cracks have already developed, only expansion joints for thermal movement need be addressed.

These coatings are cement-based products containing finely graded siliceous aggregates that are nonmetallic. Pigments are added for color; proprietary chemicals are added for integral waterproofing or water repellency. An integral bonding agent is added to the dry mix, or a separate bonding agent liquid is provided to add to the dry packaged material during mixing. The cementitious composition allows use in both above- and below-grade applications. See Fig. 3.7, for a typical above-grade cementitious application.

Since these products are water-resistant, they are highly resistant to freeze–thaw cycles; they eliminate water penetration that might freeze and cause spalling. Cementitious coatings have excellent color retention and become part of the substrate. They are also non-chalking in nature.

Color selections, such as white, that require the used of white Portland cement, increase material cost. Being cementitious, the product requires job-site mixing, which should be carefully monitored to ensure proper in-place performance characteristics of coatings. Also, different mixing quotients will affect the dried finish coloring, and if each batch is not mixed uniformly, different finish colors will occur.

Cementitious properties

Cementitious coatings have excellent compressive strength, ranging from 4000 to 6000 lb/in^2 after curing (when tested according to ASTM C-109). Water absorption rates of cementitious materials are usually slightly higher than elastomeric coatings. Rates are acceptable for waterproofing, and range from 3 to 5 percent maximum water absorption by weight (ASTM C-67).

FIGURE 3.7 Spray application of cementitious waterproofing. (*Courtesy of Vandex*)

Cementitious coatings are highly resistant to accelerated weathering, as well as being salt-resistant. However, acid rain (sulfate contamination) will deteriorate cementitious coatings as it does other masonry products.

Cementitious coatings are breathable, allowing transmission of negative water vapor. This avoids the need for completing drying of substrates before application, and the spalling that is caused by entrapped moisture. These products are suitable for the exterior of planters, undersides of balconies, and walkways, where negative vapor transmission is likely to occur. Cementitious coatings are also widely used on bridges and roads, to protect exposed concrete from road salts, which can damage reinforcing steel by chloride attack.

Cementitious installations

Water entering masonry substrates causes brick to swell, which applies pressure to adjacent mortar joints. The cycle of swelling when wet, and relaxing when dry, causes mortar joint deterioration. Cementitious coating application prevents water infiltration and the resulting deterioration. However, coatings alter the original facade aesthetics, and a building owner or architect may deem them not acceptable.

Cementitious coatings are only used on masonry or concrete substrates, unlike elastomeric coatings that are also used on wood and metal substrates. Cementitious coating use includes applications to poured-in-place concrete, precast concrete, concrete block units, brick, stucco, and cement plaster substrates (Figure 3.8). Once applied, cementitious coatings bond so well to a substrate that they are considered an integral part of the substrate rather than a film protection such as an elastomeric coating.

Typical applications besides above-grade walls include swimming pools, tunnels, retention ponds, and planters (Figure 3.9). With Environmental Protection Agency (EPA) approval, these products may be used in water reservoirs and water treatment plants. Cementitious coatings are often used for finishing concrete, while at the same time providing a uniform substrate coloring.

An advantage with brick or block wall applications is that these substrates do not necessarily have to be tuck-pointed before cementitious coating application. Cementitious coatings will fill the voids, fissures, and honeycombs of concrete and masonry surfaces, effectively waterproofing a substrate (Figure 3.10).

When conditions require, complete coverage of the substrate by a process called *bag*, or *face*, *grouting* of the masonry is used as an alternative. In this process, a cementitious coating is brush applied to the entire masonry wall. At an appropriate time, the cementitious coating is removed with brushes or burlap bags, again revealing the brick and mortar joints. The only coating material left is that in the voids and fissures of masonry units and mortar joints. Although costly, this is an extremely effective means of waterproofing a substrate, more effective only than tuck-pointing.

Complete cementitious applications provide a highly impermeable surface and are used to repair masonry walls that have been sandblasted to remove existing coatings and walls that are severely deteriorated. Cementitious applications effectively preserve a facade while making it watertight. *Bag grouting* application adds only a uniformity to substrate color; colored cementitious products can impart a different color to existing walls if desired. *Mask grouting* is similar to bag grouting. With mask grouting applications, existing masonry units are carefully taped over, exposing only mortar joints. The coating material is brush-applied

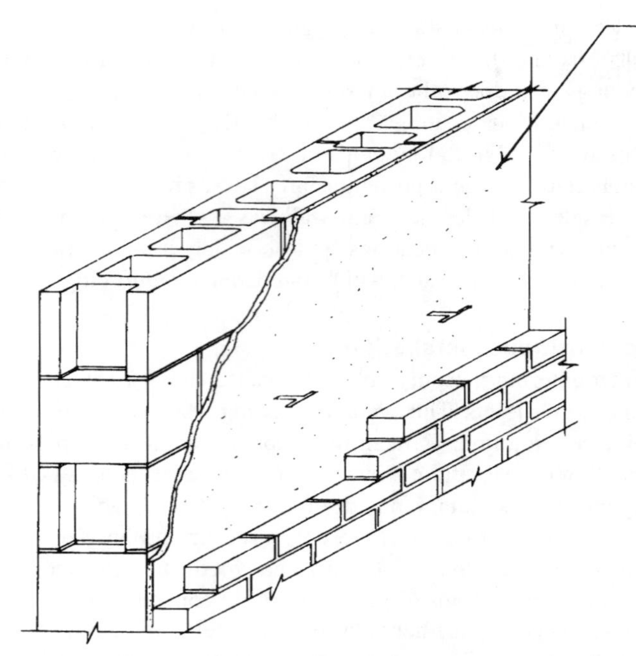

FIGURE 3.8 Block cavity wall waterproofing using cementitious waterproofing. (*Courtesy of Anti-Hydro International, Inc.*)

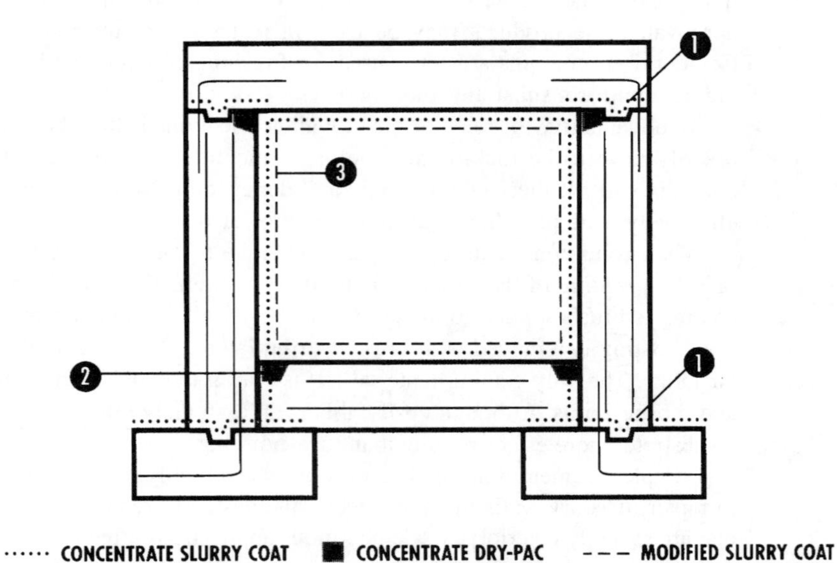

FIGURE 3.9 Typical detailing of tunnel waterproofing applicable to above-grade applications. (*Courtesy of Xypex*)

to exposed joints, then cured. Tape is then removed from the masonry units, leaving behind a repaired joint surface with no change in wall facade color.

The thickness of coating added to mortar joints is variable but is greater when joints are recessed. This system is applicable only to substrates in which the masonry units themselves, such as brick, are nondeteriorated and watertight, requiring no restoration.

Texture is easily added to a cementitious coating, either by coarseness of aggregate added to the original mix or by application methods. The same cementitious mix applied by roller, brush, spray, hopper gun, sponge, or trowel results in many different texture finishes. This provides an owner or designer with many texture selections while maintaining adequate waterproofing characteristics. A summary of the major advantages and disadvantages of cementitious coatings are given in Table 3.12.

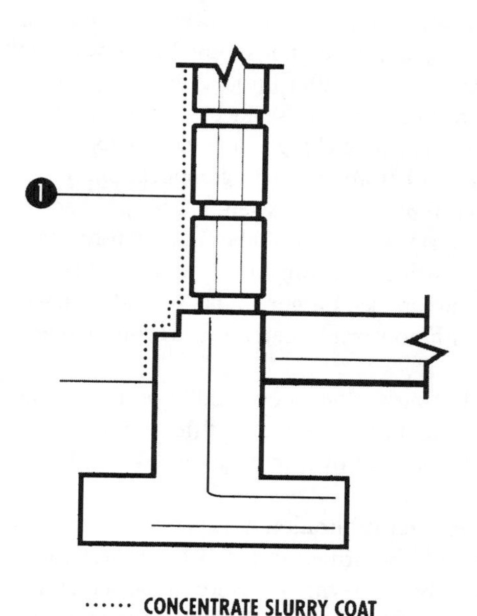

FIGURE 3.10 Waterproofing concrete block envelope with cementitious coating. (*Courtesy of Xypex*)

In certain instances, such as floor–wall junctions, it is desirable first to apply the cementitious coating to a substrate, and then to fill the joint with sealant material in a color that matches the cementitious coating. The coating will fully adhere to the substrate and is compatible with sealant materials. It is also possible first to apply cementitious coating to substrates, then to apply a sealant to expansion joints, door, and window penetrations, and other joints. This is not possible with clear sealers nor recommended with elastomeric coatings, due to bonding problems.

Cementitious coatings are a better choice over certain substrates, particularly concrete or masonry, than clear sealers or elastomeric coatings. This is because cementitious coatings have better bonding strength, a longer life cycle, lower maintenance, and less attraction of airborne contaminants. Provided that adequate means are incorporated for thermal and structural movement, cementitious coatings will function satisfactorily for above- and below-grade waterproofing applications.

TABLE 3.12 Cementitious Coating Properties

Advantages	Disadvantages
Excellent bonding capability	No movement capability
Applicable to both above-and below-grade installations	Difficult to control uniform color and texture
Excellent weathering capabilities	High degree of expertise required for installation
Numerous textures and colors available	Not resistant to acid rain and other contaminants
Can eliminate need for tuck-pointing	Not applicable over wood or metal substrates

CEMENTITIOUS COATING APPLICATION

For adequate bonding to substrates, surfaces to receive cementitious coatings should be cleaned of contaminants including dirt, efflorescence, form-release agents, laitance, residues of previous coatings, and salts. Previously painted surfaces must be sandblasted or chemically cleaned to remove all paint film.

Cementitious coating bonding is critical to successful in-place performance. Therefore, extreme care should be taken in preparing substrates for coating application. Sample applications for bond strength should be completed if there is any question regarding the acceptability of a substrate, especially with remedial waterproofing applications.

Poured-in-place or precast concrete surfaces should be free of all honeycombs, voids, and fins. All tie holes should be filled before coating application with nonshrink grout material as recommended by the coating manufacturer. Although concrete does not need to be cured before cementitious coating application, it should be set beyond the green stage of curing. This timing occurs within 24 hours after initial concrete placement.

With smooth concrete finishes, such as precast, surfaces may need to be primed with a bonding agent. In some instances a mild acid etching can be desirable, using a muriatic acid solution and properly rinsing substrates before the coating application. Some manufacturers require a further roughing of smooth finishes, such as sandblasting, for adequate bonding.

On masonry surfaces, voids in mortar joints should be filled before coating installation. With both masonry and concrete substrates, existing cracks should be filled with a dry mix of cementitious material sponged into cracks. Larger cracks should be sawn out, usually to a 3/4 inch minimum, and packed with nonshrink material as recommended by the coating manufacturer.

Moving joints must be detailed using sealants designed to perform under the expected movement. These joints include thermal movement and differential movement joints. The cementitious material should not be applied over these joints as it will crack and "alligator" when movement occurs.

If cracks are experiencing active water infiltration, this pressure must be relieved before coating is applied. Relief holes should be drilled in a substrate, preferably at the base of the wall, to allow wicking of water, thus relieving pressure in the remainder of work areas during coating application. After application and proper curing time (approximately 48–72 hours), drainage holes may then be packed with a nonshrink hydraulic cement material and finished with the cementitious coating.

After substrate preparations are completed and just before application, substrates must be wetted or dampened with clean water for adequate bonding of the coating. Substrates must be kept continually damp in preparation for application. The amounts of water used are dependent on weather and substrate conditions. For example in hot, dry weather, substrates require frequent wettings. Coatings should not be applied in temperatures below 40°F or in conditions when the temperature is expected to fall below freezing within 24 hours after application.

Cementitious coatings should be carefully mixed following the manufacturer's recommended guidelines concerning water ratios. Bonding agents should be added as required with no other additives or extenders, such as sand, used unless specifically approved by the manufacturer. With smooth surfaces such as precast concrete, an additional bonding agent is required.

Cementitious coatings may be applied by brush, trowel, or spray. Stiff, coarse, or fiber brushes are used for application. Brush applications require that the material be scrubbed into a substrate, filling all pores and voids. Finish is completed by brushing in one direction for uniformity.

Spray applications are possible by using equipment designed to move the material once mixed. Competent mechanics trained in the use of spray equipment and technique help ensure acceptable finishes and watertightness (Figure 3.11).

Trowel applications are acceptable for the second coat of material. Due to the application thickness of this method, manufacturers recommend that silica sand be added to the mix in proper portions. The first coats of trowel applications are actually brush applications that fill voids and pores. Finish trowel coats can be on a continuum from smooth to textured. Sponge finishing of the first coat is used to finish smooth concrete finishes requiring a cementitious application.

With textured masonry units such as split face or fluted block, additional material is required for effective waterproofing. On this type of finish, spraying or brush applications are the only feasible and effective means.

The amount of material required depends upon the expected water conditions. Under normal waterproofing requirements, the first coat is applied at a rate of 2 pounds of material per square yard of work area. The finish coat is then applied at a coverage rate of 1 lb/yd^2. In severe water conditions, such as below-grade usage with water-head pressures, materials are applied at 2 lb/yd^2. This is followed by a trowel application at 2 lb/yd^2. Clean silica sand is added to the second application at 25 pounds of silica to one bag, 50 pounds, of premixed cementitious coating.

FIGURE 3.11 Spray application of cementitious membrane on negative side. (*Courtesy of LBI Technologies*)

With all applications, the second material coat should be applied within 24 hours after applying the first coat. Using these application rates, under normal conditions, a 50-pound bag of coating will cover approximately 150 ft^2 (1 $lb/yd^2$2, first coat; 2 lb/yd^2, second coat). The finish thickness of this application is approximately 1/8 in.

Trying to achieve this thickness in one application, or adding excessive material thickness in one application, should not be attempted. Improper bonding will result, and material can become loose and spall. To eliminate mortar joint shadowing on a masonry wall being visible through the coating, a light trowel coat application should be applied first, followed by a regular trowel application.

The cementitious coating beginning to roll or pull off a substrate is usually indicative of the substrate being too dry; redampening with clean water before proceeding is necessary. Mix proportions must be kept constant and uniform, or uneven coloring or shadowing of the substrate will occur.

After cementitious coatings are applied they should be cured according to the manufacturer's recommendations. Typically, this requires keeping areas damp for 1–3 days. In extremely hot weather, more frequent and longer cure times are necessary to prevent cracking of the coating. The water cure should not be done too soon after application, as it may ruin or harm the coating finish. Chemical curing agents should not be used or added to the mix unless specifically approved by the coating manufacturer.

Typically, primers are not required for cementitious coating applications, but bonding agents are usually added during mixing. In some cases, if substrates are especially smooth or previous coatings have been removed, a direct application of the bonding agent to substrate surfaces is used as a primer. If there is any question regarding bonding strength, samples should first be applied both with and without a bonding agent and tested before proceeding with the complete application.

Cementitious coatings should not be applied in areas where thermal, structural, or differential movement will occur. Coatings will crack and fail if applied over sealant in control or expansion joints. Cementitious-based products should not be applied over substrates other than masonry substrates such as wood, metal, or plastics,

ELASTOMERIC COATINGS

Paints and elastomeric coatings are similar in that they always contain three basic elements in a liquid state: pigment, binder, and solvent. In addition, both often contain special additives such as mildew-resistant chemicals. However, paints and coatings differ in their intended uses.

Paints are applied only to add decorative color to a substrate. Coatings are applied to waterproof or otherwise protect a substrate. The difference between clear sealers and paints or coatings is that sealers do not contain the pigments that provide the color of paints or coatings.

Solvent is added to paints and coatings to lower the material viscosity so it can be applied to a substrate by brush, spray, or roller. The binder and solvent portion of a paint or coating is referred to as the *vehicle*. A coating referred to as *100 percent solids* is merely a binder in a liquid state that cures, usually moisture cured from air humidity, to a seamless film upon application. Thus it is the binder portion, common to all paints and coatings, that imparts the unique characteristics of the material, differentiating coatings from paints.

Waterproof coatings are classified generically by their binder type. The type of resin materials added to the coating imparts the waterproofing characteristics of the coating material. Binders are present in the vehicle portion of a coating in either of two types. An emulsion occurs when binders are dispersed or suspended in solvent for purposes of application. Solvent-based materials have the binder dissolved within the solvent.

The manner in which solvents leave a binder after application depends upon the type of chemical polymer used in manufacturing. A thermoplastic polymer coating dries by the solvent evaporating and leaving behind the binder film. This is typical of water-based acrylic elastomeric coatings used for waterproofing. A thermosetting polymer reacts chemically or cures with the binder and can become part of the binder film that is formed by this reaction. Examples are epoxy paints, which require the addition and mixing of a catalyst to promote chemical reactions for curing the solvent.

The catalyst prompts a chemical reaction that limits application time for these materials before they cure in the material container. This action is referred to as the "pot life" of material (workability time). The chemical reactions necessary for curing create thermosetting polymer vehicles that are more chemically resistant than thermoplastic materials. Thermosetting vehicles produce a harder film and have an ability to contain higher solids content than thermoplastic materials.

Resins used in elastomeric coatings are breathable. They allow moisture-vapor transmission from the substrate to escape through the coating without causing blisters in the coating film. This is a favorable characteristic for construction details at undersides of balconies that are subjected to negative moisture drive. Thermosetting materials such as epoxy paints are not breathable. They will blister or become unbonded from a substrate if subjected to negative moisture drive.

Resins

Elastomeric coatings are manufactured from acrylic resins with approximately 50 percent solids by volume. Most contain titanium dioxide to prevent chalking during weathering. Additional additives include mildewcides, alkali-resistant chemicals, various volume extenders to increase solids content, and sand or other fillers for texture.

Resins used in waterproofing coatings must allow the film to envelop a surface with sufficient dry film millage (thickness of paint measured in millimeters) to produce a film that is watertight, elastic, and breathable. Whereas paints are typically applied 1–4 mil thick, elastomeric coatings are applied 10–20 mil thick.

It is this thickness (with the addition of resins or plasticizers that add flexibility to the coating) that creates the waterproof and elastic coating, thus the term *elastomeric coating*. Elastomeric coatings have the ability to elongate a minimum of 300 percent at dry millage thickness of 12–15 mil. Elongation is tested as the minimum ability of a coating to expand and then return to its original shape with no cracking or splitting (tested according to ASTM D-2370). Elongation should be tested after aging and weathering to check effectiveness after exposure to the elements.

Elastomeric coatings are available in both solvent-based and water-based vehicles. Water-based vehicles are simpler to apply and not as moisture-sensitive as the solvent-based vehicles. The latter are applied only to totally dry surfaces that require solvent materials for cleanup.

Typical properties of a high-quality, waterproof, and elastic coating include the following:

- Minimum of 10-mil dry application
- High solids content (resins)
- Good ultraviolet weathering resistance
- Low water absorption, withstanding hydrostatic pressure
- Permeability for vapor transmission
- Crack-bridging capabilities
- Resistance to sulfites (acid rain) and salts
- Good color retention and low dirt pickup
- High alkali resistance

Acrylic coatings are extremely sensitive to moisture during their curing process, taking up to 7 days to cure. Should the coating be subjected to moisture during this time, it may reemulsify (return to liquid state). This becomes a critical installation consideration whenever such coatings are used in a horizontal or slightly inclined surface that might be susceptible to ponding water.

Elastomeric coating installations

Elastomeric coatings, which are used extensively on stucco finish substrates and exterior insulation finish systems (EIFS), are also used on masonry block, brick, concrete, and wood substrates. Some are available with asphalt primers for application over asphalt finishes. Others have formulations for use on metal and sprayed urethane foam roofs.

Elastomeric coatings are also successfully used over previously painted surfaces. By cleaning, preparing the existing surface, repairing cracks (Fig. 3.12), and priming, coatings can be used to protect concrete and masonry surfaces that have deteriorated through weathering and aging (Fig. 3.13).

Proper preparation, such as tuck-pointing loose and defective mortar joints and injecting epoxy into cracks, must be completed first. In single-wythe masonry construction, such as split-face block, applying a cementitious block filler is necessary to fill voids in the block before applying elastomeric coating for effective waterproofing.

Aesthetically, coatings are available in a wide range of textures and are tintable to any imaginable color. However, deep, dark, tinted colors may fade, or pigments added for coloring may bleed out creating unsightly staining. Heavy textures limit the ability of a coating to perform as an elastomeric due to the amount of filler added to impart texture. Because elastomeric coatings are relatively soft materials (lower tensile strength to impart flexibility), they tend to pick up airborne contaminants. Thus lighter colors, including white, may get dirty quickly.

Uniform coating thickness is critical to ensure crack bridging and thermal movement capabilities after application. Applicators should have wet millage gages for controlling the millage of coating applied. Applications of elastomeric coatings are extremely labor-sensitive. They require skilled application of the material. In addition, applicators must transition coating applications into adjacent members of the building envelope, such as window frames and flashings, for effective envelope waterproofing. (See Table 3.13.)

FIGURE 3.12 Preparation of substrate including crack repair prior to elastomeric coating application. (*Courtesy of Coastal Construction Products*)

FIGURE 3.13 Application of elastomeric coating. (*Courtesy of Coastal Construction Products*)

TABLE 3.13 Elastomeric Coating Properties

Advantages	Disadvantages
Excellent elastomeric and crack-bridging capability	Uniform application thickness difficult to control
Wide range of colors and textures available	Life cycle shorter than cementitious
Breathable	No below-grade usage
Applicable over wood and metal substrates	Masonry substrates may require extensive repairs before application
Resistant to acid rain and other pollutants	May fade over time

ELASTOMERIC COATING APPLICATION

Successful application of elastomeric coatings depends entirely on proper substrate preparation. Although they are effective waterproof materials, they should not be applied over cracks, voids, or deteriorated materials, as this will prevent cohesive waterproofing of the building envelope. Coatings chosen must be compatible with any existing coatings, sealants, or patching compounds used in crack repairs. Coating manufacturers have patching, sealing, and primer materials, all compatible with their elastomeric coating.

Applying elastomeric coating requires applicator knowledge beyond a typical paint job. Most painting contractors do not have the experience or knowledge to apply these coatings.

Existing substrates must be cleaned to remove all dirt, mildew, and other contaminants. This is accomplished by pressure-cleaning equipment with a minimum capability of 1500 lb/in^2 water pressure. All grease, oils, and asphalt materials must be removed completely.

Mildew removal with chlorine should be done where necessary. Chemical cleaning is also necessary to remove traces of release agents or incompatible curing agents. If chemicals are used, the entire surface should be rinsed to remove any chemical traces that might affect the coating bonding.

Previously painted substrates should have a duct-tape test for compatibility of the elastomeric coating application. A sample area of coating should be applied over existing materials and allowed to dry. Then duct tape should be sealed firmly to the substrate then pulled off quickly. If any amount of coating comes off with the tape, coatings are not properly adhering to existing materials. In that case, all existing coatings or paints must be removed to ensure adequate bonding. No coating can perform better than the substrate to which it is applied, in this case a poorly adhered existing coating. Either excessively chalky coatings must be removed or a primer coat applied. Primers will effectively seal the surface for proper bonding to a substrate.

High-alkaline masonry substrates must be checked for a pH rating before installation. The pH rating is a measure of substrate acidity or alkalinity. A rating of 7 is neutral, with higher ratings corresponding to higher alkaline substrates. A pH of more than 10 requires following specific manufacturer's recommendations. These guidelines are based upon the alkali resistance of a coating and substrate pH.

Surface preparations of high-alkali substrates include acid washing with 5 percent muriatic acid or primer application. In some cases, extending curing time of concrete or stucco substrates will effectively lower their pH. Immediately after application stucco has a high pH, but it has continually lower pH values during final curing stages. New stucco should cure for a minimum of 30 days, preferably 60–90 days, to lower the pH. This also allows shrinkage and thermal cracks to form and be treated before coating application.

Sealant installation should be completed before applying elastomeric coating to prevent joint containment by the coating. This includes expansion and control joints, perimeters of doors and windows, and flashings. Small nonmoving cracks less than 1/16 inch wide require filling and overbanding 2 inch wide with a brushable or knife-grade sealant material (Fig. 3.14).

Cracks exceeding 1/16 inch that are also nonmoving joints should be sawn out to approximately a 1/4-inch width and depth and filled with a knife-grade sealant, followed by overbanding approximately 4 inches wide (see Fig. 3.15). Changes in direction should be reinforced as shown in Fig. 3.16.

Overbanding (bandage application of a sealant) requires skilled craftspeople to featheredge banding sides to prevent telescoping of patches through the coating. Thick, unfeathered applications of brushable sealant will show through coating applications, providing an unacceptable substrate appearance.

Large cracks over 1/2 inch wide that are nonmoving, such as settlement cracks, should be sawn out, and proper backing materials applied before sealant installation (Fig. 3.17). Fiberglass mesh in 4-inch widths can be embedded into the brushable sealant for additional protection.

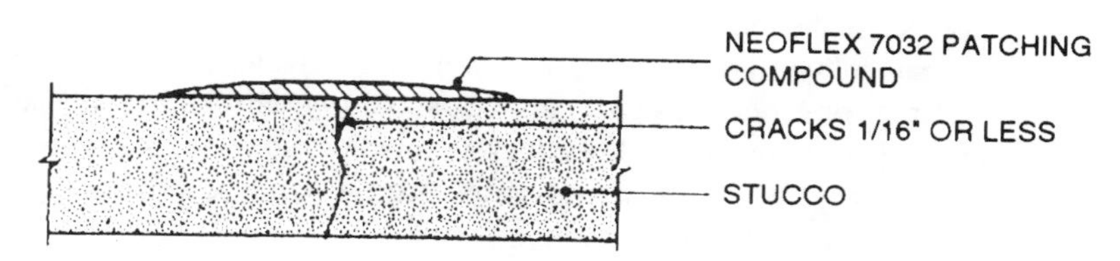

FIGURE 3.14 Crack repair, under 1/16 inch, for elastomeric substrate preparation. (*Courtesy of Neogard*)

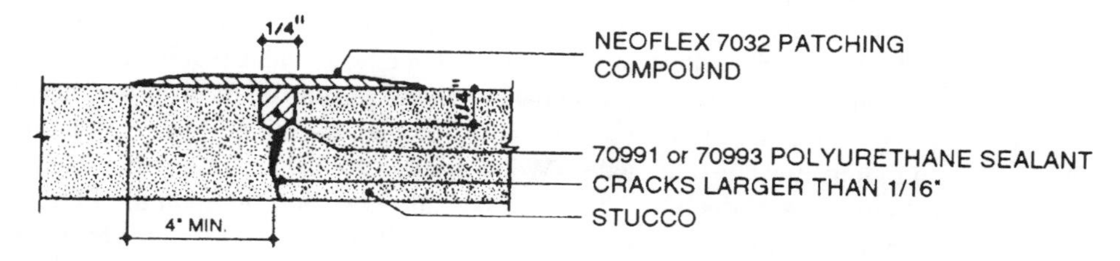

FIGURE 3.15 Crack repair, over 1/16 inch, for elastomeric substrate preparation. (*Courtesy of Neogard*)

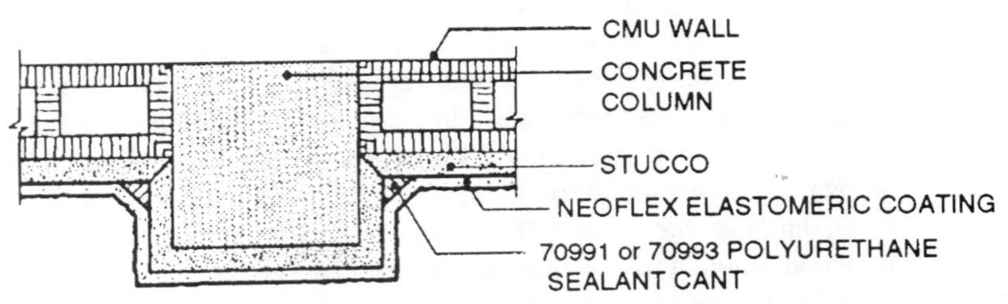

FIGURE 3.16 Changes in envelope plane require detailing prior to elastomeric application. (*Courtesy of Neogard*)

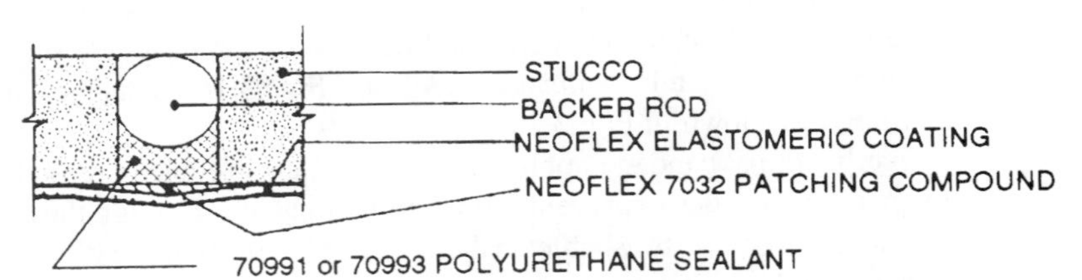

FIGURE 3.17 Large movement crack or joint repair for elastomeric coatings. (*Courtesy of Neogard*)

Joints that are expected to continue moving, such as joints between dissimilar materials, should be sealed using guidelines set forth in Chap. 4. These joints should not be coated over, since the movement experienced at these joints typically exceeds the elastomeric coating capability. In such cases, the coating will alligator and develop an unsightly appearance.

Brick or block masonry surfaces should be checked for loose and unbonded mortar joints. Faulty joints should be tuck-pointed or sealed with a proper sealant. With masonry applications, when all mortar joints are unsound or excessively deteriorated, all joints should be sealed before coating.

Additionally, with split-face block, particularly single-wythe construction, a cementitious block filler should be applied to all cavities and voids. This provides the additional waterproofing protection that is necessary with such porous substrates. On previously painted split face construction, an acrylic block filler may be used to prepare the surface.

All sealants and patching compounds must be cured before coating application; if this is not done, patching materials will mildew beneath the coating and cause staining. For metal surfaces, rusted portions must be removed or treated with a rust inhibitor, then primed as recommended by the coating manufacturer. New galvanized metal should also be primed.

Wood surfaces require attention to fasteners that should be recessed and sealed. Laps and joints must also be sealed. Wood primers are generally required before coating application. The success of an elastomeric coating can depend upon use of a proper primer for specific conditions encountered. Therefore, it is important to refer to manufacturer guidelines for primer usage.

Elastomeric coatings are applied by brush, roller, or spray after proper mixing and agitating of the coating (see Fig. 3.18). Roller application is preferred, as it fills voids and crevices in a substrate. Long nap rollers should be used with covers having a $^3/_4$–$^1/_2$-inch nap. Elastomeric coatings typically require two coats to achieve proper millage. The first application must be completely dried before the second coat is applied.

Spray applications require a mechanic properly trained in the crosshatch method. This method applies coating by spraying vertically and then horizontally to ensure uniform coverage. Coatings are then back-rolled with a saturated nap roller to fill voids and crevices.

Brushing is used to detail around windows or protrusions, but it is not the preferred method for major wall areas. When using textured elastomeric coatings, careful application is extremely important to prevent unsightly buildup of texture by rolling over an area twice. Placing too much pressure on a roller nap reduces the texture applied and presents an unsightly finish. Textured application should not be rolled over adjacent applications, as roller seams will be evident after drying.

Coatings, especially water-based ones, should not be applied in temperatures lower than 40°F and should be protected from freezing by proper storage. Manufacturers do not recommend application in humidity over 90 percent. Application over excessively wet substrates may cause bonding problems. In extremely hot and dry temperatures, substrates are misted to prevent premature coating drying. Complete curing takes 24–72 hours; coatings are usually dry to the touch and ready for a second coat in 3–5 hours, depending on the weather.

Coverage rates vary depending upon the substrate type, porosity of the substrate, and millage required. Typically, elastomeric coatings are applied at 100–150 ft^2/gal per coat, for a net application of 50–75 ft^2/gal. This results in a dry film thickness of 10–12 mil.

Elastomeric coatings should not be used in below-grade applications where they can reemulsify and deteriorate, nor are they designed for horizontal surfaces subject to traffic.

FIGURE 3.18 Elastomeric coating application after preparatory work is completed. (*Courtesy of Innovative Coatings*)

Horizontal areas such as copings or concrete overhangs should be checked for ponding water that may cause debonding and coating reemulsification (Fig. 3.19).

DECK COATINGS

Several choices are available for effective waterproofing of horizontal portions of a building envelope. Several additional choices of finishes or wearing surfaces over this waterproofing are also available. Liquid-applied seamless deck coatings or membranes are used where normal roofing materials are not practical or acceptable. Deck coatings may be applied to parking garage floors, plaza decks, balcony decks, stadium bleachers, recreation roof decks, pool decks, observation decks, and helicopter pads. In these situations, waterproof coatings occupy areas beneath the decks and provide wearing surfaces acceptable for either vehicular or pedestrian traffic. These systems do not require topping slabs or protection such as tile pavers to protect them from traffic.

Deck coatings make excellent choices for remedial situations where it is not possible to allow for the addition of a topping slab or other waterproofing system protection. Deck coatings are installed over concrete, plywood, or metal substrates, but should not be installed over lightweight insulating concrete.

Deck coatings are also used to protect concrete surfaces from acid rain, freeze–thaw cycles, and chloride ion penetration, and to protect reinforcing steel.

In certain situations, deck coatings are not specifically installed for their waterproofing characteristics but for protection of concrete against environmental elements. For example,

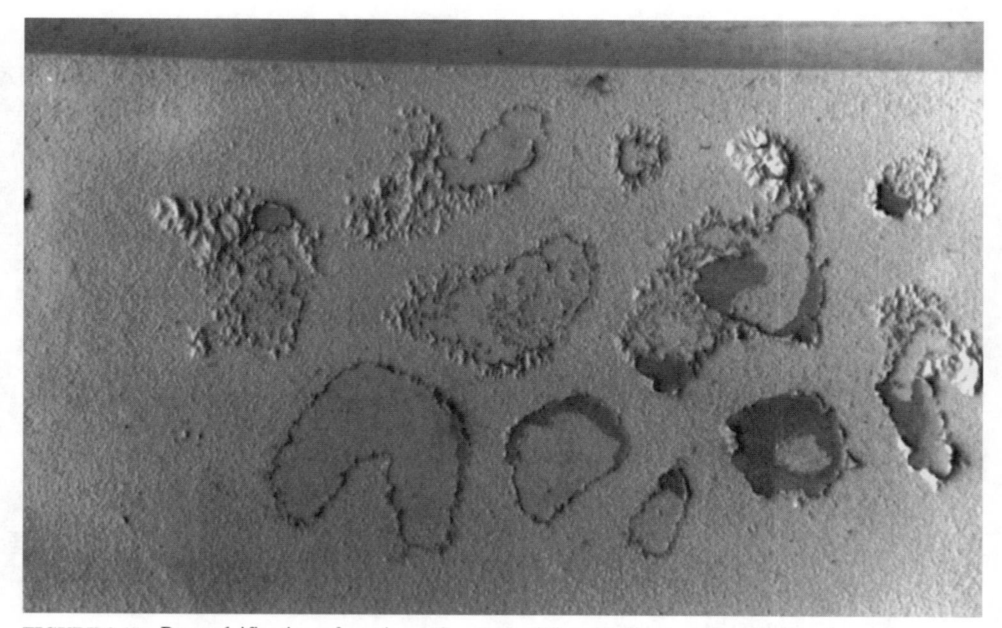

FIGURE 3.19 Reemulsification of coating. (*Courtesy of Coastal Construction Products*)

whereas deck coatings on the first floor of a parking garage protect occupied offices on ground level, they also protect concrete against road salts and freeze–thaw cycles on all other levels. In these situations, coatings are installed to prevent unnecessary maintenance costs and structural damage during structure life-cycling.

Deck coatings are usually installed in two- to four-step applications, with the final coat containing aggregate or grit to provide a nonslip wearing surface for vehicular or foot traffic. Aggregate is usually broadcast into the final coat either by hand seeding or by mechanical spray such as sandblast equipment. Aggregates include silica sand, quartz carbide, aluminum oxide, or crushed walnut shells. The softer, less harsh silica sand is used for pedestrian areas; the harder-wearing aggregate is used for vehicular traffic areas. The amount of aggregate used varies, with more grit concentrated in areas of heavy traffic such as parking garage entrances or turn lanes.

Due to the manufacturing processes involved, deck coatings are available in several standard colors but usually not in custom colors. A standard gray color is recommended for vehicular areas because oils and tire trackings will stain lighter colors. Some manufacturers allow their coatings to be color-top-coated with high-quality urethane coatings, if a special color is necessary, but only in selected cases and not in vehicular areas.

Deck coatings are supplied in two or three different formulations for base, intermediate, and wearing coats. Base coats are the most elastomeric of all formulations. Since they are not subject to wear, they do not require the high tensile strength or impact resistance that wearing layers require. Lower tensile strength allows a coating to be softer and, therefore, to have more elastomeric and crack-bridging characteristics than topcoats. As such, base coats are the waterproof layer of deck-coating systems.

Top and intermediate coats are higher in tensile strength and are impact-resistant to withstand foot or vehicular traffic. However, the various coating layers must be compatible and sufficiently similar to base coat properties not to crack or alligator as a paint applied over an elastomeric coating might. This allows base coatings to move sufficiently to bridge cracks that develop in substrates without cracking topcoats.

Adding grit or aggregate in a coating further limits movement capability of topcoats. The more aggregate added, the less movement topcoats can withstand, further restricting movement of base coats.

Deck coatings are available in several different chemical formulations. They are differentiated from clear coatings, which are penetrating sealers, in that they are film-forming surface sealers. Deck coating formulations include the following:

- Acrylics
- Cementitious coatings
- Epoxy
- Asphalt overlay
- Latex
- Neoprene
- Hypalon
- Urethane

- Modified urethane
- Sheet systems

Acrylics

Acrylics are not waterproof coatings, but act as water-repellent sealers. Their use is primarily aesthetic, to cover surface defects and cracking in decks. These coatings have low elastomeric capabilities; silica aggregate is premixed directly into their formulations, which further lowers their elastic properties. These two characteristics prevent acrylics from being true waterproof coatings.

The inherent properties of acrylics protect areas such as walkways or balconies with no occupied areas beneath from water and chloride penetration. In addition to concrete substrates, acrylics are used over wood or metal substrates, provided that recommended primers are installed. Acrylics are also used at slab-on-grade areas where urethane coatings are not recommended.

Sand added in acrylic deck coatings provides excellent antislip finishes. As such, they are used around pools or areas subject to wet conditions that require protection against slips and falls. Acrylics are not recommended for areas subject to vehicular traffic. Some manufacturers allow their use over asphaltic pavement subject only to foot traffic, for aesthetics and a skid-resistant finish. (See Table 3.14.)

Cementitious

Cementitious deck coatings are used for applications over concrete substrates and include an abrasive aggregate for exposure to traffic. These materials are supplied in prepacked and premixed formulations requiring only water for mixing. Cementitious coatings are applied by trowel, spray, or squeegee, the latter being a self-leveling method.

Cementitious systems contain proprietary chemicals to provide necessary bonding and waterproofing characteristics. These are applied to a thickness of approximately 1/8 inch and will fill minor voids in a substrate. A disadvantage of cementitious coatings, like below-grade cementitious systems, is their inability to withstand substrate movement or cracking. They are one-step applications, with integral wearing surfaces, which require no primers and are applicable over damp concrete surfaces.

Modified acrylic cementitious coatings are also available. Such systems typically include a reinforcing mesh embedded into the first coat to improve crack-bridging capabilities. Acrylics are added to the basic cement and sand mixture to improve bonding and performance characteristics.

Cementitious membrane applications include the dry-shake and power-trowel methods previously discussed in Chapter 2. Successful applications depend on properly designed, detailed,

TABLE 3.14 Acrylic Deck-Coating Properties

Advantages	Disadvantages
Ease of application	Not a complete waterproof system
Aggregate is integral with coating	No movement capability
Slab-on-grade applications	Not resistant to vehicular traffic

and installed allowances for movement, both thermal and differential. For cementitious membranes to be integrated into a building envelope, installations should include manufacturer-supplied products for cants, patching, penetrations, and terminations. (See Table 3.15.)

Epoxy

As with acrylics, epoxy coatings are generally not considered true waterproof coatings. They are not recommended for exterior installations due to their poor resistance to ultraviolet weathering. Epoxy floor coatings have very high tensile strengths, resulting in low elastomeric capabilities. These coatings are very brittle and will crack under any movement, including thermal and structural.

Epoxy coatings are used primarily for interior applications subject to chemicals or harsh conditions such as waste and water treatment plants, hospitals, and manufacturing facilities. For interior applications not subject to movement, epoxy floor coatings provide effective waterproofing at mechanical room floor, shower, and locker room applications. Epoxy coatings are available in a variety of finishes, colors, and textures, and may be roller- or trowel-applied.

Epoxy deck coatings are also used as top coats over a base-coat waterproof membrane of urethane or latex. However, low-movement capabilities and brittleness of epoxy coatings limit elastomeric qualities of waterproof top coats. (See Table 3.16.)

Asphalt

Asphalt overlay systems provide an asphalt wearing surface over a liquid-applied membrane. The waterproofing base coat is a rubberized asphalt or latex membrane that can withstand the heat created during installation of the asphalt protective course. Both the waterproof membrane and the asphalt layers are hot-applied systems.

Asphalt layers are approximately 2 inches thick. These systems have better wearing capabilities due to the asphaltic overlay protecting the waterproof base coating.

The additional weight added to a structure by these systems must be calculated to ensure that an existing parking garage can withstand the additional dead loads that are created. Asphalt severely restricts the capability of the waterproof membrane coating to bridge

TABLE 3.15 Cementitious Deck-Coating Properties

Advantages	Disadvantages
Excellent bonding to concrete substrates	No movement capabilities
Good wearing surfacing	Not applicable over wood or metal
Dry-shake and power-trowel applications	Not resistant to acid rain and other contamination

TABLE 3.16 Epoxy Deck-Coating Properties

Advantages	Disadvantages
Excellent chemical resistance	Brittle; no movement capability
High tensile strength	Trowel application
Variety of finishes, colors, and textures	Not for exterior applications

cracks or to adjust to thermal movement. Additionally, it is difficult to repair the waterproofing membrane layers once the asphalt is installed. There is no way to remove overlays without destroying the base coat membrane. Asphaltic systems are not recoatable. For maintenance, they must be completely removed and reinstalled. (See Table 3.17.)

TABLE 3.17 Asphalt Deck-Coating Properties

Advantages	Disadvantages
Protection of membrane by asphalt overlay	Weight added to structure
Longer wearing capability	Movement capability restricted
Thickness of applied system	Inaccessibility for repairs

Latex, neoprene, hypalon

Deck coatings are available in synthetic rubber formulations, including latex, neoprene, neoprene cement, and hypalon. These formulations include proprietary extenders, pigments, and stabilizers. Neoprene derivatives are soft, low-tensile materials and require the addition of a fabric or fiberglass reinforcing mesh. For traffic-wear resistance, this reinforcing mesh enhances in-place performance properties such as elongation and crack-bridging capabilities. Reinforcing requires that the products be trowel applied rather than roller or squeegee applied.

Trowel application and a finish product thickness of approximately 1/4 inch increase the in-place costs of these membranes. They also require experienced mechanics to install the rubber derivative systems. Trowel applications, various derivatives, and proprietary formulations provide designers with a wide range of textures, finishes, and colors.

Rubber compound coatings have better chemical resistance than most other deck-coating systems. They are manufactured for installation in harsh environmental conditions such as manufacturing plants, hospitals, and mechanical rooms. They are appropriate in both exterior and interior applications.

Design allowances must be provided for finished application thickness. Deck protrusions, joints, wall-to-floor details, and equipment supports must be flashed and reinforced for membrane continuity and watertightness. Certain derivatives of synthetic rubbers become brittle under aging and ultraviolet weathering, which hinders waterproofing capabilities after installation. Manufacturer's literature and applicable test results should be reviewed for appropriate coating selection. (See Table 3.18.)

Urethanes

Urethane deck coatings are frequently used for exterior deck waterproofing. These are available for both pedestrian and vehicular areas in a variety of colors and finishes. Urethane systems include aromatic, aliphatic, and epoxy-modified derivatives and formulations.

TABLE 3.18 Latex, Neoprene, and Hypalon Deck-Costing Properties

Advantages	Disadvantages
Excellent chemical resistance	Trowel application required
Good aging and weathering	Fabric reinforcement required
Good wear resistance	High cost

Aliphatic materials have up to three times the tensile strength of aromatics but only 50 percent of aromatic elongation capability. Many manufacturers use combinations of these two materials for their deck-coating systems. Aromatic materials are installed as base coats for better movement and recovery capabilities; aliphatic urethane top coats make for better weathering, impact resistance, and ultraviolet resistance.

Epoxy urethane systems are also used as top coat materials. These modified urethane systems provide additional weathering and wear, while still maintaining necessary waterproofing capabilities.

Urethane coatings are applied in two or more coats, depending upon the expected traffic wear. Aggregate is added in the final coating for a nonslip wearing surface. An installation advantage with urethane systems is their self-flashing capability. Liquid-applied coatings by brush application are turned up adjoining areas at wall-to-floor junctions, piping penetrations, and equipment supports and into drains.

Urethane coatings are manufactured in self-leveling formulations for applications control of millage on horizontal surfaces. Nonflow or detailing grades are available for vertical or sloped areas. The uncured self-leveling coating is applied by notched squeegees to control thickness on horizontal areas. At sloped areas, such as the up and down ramps of parking garages or vertical risers of stairways, nonflow material application ensures proper millage. If self-leveling grade is used in these situations, material will flow downward and insufficient millage at upper areas of the vertical or sloped portions will occur.

Nonflow liquid material is used to detail cracks in concrete decks before deck-coating application. Cracks wider than 1/16 inch, which is the maximum width that urethane materials bridge without failure, are sawn out and sealed with a urethane sealant. This area is then detailed 4 inches wide with nonflow coating.

In addition, urethane coatings are compatible with urethane sealants used for cants between vertical and horizontal junctions, providing a smooth transition in these and other changes of plane. This is similar to using wood cants for roof perimeter details (see Table 3.19).

Sheet systems

While they do not fit the description of a deck coating per se, there are balcony and deck waterproofing systems that are available in sheet materials that provide waterproofing capabilities. There are a variety of systems available, including those that require the sheet embedded in a trowel- or spray-applied acrylic or resin material, and those that are act as a complete system.

The latter is a vinyl product, similar to a typical interior vinyl flooring product with the exception that the product is improved to withstand exterior weathering and of course water infiltration. The system is vulnerable for leakage at the seams, following the 90%/1% principle. If seaming is adequately addressed, including the necessary vertical

TABLE 3.19 Urethane Deck-Coating Properties

Advantages	Disadvantages
Excellent crack-bridging capability	Limited color selection
Simple installations	Low chemical resistance
Expanded product line	Maintenance required with heavy traffic

turn-ups, the product can be an effective barrier system. These systems make excellent candidates for remedial application, as they can hide considerably more substrate imperfections than the liquid systems discussed previously. These systems can also be applied to wood substrates and make excellent choices for residential applications including apartment projects.

Many systems combine the properties of the liquid-applied systems with sheet good reinforcing for "belt and suspenders" protection. The limiting factor is cost, as the more material and layers a system requires for effectiveness, the more the final in-place cost rises. Table 3.20 summaries the advantages and disadvantages of using sheet systems for waterproofing applications.

DECK-COATING CHARACTERISTICS

Deck coatings bond directly to concrete, wood, or metal substrates. This prevents lateral movement of water beneath the coatings, as is possible with sheet good systems. Once cured, coatings are nonbreathable and blister if negative vapor drive is present. This is the reason deck coatings, with the exception of acrylic and epoxies, are not recommended for slab-on-grade applications. Specifically, moisture in soils is drawn up into a deck by capillary action, causing blistering in applied deck coatings. In the same manner, blistering occurs in deck coatings applied on upper deck portions of sandwich-slab membranes due to entrapped moisture and negative vapor drive. In both cases, an epoxy vapor barrier prime coat should be installed to protect deck-coating systems from being subjected to this vapor drive.

Physical properties of deck coatings vary as widely as the number of systems available. Important considerations to review when choosing a coating system include tensile strength, elongation, chemical resistance, weathering resistance, and adhesion properties. Different installation types, expected wearing, and weathering conditions require different coating types.

High tensile strength is necessary when a coating is subject to heavy wear including vehicular traffic or forklift traffic at loading docks. Tensile strengths of some deck coatings exceed 1000 lb/in^2 (tested according to ASTM D-412) and are higher for epoxy coatings. This high tensile strength reduces the elongation ability of coatings.

Elongation properties range from 200 percent (for high-tensile-strength top coats) to more than 1000 percent (for low-tensile-strength base coats). For pedestrian areas where impact resistance and heavy wear is not expected, softer, higher elongation aromatic urethanes are used. Sun decks subject to impact from lawn chairs and tables would be better served by a coating between the extremes of high and low tensile strength.

Table 3.20 Sheet Systems

Advantages	Disadvantages
Uniformity of material application	Difficult applications in small areas
Forgiving of substrate flaws	Repairs difficult
Applicable over wood substrates	Water can travel under sheet systems

Chemical resistance can be an important consideration under certain circumstances. Parking garage decks must have coatings resistant to road salts, oil, and gasoline. A pedestrian sun deck may be subjected to chlorine and other pool chemicals. Testing for chemical resistance should be completed according to recognized tests such as ASTM D-471.

Weathering resistance and ultraviolet resistance are important to coatings exposed to the elements such as on upper levels of a parking garage. These areas should be protected by the ultraviolet-resistant properties of coatings such as an aliphatic urethane. Weathering characteristics can be compared with accelerated weathering tests such as ASTM D-822. Other properties to consider on an as-needed basis include adhesion tests, solvent odor for interior uses, moisture vapor transmission, and fire resistance.

Once installed, the useful life of deck coatings depends upon proper maintenance as well as traffic wear. Heavily traveled parking garage decks and loading docks will wear faster than a seldom-used pedestrian deck area. To compensate, manufacturers recommend a minimum of one to as many as three additional intermediate coat applications. Additional aggregate is also added for greater wear resistance (Fig. 3.20).

With proper installation, deck coatings should function for upward of 5 years before requiring resealing. Resealing entails cleaning, patching existing coatings as required, reapplying top coatings, and, if required, adding intermediate coats at traffic lanes. Proper maintenance prevents coatings from being worn and exposing base coatings that cannot withstand traffic or exposure.

Exposed and unmaintained deck-coating systems require complete removal and replacement when repairs become necessary. Chemical spills, tears or ruptures, and improper usage must also be repaired to prevent unnecessary coating damage. Maintaining the top coat or wearing surface properly will extend the life cycle of a deck-coating system indefinitely.

Deck coatings are also effective in remedial waterproofing applications. If a sandwich-slab membrane installed during original construction becomes ineffective, a deck coating can be installed over the topping slab provided proper preparatory work is completed. Deck coatings can also be successfully installed over quarry and other hard-finish tile surfaces, precast concrete pavers, and stonework. With any special surfacing installation, proper adhesive tests and sample applications should be completed.

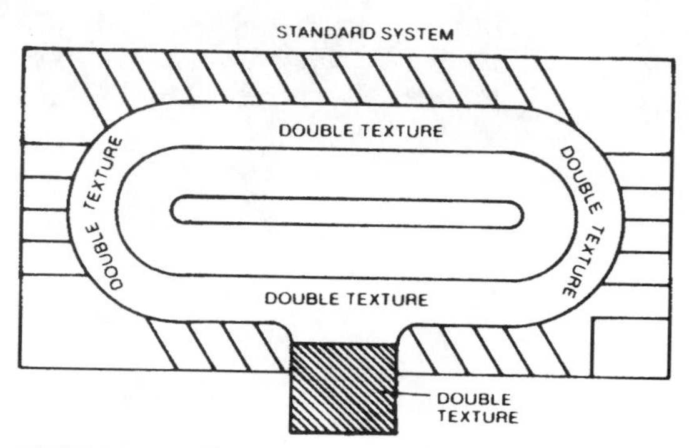

FIGURE 3.20 Suggested aggregate texture layout for maximum protection of deck coating. (*Courtesy of General Polymers*)

DECK-COATING APPLICATION

When deck coatings are being applied at a job site, it is the only time when golf shoes are mandatory attire! Liquid deck coatings are required to be squeegee-applied to ensure sufficient and uniform milage. The millage rate is too thick for spray applications, which cannot also provide the uniform thickness required.

During application the squeegee is pushed, not pulled, to prevent the blade end of the squeegee from being pulled down too hard against the substrate and applying too thin a wet millage of material. Pushing of the squeegee blade maintains the blade in an upright position and a uniform millage application.

This pushing of the squeegee requires that the mechanic walk through the applied material, thus requiring the golf shoes so as not to damage the installation and have shoes stick to the wet membrane. The material self-levels after installation, so that any minute impressions the golf-shoe spikes leave are quickly covered by the material.

Most deck coatings also require that the material be immediately back-rolled after initial squeegee application to further ensure uniformity of millage. These applicators must also wear the golf shoes, as do those applying the aggregate that forms the wearing surface in the top coat applications. Figure 3.21 pictures the application process of a typical deck coating application.

Substrate adhesion and proper substrate finishing are critical for successful deck-coating applications. In general substrates must be clean, dry, and free of contaminants. Concrete substrates exhibiting oil or grease contamination should be cleaned with a biodegradable degreaser such as trisodium phosphate. Contaminants such as parking-stall stripe paint

FIGURE 3.21 Deck-coating application. Note the pushing of squeegees in the background, back-rolling of coating and spreading of the aggregate by hand in the foreground. All crew members are wearing golf shoes. (*Courtesy of Coastal Construction Products*)

should be removed by mechanical grinder or sandblasting (Figs. 3.22 and 3.23).

For new concrete substrates, a light broom finish is desirable. Surface laitance, fins, and ridges must be removed. Honeycomb and spalled areas should be patched using an acceptable nonshrink grout material.

Coatings should not be applied to exposed aggregate or reinforcing steel. If present, these areas should be properly repaired. Concrete surfaces, including patches, should be cured a minimum of 21 days before coating application. Use of most curing compounds is prohibited by coating manufacturers, since resins contained in curing compounds prevent adequate adhesion. If present, substrates require preparatory work, including sandblasting, or acid etching with muriatic acid. Water curing is desirable, but certain manufacturers allow use of sodium silicate curing agents.

Substrate cracks must be prepared before coating application (Fig. 3.24). Cracks less than 1/16 inch wide should be filled and detailed with a 4-in band of nonflow base coat. Larger cracks, from 1/2 inch to a maximum of 1-in width, should be sawn out and filled with urethane sealant (Figure 3.25). Moving joints should have proper expansion joints installed with coating installed up to but not over these expansion joints. Refer to Figures 3.26 and 3.27 for typical expansion joint detailing.

Substrates should be sloped, to drain water toward scuppers or deck drains. Plywood surfaces should be swept clean of all dirt and sawdust. Plywood should be of A-grade only, with tongue and groove connections (Figure 3.28). Only screw-type fasteners should be

FIGURE 3.22 Mechanical removal of contaminants. (*Courtesy of Coastal Construction Products*)

FIGURE 3.23 Substrate has been scarified prior to deck-coating application to ensure proper adhesion. (*Courtesy of Coastal Construction Products*)

FIGURE 3.24 Crack repair prior to deck-coating application. (*Courtesy of Coastal Construction Products*)

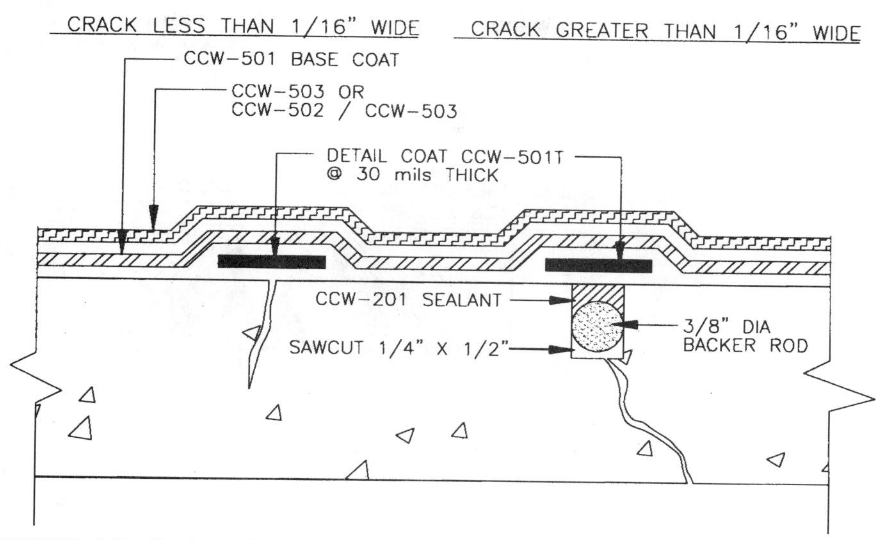

FIGURE 3.25 Crack repair detailing for deck-coating applications. (*Courtesy of Carlisle Corporation*)

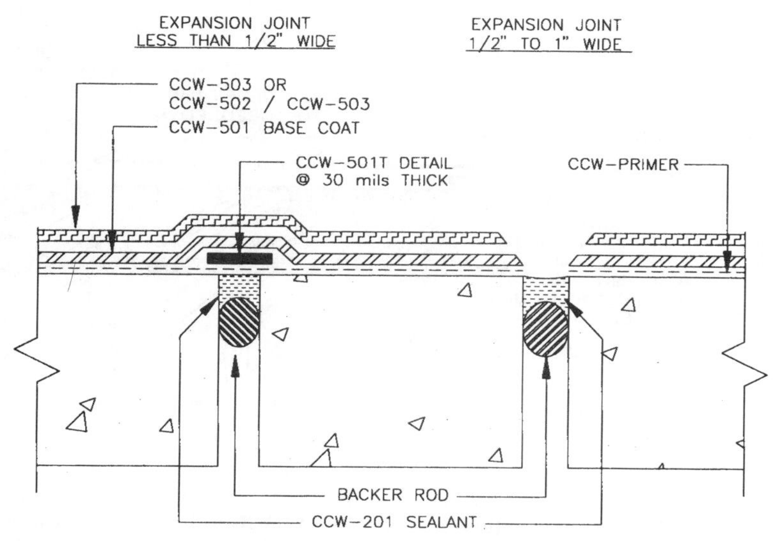

FIGURE 3.26 Expansion joint detailing for deck-coating applications. (*Courtesy of Carlisle Corporation*)

used, and they should be countersunk. The screw head is filled with a urethane sealant and troweled flush with the plywood finish (Figure 3.29). As these coatings are relatively thin, 60–100 mil dry film, their finish mirrors the substrate they are applied over. Therefore, if plywood joints are uneven or knots or chips are apparent in the plywood, they also will be apparent in the deck-coating finish.

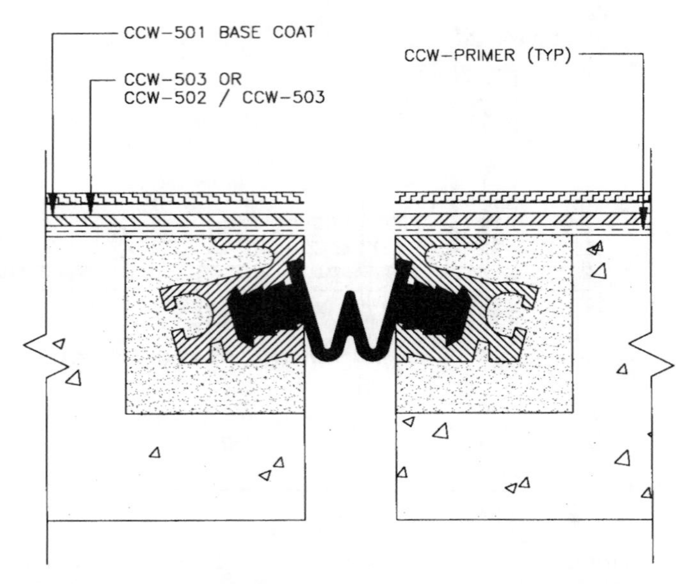

FIGURE 3.27 Mechanical expansion joint detailing for deck-coating applications. (*Courtesy of Carlisle Corporation*)

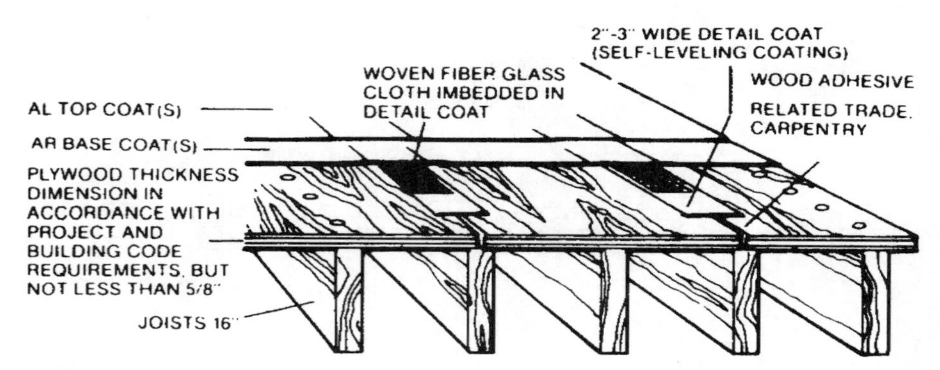

FIGURE 3.28 Plywood deck preparation for deck-coating applications. (*Courtesy of General Polymers*)

Metal surfaces require sandblasting or wire-brush cleaning, then priming immediately afterward (Figure 3.30). Aluminum surfaces also require priming (Figure 3.31). Other substrates such as PVC, quarry tile, and brick pavers should be sanded to roughen the surface for adequate adhesion. Sample test areas should be completed to check adhesion on any of these substrates before entire application.

For recoating over previously applied deck coatings, existing coatings must be thoroughly cleaned with a degreaser to remove all dirt and oil. Delaminated areas should be cut out and patched with base coat material. Before reapplication of topcoats, a solvent is applied to reemulsify existing coatings for bonding of new coatings.

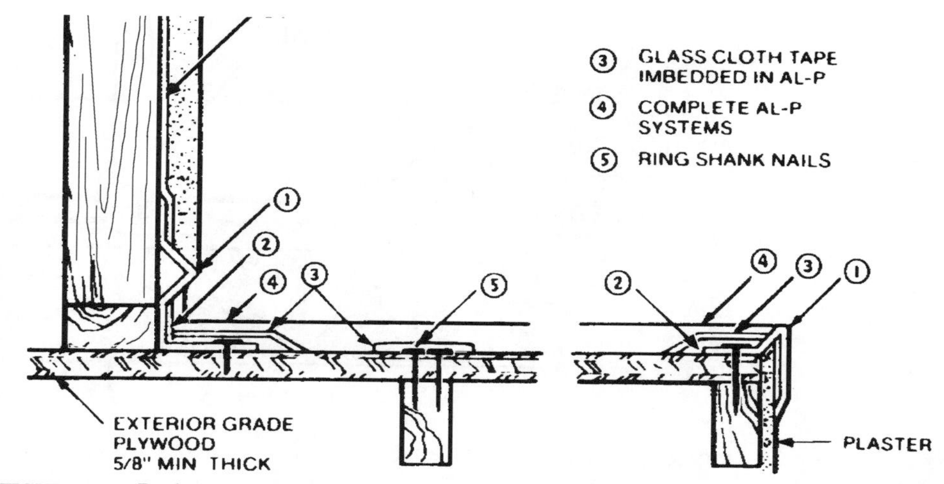

FIGURE 3.29 Deck-coating details for plywood deck applications. (*Courtesy of General Polymers*)

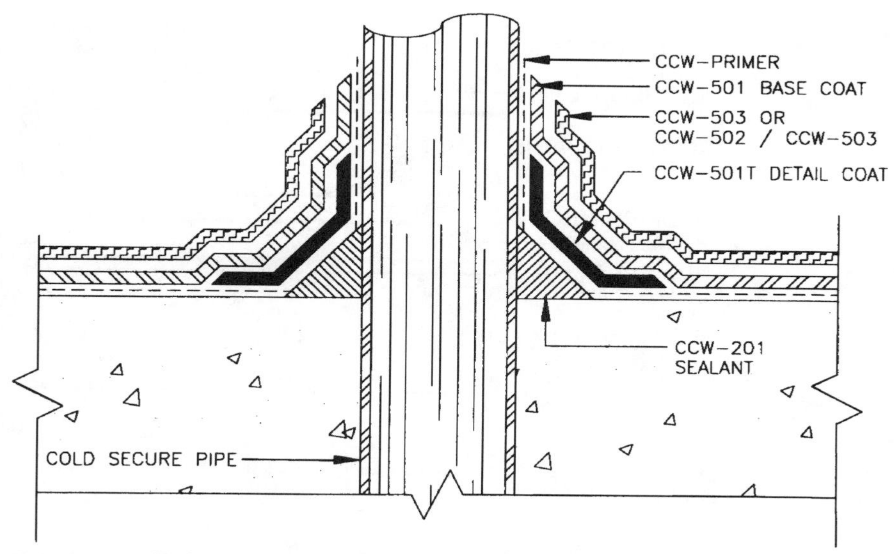

FIGURE 3.30 Typical penetration detail for deck coatings. (*Courtesy of Carlisle Corporation*)

All vertical abutments and penetrations should be treated by installing a sealant cove, followed by a detail coat of nonflow material (Figures 3.32, 3.33, and 3.34). If a joint occurs between changes in plane such as wall-to-floor joints, an additional detail coat is added or reinforcement. Figures 3.35 and 3.36 show typical installation procedures for this work.

With new construction, detail coats of base coat membrane are turned up behind the facing material (e.g., brick cavity wall), followed by coating and detailing to the facing material. This allows for double protection in these critical envelope details. At doors or

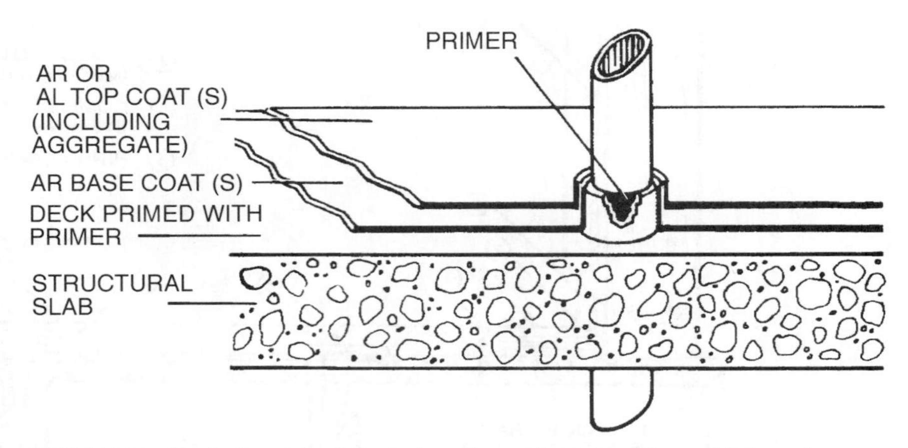

FIGURE 3.31 Handrail post detail for deck coatings. (*Courtesy of General Polymers*)

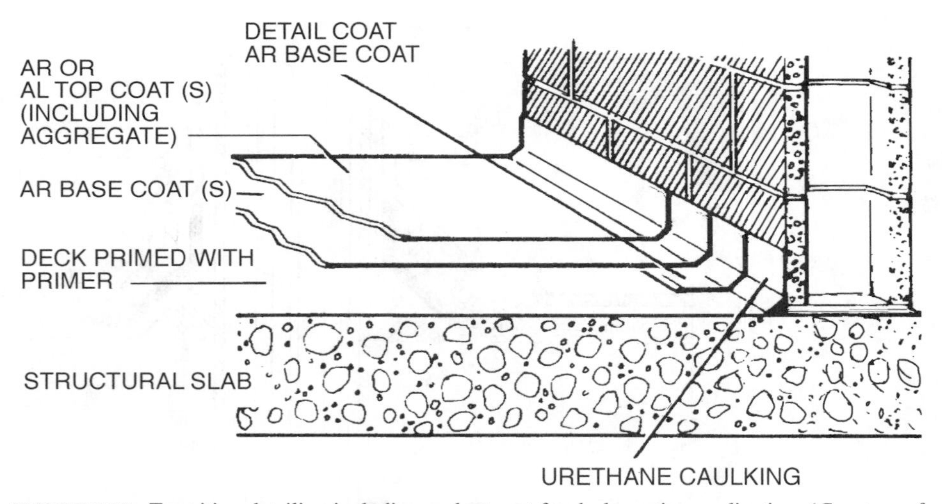

FIGURE 3.32 Transition detailing including sealant cant for deck-coating application. (*Courtesy of General Polymer*)

sliding glass doors, coating is installed beneath thresholds before installation of doors. Figure 3.37 shows application procedures at a deck drain.

For applications over topping slabs with precast plank construction, such as double-T, a joint should be scored at every T-joint. These joints are then filled with sealant and a detail coat of material is applied allowing for differential and thermal movement. Refer to Figures 3.38 and 3.39 for typical installation detail at these areas.

Base coats are installed by notched squeegees for control of millage, typically 25–40 mil dry film, followed by back-rolling of materials for uniform millage thickness (Fig. 3.40).

FIGURE 3.33 Sealant cant installed at vertical termination prior to deck-coating application. (*Courtesy of Coastal Construction Products*)

FIGURE 3.34 Detail coat of deck coating installed at vertical transitioning prior to deck-coating application. (*Courtesy of Coastal Construction Products*)

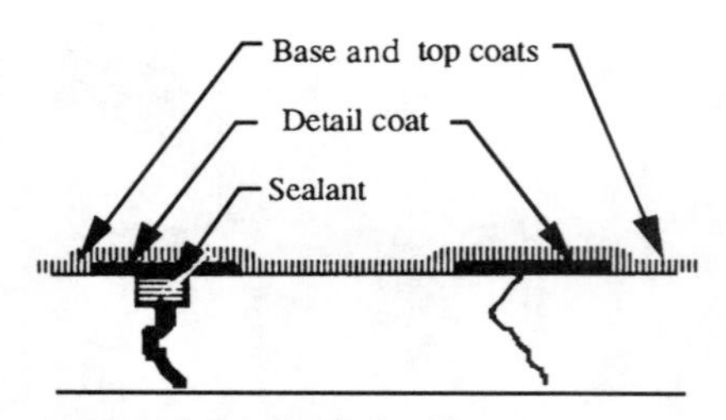

FIGURE 3.35 Crack detailing.

Following initial base-coat curing, within 24 hours intermediate coats, topcoats, and aggregates are installed (Fig. 3.41).

Aggregate, silica sand, and silicon carbide are installed in intermediate or final topcoats or possibly both in heavy traffic areas. On pedestrian decks grit is added at a rate of 4–10 lb square (100 ft^2) of deck area. In traffic lanes, as much as 100–200 pounds of aggregate per square is added.

Aggregate is applied by hand seeding (broadcasting) or by mechanical means (Fig. 3.42). If aggregate is added to a topcoat, it is back-rolled for uniform thickness of membrane and grit distribution. With installations of large aggregate amounts, an initial coat with aggregate fully loaded is first allowed to dry. Excess aggregate is then swept off, and an additional topcoat is installed to lock in the grit and act as an additional protective layer. See Fig. 3.43 for aggregate comparisons.

Intermediate coats usually range in thickness from 10–30 mil dry film, whereas top coats range in thickness from 5–20 mil. Refer to Figs. 3.44 and 3.45 for typical millage requirement. Final coats should cure 24–72 hours before traffic is allowed on the deck, paint stripping is installed, and equipment is moved onto the deck. Approximate coverage rates for various millage requirements are shown in Table 3.21. Trowel systems are applied to considerably greater thickness than liquid-applied systems. Troweled systems range from $1/8$–$1/4$ inch total thickness, depending upon the aggregate used.

Other than applications of acrylic coatings, manufacturers require primers on all substrates for improved membrane bonding to substrates. Primers are supplied for various substrates, including concrete, wood, metal, tile, stone, and previously coated surfaces. Additionally, priming of aggregate or grit is required before its installation in the

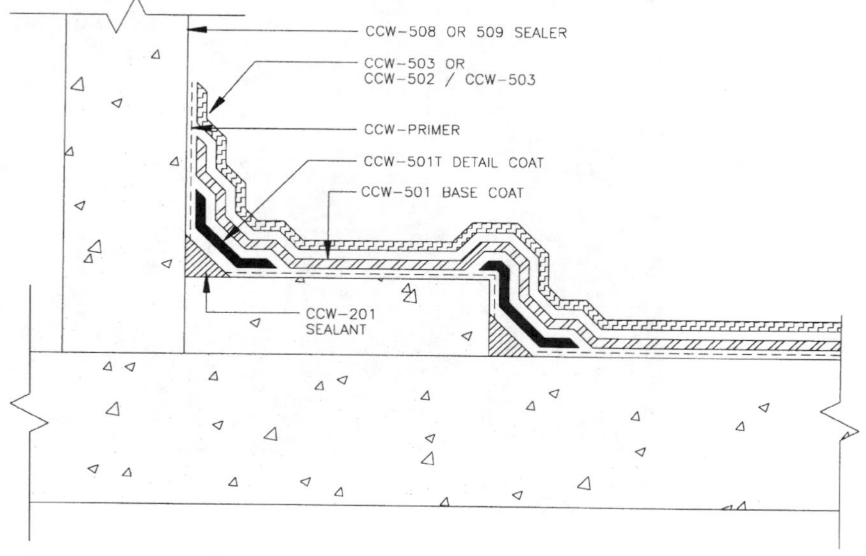

FIGURE 3.36 Detailing for vertical transitions. (*Courtesy of Carlisle Corporation*)

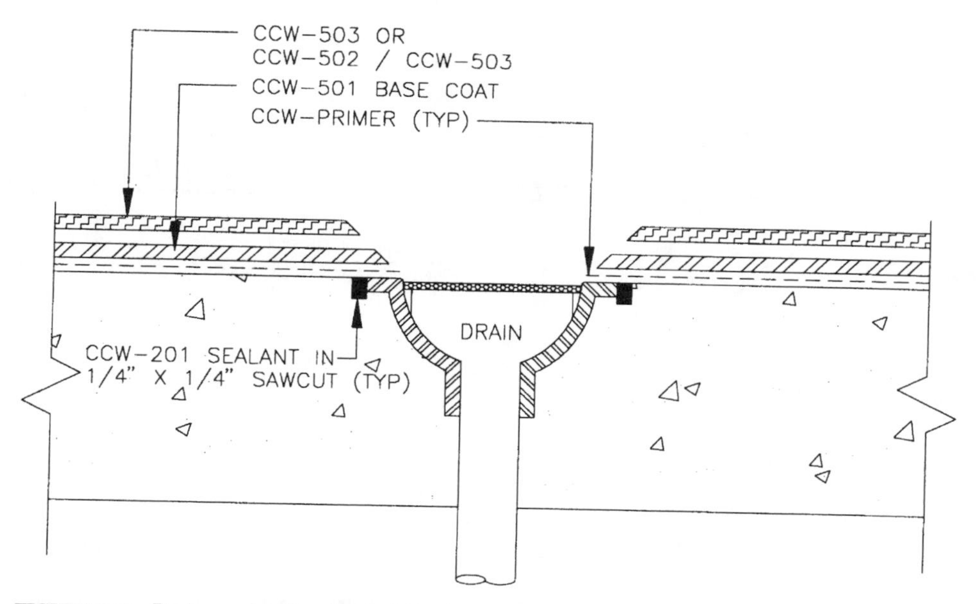

FIGURE 3.37 Drainage detailing for deck coating. (*Courtesy of Carlisle Corporation*)

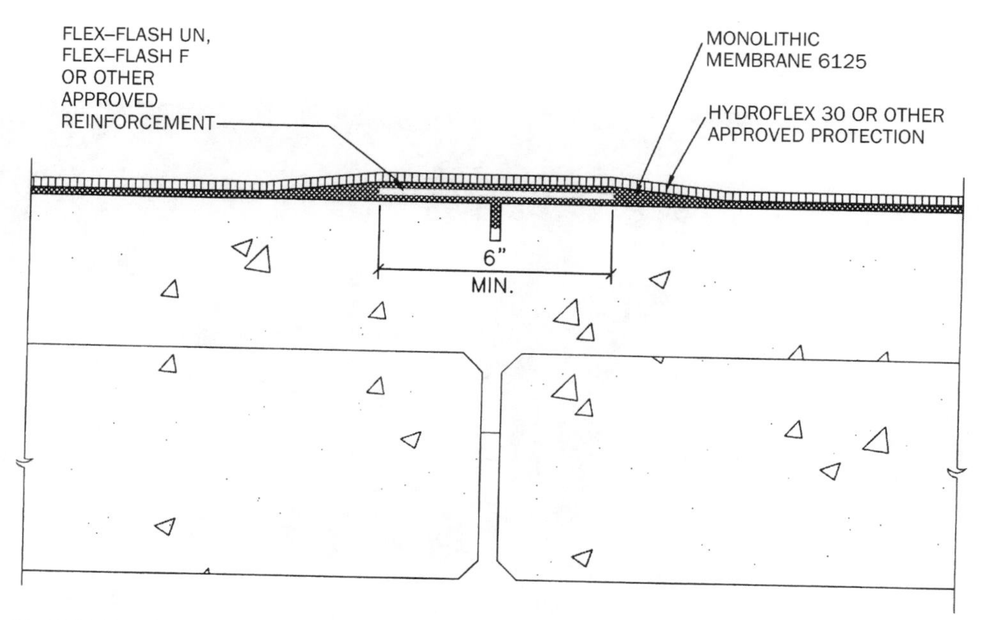

FIGURE 3.38 Precast or Double-T topping detailing for deck-coating applications. (*Courtesy of American Hydrotech*)

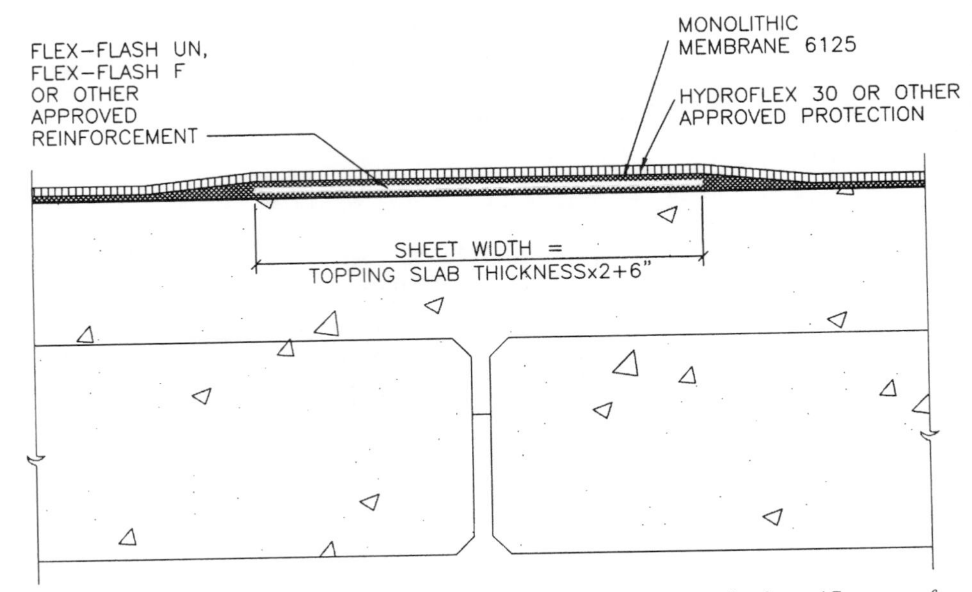

FIGURE 3.39 Precast or Double-T topping detailing for deck-coating applications. (*Courtesy of American Hydrotech*)

FIGURE 3.40 Squeegee application and back-rolling of deck coating. (*Courtesy of Coastal Construction Products*)

FIGURE 3.41 Aggregate being applied into top coat, followed by back rolling. (*Courtesy of Coastal Construction Products*)

FIGURE 3.42 Mechanical broadcasting of aggregate into deck coating. (*Courtesy of Coastal Construction Products*)

coating. Some primers must be allowed to dry completely (concrete); others must be coated over immediately (metal). In addition to primers, some decks may require an epoxy vapor barrier to prevent blistering from negative vapor drive.

Because of the volatile nature and composition of deck-coating materials, they should not be installed in interior enclosed spaces without adequate ventilation. Deck coatings are highly flammable, and extreme care should be used during installation and until fully

FIGURE 3.43 Surface priming after crack preparatory work has been completed. (*Courtesy of Western Group*)

cured. Deck coating requires knowledgeable, trained mechanics for applications, and manufacturer's representatives should review details and inspect work during actual progress.

Figure 3.46 demonstrates proper deck coating application, and Fig. 3.47 demonstrates the various stages of deck-coating application.

CLEAR DECK SEALERS

Although similar to vertical surface sealers, clear horizontal sealers require a higher percentage of solids content to withstand the wearing conditions encountered at horizontal areas. Decks are subject to ponding water, road salts, oils, and pedestrian or vehicular traffic. Such in-place conditions require a solids content of 15 to 30 percent, depending on the number of application steps required. Typically, two coats are required for lower solids material and one coat for 30 percent solids material. In addition, complete substrate saturation is required rather than the spray or roller application suitable for vertical installations.

Clear wall sealers differ from elastomeric coatings in much the same way that clear deck sealers differ from deck coatings. Clear deck sealers cannot bridge cracks in a substrate, whereas most deck coatings bridge minimum cracking. Clear sealers can be applied only over concrete substrates, whereas deck coatings can be applied over metal and wood substrates. Clear sealers are penetrating systems, whereas deck coatings are surface sealers.

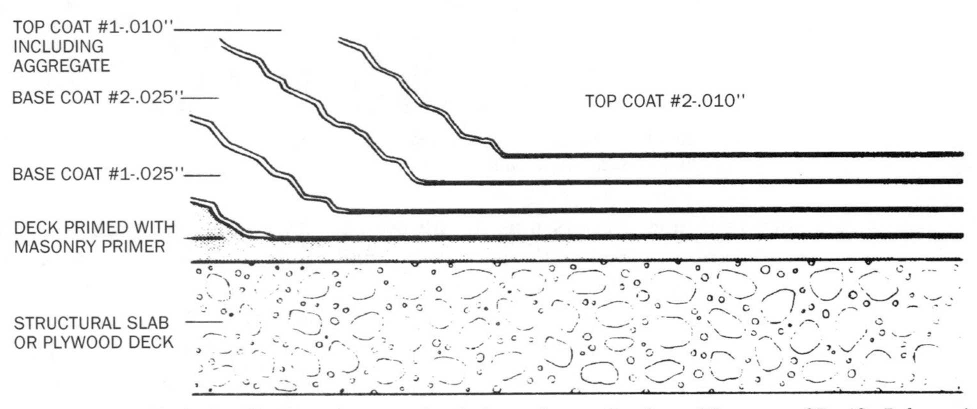

FIGURE 3.44 Typical millage requirements for deck-coating applications. (*Courtesy of Pacific Polymers*)

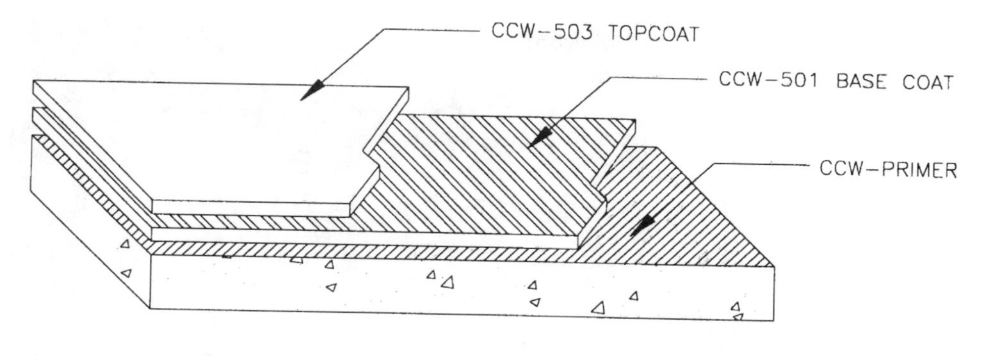

CCW-5013 PEDESTRIAN TRAFFIC DECK COATING

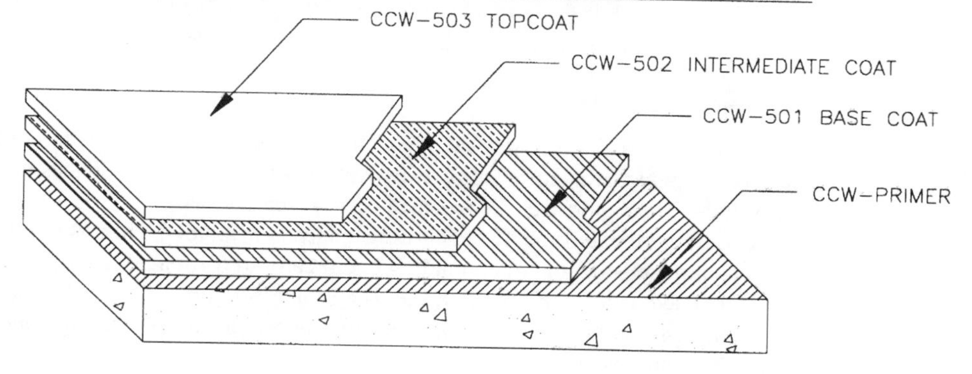

CCW-5123 VEHICULAR TRAFFIC DECK COATING

FIGURE 3.45 Typical millage requirements for deck-coating applications. (*Courtesy of Carlisle Corporation*)

TABLE 3.21 Approximate Coverage Rates for Liquid-Applied Deck Coatings

Dry millage	Coverage (ft^2/gal)
5	250
10	130
15	72
20	55
25	44
30	36
50	27

Properly used and detailed into other building envelope components, sealers provide the protection necessary for many deck applications. In addition, deck sealers are frequently used to protect concrete surfaces from chloride attack and other damaging substances such as acid rain, salts, oils, and carbons. By preventing water penetration, substrates are protected from the damaging effects of freeze–thaw cycles.

FIGURE 3.46 Deck-coating application; note crack detailing. (*Courtesy of Karnak*)

Unlike clear sealers for vertical applications, the chemical composition of horizontal deck sealers is limited. It includes silicone derivatives of siloxanes and silanes and clear urethane derivatives. The majority of products are siloxane-based.

A sodium silicate type of penetrating sealer is available. This material reacts with the free calcium salts in concrete, bonding chemically to form a dense surface. The product is typically used as a floor hardener, not as a sealer. Sodium silicates do not have properties that sufficiently repel water and the chlorides necessary for protecting concrete exposed to weathering and wear.

To ensure sealer effectiveness to repel water, test results such as ASTM C-642, C-67, or C-140 should be reviewed. Reduction of water absorption after treatment should be over 90 percent and preferably over 95 percent. Additionally, most sealers are tested for resistance to chlorides to protect reinforcing steel and structural integrity of concrete. Tests for chloride absorption include AASHTO 259 and NCHRP 244. Effective sealers will result in reductions of 90 percent or greater.

Penetration depth is an important consideration for effective repellency and concrete substrate protection. As with vertical sealers, silanes with smaller molecular structure penetrate deepest, up to $^1/_2$ in. Siloxanes penetrate to a depth of approximately $^1/_4$–$^3/_8$ in. Urethanes, containing higher solids content, penetrate substrates approximately $^1/_8$ in.

Silicone derivative sealers react with concrete and atmospheric humidity to form a chemical reaction bonding the material to a substrate. This provides the required water repellency. Substrates can be slightly damp but not saturated for effective sealer penetration. Over dense, finished concrete, such as steel-troweled surfaces, acid etching may be required.

Since sealers are not completely effective against water-head pressures and do not bridge cracks, proper detailing for crack control, thermal and differential movement, and detailing into other envelope components must be completed. Expansion joints, flashings, and counterflashings should be installed to provide a watertight transition between various building envelope components and deck sealers.

Clear deck sealers are often chosen for application on balconies and walkways above grade (not over occupied spaces) as well as for parking garage decks. In the latter, the

FIGURE 3.47 Deck-coating application; note prepared deck in foreground, base coat applied in middle, and finished top coat in background of parking garage. (*Courtesy of Coastal Construction Products*)

upper deck or lower decks, which cover occupied areas, are sealed with deck coatings, while intermediate decks are sealed with clear sealers. (See Table 3.22.)

CLEAR DECK SEALER APPLICATION

Clear deck sealers are penetrants, and it is critical for concrete surfaces to be prepared to allow proper penetration and bonding. A light broom finish is best for proper penetration; smooth, densely finished concrete should be acid-etched. Test applications of sealers are recommended by manufacturers to check compatibility, penetration, and effectiveness for desired results. Concrete must be completely cured, and only water-curing or dissipating, resin-curing agents are recommended. Primers are not required with clear deck sealers.

TABLE 3.22 Clear Deck Sealer Properties

Advantages	Disadvantages
Cost	Not completely waterproof
Ease of application	No crack-bridging capability
Penetrating applications	Not for wood or metal substrates

Deck sealers should be applied directly from the manufacturer's containers in the provided solids contents, and not diluted in any manner. Application is by low-pressure spray equipment (Fig. 3.48), deep-nap rollers (Fig. 3.49), or squeegees. The concrete should be thoroughly saturated with the material. Brooming of material to disperse ponding collection for even distribution and penetration is then required. Concrete porosity will determine the amount of material required for effective treatment, typically ranging from 100–150 ft^2/gal of material.

Adjacent metal, glass, or painted surfaces of the building envelope should be protected from sealer overspray. All sealant work should be completed before sealer application to prevent joint contamination that causes disbonding of sealants.

Deck sealers are extremely toxic and should not be applied in interior or enclosed areas without adequate ventilation. Workers should be protected from direct contact with materials (Fig. 3.50). Sealers are flammable and should be kept away from open flame and extreme heat.

PROTECTED MEMBRANES

With certain designs, horizontal above-grade decks require the same waterproofing protection as below-grade areas subjected to water table conditions. At these areas, membranes

FIGURE 3.48 Horizontal spray application of clear deck sealer. *(Courtesy of Sivento)*

FIGURE 3.49 Flood coat application of clear deck sealer. (*Courtesy of Saver Systems*)

are chosen in much the same way as below-grade applications. These installations require a protection layer, since these materials cannot be subjected to traffic wear or direct exposure to the elements. As such, a concrete topping slab is installed over the membrane, sandwiching the membrane between two layers of concrete; hence the name sandwich-slab membrane. Figure 3.51 details a typical sandwich-slab membrane.

In addition to concrete layers, other forms of protection are used, including wood decking, concrete pavers (Fig. 3.52), natural stone pavers (Fig. 3.53), and brick pavers (3.54). Protected membranes are chosen for areas subjected to wear that deck coatings are not able to withstand, for areas of excessive movement, and to prevent the need for excess maintenance. Although they cost more initially due to the protection layer and other detailing required, sandwich membranes do not require the in-place maintenance of deck coatings or sealers.

Protected membranes allow for installation of insulation over waterproof membranes and beneath the topping layer (Fig. 3.55). This allows occupied areas beneath a deck to be

FIGURE 3.50 Deck seal application; note the crew's protective wear. (*Courtesy of Coastal Construction Products*)

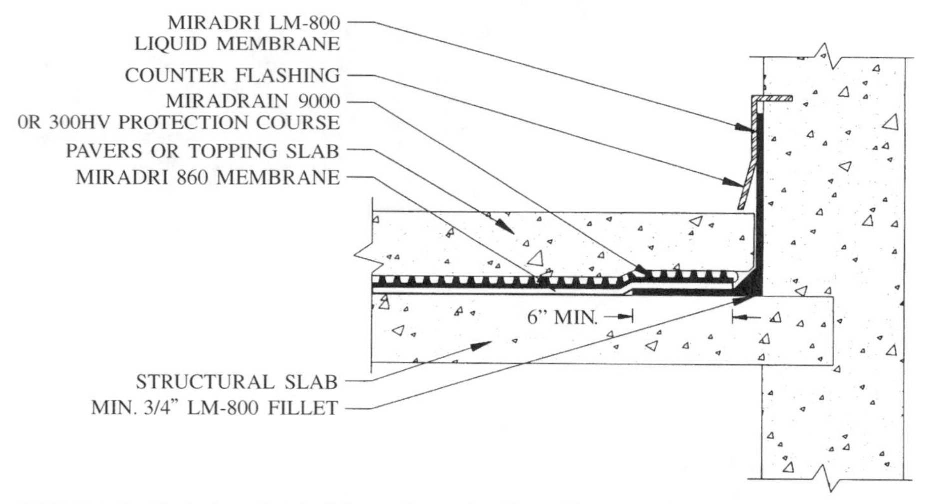

FIGURE 3.51 Typical sandwich-slab membrane detailing. (*Courtesy of TC MiraDRI*)

insulated for environmental control. All below-grade waterproofing systems, with the exception of hydros clay and vapor barriers, are used for protected membranes above grade. These include cementitious, fluid-applied, and sheet-good systems, both adhering and loose-laid. Additionally, hydros clay systems have been manufactured attached to sheet-good membranes, applicable for use as protected membrane installations.

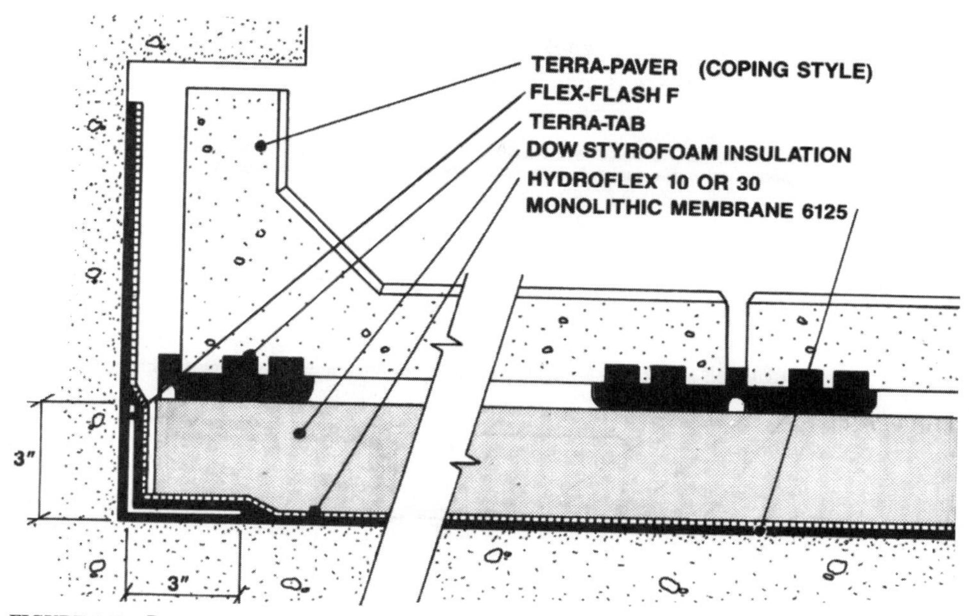

FIGURE 3.52 Protected membrane application using concrete pavers. (*Courtesy of American Hydrotech*)

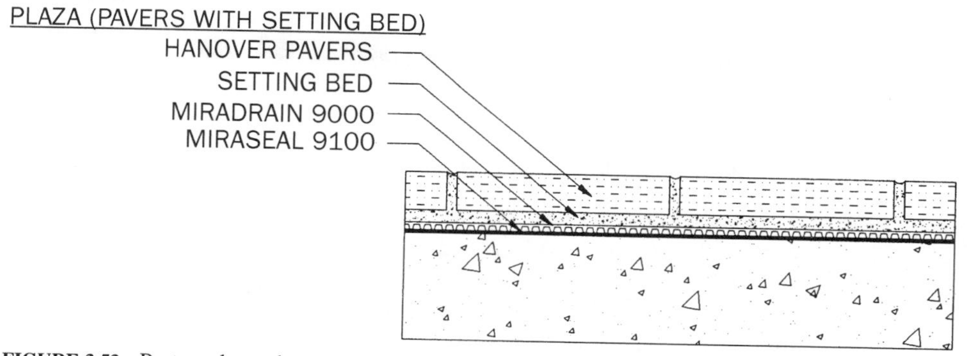

FIGURE 3.53 Protected membrane application using stone pavers. (*Courtesy of TC MiraDRI*)

DRAINAGE REQUIREMENTS

Protected membranes are used for swimming pool decks over occupied areas, rooftop pedestrian decks, helicopter landing pads, parking garage floors over enclosed spaces, balconies, and walkways. Sandwich membranes should not be installed without adequate provision for drainage at the membrane elevation; this allows water on the topping slab, as well as water that penetrates the protection layer onto the waterproof membrane, to drain (Fig. 3.56). If this drainage is not allowed, water will collect on a membrane and lead to numerous problems, including freeze–thaw damage, disbonding, cracking of topping

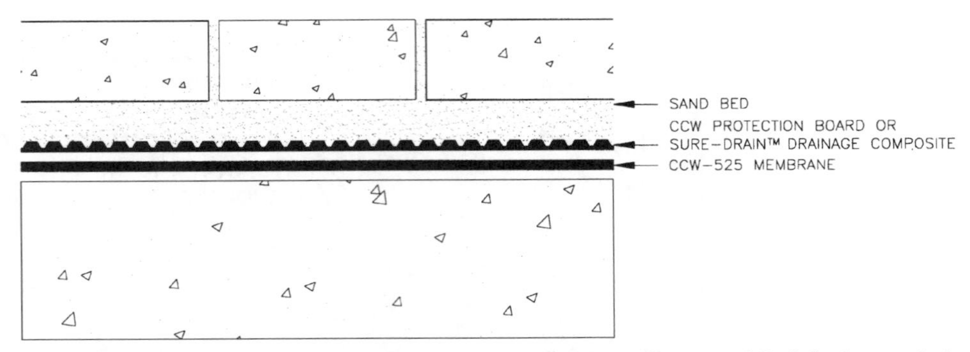

FIGURE 3.54 Protected membrane application using brick pavers. (*Courtesy of Carlisle Corporation*)

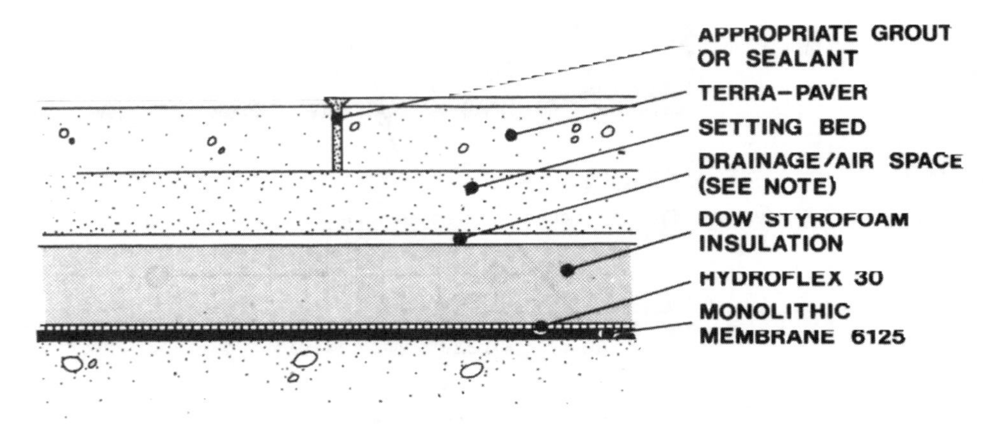

FIGURE 3.55 Insulation layer in protected membrane application. (*Courtesy of American Hydrotech*)

slabs, and deterioration of insulation board and the waterproof membrane. Refer to Figs. 3.57 and 3.58 for an example of these drainage requirements.

For the best protection of the waterproofing membrane, a drainage layer should be installed that directs water to dual drains or terminations of the application. Water that infiltrates through the topping slab can create areas of ponding water directly on top of the membrane even if the structural slab is sloped to drains. This ponding can be created by a variety of causes, including imperfections of the topping slab and protection layer, dirt, and debris.

To prevent this from occurring and to ensure that the water is removed away from the envelope as quickly as possible, a premanufactured drainage mat should be installed on top of the waterproofing membrane. The drainage layer can also be used in lieu of the protection layer.

The drainage system is the same basic system as manufactured for below-grade applications and discussed in Chapter 2. However, the sandwich membrane drainage systems have one major difference: they are produced with sufficient strength to prevent crushing of the material when traffic, foot or vehicular, is applied after installation. A typical drainage mat is shown in Figure 3.59.

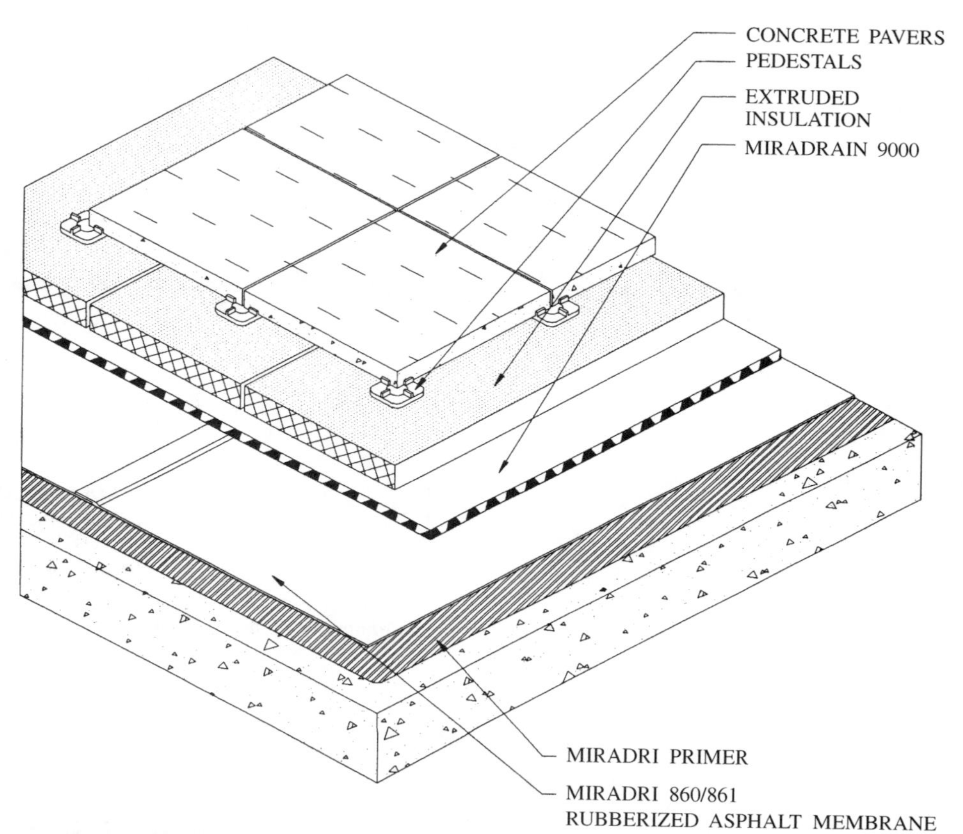

FIGURE 3.56 Prefabricated drainage layer in sandwich application. Note the insulation is spaced to permit drainage also. (*Courtesy of TC MiraDRI*)

It is imperative that termination detailing be adequately included to permit the drainage or weeping of water at the edges or perimeter of the sandwich slab installation. This is usually provided by installing an edge-weep system and counter-flashing, as shown in Figure 3.60. Or if the structural slab is sloped to drain towards the edges of the slab, a drain and gutter system should be provided as shown in Figure 3.61.

Note that in each of these details the drainage is designed to weep water directly at the pre-fabricated drainage board level. The drainage board should be installed so that the channels created are all aligned and run towards the intended drainage. The entire purpose of the various drainage systems in a sandwich-slab application (drainage mat, deck drains, and edge drainage systems) can be entirely defeated if the pre-fabricated drainage board is not installed correctly.

DETAILING SANDWICH MEMBRANES

Expansion or control joints should be installed in both the structural slab portion and the protection layer. Providing for expansion only at the structural portion does not allow for thermal

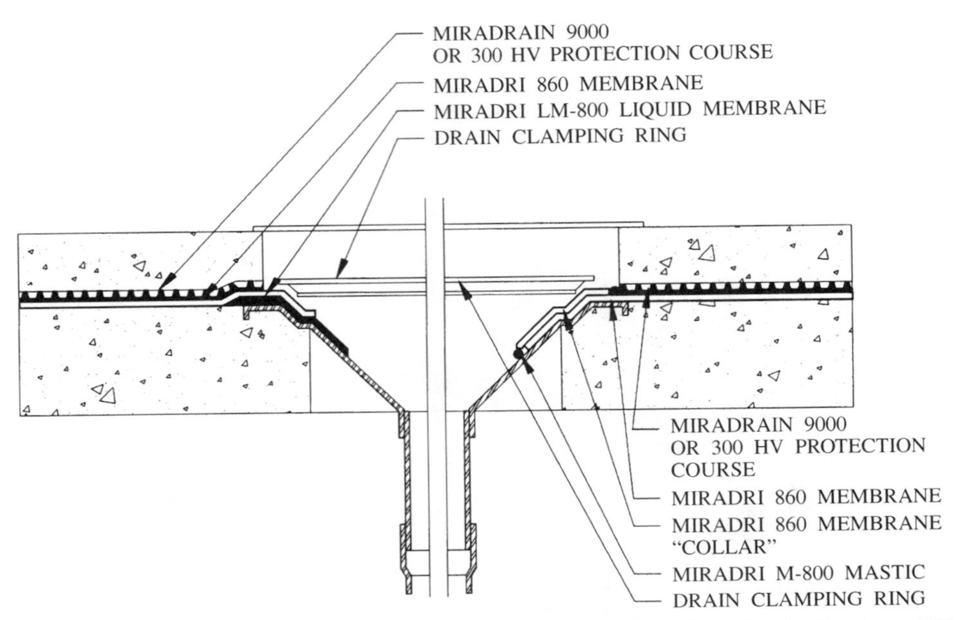

FIGURE 3.57 Dual drain installed for proper drainage of protected membrane level. (*Courtesy of TC MiraDRI*)

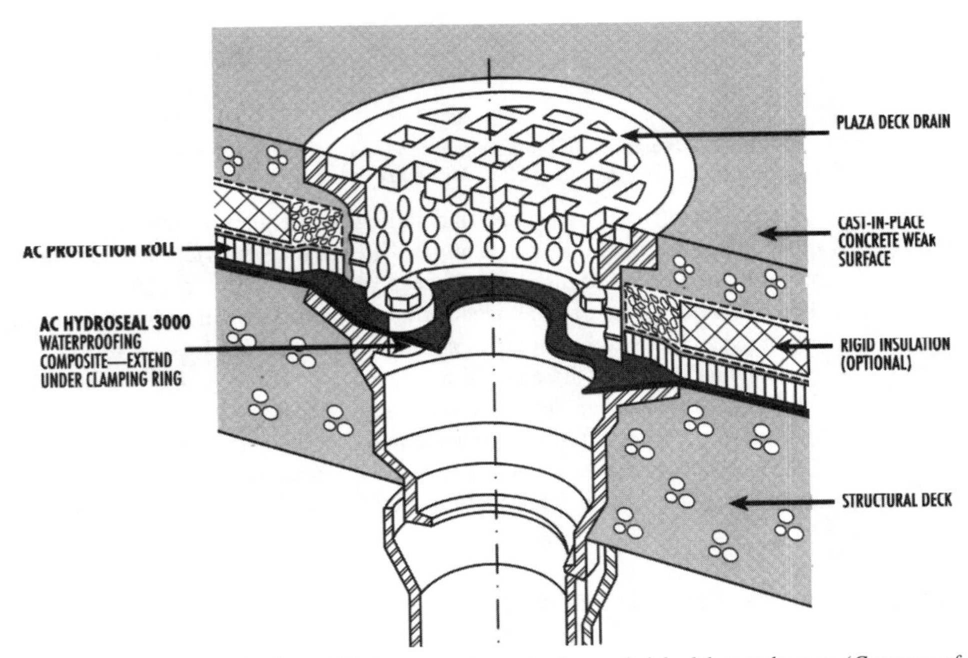

FIGURE 3.58 Schematic view of drainage requirements for sandwich-slab membranes. (*Courtesy of NEI Advanced Composite Technology*)

FIGURE 3.59 Premanufactured drainage for sandwich-slab construction. (*Courtesy of American Hydrotech*)

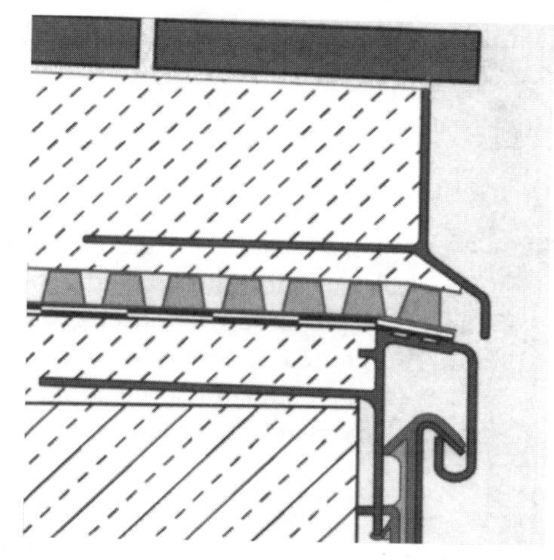

FIGURE 3.60 Topping slab with tile perimeter edge detail that permits drainage. Note drainage board over membrane, and provision to weep water out of envelope at edge. (*Courtesy of Schluter Systems*)

or structural movement of the topping slab. This can cause the topping slab to crack, leading to membrane deterioration. Refer to Figures 3.62 and 3.63 for proper detailing.

Membranes should be adhered only to the structural deck, not to topping layers, where unnecessary stress due to differential movement between the two layers will cause membrane failure.

Waterproof membranes should be adequately terminated into other building envelope components before applying topping and protection layers. The topping is also tied into the envelope as secondary protection. Control or expansion joints are installed along topping slab perimeters where they abut other building components, to allow for adequate movement (Fig. 3.64). Waterproof membranes at these locations are turned up vertically, to prevent water intrusion at the protection layer elevation. Refer to Fig. 3.65 for a typical design at this location.

When pavers are installed as the protection layer, pedestals are used to protect the membrane from damage. Pedestals allow leveling of pavers, to compensate for elevation deviations in pavers and structural slabs (Figure 3.66).

At areas where structural slabs are sloped for membrane drainage, pavers installed directly over the structural slab would be unlevel and pose a pedestrian hazard. Pedestals allow paver elevation to be leveled at these locations. Pedestals are manufactured to allow four different leveling applications, since each paver typically intersects four pavers, each of which may require a different amount of shimming (Figure 3.67).

If wood decking is used, wood blocking should be installed over membranes so that nailing of decking into this blocking does not puncture the waterproofing system. Blocking should run with the structural drainage design so that the blocking does not prevent water draining.

Tile applications, such as quarry or glazed tile, are also used as decorative protection layers with regular setting beds and thin-set applications applied directly over membranes.

With thin-set tile installations, only cementitious or liquid-applied membrane systems are used, and protection board is eliminated. Tile is bonded directly to the waterproof membrane.

Topping slabs must have sufficient strength for expected traffic conditions. Lightweight or insulating concrete systems of less than 3000 lb/in^2 compressive strength are not recommended. If used in planting areas, membranes should be installed continuously over a structural deck and not terminated at the planter walls and restarted in the planter. This prevents leakage through the wall system bypassing the membrane. See Fig. 3.68 for the differences in these installation methods. Fig. 3.69 represents a typical manufacturer detail for a similar area.

Using below-grade membranes for above-grade planter waterproofing is very common, especially on plaza decks. While these decks themselves are often waterproofed using the techniques described in this chapter, the planter should in itself be made completely waterproof to protect the building envelope beneath or adjacent to the planter.

Figures 3.70 and 3.71 detail the typical application methods of waterproofing above-grade planter areas. Note that each of these details incorporates the use of drainage board to drain water towards the internal planter drain. Since these areas are watered frequently, drainage is imperative, in this case, not only for waterproofing protection but also for the health of the vegetation planted in the planter.

Figure 3.72 shows the application of liquid membrane to planter walls as does Figure 3.73. In the latter note how difficult the use of a sheet good system would be in this particular application. Whenever waterproofing above-grade planters with tight and numerous changes-in-plane or direction, liquid applied membranes are preferred over sheet good systems as the preferred "idiot-proof" application. The continual cutting of sheets in these smaller applications results in a corresponding number of

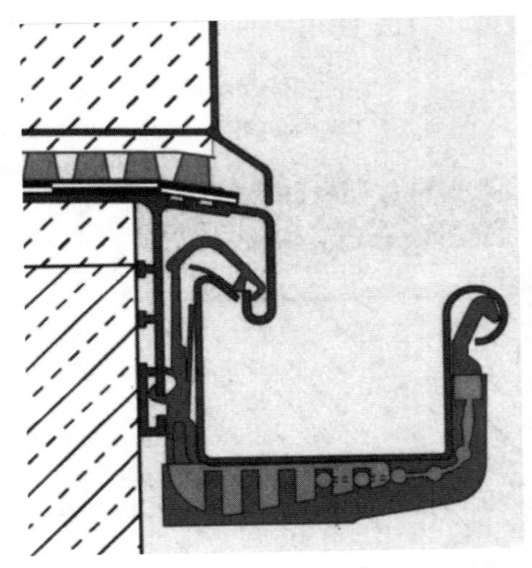

FIGURE 3.61 Drainage system detailed into gutter system. (*Courtesy of Schluter Systems*)

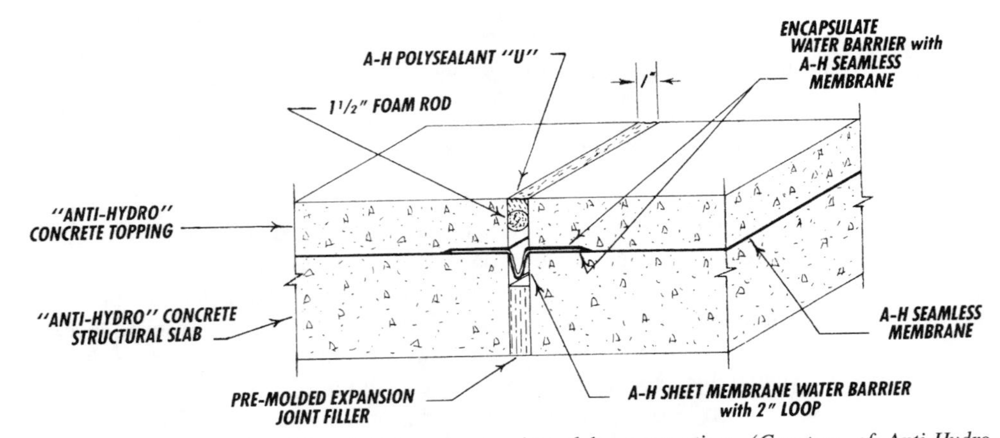

FIGURE 3.62 Expansion joint detailing for topping slab construction. (*Courtesy of Anti-Hydro International, Inc.*)

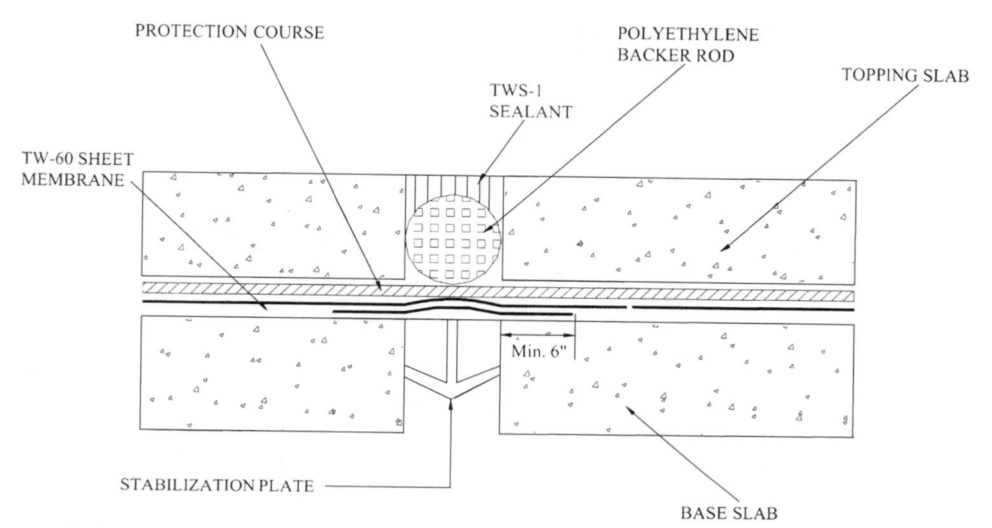

FIGURE 3.63 Expansion joint detailing for topping slab construction. (*Courtesy of Tamko Waterproofing*)

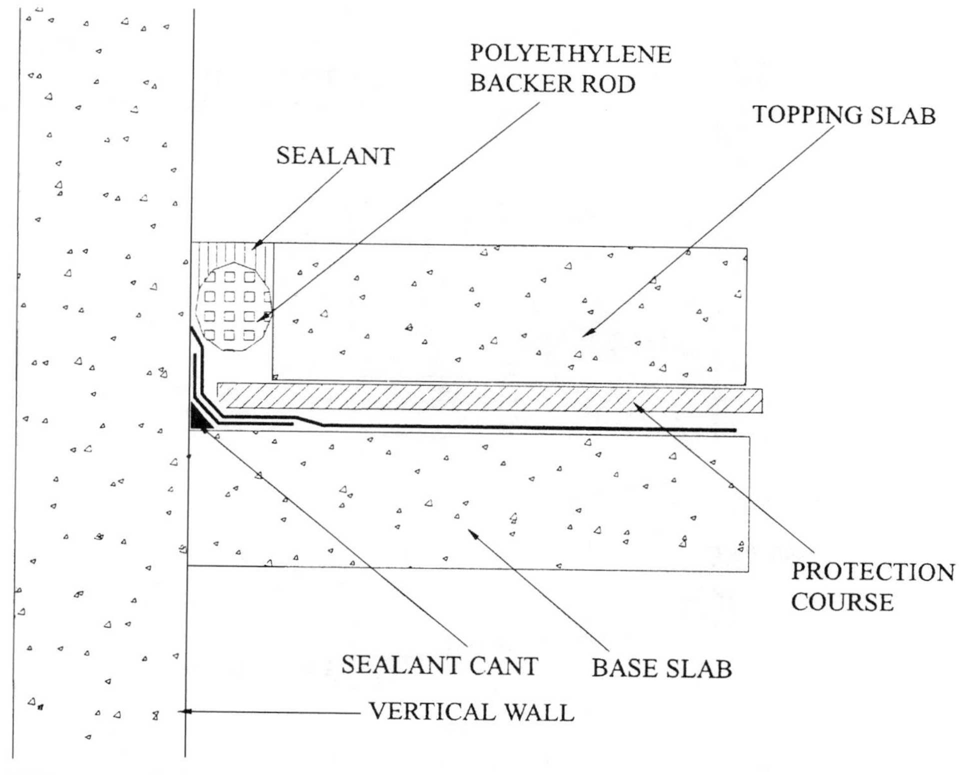

FIGURE 3.64 Perimeter expansion joint detailing for sandwich-slab membranes. (*Courtesy of Tamko Waterproofing*)

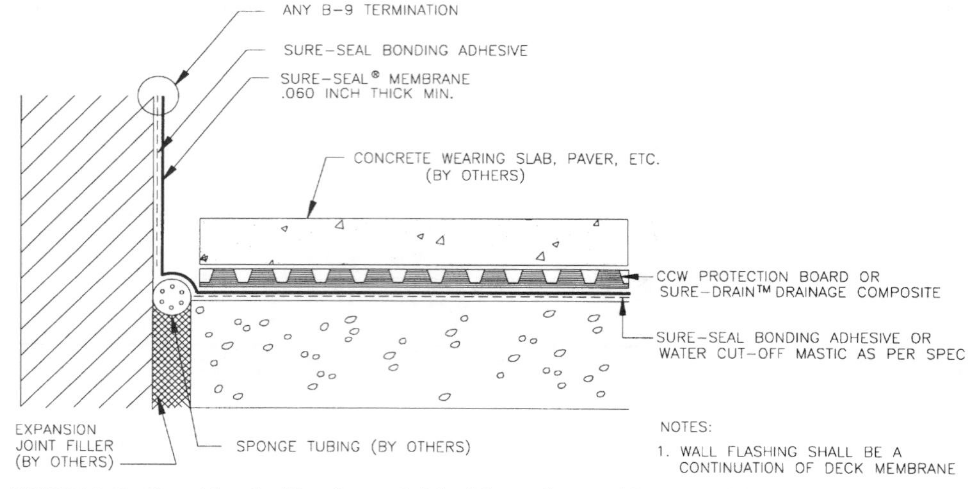

FIGURE 3.65 Transition detailing for sandwich-slab membranes. (*Courtesy of Carlisle Corporation*)

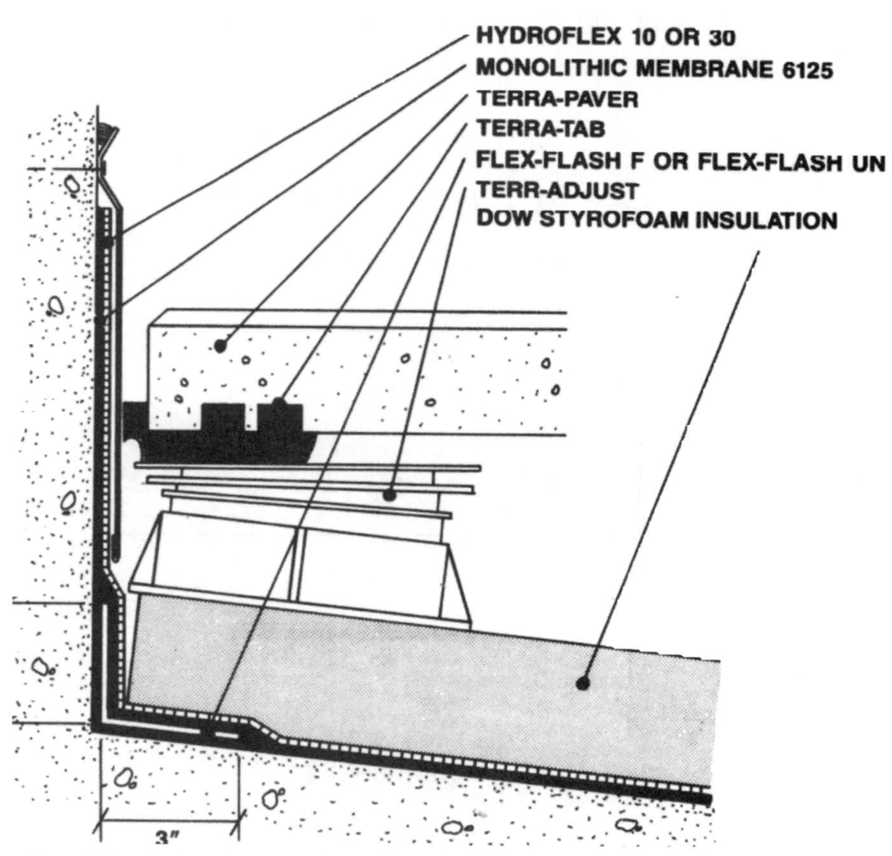

FIGURE 3.66 Pedestals permit the leveling of the walking surface on sloped structural decks using sandwich-slab membranes. (*Courtesy of American Hydrotech*)

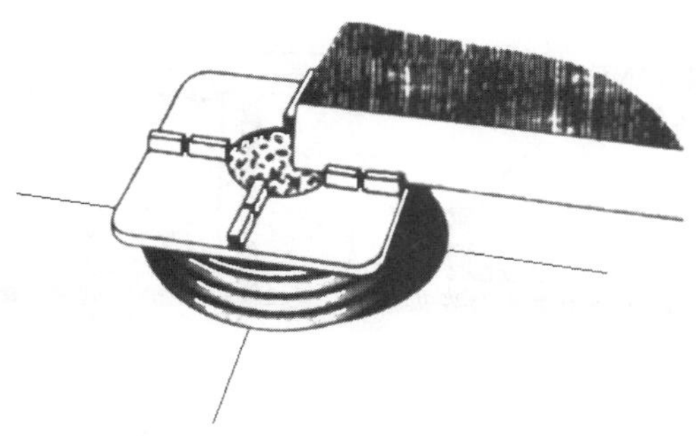

FIGURE 3.67 Pedestal detail. (*Courtesy of American Hydrotech*)

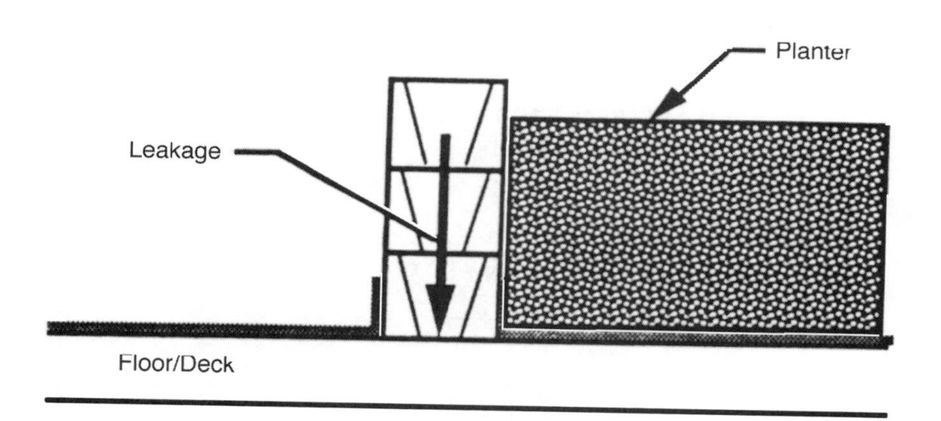

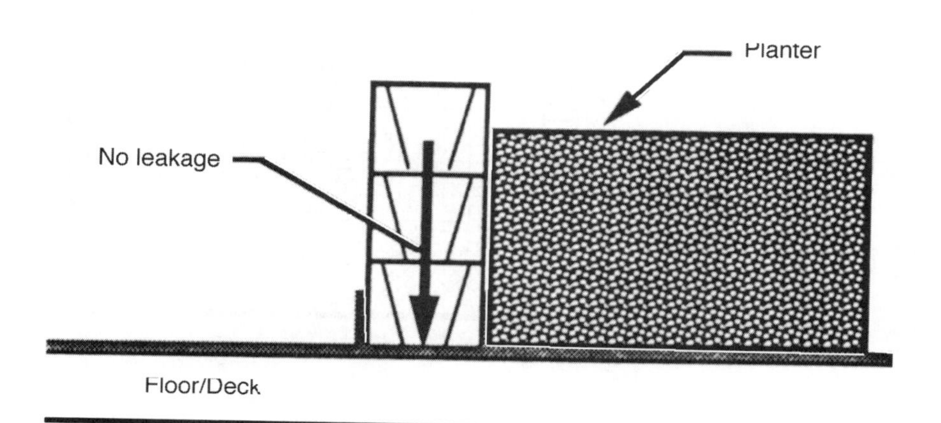

FIGURE 3.68 Planter detailing for split-slab membrane.

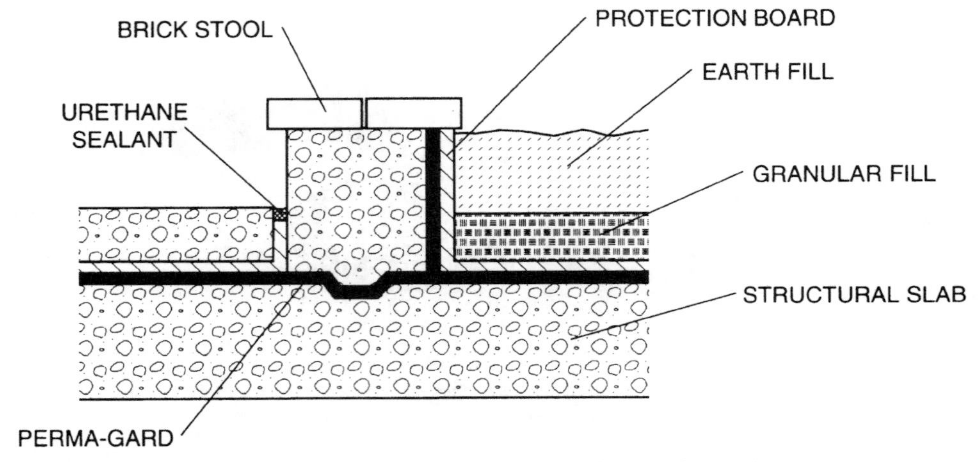

FIGURE 3.69 Typical detail for planters on decks. (*Courtesy of Neogard*)

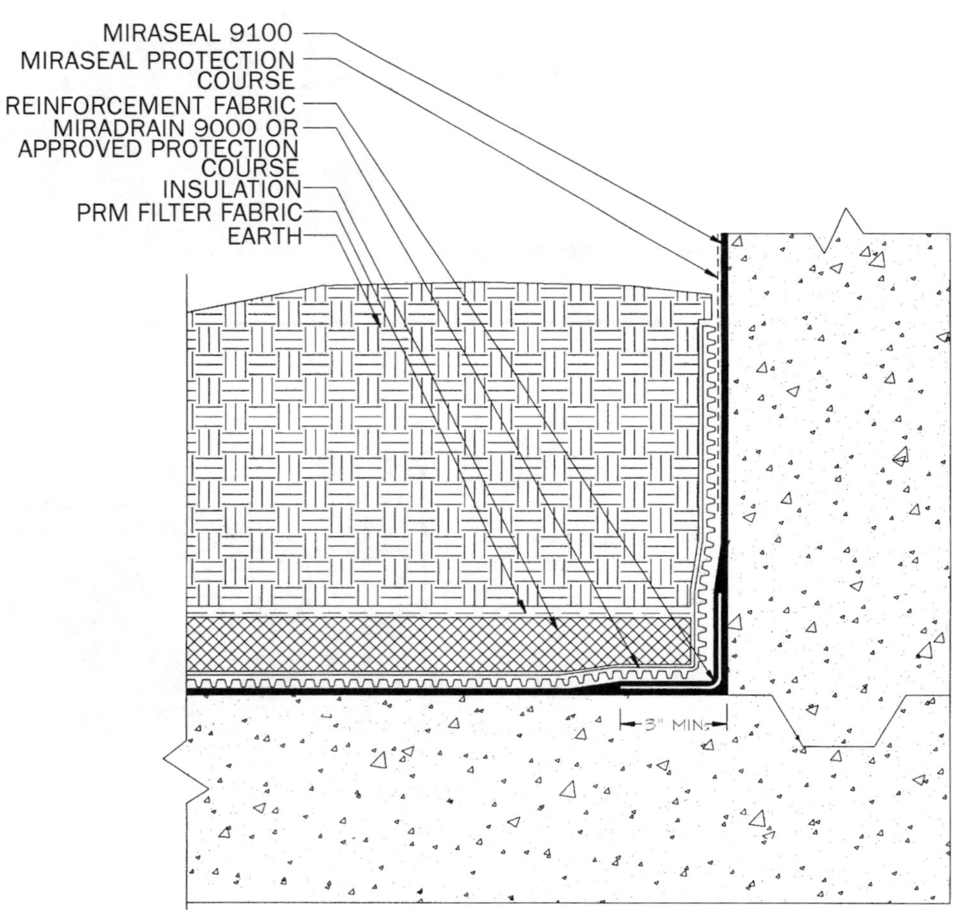

FIGURE 3.70 Typical detailing for above-grade planter areas. (*Courtesy of TC MiraDRI*)

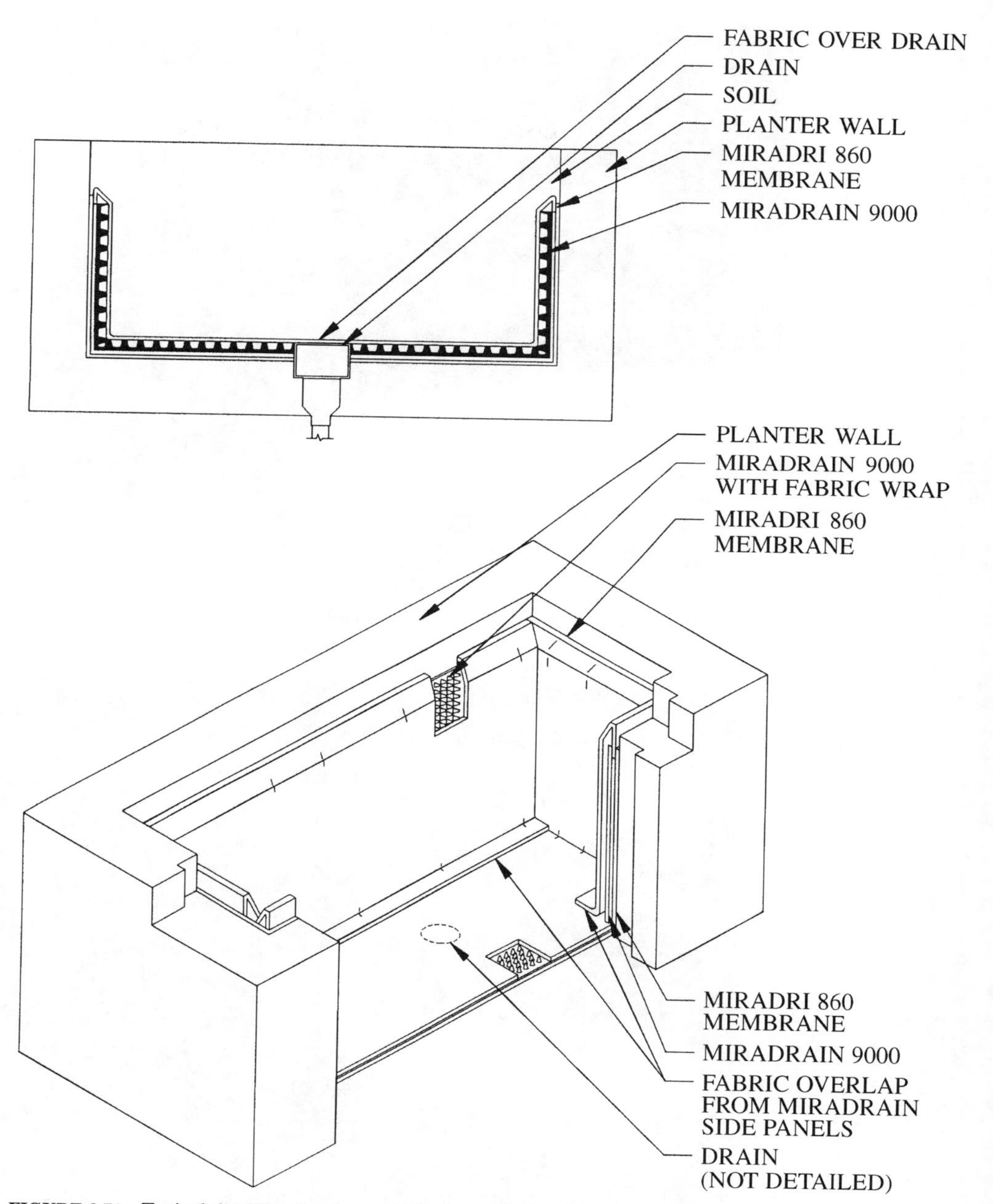

FIGURE 3.71 Typical detailing for above-grade planter areas. (*Courtesy of TC MiraDRI*)

FIGURE 3.72 Application of fluid membrane in above-grade planter. (*Courtesy of LBI Technologies*)

FIGURE 3.73 Liquid membrane application in planter. Note how difficult this application would be for a sheet system. (*Courtesy of American Hydrotech*)

seams that emphasize the 90%/1% principle. Liquid applications are seamless and can prevent the problems associated with sheet-good installation in small planter areas.

Selection of a protected system should be based on the same performance criteria as those for materials used with below-grade applications. For example, cementitious systems are rigid and do not allow for structural movement. Sheet goods have thickness controlled by premanufacturing but contain seams; liquid-applied systems are seamless but millage must be controlled. (See Table 3.23.) Refer to Chap. 2 on below-grade systems for more specific information.

TABLE 3.23 Protected Membrane Properties

Advantages	Disadvantages
Excellent elastomeric capabilities	No access for repairs
Topping provides protection	Requires drainage at membrane level
No maintenance	No remedial applications

FIGURE 3.74 Drainage system for sandwich-slab application. (*Courtesy of TC MiraDRI*)

PROTECTED MEMBRANE APPLICATION

Guidelines for protected membrane installation are the same as for below-grade materials. Concrete surfaces must be clean and free of all lattice, dirt, and oils and must be properly cured. Most systems prohibit the use of curing agents or form-release agents.

Applications follow below-grade installation methods (see Figs. 3.74, 3.75, and 3.76). Refer to Chapter 2 for specific installation details. With protected membrane installations,

FIGURE 3.75 Spray application of liquid-protected membrane application. (*Courtesy of LBI Technologies*)

FIGURE 3.76 Liquid-membrane roller application. (*Courtesy of American Hydrotech*)

FIGURE 3.77 Application of protection board prior to topping slab placement. (*Courtesy of American HydroTech*)

FIGURE 3.78 Primer application for sandwich-slab membrane installation. (*Courtesy of American Hydrotech*)

TABLE 3.24 Comparative Characteristics of Horizontal Waterproofing Systems

Membrane system	Advantages	Disadvantages
Deck coating	No protection or topping	Maintenance required
Ease of repairs	Control of millage	
Crack-bridging capabilities	Limited color selections	
Clear sealers	Single-step installation	No crack-bridging capability
No protection required	Not completely waterproof	
No grit or aggregate	Highly volatile materials	
Membranes	No maintenance	Protection layer required
Crack-bridging capabilities	Interslab drainage required	
Applicable over wood and metal	No direct access for repairs	

high-density insulation board can be used in place of protection board, Fig. 3.77. In these applications, insulation serves two purposes—protecting the waterproofing during placement of topping and providing insulation value for occupied spaces. Insulation must have sufficiently high density and compressive strength to withstand the weight of expected live and dead loads to protect cracking of toppings.

Adequate allowances should be made for movement in toppings, such as control joints or expansion joints. A water test, by complete flooding of the waterproofing system, must be completed before topping installation. This prevents unnecessary repairs to waterproofing after topping installation, should leaks later be discovered.

Most sheet goods and liquid-applied systems require primers (Figure 3.78), whereas cementitious systems do not. Waterproof membranes should not be installed over lightweight concrete decks such as insulating concrete. These lightweight mixes have insufficient strength for adequate substrate usage.

Detailing at penetrations is handled the same way below-grade installations require. See Fig. 3.79 and 3.80 for detailing at penetrations through both structural and topping slab.

HORIZONTAL WATERPROOFING SUMMARY

Advantages and disadvantages of exposed deck coatings, sealers, and protected membranes are summarized in Table 3.24. Once chosen it is important for proper detailing or terminating into other elements of the envelope to be done with flashings, counterflashings, control joints, or reglets for weatherproof integrity of all systems.

CIVIL STRUCTURE WATERPROOFING

For civil construction projects such as bridges, highway overpasses, and other similar concrete structures, waterproofing is often applied more for protection of the substrate and embedded reinforcing steel than for water infiltration to underlying spaces. Waterproofing is applied to protect against freeze–thaw cycles, road salt intrusion into the concrete that attacks the reinforcing, and water infiltration into the substrate that brings along other contaminants such as acid rain that can substantially reduce the life cycle of a structure.

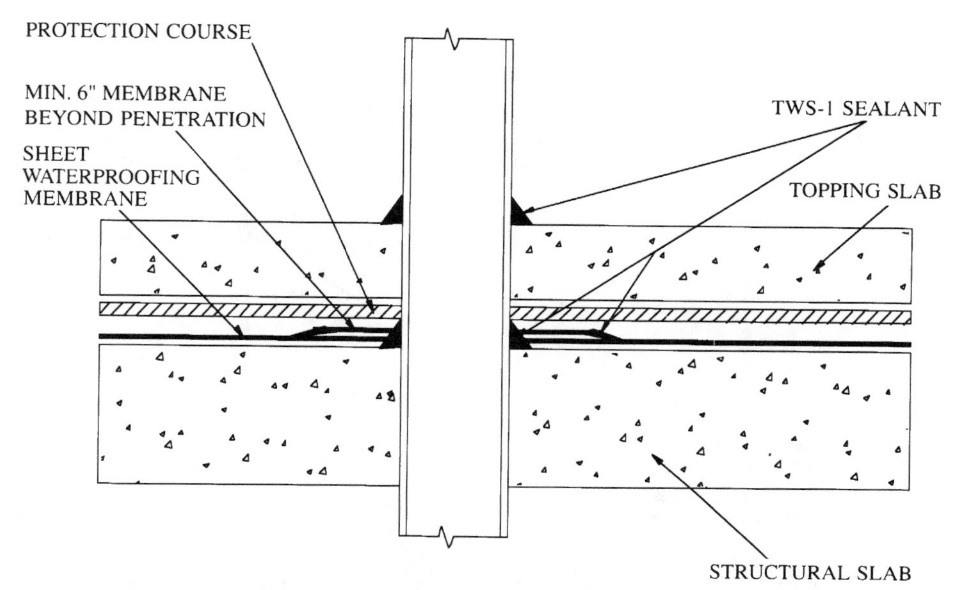

FIGURE 3.79 Sandwich-slab membrane penetration detailing. (*Courtesy of Tamko Waterproofing*)

Typically, waterproofing systems used on these types of structures is limited to cementitious or clear repellents, though some elastomeric coatings and sheet systems can be used. Figures 3.81 and 3.82 detail the typical application of a cementitious coating to a bride structure.

On other civil projects including tunnels and water reservoir, most of the below-grade and above-grade systems are used as appropriate, Figure 3.83. The application methods are also the same as addressed in the previous sections and chapter. For instance, a water retaining structure above grade might use negative-side cementitious coatings, while any below-grade portions of the structure might be protected by positive-side systems including clay, liquid membranes, or sheet goods.

Figure 3.84 shows a liquid membrane being sprayed on the interior of a water-retaining civil structure. Since it is a negative application, the membrane would have to be acceptable for potable water containment.

Landfill, pond linings, earthen reservoirs and other environmental protection projects are increasingly using standard and modified waterproofing systems to provide the necessary protection. The waterproofing systems in these situations provide protection from erosion in ponds, leaching from contaminant placed in landfills, and other nontraditional uses for waterproofing systems. Figure 3.85 shows the use of a drainage matting that is used to collect the discharge from the landfill and divert it into receptor drains that can direct the water to facilities designed to treat the water before it is discharged.

This traditional waterproofing system, while being used to prevent water intrusion in this case back into the earth, is also providing a nontraditional use as a collector system for the landfill water runoff so that it can be properly treated. Figure 3.86 details the use of the prefabricated drainage mat on earth structures.

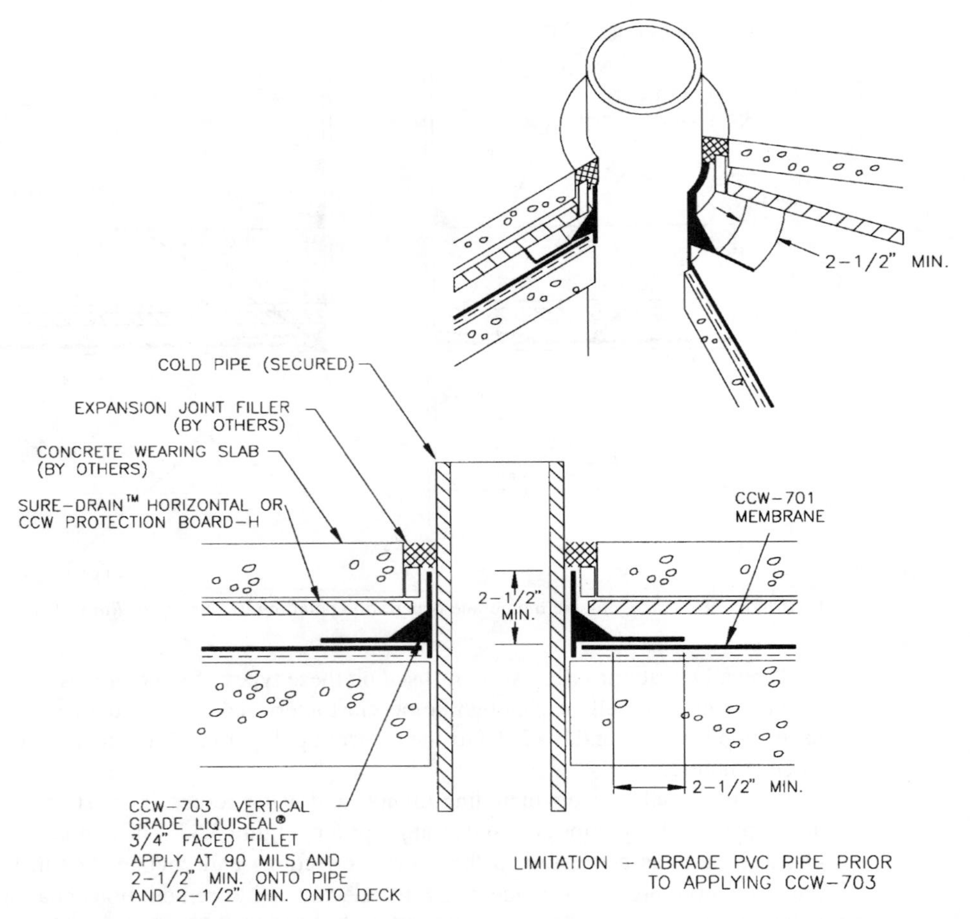

FIGURE 3.80 Pipe penetration detailing for topping slab construction. (*Courtesy of Carlisle Corporation*)

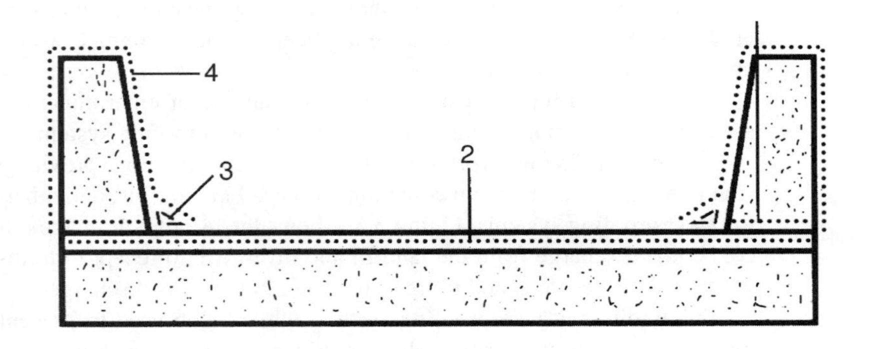

FIGURE 3.81 Cementitious waterproofing system for civil project. (*Courtesy of Xypex*)

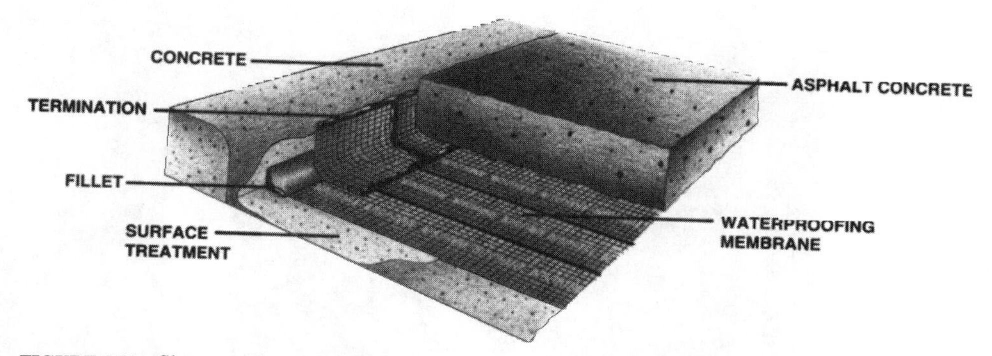

FIGURE 3.82 Sheet membrane application for civil project. (*Courtesy of Grace Construction Products*)

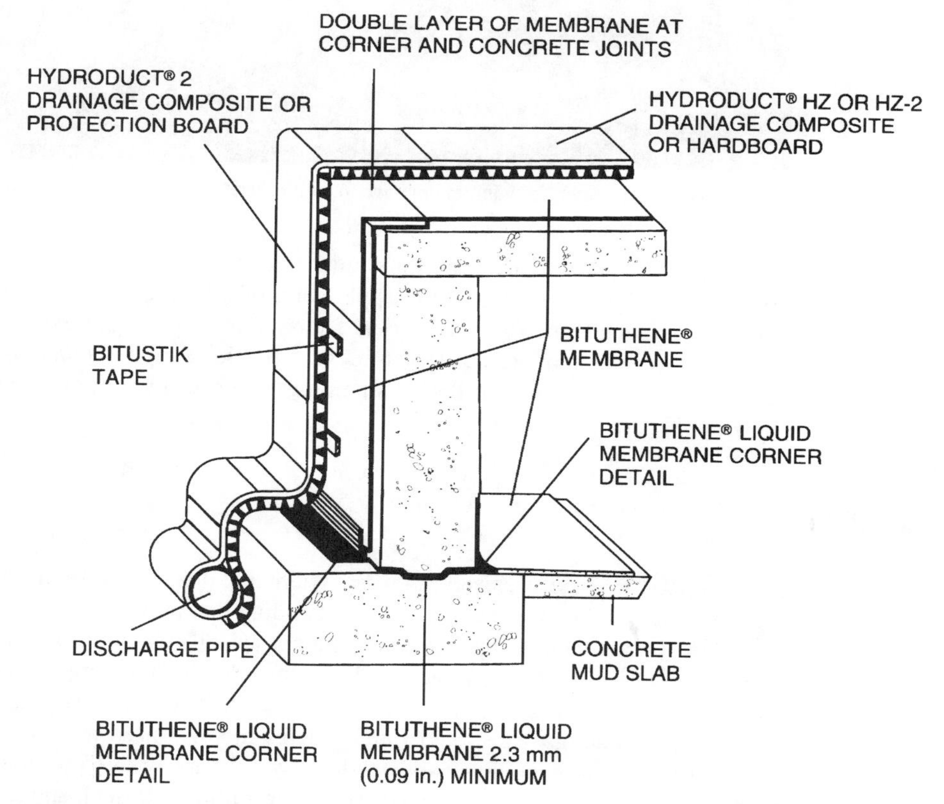

FIGURE 3.83 Sheet system for below-grade civil project. (*Courtesy of Grace Construction Products*)

FIGURE 3.84 Application of liquid membrane to civil project. (*Courtesy of LBI Technologies*)

Many other systems are used on similar site structures, including sheet systems as pond liners. Figure 3.87 shows the application of a liquid membrane to a fabric scrim material to provide an effective barrier. The system can be used exposed as detailed in Fig. 3.88 for a basin or similar structure, or if traffic is applied (such as a landfill using heavy trucks and equipment during dumping), a protective topping can be applied as shown in Figure 3.89.

ROOFING

Roofing is not defined as a waterproofing system, but it does form an integral part of building envelopes. Roofing is that portion of a building that prevents water intrusion into horizontal or slightly inclined elevations. Although typically exposed to the elements, roofing systems can also be internal or sandwiched between other building components.

Several waterproofing systems perform as excellent roofing systems, including fluid-applied deck coatings. Sheet and fluid systems are also used as sandwich or protected membranes as roofing components. All these systems allow roof or horizontal portions of a structure to be used for pedestrian or vehicular traffic. With such applications, insulation must be placed on the envelope's interior.

Waterproofing membranes are also used as the membrane portion of inverted roofing. These systems apply membranes directly to deck substrates, with insulation and ballast over the membrane to protect it from weathering, including ultraviolet rays.

This section does not provide detailed coverage of all available roofing systems that are not actual waterproofing materials; it highlights commonly used systems, and their relationship and use in above-grade envelopes are presented. In addition, Chapter 8 covers transition and termination detailing of roofing systems along with other envelope components.

The most commonly used roofing systems in building construction include the following:

- Built-up roofing
- Single-ply systems
- Modified bitumens
- Sprayed in-place urethane foam
- Metal roofing
- Protected and inverted membranes
- Deck coatings

FIGURE 3.85 Sheet system used for application directly over compacted grade (landfill) project. (*Courtesy of TC MiraDRI*)

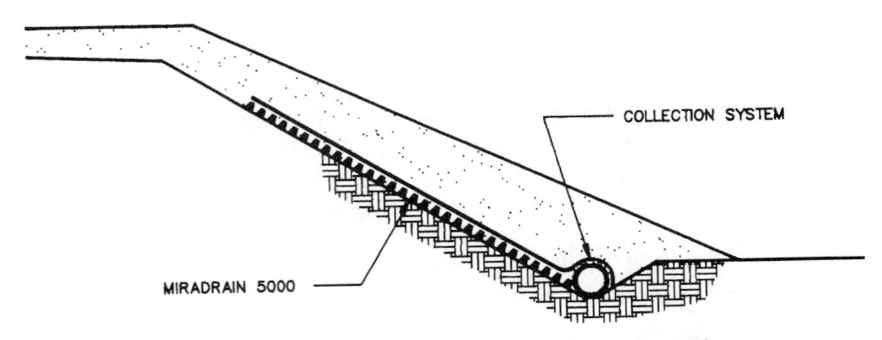

FIGURE 3.86 Landfill waterproofing detailing. (*Courtesy of TC MiraDRI*)

FIGURE 3.87 Liquid membrane used on civil project. (*Courtesy of C.I.M. Industries*)

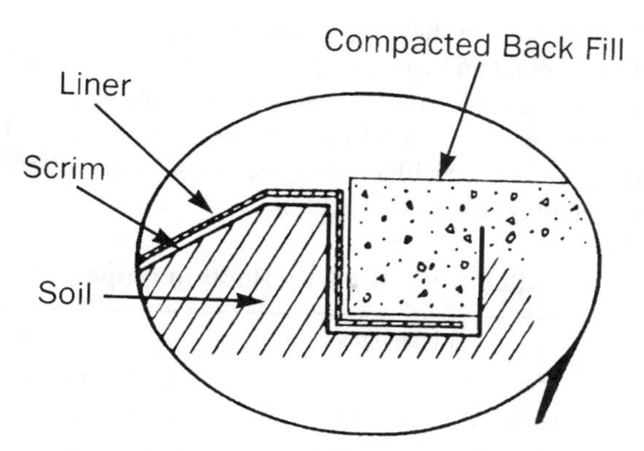

FIGURE 3.88 Basin termination waterproofing detail. (*Courtesy of C.I.M. Industries*)

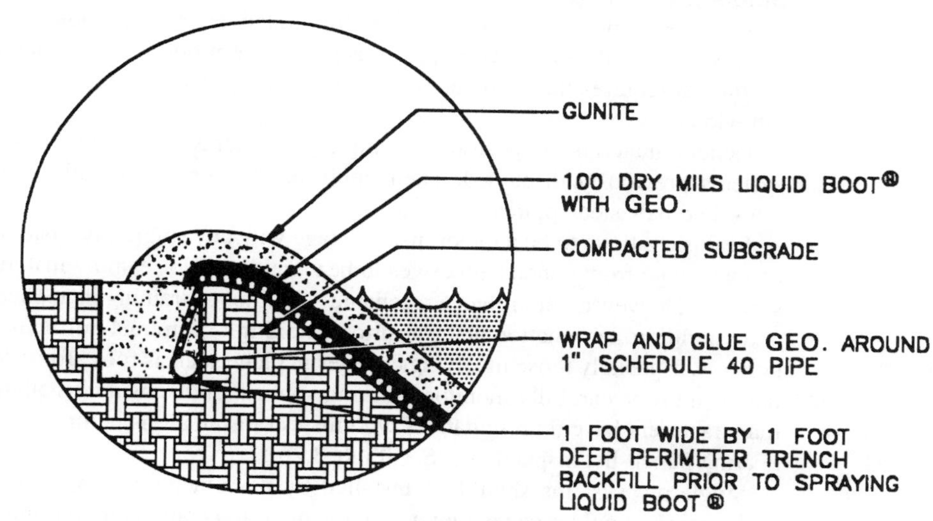

FIGURE 3.89 Sandwich system used for civil construction on-grade. (*Courtesy of LBI Technologies*)

Built-up roofing

The oldest system still commonly used today, built-up roofing (BUR), derives it name from the numerous layers of felts and bitumens applied to a substrate. Bitumens used as roofing membranes include coal tar and asphalt bitumens. By virtually adding layer upon layer, this type of system eventually covers and waterproofs a substrate and associated termination and transition details.

In the past, the quality of the roofing felts used was often inadequate for in-place service conditions. As technology advanced with other systems, BUR use declined. Improvements in materials now used for felts, including fiberglass, has led to reacceptance of the built-up roofing system. Field labor safety and field quality control of actual installations do, however, present problems in most BUR applications, particularly in confined and tight areas containing intricate termination and transition detailing.

Built-up systems have poor elongation properties, with coal-tar systems being very brittle. Both asphalt and coal-tar systems do have high tensile strength. As with other sheet systems, a major disadvantage with built-up roofing is that it allows any water infiltration to travel transversely until it finds a path to interior areas. This makes leakage causes difficult to determine, in particular if they are being caused by rooftop equipment or associated transition detailing. (See Table 3.25.)

TABLE 3.25 Built-Up Roofing Properties

Advantages	Disadvantages
Material cost	Material quality
High tensile strength	Low elongation properties
Multiple layers of protection	Difficult construction and safety conditions

Single-ply roofing

Single-ply roofing systems were derived from relatively new technology for use with roofing envelope applications. Used previously for waterproofing, their adaptation to exposed conditions requires that a membrane be resistant against exposure to weathering such as ultraviolet rays.

Generic material compositions of single-ply systems are as numerous as waterproofing systems previously discussed. Their applications range from fully adhered systems to loose-laid ballasted applications.

Seams continue to be a major disadvantage with any single-ply system. Seaming applications range from contact adhesives to heat welding. No seaming material or application system is, however, better than a mechanic's abilities or training in application procedures.

In addition, termination and transition detailing is extremely difficult using single-ply systems, especially those involving changes in plane such as roof protrusions. Seam installations must be carefully monitored during application to ensure installation quality. Most manufacturers have representatives who inspect installations before architectural or engineering punch-list inspections. (See Table 3.26.)

Single-ply systems should be installed by manufacturer's approved applicators, and each project should receive a joint manufacturer and contractor warranty. Guarantees and warranties are discussed in Chapter 9.

Modified bitumen

Modified bitumen systems are available in a variety of materials and application types. These include hot-applied liquid membranes and cold-applied systems with protective top coats. Systems are available with or without fabric or felt reinforcing.

TABLE 3.26 Single-Ply Roofing Properties

Advantages	Disadvantages
Manufactured quality control of materials	Multiple seams
Weathering durability	Termination and transition detailing
Selection of materials available	Patching and repairs

Most systems allow for seamless applications; this makes termination and transitional detailing easily detailed within the roofing installation. Modified bitumens are also used as protected membranes with inverted roofing systems. They are typically manufactured from a basic asphaltic product with added plasticizers and proprietary additives. This provides better performance characteristics as compared to basic asphaltic systems.

Generally, modified bitumens are not resistant to heavy pedestrian traffic. Adequate measures must be taken to provide for walkway pads or other protection. As with single-ply systems roofing, applicators are approved and trained by material manufacturers. (See Table 3.27.)

TABLE 3.27 Modified Bitumen Properties

Advantages	Disadvantages
Increased performance properties	Thickness control of applications
Seamless applications	Temperature control of hot-applied systems
Termination and transition detailing	Safety concerns

Metal roofing

Metal roofing systems on many building envelopes are used as decorative highlights for small portions of the entire roofing area. Often, metal systems such as copper domes are used for aesthetic purposes.

Complete metal roofing systems are now used regularly, especially on low-rise educational facilities and warehouse-type structures. Metal roofing systems are available in a variety of compositions, from copper to aluminum. They also include various manufactured preengineered systems.

Termination and transition detailing are difficult with metal systems, particularly when large amounts of rooftop equipment are installed. Additionally, metal roofs are not recommended for flat or minimally sloped areas that frequently occur on building envelopes.

Metal or sheet flashings are typically used for transitional detailing. This makes round protrusions and sloped areas subject to problems in detailing and possible water infiltration. If used in proper situations and expertly installed, however, certain roof systems such as copper domes will far outlast any other type of roofing installation. (See Table 3.28.)

Sprayed urethane foam roofing

Sprayed urethane foam roofing systems consist of high-density urethane foam coated with an elastomeric roof coating. The foam is of sufficient density to withstand minor foot traffic. The elastomeric coating, although similar to that used for vertical envelope waterproofing, must be able to withstand ponding or standing water.

TABLE 3.28 Metal Roofing Properties

Advantages	Disadvantages
Aesthetics	Termination and transition detailing
Durability	Not for flat or minimally sloped roofs
Life-cycle weathering	Cost

A major advantage of foam roofs is their seamless application, particularly with remedial or reroofing applications. Foam roofs can be installed over many types of existing failed or leaking roof envelopes, including built-up and single-ply roofing. Minimal preparation work is required when applying urethane foam roofs in these situations.

The urethane foam portion adds substantial insulation value to a roof, depending on foam thickness. Foams have an insulation R value of approximately 7 inches of foam insulation. Urethane foam can be installed in various thicknesses and sloped to provide drainage where none currently exists. Typically, foam roofs are installed from 1/2–6 inch thick.

After the urethane foam is installed it must be protected not from water, as it is waterproof, but from ultraviolet weathering. Foam left exposed is initially waterproof, but ultraviolet weathering will eventually degrade the foam until leakage occurs.

Thus, coating is applied to provide weathering protection for transitions and termination detailing. Coatings allow the systems to be self-flashing around roof protrusions and similar details. Foam roofs are installed in a completely seamless fashion, and their spray application makes termination and transition detailing relatively simple.

A major disadvantage with foam roofing systems is their reliance on 100 percent job-site manufacturing. Foam is supplied in two-component mixes that must be carefully mixed proportionally, heated to proper temperatures, and correctly sprayed in almost perfect weather. Any moisture on a roof, even high humidity and condensation, will cause foam to blister, as urethane foam does not permit vapor transmission. (See Table 3.29.)

TABLE 3.29 Sprayed Urethane Foam Roofing Properties

Advantages	Disadvantages
Applications for reroofing	Quality-control problems caused by weather conditions
Seamless	Completely job-site manufactured system
Termination and transition detailing	Safety concerns during application

Protected and inverted membranes

Sheet systems and fluid membranes used in below- and above-grade waterproofing have been successfully used for protected, sandwiched, and inverted roofing systems. These materials are identical to those previously discussed under the protected membrane section of this chapter.

Using protected membranes allows the envelope roofing portion to be used for other purposes including tennis courts and pedestrian areas. A roof area can also be used for vehicular parking when necessary structural provisions are provided.

Protrusions, particularly HVAC and electrical, are difficult to waterproof since they penetrate both structural and topping slab portions. If used, all protrusions and similar detailing should be in place and detailed before membrane installation. After the topping or protection slab is in place, protection layers should be detailed for additional protection, including movement allowances. Drainage should be provided at both topping and structural slab elevations to ensure that water is shed as quickly as possible.

A major disadvantage with these systems is their difficulty in finding and repairing leakage, since the membrane is inaccessible. These systems require that all applications be completely flood tested after membrane installation and before topping protection is installed to prevent unnecessary problems. (See Table 3.30.)

TABLE 3.30 Protected and Inverted Roofing Properties

Advantages	Disadvantages
Protection from vehicular and pedestrian damage	Difficult to repair
Protected from weathering	Drainage problems occur frequently
Provide multiple use of roof areas	Subject to insulation deterioration

Deck coatings for roofing

Among the simplest and most successful but most underused roofing systems are deck-coating systems. These materials are used primarily for waterproofing pedestrian and vehicular decks as previously discussed in this chapter. Deck coatings applied as roofing systems provide seamless applications, including terminations and transitions, and are completely self-flashing. They can be applied over wood, metal, and concrete substrates.

The only major disadvantage in roofing application of deck coating is that insulation must be installed to the underside of the exposed envelope portion. If this is possible, deck coatings provide numerous benefits when installed as roofing systems.

Since deck coatings adhere completely to a substrate, water cannot transverse longitudinally beneath the system. Therefore, should leakage occur, it will be directly where a membrane has failed and easily determinable. These membranes are resistant to pedestrian and vehicular traffic, providing resistance to abuse that most other roofing systems cannot.

Repairs are easily completed by properly repairing and recoating an affected area. Any equipment changes, roof penetrations, and so forth, can be made after the initial roof installation. These repaired areas also become seamless with the original application.

Deck coatings are applied directly from manufacturer's containers in a liquid. This application eliminates the need to heat materials, seam, and use spray application, which are required with other roofing systems. By providing the simplest installation procedure, deck coatings eliminate most human error and result in successful roofing systems for building envelopes. (See Table 3.31.)

ROOFING INSTALLATION

Mechanical equipment, plumbing stacks, and electrical penetrations are often poorly detailed, presenting almost impossible conditions in which to install roofing materials. The

TABLE 3.31 Deck Coatings Used for Roofing Properties

Advantages	Disadvantages
Seamless applications	Insulation must be placed on the underside of the deck
Seamless termination and transition detailing	Control of millage thickness
Allows roof areas to be used for different purposes	Subject to blistering by negative moisture drive

difficult areas include inaccessible places beneath mechanical equipment, and electrical conduit protrusions too close to adjacent equipment to properly flash roof transition materials.

By limiting the number of roof penetrations, providing areas large enough to install transition detailing, and ensuring minimum heights above roof for termination, detailing will limit common 90 percent leakage problems. All equipment should be placed on concrete curbs a minimum of 8 inches above roofing materials. Wood used for curbs can rot and eventually damage the roof. This minimal height provides sufficient areas to transition, and terminates roofing materials properly into equipment that becomes part of the building envelope.

Curbs should be placed to complement roof drainage and not block it. Roofing membranes should extend both under the curb and over it, completely enveloping it to prevent leakage.

Any conduits or drains running to and from equipment should be raised off the roof so as not to prevent drainage and damage to roof membranes. Any rooftop-mounted equipment such as balustrades, signs, and window-washing equipment should be placed on curbs. Equipment fasteners used at curb detailing, as well as the equipment itself, should be waterproofed to prevent transition and termination water infiltration.

Roof penetrations and all protrusions, such as electrical conduits, should be kept in as absolute a minimum of groupings as possible. Roofs should not be used as penthouse areas for electrical and mechanical equipment nor storage areas for excess equipment. Too often equipment is placed on a roof in groupings that make maintenance and drainage impossible. Further, any equipment added after roofing completion should be reviewed by the material manufacturer and roofing contractor to ensure that warranties are not affected by the installation.

Roofs should be tested for adequate drainage before membrane installation. Once rooftop equipment has been installed, drainage should be checked and adjustments made where necessary. After roofing is installed, it is too late to repair areas of ponded water. Roof drains must be placed at the lowest elevations of the roof and not be obstructed.

All related roof envelope equipment should be tested for watertightness after installation. Roofing envelope portions are often damaged by equipment that allows water or condensation to bypass roof membranes and to enter directly into interior areas.

Sealants should not be used excessively as termination or transition detailing anywhere within the roofing envelope. Sealants typically have a much shorter life cycle than roofing membranes. Sealants then become a maintenance problem, since when not properly attended they create leakage.

ROOFING SUMMARY

As with a complete envelope, typically it is not roofing materials or systems that directly cause water infiltration; it is the 1 percent of a roofing envelope portion. This 1 percent includes termination and transition details including flashings, protrusions, and mechanical supports that typically occur within a roofing application.

Roofing envelope installation often involves more subcontractors and trades than any other building envelope portion; these people range from sheet metal mechanics to window-washing-equipment installers. This extreme multiple-discipline involvement call for the utmost care in detailing and installing termination and transition details, not only between various rooftop components but transitions between roofing and other building envelope components of the envelope. These transition details are covered fully in Chapter 8.

VAPOR BARRIERS

Vapor barriers are used in above-grade construction to prevent moisture vapor transmission between interior and exterior areas. In winter conditions, warm, moist interior air is drawn outward to the drier outside air by the difference in vapor pressures (negative vapor drive). In summer, moisture vapor travels from moist and warm outside air to cool and dry interior areas (positive vapor drive).

Vapor barriers or retarders are not waterproofing materials but are used as part of wall assemblies to prevent vapor transmission and allow this vapor to condense into liquid form. Vapor barriers are most useful in hot tropical areas where vapor transmission into air-conditioned areas can be so severe that mold and mildew frequently form on exterior walls. This problem is often mistaken for water leakage or infiltration when it is not. Attempts to repair, including applying breathable coatings to an envelope (e.g., elastomeric coatings), will not solve this problem.

Should a nonbreathable coating be applied to an envelope under these conditions, however, coating blistering and disbonding will occur when negative vapor drives occur. This requires that vapor retarders or barriers be applied to the interior or warm side of insulated areas. In tropical areas, the barrier is placed on exterior sides to prevent condensation or vapor from wetting insulation caused by positive drive. In most areas, barriers are placed on interior sides of insulation due to the predominance of negative vapor drive.

Vapor barriers are commonly available in polyethylene sheets or aluminum foil sheets on laminated reinforced paper. Sheets must be applied with seams lapped and sealed to prevent breaks in the barrier.

A vapor barrier's performance is measured in perms (permeability). This is the measure of vapor transmitting through a particular envelope material or component. Materials such as masonry block have high permeability, whereas polyethylene materials have very low permeability. Glass is an example of a barrier. Moisture collects and condenses on glass because it cannot pass through the glass.

Although vapor barriers are not used as waterproofing systems, they can affect the selection and use of waterproofing materials for use on an envelope. If negative vapor drive is possible (winter conditions), it is necessary for permeable waterproof materials to be used to allow this moisture to pass without damaging the waterproofing material by blistering or delaminating.

INTERIOR WATERPROOFING APPLICATIONS

Most of the waterproofing systems described in the previous sections and Chapter 2 are applicable for interior spaces including showers, kitchens and specialty areas such as steam or locker rooms, laboratories, and mechanical rooms. Liquid-applied membranes and sheet goods are most commonly used. Clay systems are not applicable for interior spaces; cementitious systems, however, make excellent choices, since these areas are not subject to thermal movement (the one major disadvantage with cementitious product is that they do not move under thermal expansion or contraction).

Shower and bath areas are common areas requiring waterproofing anytime they occur above occupied spaces. Figure 3.90 details the application of a liquid-applied membrane with the tile and setting bed acting as the protection or wearing surface for the membrane. Note that the material is applied over a concrete substrate. While liquid membranes can be applied over plywood, it is often preferable to use a sheet or cementitious systems in this situation since the plywood joints are not sufficiently tight to apply liquid membranes. Figure 3.91 details a sheet membrane installation and Fig. 3.92 details a cementitious and liquid combination system.

These same systems can be used in kitchen or any other similar interior space where a finished floor that acts as a protection layer for the membrane is applied. Usually this finish floor surface is a tile that works well in wet room areas. Figure 3.93 details the application of a sheet application for the kitchen area. Note that the membrane is applied continuously under the finished floor, including the interior partitions. This is a critical detail to prevent the 90%/1% principal from applying to interior spaces as well. Running the membrane continuously under the interior partitions prevents water from entering the wall and bypassing the membrane in the same manner as described for planters on plaza decks in the previous section. Figure 3.94 details such a sandwich-slab application for interior projects.

In any area where an expansion joint is required, the joint should be detailed in the same manner as sandwich-slab construction. The joint should occur on the finished floor as well as the substrate on which the membrane is applied. Note this detailing in Fig. 3.95.

If necessary, these waterproofing systems can be used to completely envelope a room, as is the case with the steam room detailing shown in Fig. 3.96. For any interior application, the same installation precautions should be used as with above- or below-grade applications. Specific guidelines for interior applications should be requested from the manufacturers; for instance, whether or not thin-set tile can be used directly over the membrane.

Thin set tile adhering directly to a waterproofing membrane obviously deters from its movement capability. However, since the movement experienced inside a building is not as much as on an exterior envelope portion, most manufacturers will accept a thin-set application for interior areas.

A major problem with many interior waterproofing application is the amount of traffic from different trades that might occur on a waterproof membrane before it is protected with the finished floor or wall surfacing. For example, after a membrane is installed, various trades might be required in the room, such as electrical or mechanical, before the finished surfaces can be applied. As such, it is imperative that protection be provided on the waterproofing membrane until such time as the finished floor or wall surfaces can be applied properly over the membrane.

The general contractor should carefully examine the membrane before application of the finish systems, to ensure that the waterproofing membrane is intact and has not been damaged by PPes over the membrane and to verify that it functions properly. This will prevent costly repairs such as removal and replacement of tile to fix an interior membrane application. Flood tests consist of temporarily damming the drains in the room, flooding the deck with

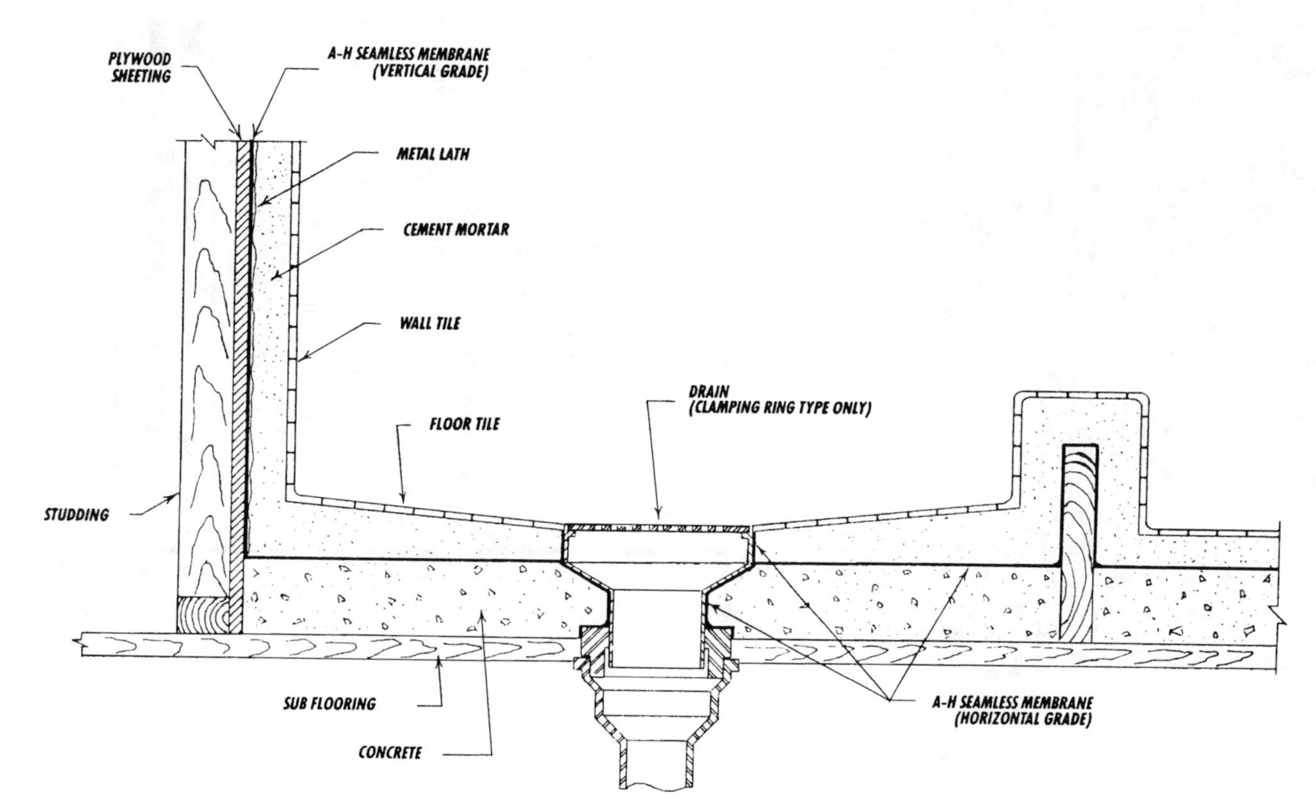

FIGURE 3.90 Interior waterproofing detailing for shower or wet room area. (*Courtesy of Anti-Hydro International, Inc.*)

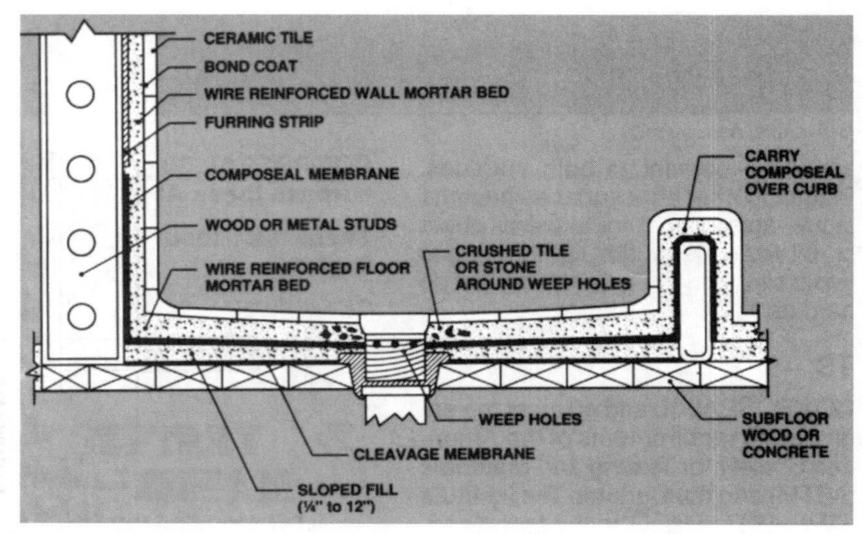

FIGURE 3.91 Sheet system waterproofing for interior waterproofing. (*Courtesy of Composeal*)

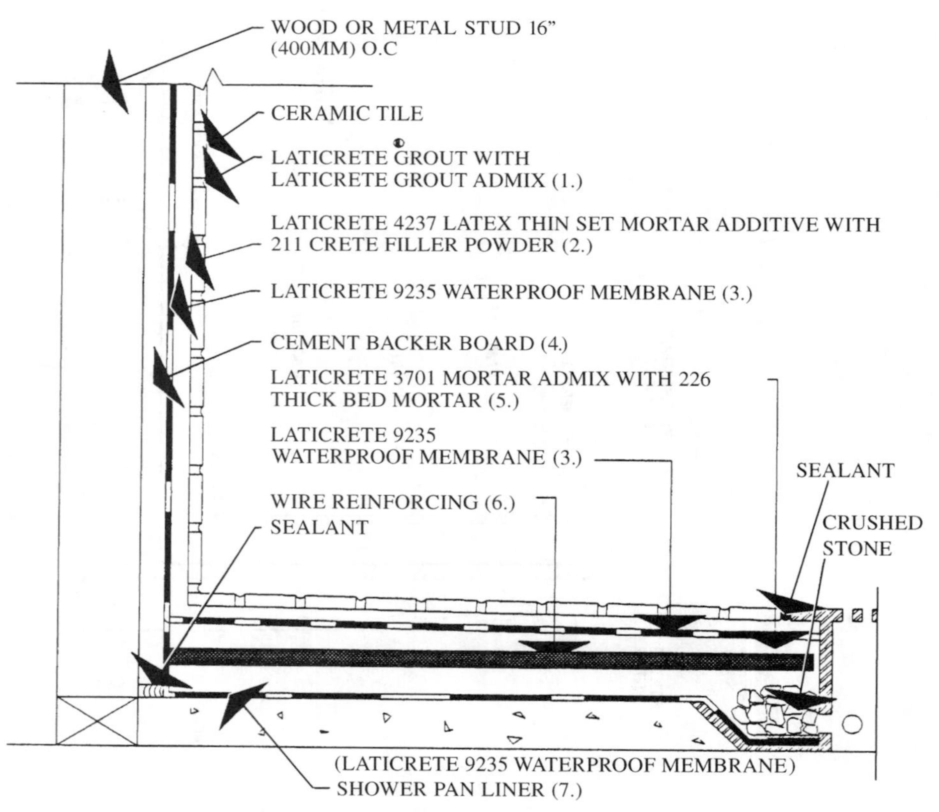

FIGURE 3.92 Cementitious and fluid-applied waterproofing system for interior areas. (*Courtesy of Laticrete*)

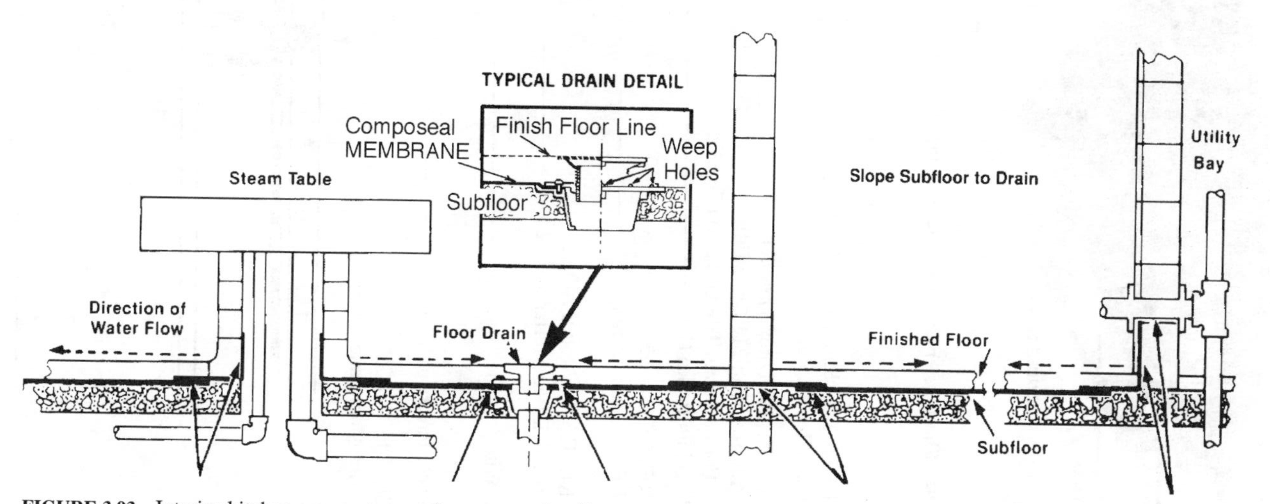

FIGURE 3.93 Interior kitchen area waterproofing system detailing. (*Courtesy of Composeal*)

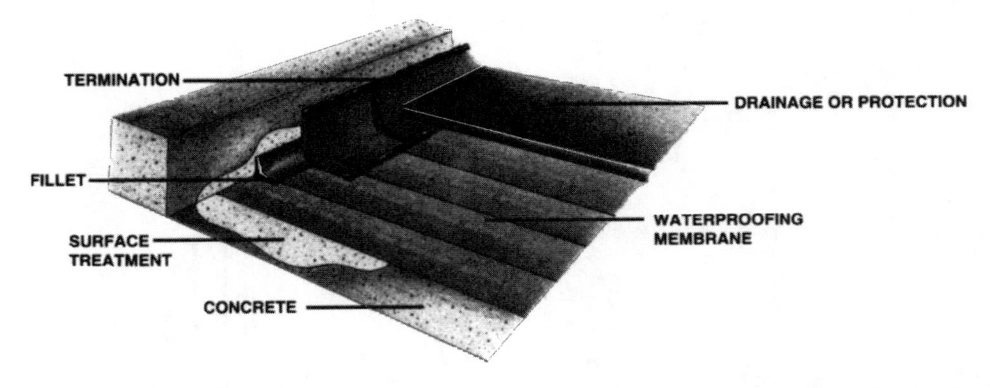

FIGURE 3.94 Sheet system interior waterproofing system. (*Courtesy of Grace Construction Products*)

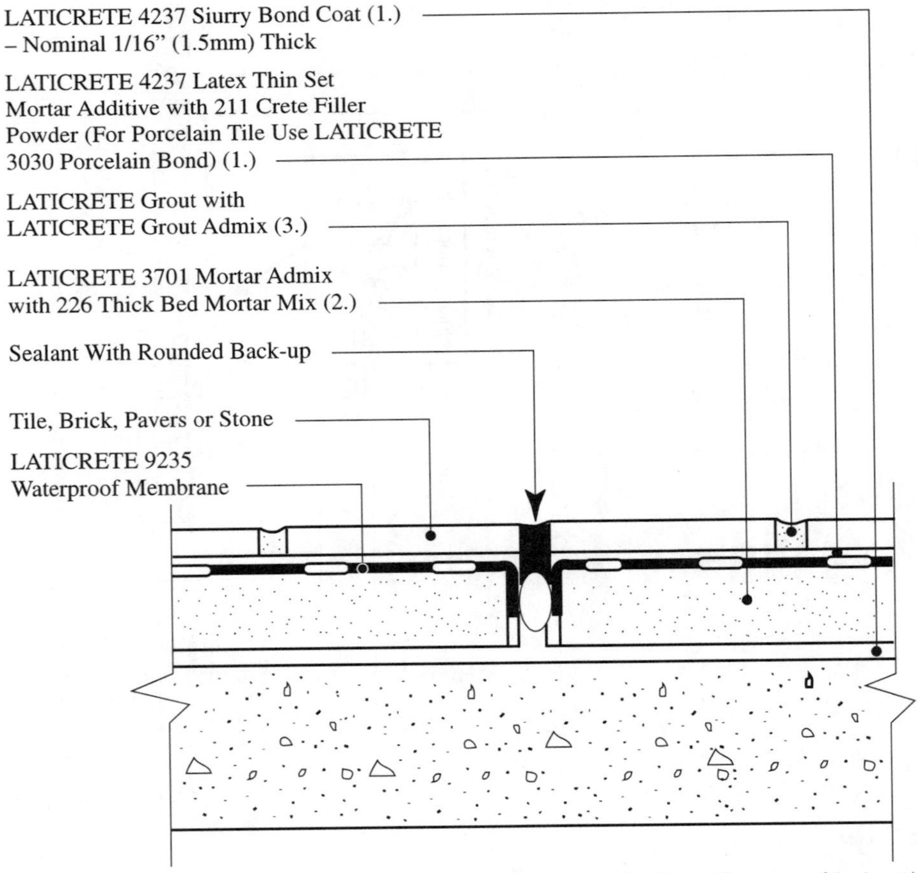

FIGURE 3.95 Interior expansion joint detailing for membrane applications. (*Courtesy of Laticrete*)

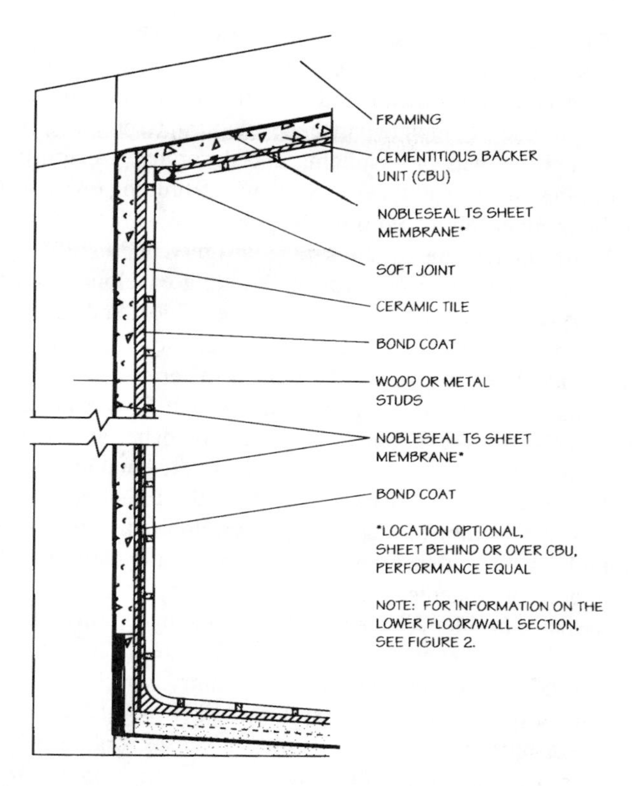

FIGURE 3.96 Interior steam room waterproofing detailing. (*Courtesy of Nobel Company*)

approximately one inch of water and letting it stand for 24 hours to ensure that the membrane holds water.

EXTERIOR INSULATED FINISH SYSTEMS

While an exterior insulated finish system (EIFS) is not exclusively a waterproofing product, it is sold as a cladding material that is waterproof. It is EIFS's checkered past involving water infiltration and related envelope damage that requires a section devoted to these systems.

EIFS systems were first commercially used in Germany as a means to add insulation to existing concrete buildings that were originally constructed without any insulation. The systems combined excellent insulating ratings and a decorative finish at minimal costs and needing little craft capability to install. They were not used or installed for their waterproofing capability. The EIFS products were introduced in the United States in the late 1960s and during the energy crisis of the 70s became a popular choice to add energy efficiency to building envelopes. (EFIS is also referred to as *synthetic stucco systems*, particularly for residential applications).

The original systems included an insulation board, base coat with or without fiber mesh reinforcing, and a decorative finish coat of polymer-modified coatings. Little emphasis was placed on termination and transition detailing, although designers expected the product to provide a complete barrier envelope waterproofing system. While the product's original emphasis was on insulating and decorative capability, it soon became a standalone cladding system in all types of structures. Building owners that loved the cost effectiveness of the product reinforced this acceptance.

Unfortunately, the systems were designed and used in applications that would perform poorly as a barrier system for the building envelope. Excluding the insulation thickness, these systems are applied at 1/8 to 1/4 inch thick, and this thin application created multiple problems related to termination and transitioning detailing. For example, early applications included areas where the sealant material was applied into joints a minimum of 1/4 inch thick when the coating was only 1/8 inch. This resulted in the sealant material attempting to adhere to the incompatible insulation material, eventually resulting in a loss of adhesion, failure of the joint, and resulting leakage and envelope damage.

Damage was further exasperated by the finish coatings not being permeable, so that once water or moisture entered the envelope, diverter systems were not included to exit water back out to the exterior. The trapped water then caused wood rot, and rusting of other envelope components.

Leakage problems became common and in certain areas of the country class-action lawsuits were filed against the manufacturers (e.g., New Hanover, North Carolina, case 96 CVS 0059) and in some areas of the country local building departments (e.g., Wilmington, North Carolina) banned the systems. EFIS problems became so prevalent in the home building industry that a consumer advocacy organization was formed (Stucco Home Owners Committee [SHOC]) to share information about the systems, in particular the problems related to water damage and envelope damage. Further complicating the situation was the fact that insurers began to exclude coverage for EIFS systems.

There is no reason not to assume that these problems are directly related to the 90%/1% and 99% principles presented in Chapter 1, which state that the majority of leakage is attributable to problems in transition and termination detailing, not to the material or system itself. To alleviate the problems, manufacturers, designers and contractors have responded with better detailing and installation practices to enable EIFS systems to perform properly and successfully as an envelope component with sufficient waterproofing capability. One only has to visit Las Vegas, Nevada and carefully view the multitude of envelope designs capable with EIFS systems to understand their continued popularity with designers and building owners.

While the thinness of application presents unique challenges in designing and installing proper termination and transition detailing, the cost effectiveness of the system for building owners offsets any negative connotations of the required intricate detailing. EIFS systems require the same adherence to the 90%/1% principle as any other envelope component or system.

The only major change to occur in EIFS systems in responding to these leakage problems is the increased usage of EIFS diverter systems as opposed to EIFS barrier systems. Acknowledging that some water or moisture is likely to enter the systems for whatever reason (including the 90%/1 % principle), provisions can be provided to drain this water back out to the exterior. In addition this drainage permits the release of entrapped moisture vapor, eliminating the possibility of wood rot and other damage to envelope structural components. These systems are referred to as *water-managed* or *drainable EIFS systems*. Some local building codes are now adopting the requirement to use these diverter systems and prohibiting the use of barrier systems.

EIFS waterproof installations

The intent of this section is not to provide general installation practices for EIFS systems, but to highlight the application techniques and proper detailing for terminations and transitions that provide a quality watertight finished product. Since each system is unique, consulting with EIFS manufacturers and reviewing their specifications and details before completing any installation is necessary to verify that these suggestions are the most appropriate to ensure watertight applications.

The waterproofing recommendations provided here are all based on the use of water-managed or diverter systems. No attempt is made to review the barrier systems, and caution should be applied whenever a barrier system is selected for installation. While the barrier systems may function properly in dry climates such as Nevada, areas with wet and humid conditions such as Florida should include the additional protection provided by a drainage system.

Terminations

All terminations of any EIFS installation should never have the system touching any horizontal envelope component such as roofs, sidewalks, and balconies. The finished material is not designed to function under standing water and therefore all terminations, particular those near wet areas such as sidewalks and roofs, must terminate sufficiently above the horizontal transition. Never should an EIFS system be used in a below-grade application.

Manufacturer requirements vary, but a minimum of 2 inches above the horizontal plane should be used as a guideline for termination detailing. Figure 3.97 shows a typical termination detailing at a roof parapet, with the manufacturer requiring an 8-inch minimum spacing from the roof surface. Note the cant strip used at the roof-to-wall transition to ensure that no standing water is permitted near the EIFS system.

At any termination the system must be backwrapped or protected by some means to eliminate any exposed edges of insulation that would permit moisture or water to enter the system and bypass the drainage protection. The backwrapping can be the reinforcing mesh and base coats, flashing, or some type of edge trim supplied by the manufacturer. In addition, the termination detailing requires a weep or drainage capability for completing the water diverter capability.

Figure 3.98 details the use of backwrapping with the mesh and base coat, with the flashing portion of the water diverter components completing the termination detail. Figure 3.99 details a termination onto an adjacent concrete substrate; the manufacturer requires only a ¾-inch distance off the horizontal plane but with an added self-adhering flashing membrane material behind the system completed with a sealant joint for additional protection. Note that the weep base system is on top and in front of these other termination protections. In this detail, the manufacturer does not require an edge wrap for the material adjacent to the weep flashing.

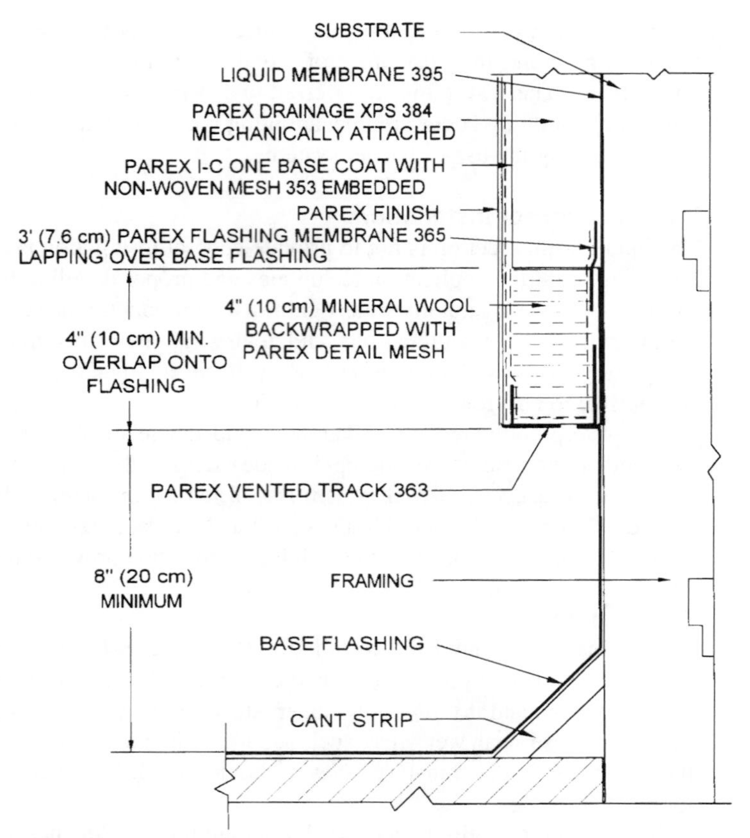

FIGURE 3.97 EIFS Termination detailing above roof. Note the cant detail to shed water away from the envelope. (*Courtesy of Parex*)

Drainage Systems

All EIFS diverter systems provide for drainage system capability that consists of a moisture barrier applied to the substrate, with associated flashing and weep mechanisms to redirect entering water back to the exterior. Even with these drainable systems, water and moisture can become trapped if the insulation board is adhered directly to the moisture barrier especially around fasteners and seams. It is recommended as additional protection that a drainable insulation board or some type of drainage mat be used in all applications to ensure that entering water is quickly and adequately drained away from the envelope.

Figure 3.100 details a typical drainage insulation board that is recommended for use with EIFS water-managed systems. Figure 3.101 details a complete schematic detail of an installation using a waterproof membrane over the substrate and grooved insulation board attached to the substrate at the high points to permit the proper drainage in the channels. The system allows entering water to be diverted to the exterior through the manufacturer's vented track at the edge termination. Figure 3.102 details a similar system using mechanical fasteners for the insulation.

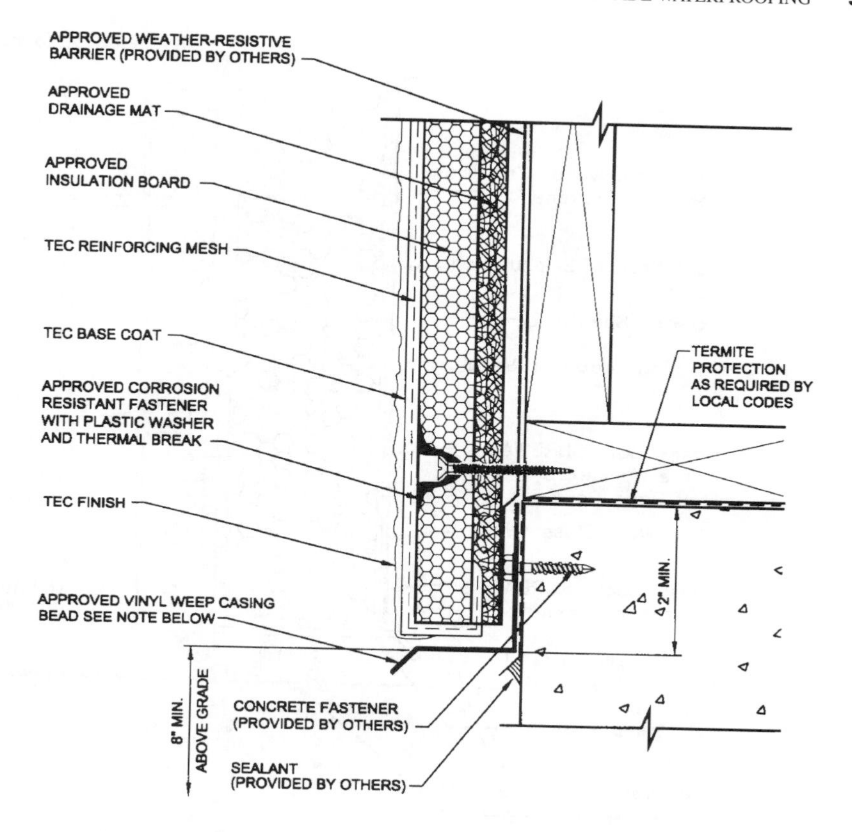

FIGURE 3.98 EIFS Termination detailing including backwrapping and flashing. (*Courtesy of TEC Specialty Products, Inc.*)

Figure 3.103 details a drainable system using flashing and weeps that also act as an expansion joint for the system. Figure 3.104 details the use of a drainage board applied behind the insulation board to effect the proper drainage capability.

Transitions

As with all waterproofing and envelope cladding systems, much of the leakage attributable to EIFS systems can be traced to the transition detailing. Transitions between EIFS systems and other envelope components must be carefully detailed and installed, due to the thinness of the actual EIFS base and finish coats that average $1/8$ to $1/4$ inch thick.

Figure 3.105 details the transition of EIFS system to masonry cladding. Note that the details include a vented track for the drainage system that incorporates flashing and sealant with weeps above the flashing that is used as a transition system. The flashing is also sealed underneath to the masonry for additional waterproofing protection. Figure 3.106 details the transition of EIFS to a coping cap. A moisture barrier is used beneath the cap flashing that overlaps the EIFS on both sides of the wall. Sealant is used to transition the flashing directly

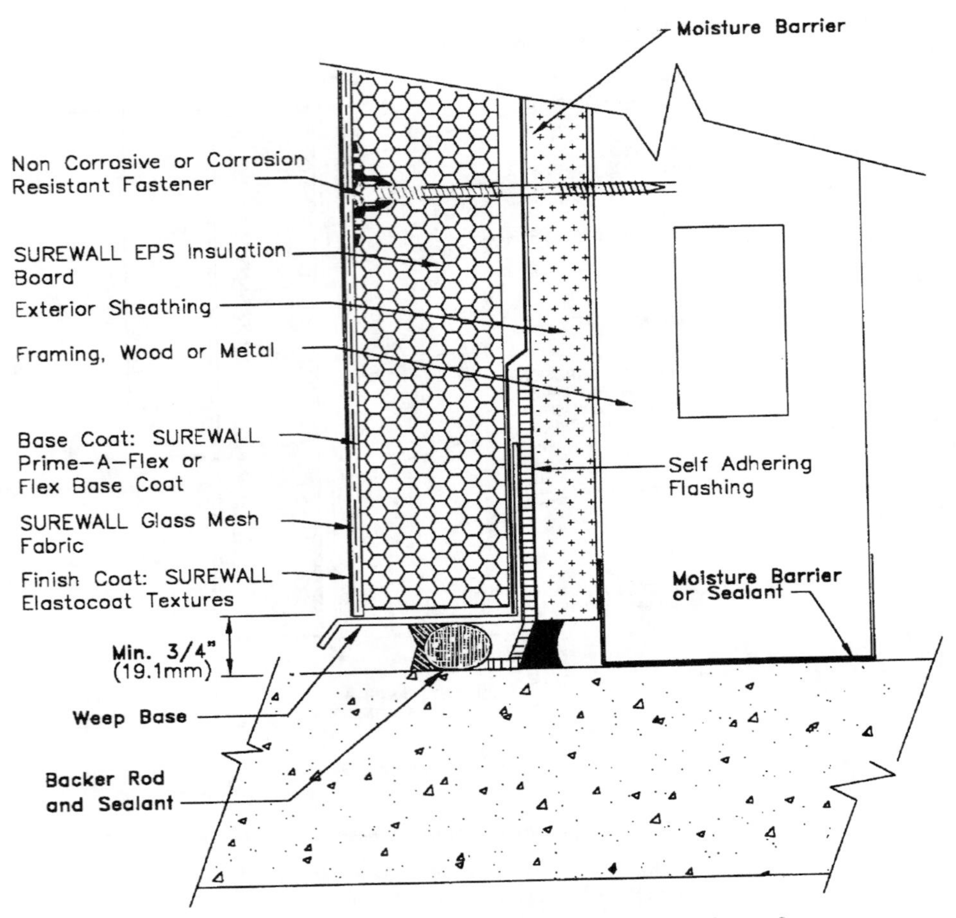

FIGURE 3.99 Termination to horizontal envelope component. (*Courtesy of Bonsal*)

to the EFIS system. Note the weep base at the roof side for the EIFS system and the additional sealant cant transition from the EIFS to the roof flashing.

Figure 3.107 details a vertical transition of EIFS to masonry facade components. Note that a double seal is used at the expansion joint that becomes the transition detail between the two cladding systems. The moisture or dampproofing barrier runs continuously behind both the masonry and EFIS system. Excellent detailing, but it is important to recognize that there would be an obvious problem with contractural and installation responsibility for the dampproofing. Would the mason or the EIFS applicator install the dampproofing shown running continuously? If the mason installs the product and leakage occurs behind the EIFS system, who would be held responsible, since the mason likely would not have supplied a warranty for the EIFS system? Such details, while providing excellent waterproofing protection, are often not completed in the field properly, resulting in leakage that could easily have been prevented. This topic of responsibility for transition detailing is presented in Chapter 8 for all envelope components.

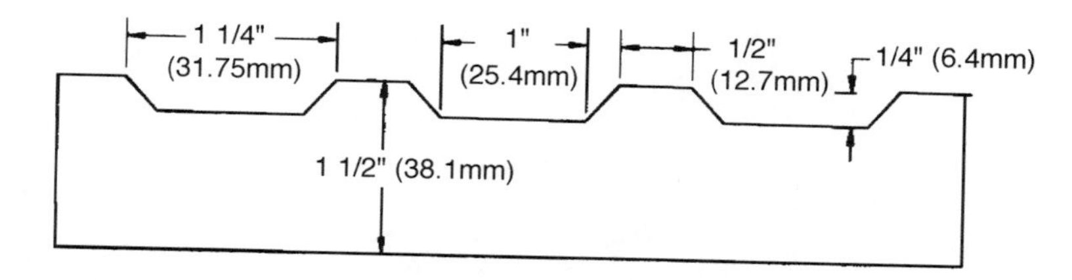

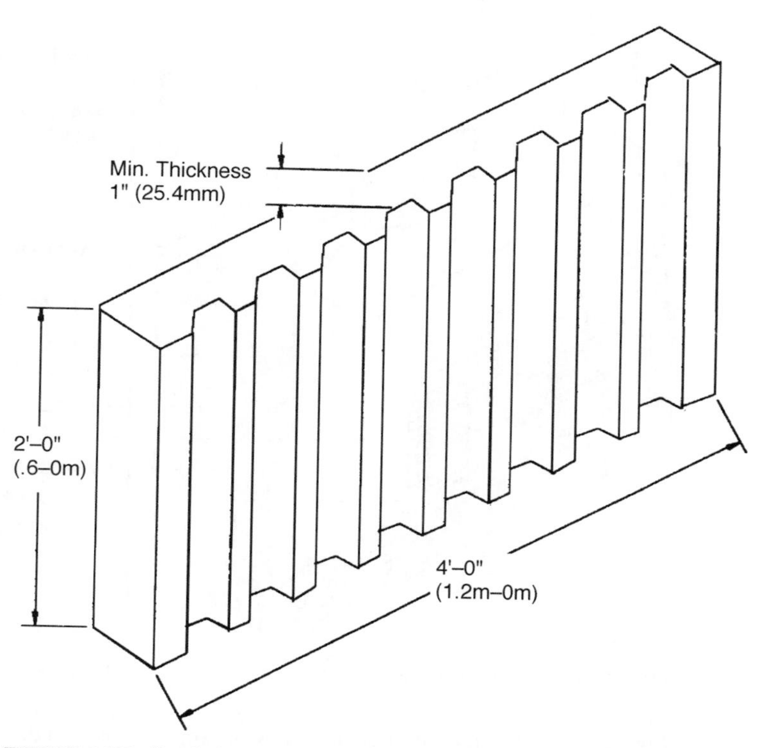

FIGURE 3.100 Insulation used for draining EIFS divertor systems. (*Courtesy of Bonsal*)

Figure 3.108 provides a common EIFS-to-window head transition detail. Note that the manufacturer is showing that the transitions systems (flashing and sealants) are to be provided by others. This could present problems in the field if the general contractor has not properly assigned this work to subcontractors that recognize the importance of the detailing to ensure a watertight application. Should the contractor assign the responsibility of the flashing to a sheet metal contractor, the sealant to the waterproofer, and leakage occurs at this window, four different subcontractors could be involved in this simple transition detail. This invariably leads to "finger-pointing" or passing the blame, while the building owner continues to suffer leaks and damage to the envelope. Again, refer to Chapter 8 for a review of methods including "following the barrier" to prevent such situations.

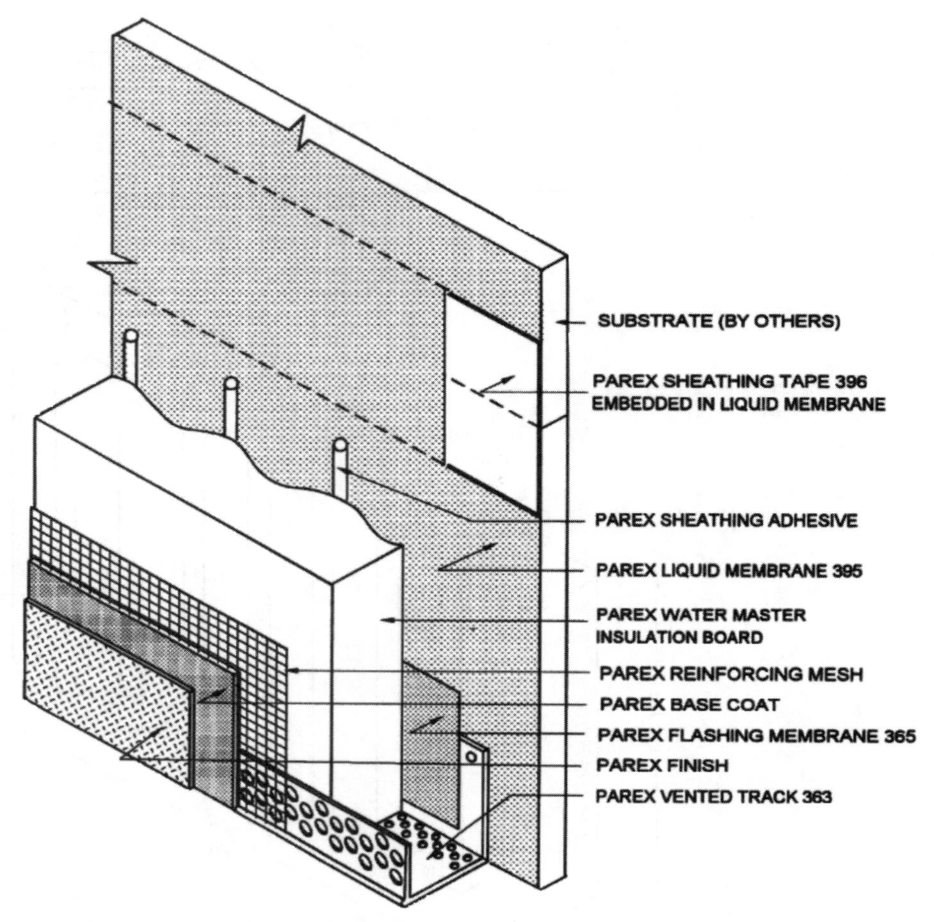

FIGURE 3.101 Schematic detail for drainage requirements of EIFS system. (*Courtesy of Parex*)

Figure 3.109 provides a schematic view of proper transition detailing at a punch window. Figure 3.110 details the transitioning at a typical mechanical penetration. Note that since the penetration is round, the manufacturer has not terminated the flashing and weep system above the duct. Water entering the EIFS cladding would travel around the duct and down to where the flashing weeps the water back to the exterior. This makes the membrane flashing applied around the duct critical to the watertightness of this detail, since water passing through the EIFS system can bypass the sealant material and enter into the building if the membrane flashing is not installed properly. Again an excellent waterproofing detail, but the actual field application would likely result in upholding the 90%/1% principle.

Sealants

Expansion and control joints in EIFS claddings can be difficult to detail and waterproof due to the thinness of the EIFS coatings. All EIFS manufacturers recommend that the sealant be applied or adhered to the base coat only and not the topcoat. The topcoat of most

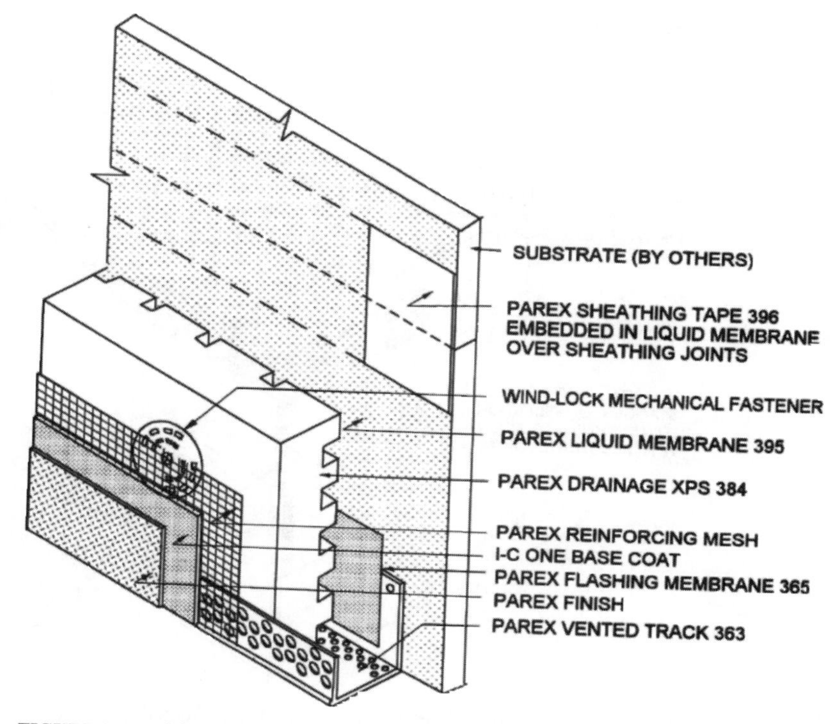

FIGURE 3.102 Mechanical-fastened EIFS drainage system. (*Courtesy of Parex*)

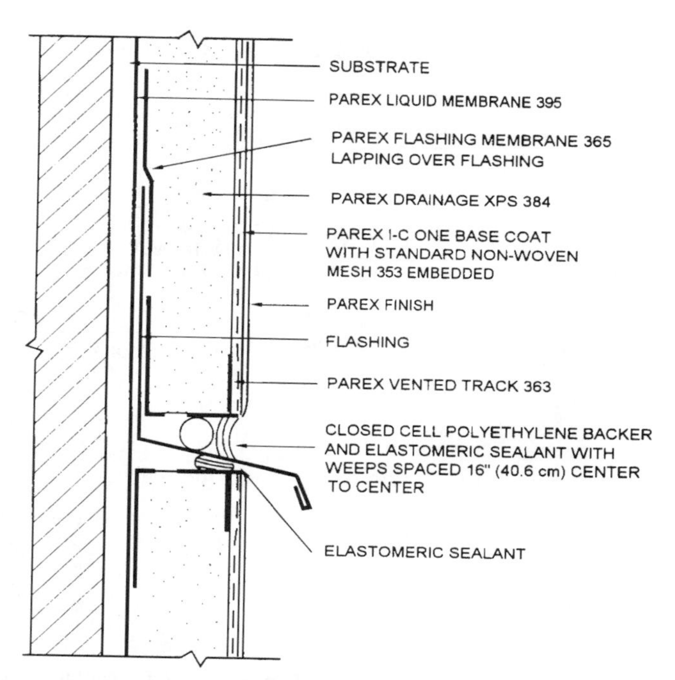

FIGURE 3.103 Divertor component, including flashing and weeps, of EIFS systems. (*Courtesy of Parex*)

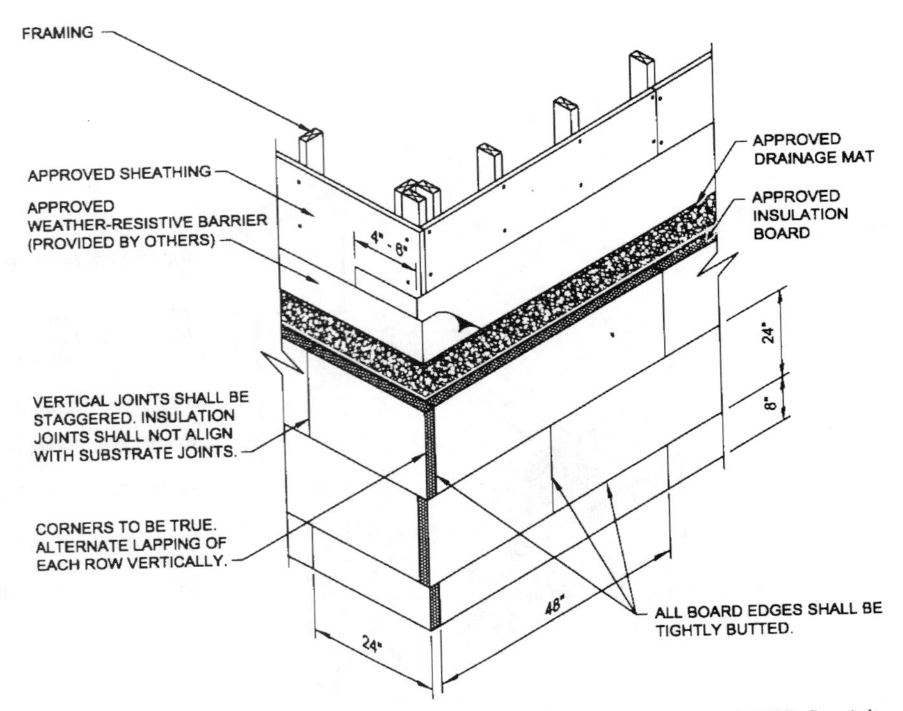

FIGURE 3.104 EIFS drainage board application detailing. (*Courtesy of TEC Specialty Products, Inc.*)

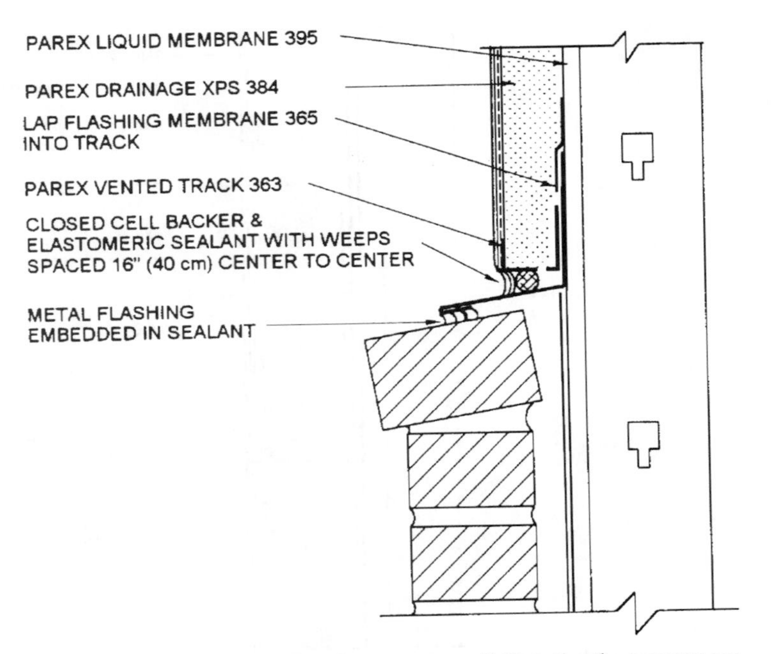

FIGURE 3.105 Transition of EIFS to masonry wall. Note that the masonry cap is sloped, to enhance drainage away from the envelope. (*Courtesy of Parex*)

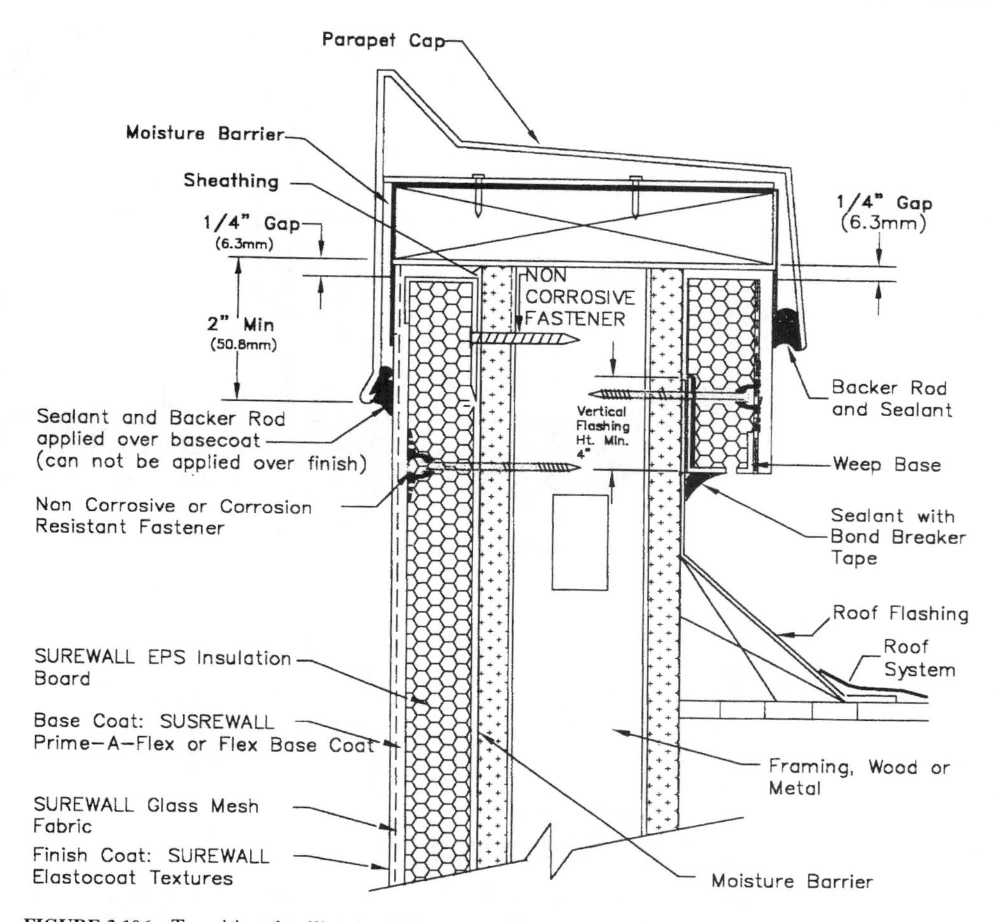

FIGURE 3.106 Transition detailing at parapet wall and coping for EIFS systems. (*Courtesy of Bonsal*)

EIFS systems is of a much softer material, to permit movement in thermal expansion and contraction. This means the sealant is being applied to a substrate that is only 1/8-inch thick in most applications. In high-movement joints, sealants with high adhesive strength could actually rip the base coat away from the substrate, resulting in leakage.

All EIFS manufacturers require that the base coat be wrapped entirely into the expansion joint to provide a sufficient substrate for sealant application. To keep the sealant from adhesion failure, manufacturers recommend that only low modulus sealants be used. The sealants recommended include one component silicone and two components urethane material (refer to Chapter 4 for additional information).

For the best detailing of expansion and control joints with EIFS systems, a dual sealant system is recommended as presented in Chapter 5. Figure 3.111 details the use of dual seal systems, an expanding foam material beneath a recommended sealant. Figure 3.112 details the use of a manufactured dual seal that is installed using field-applied silicone cants of sealant to the edges of the joint as discussed in Chapter 5.

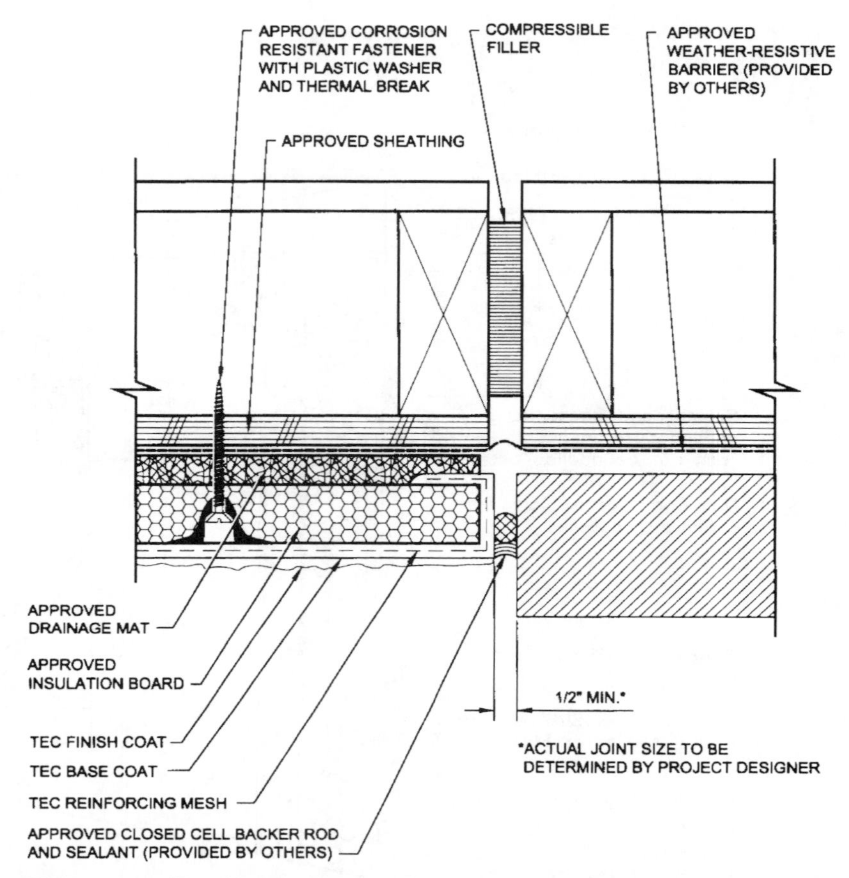

FIGURE 3.107 Transition detail from EIFS to masonry. Note the continuous dampproofing behind both systems. (*Courtesy of TEC Specialty Products, Inc.*)

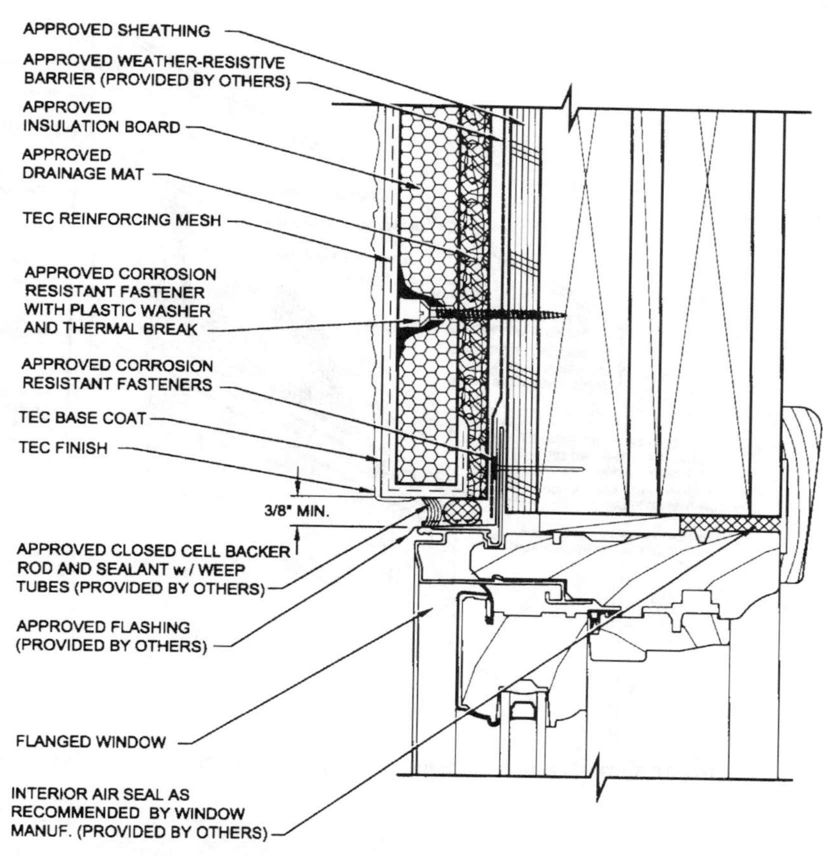

FIGURE 3.108 Transition detail from EIFS to punch window or curtain wall. Note the windows are properly recessed for added protection. (*Courtesy of TEC Specialty Products, Inc.*)

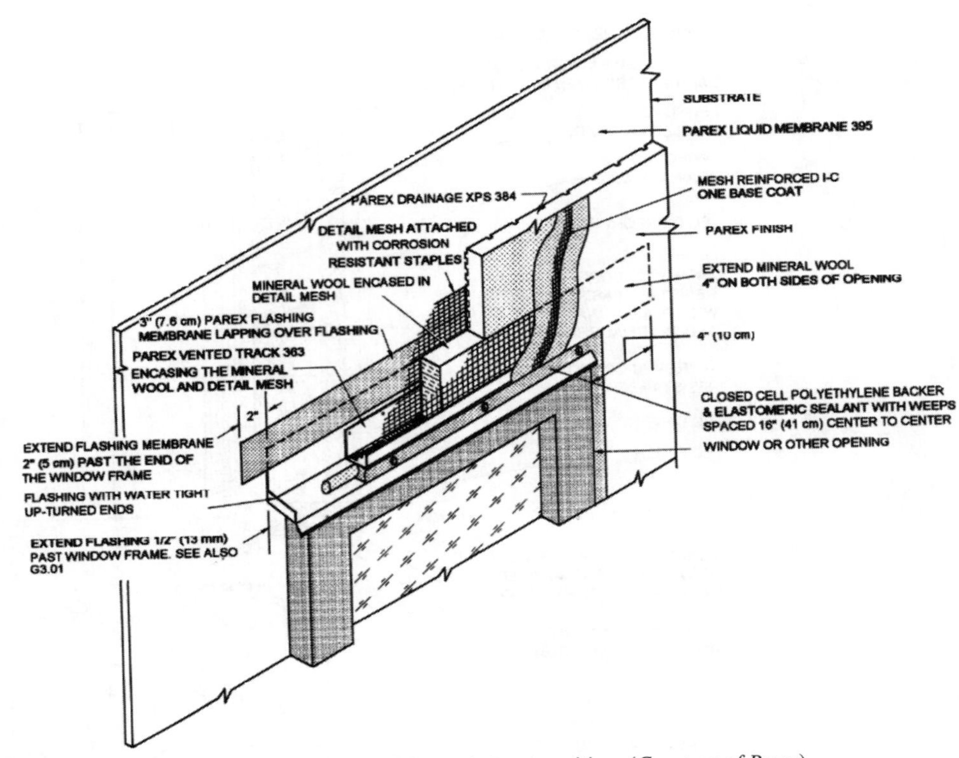

FIGURE 3.109 Schematic detailing of EIFS-to-window transition. (*Courtesy of Parex*)

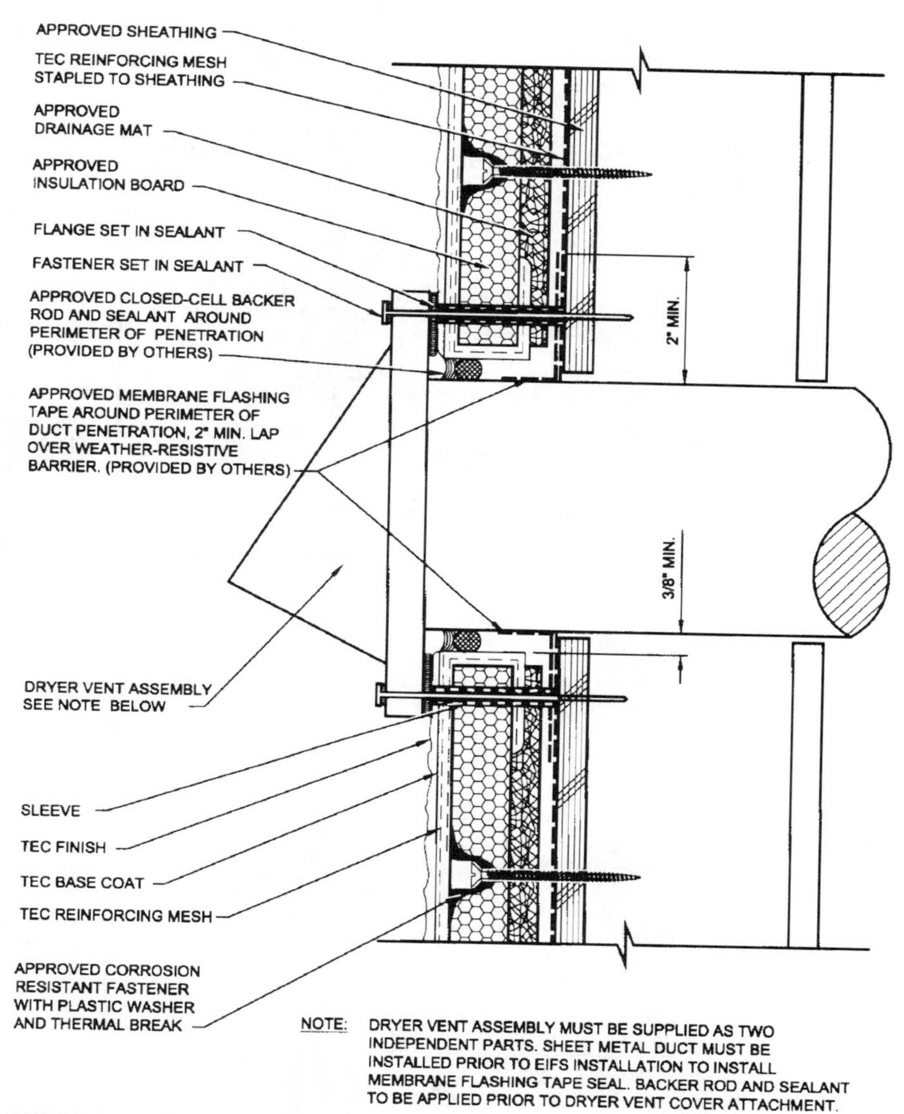

FIGURE 3.110 Typical penetration detailing for EIFS system. (*Courtesy of TEC Specialty Products, Inc.*)

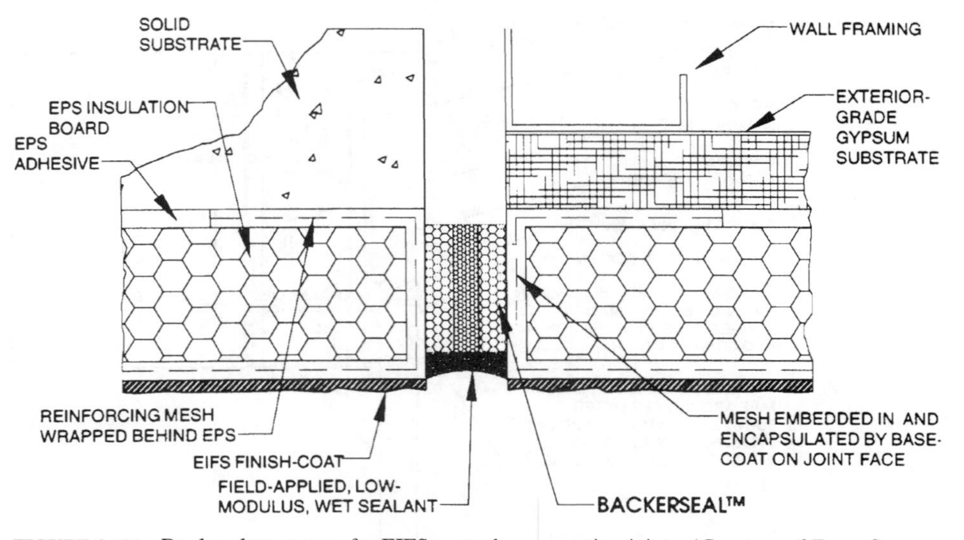

FIGURE 3.111 Dual sealant system for EIFS control or expansion joints. (*Courtesy of Emseal*)

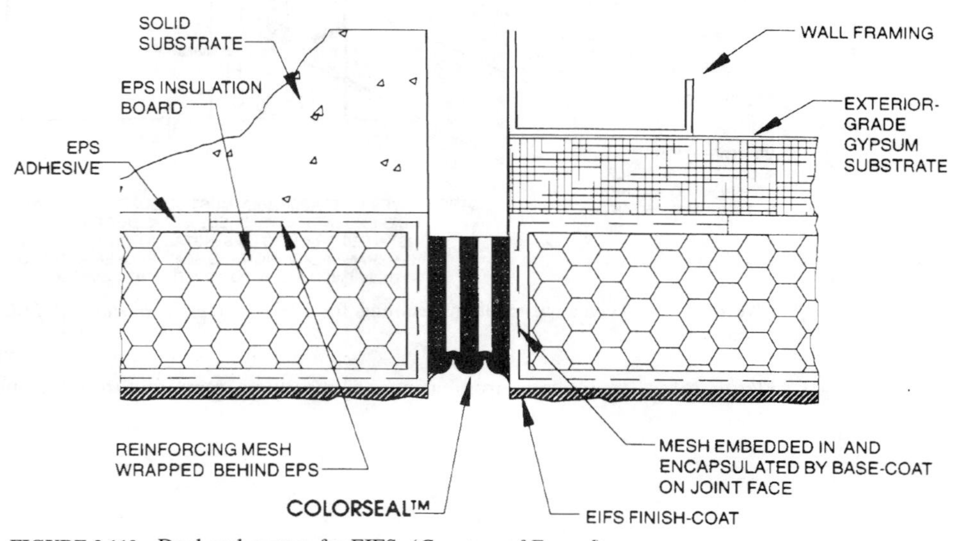

FIGURE 3.112 Dual seal system for EIFS. (*Courtesy of Emseal*)

CHAPTER 4
SEALANTS

INTRODUCTION

Sealants are not only the most widely used waterproofing materials, but also the most incorrectly used. Although sealants are a relatively minor cost item, they constitute a major function in a building's life cycle. Applied from below grade to roof areas, and used as components of complete waterproofing systems and for detailing junctures and terminations, sealants act as direct waterproofing barriers. As such, sealants are important in constructing successful watertight building envelopes. Sealants are also used to prevent air from infiltrating in or out of a building. Sealants thus have a dual weatherproofing role, with waterproofing as the primary role and environmental control as the secondary role.

Practically every building's exterior skin requires sealants for weathertightness. Junctures of dissimilar materials or joints installed to allow for structural or thermal movement require sealants to maintain envelope effectiveness. Below grade, sealants are used for sealing expansion joints, junctures, or terminations of waterproofing compounds and protrusions. Above-grade applications include sealing joints between changes in building facade materials, window and door perimeters, and expansion and control joints. Sealants are also used to detail numerous joints, including flashings and copings that act as termination or transition details.

Since sealants are a minor portion of overall construction scope, they receive a comparable amount of effort in their design and installation. Yet because they are a first-line defense against water infiltration, sealant failures can cause an unequal proportion of problems and resulting damage (Fig. 4.1).

SEALANT/CAULKING/GLAZING

Technologically, sealants have advanced dramatically from the white stuff in a tube, and a clear differential should be made between caulking, sealants, and glazing materials. *Caulking* refers to interior applications, to products manufactured for interior use and installed by paint contractors. These materials are usually painted after installation. Caulking is installed as a filler between dissimilar materials in an interior controlled environment not subject to thermal or structural movement. Therefore, caulking does not require the performance materials that exterior high-movement joints do. *Sealants* are exterior applications using high-performance materials (e.g., silicones), which are typically colored rather than painted and are applied by waterproofing contractors.

FIGURE 4.1 Failure of sealant. (*Courtesy of Coastal Construction Products*)

Sealants are also differentiated from *glazing materials*, which are considerably higher in tensile strength. This higher tensile strength produces lower elongation capabilities than sealants or caulking possess. Glazing materials are used in construction of window panels or curtain walls where higher strength is more important than movement capability. This strength (tensile) is referred to as the *modulus of elasticity*.

SEALANT INSTALLATION

Every building project involves numerous different types of building materials that require sealing to prevent either direct or indirect water infiltration. Every building component involved in a structure must be reviewed to ensure that the sealants specified are compatible with the substrate and to determine if a primer is necessary prior to sealant installation. Among the common envelope products occurring on most projects:

- Concrete, including precast with various finishes
- Masonry, and numerous formulations of mortar mixes
- Natural stone, such as granite, marble, and limestone
- Metals, including aluminum, copper, steel (galvanized, stainless, and structural)
- Wood or painted, stained or sealed

- Glass
- Plastics, including acrylics, waterproofing sheet goods, PVC, fiberglass
- Substrate finishes such as fluoropolymers on aluminum, paints, primers, and admixtures
- Roofing and waterproofing products

Each of these substrates must be reviewed and an appropriate sealant chosen to facilitate closure of the building envelope and effective watertightness. Contrary to popular belief, there is no one sealant for every purpose. Many substrates require primers before sealing; therefore, it is imperative that the sealant manufacturer be consulted to confirm that the substrate is acceptable for use with the specified product; and in addition, if needed,what type of primer is required to guarantee adhesion.

Actual selection of a sealant product is one of many important steps required to ensure successful installations, including:

- Joint design
- Material selection
- Substrate preparation
- Joint preparation and installation

Each step is critical for sealant systems to perform successfully for extended periods of time. Better sealant materials will perform for 5–10 years. But because of improper design, incorrect material choices, poor installation, or a combination of these factors, sealant joints rarely function within these time parameters.

JOINT DESIGN

Joint design failures are often attributable to improper spacing and sizing of joints. Joints are frequently arranged for aesthetic purposes, and actual calculations to determine optimum number and spacing of joints are overlooked. Precast and prefabricated panel joints are often determined by panel size of an individual precast manufacturer rather then by sound joint design.

Even if joint size requirements are actually determined, far too often panel erectors are primarily concerned with the installation of panels, with joints being used to absorb installation tolerance during erection. Often these joints end up varying greatly in width from those originally intended, with no procedures followed for maintaining proper tolerance in joint width.

Joint type

The first step in proper joint design is to determine areas and joint locations required within a building envelope. Areas of change in materials (e.g., from brick to concrete), of changes in plane, of differential movement potential (e.g., spandrel beams meeting columns), of protrusions (e.g., plumbing and ventilation equipment), and of thermal movement all must be studied to determine location requirements for joints.

Once this study is complete, design calculations must be completed to determine the type, spacing, and size of joints. Joint types include

- Expansion joints
- Control joints
- Isolation joints
- Detailing joints

Expansion joints. Expansion joints allow for movement in a structure or material that is caused by thermal expansion or contraction and other inducements such as wind loading or water absorption. Expansion joints are active dynamic movement joints that continue to move by expansion or contraction. See Fig. 4.2 for a typical masonry expansion joint.

Control joints. Control joints allow for expected cracking due to settling, curing, or separation in building materials after installation. Interior control joints, including concrete slab control joints, typically are nonmoving joints and are placed and sized for expected cracking or shrinkage only. Exterior control joints, such as brick paver joints, provide for settling as well as movement, the latter due to vehicular or pedestrian loading and expected thermal movement. These joints require more design work than interior joints, as they will become dynamic moving joints. Figure 4.3 details differences between expansion and control joints.

Masonry and mortar shrinkage after placement and curing requires that control joints be placed at appropriate locations. These joints allow for shrinkage and settlement to occur without affecting an envelope's performance. Control joint locations should include:

- Areas of change in wall height
- Junctures or transitions at columns or other structural construction

FIGURE 4.2 Expansion joint at a masonry envelope ready for sealant application.

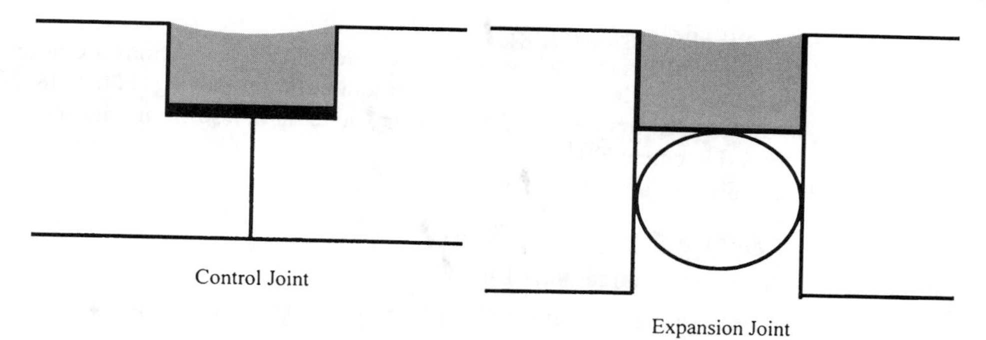

FIGURE 4.3 Difference between control and expansion joint detailing of substrate.

- Wall intersections or changes in plane, Fig. 4.4
- Areas where large openings occur in the envelope, such as above and below window openings

Isolation joints. Junctures at changes in materials require isolation joints to allow for any differential movement between two different materials. Window frame perimeters abutting facade materials and like areas of change in structural components (e.g., spandrel beam meeting brick facing material) require sealant joints because of differential movement.

Detailing joints. Detailing joints are required as a component or part of complete waterproofing systems. They are used to impart watertightness at building details such as pipe penetrations and changes in plane before the application of primary waterproofing materials.

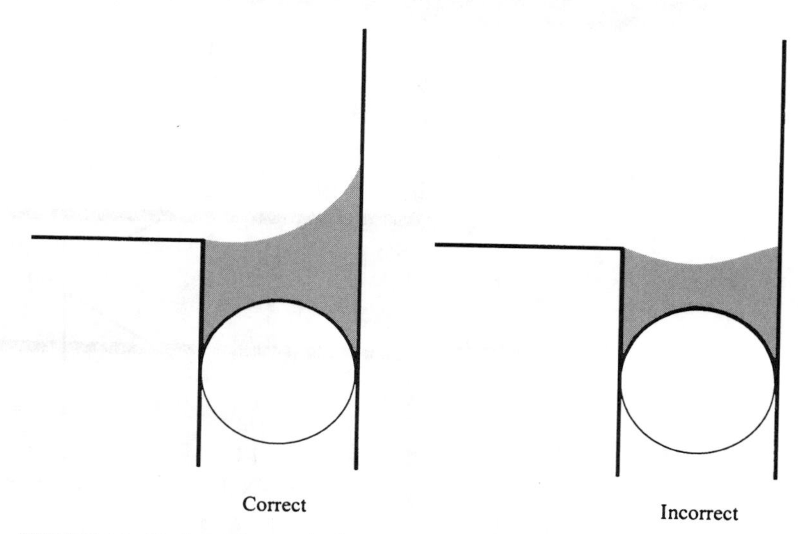

FIGURE 4.4 Horizontal to vertical transition detailing with sealant joints.

Spacing and sizing joints

Once the appropriate types of joints are determined, calculations are necessary to determine proper spacing and sizing of the required joint opening. Following are established guidelines used frequently in the industry; note that these are not meant to replace actual engineering calculations.

Basic rules for joint design, Fig. 4.5, include:

1. Joint size no smaller than 1/4 inch
2. Joint size no larger than 1 inch
3. Joint opening a minimum of four times anticipated movement 4-

Basic rules for sealant design, Fig. 4.6, include:

1. Material thickness no less than 1/4 inch
2. Joints up to 1/2 inch wide; depth of material is equal to width of material
3. Joints wider than 1/2 inch; depth of material is one-half the width
4. Maximum recommended width is 1 inch
5. Maximum depth is 1/2 inch

The *number and spacing of joints* are determined by:

1. Anticipated substrate movement, determined by coefficient of expansion
2. Length of substrate material span
3. Joint width

Design for movement is usually based on a temperature differential of 150°F. This is movement occurring in a selected material in a change of temperature of 150°F.

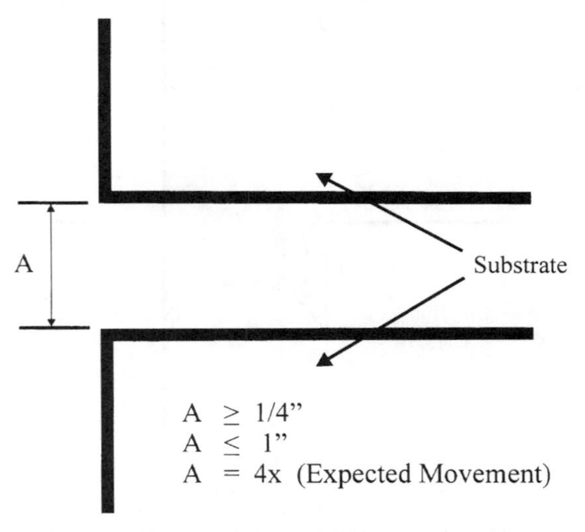

FIGURE 4.5 Design of substrate joint.

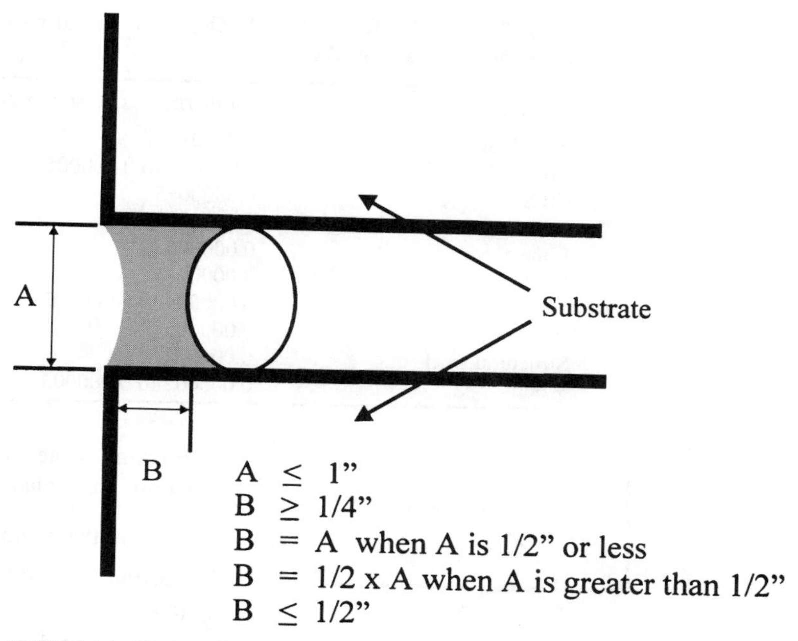

FIGURE 4.6 Design of sealant joint.

Coefficients of thermal expansion are usually expressed as inch per inch per degree Fahrenheit. To determine expected movement and resulting joint size, the coefficient of linear expansion is multiplied by temperature range, span length of material, and appropriate safety factor (usually at least a factor of 4).

The following is a typical calculation for joint design. As an example using 10-ft concrete precast panels with a coefficient of 0.000007 in/in/°F, the following would be the recommended joint design width:

$$150°\text{F} \times 0.000007 \text{ in/in/°F} \times 10 \text{ ft} \times 12 \text{ in/ft} \times 4 \text{ (safety factor)} = 0.504 = \tfrac{1}{2} \text{ in}$$

In determining the joint size necessary for moving joints located at different material intersections, materials with highest coefficients of expansion are used in calculations. However, if a material with a lower movement coefficient is spanning a greater width, these data may present a larger joint size. Therefore, it is necessary to calculate all possible combinations to determine the largest joint size necessary. Table 4.1 summarizes coefficients of thermal expansion for several common building materials.

BACKING SYSTEMS

No sealant joint can be properly installed without an appropriately sized backing system installed first. Backing materials are as important to successfully installing a joint as the sealant material itself. Unfortunately, backing systems are misused as much as sealants.

TABLE 4.1 Coefficients of Thermal Expansion for Common Envelope Materials

Material	Coefficient of thermal expansion, in/in/°F
Aluminum	0.000013
Concrete	0.000008 to 0.000005
EIFS	0.0000075
Granite	0.000005
Limestone	0.000005
Marble	0.000007
Masonry	0.000004 to 0.000003
Plate glass	0.000005
Structural steel	0.000007
Wood	0.000002 to 0.000003

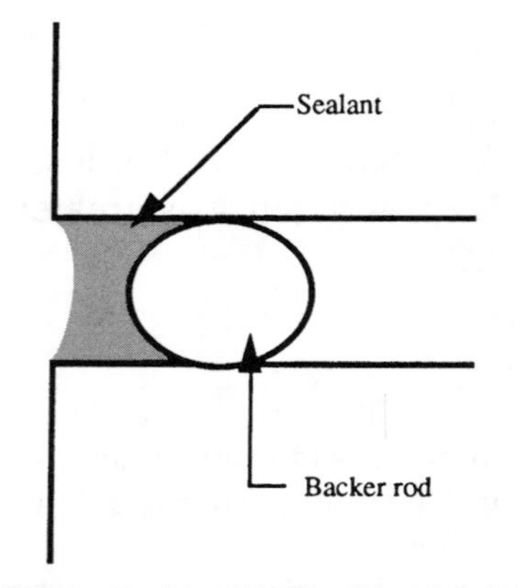

FIGURE 4.7 Backing detailing in sealant joints.

FIGURE 4.8 Backer Rod material is supplied in rolls of various sizes. (*Courtesy of Nomaco*)

Backing systems provide three critical functions for sealant joints:

1. Control the depth of sealant application
2. Prevent three-sided adhesion
3. Form the hourglass shape necessary for proper joint design

It is imperative that sealant materials be allowed to expand to their maximum capability without exerting unnecessary stress at the adhered substrate area. Thick beads of sealant are more difficult to elongate (a thick rubber band is harder to stretch than a thin band), which places more stress on the sealant. If this stress exceeds a sealant's bonding capability, adhesive failure will occur. Stress that exceeds a sealant's cohesive capability results in sealant tear failure.

To prevent failure, a backing material is inserted into joints to provide a large adhered contact area, at only two sides of the joint, with a thin bead of sealant. This is shown in Fig. 4.7. This backing material, or backer rod (Fig. 4.8) as it is commonly referred to, is of major importance in joint design and installation. Besides ensuring proper joint design, the backer rod allows applicators to monitor proper depth of material installation. Figure 4.9 shows failed material due to excessive thickness of material applied in a joint. It also provides a surface against which uncured sealant material can be tooled to force it against both sides of the joint for proper installation.

Sealants do not adhere to the backer rod, only to the joint sides. Three-sided adhesion (two sides and bottom of joint) places too much stress on sealant material in movement (elongation), causing tears that result in cohesive failure. See Fig. 4.10.

Backer-rod is round extruded foam that is installed into joints by using a much-larger-size rod diameter than the joint width. After being fitted into the joint, the rod expands tight against the sides of the joint to permit the application of sealant over it. To operate in this manner, the rod material used should be a minimum 25–50 percent larger than the joint width.

In joints not sufficiently deep for installing a backer rod, a bond breaker tape should be used, Fig. 4.11. For cant or fillet-shaped joints, either tape is used or some manufacturers now produce backer rod that comes in half or quarter-round shapes to facilitate the proper installation of sealant materials in these types of joints, Fig. 4.12 and Table 4.2. Half-round backer rod should be 20 to 40 percent larger than the joint opening width. Quarter-round material is used on cove joints, see the installation section and detail in Fig. 4.53, on whose horizontal substrate meets a vertical substrate.

There are four basic types of backing systems for sealant application:

- Closed-cell backer rod
- Open-cell backer rod
- Dual-cell backer rod
- Backer tapes

CLOSED-CELL BACKER ROD

Closed-cell backer rod is a cylindrical closed cell polyethylene foam. It is extruded to eliminate any open cell structure that can permit moisture or air transmission between the cell structures. Closed cell backer-rod is recommended for joints susceptible to the presence of moisture prior to joint sealing such as horizontal joints.

If moisture accumulates in the backer rod (possible with open-cell materials), it will prevent the sealant from curing, as moisture remulsifies sealants. Closed-cell rod is not susceptible to moisture, but it can not be used with air-cured material since it prevents curing the unexposed backside of the sealant material.

Closed-cell backer rod materials can cause sealants to improperly cure and "out-gas" if the closed-cell structure is inadvertently damaged during installation. Should the baker-rod be cut improperly or otherwise damaged and sealant then applied over it, the damaged cell release entrapped air into the uncured sealant. This causes blistering in the sealant during the curing process, Fig. 4.13. Whenever closed-cell material is used, applicators should never use sharp applicator tool that can puncture the cells and cause this out-gassing problem.

Nongassing formulations of closed-cell backer rod are available, manufactured from polyolefin material. Such formulations claim to prevent damage to sealants if accidentally punctured during installation.

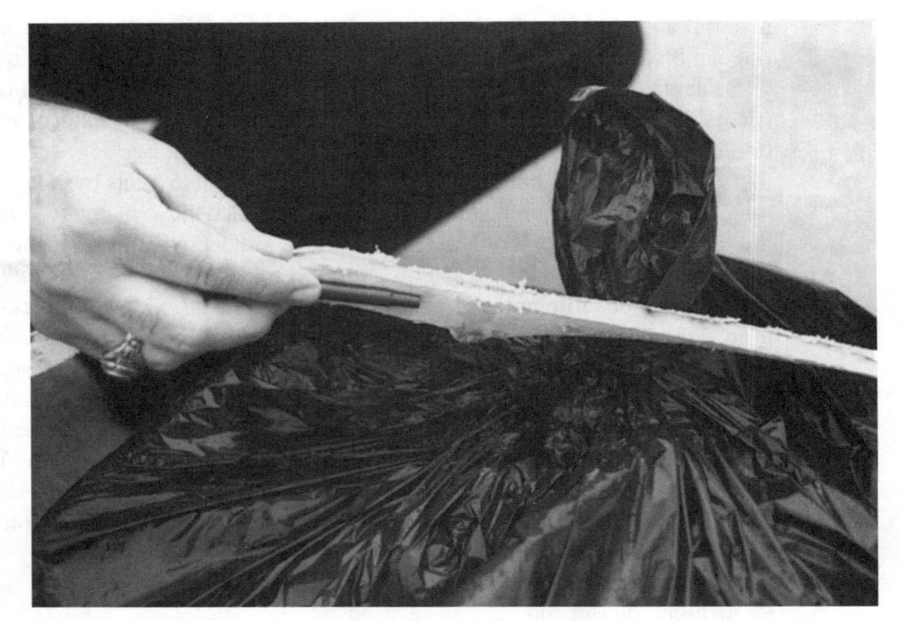

FIGURE 4.9 Excessive depth of sealant causes failure. (*Courtesy of Coastal Construction Products)*

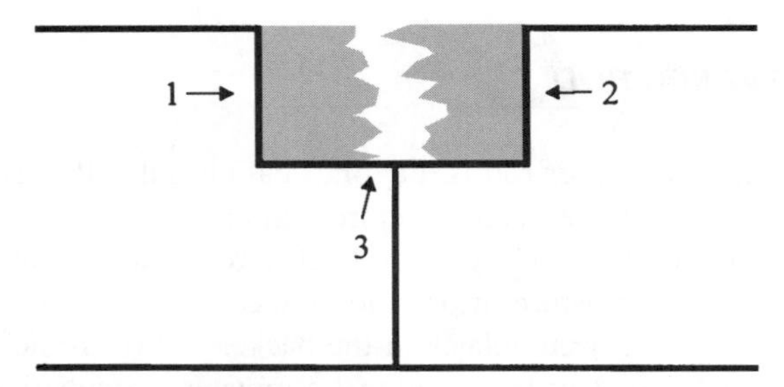

FIGURE 4.10 Three-sided adhesion results in cohesive failure.

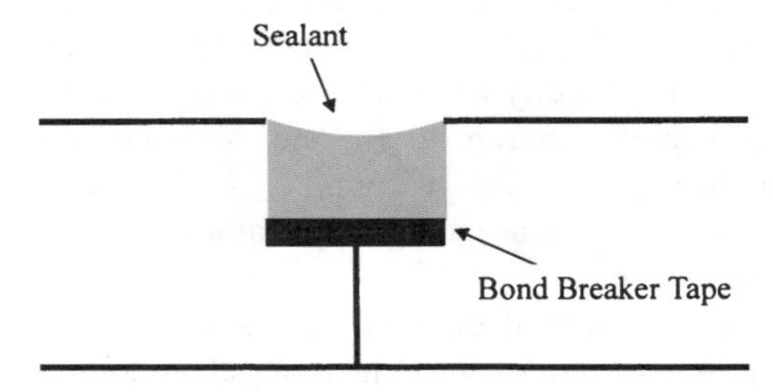

FIGURE 4.11 Use of bond breaker tape prevents three-sided adhesion.

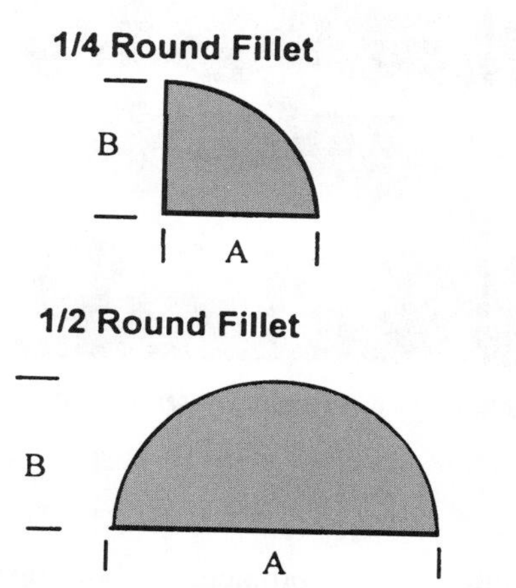

FIGURE 4.12 Backer rod is manufactured in fillet shapes for cant type applications. (*Courtesy of Nomaco*)

OPEN-CELL BACKER ROD

Open-cell backer rod is manufactured with polyethylene or polyurethane foam, but its cell structure remains open to permit vapor or air transmission. Open-cell rod can easily be compressed (like foam rubber), making it readily differentiated from closed-cell material that is not so compressible.

Open cell is specifically designed for sealant materials that are moisture- or air-cured products and not recommended for use with urethane materials or moisture-sensitive sealants. Open cell is generally used with low-modulus, slow-curing, high-performance sealant materials. Moisture curing is only applicable with airborne moisture and not with visible standing moisture that deteriorates sealant material. The open-cell material promotes curing on both sides of the joint and should be used whenever double seal joints are required to facilitate proper curing of all joint faces.

Open-cell rod is not generally used on horizontal joints, due to the probability of moisture accumulating in the joint before and immediately after sealant application that can cause the material not to cure properly. Open cell was developed to eliminate the problems associated with gassing of the joint as described above. However, the resulting open-cell structure permits moisture entrapment that can equally destroy sealant materials. Open-cell materials are typically not compatible with EIFS systems; when installing EIFS systems, the manufacturer should provide specific recommendations for sealants and the required backing materials.

Table 4.2 Typical Sizes of 1/4 and 1/2 Round Backer Rod Available. (*Courtesy of Nomaco*)

HBR®				1/4 AND 1/2 ROUND AVAILABLE			
SIZES							
1/4 ROUND				**1/2 ROUND**			
A Dimension in.—mm		B Dimension in.—mm		A Dimension in.—mm		B Dimension in.—mm	
1/4	6	1/4	6	5/8	16	5/16	8
3/8	10	3/8	10	3/4	19	3/8	10
1/2	13	1/2	13	1	25	1/2	13
3/4	19	3/4	19	1 1/4	32	5/8	16
1	25	1	25	1 1/2	38	3/4	19
1 1/4	32	1 1/4	32	2	51	1	25
1 1/2	51	1 1/2	51	2 1/2	38	1 1/4	32
2	51	2	51	3	51	1 1/2	38

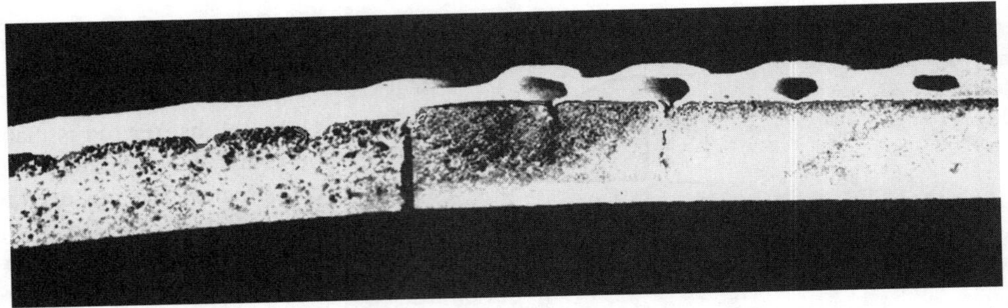

FIGURE 4.13 Outgassing of closed cell rod can cause failure of sealant material. *(Courtesy of Nomaco)*

DUAL-CELL BACKER ROD

Dual-cell rod was developed to eliminate the problems of both open- (moisture entrapment) and closed-cell rod materials (gassing). It is manufactured from extruded polyolefin foam that contains a combination of both open- and closed-cell structures (bi-cellular).

The dual-cell materials prohibit the absorption of too much moisture that can damage the sealant, but in sufficient quantities to facilitate the proper air or moisture curing of the sealant. At the same time, the dual-cell structure does not permit out-gassing when ruptured.

The dual cell is ideal for both vertical and horizontal joints and all types of sealant materials. It assists in making the sealant application on a project with various types of materials and joint design "idiot-proof" by being applicable in all situations. This eliminates the possibility of a mechanic choosing a closed-cell rod when an open-cell is required.

BACKING TAPE

Many construction joints are constructed that do not permit sufficient size or shape for the use of round backer rod materials (Figure 4.14). In these instances, a backing tape is used. The tape must act as a bond breaker, not permitting the adherence of the sealant material to the tape itself as well as the backside of the joints.

Backing tape is a specially produced tape that does not facilitate adherence to sealants. It is not similar to masking tape or duct tape, which are not acceptable products to use as backing tape. Backing tape is typically of polyethylene composition, 12–20 mil thick and provided in several widths, although cutting of the tape to the exact width of the joint is required.

It is extremely important that the tape not be so narrow that the sealant is allowed to adhere on the backside of the joint (Fig. 4.15). Tape should also not be so wide that it turns up joint sides, preventing proper adhesion (Fig. 4.16). A proper fit can be accomplished by cutting installed tape that is slightly larger than the joint's width along the joint edges with a razor knife and removing excess tape.

JOINT DETAILING

The relatively insignificant cost of sealants compared to the overall cost of a specific building project should facilitate the use of superior joint detailing to eliminate the common

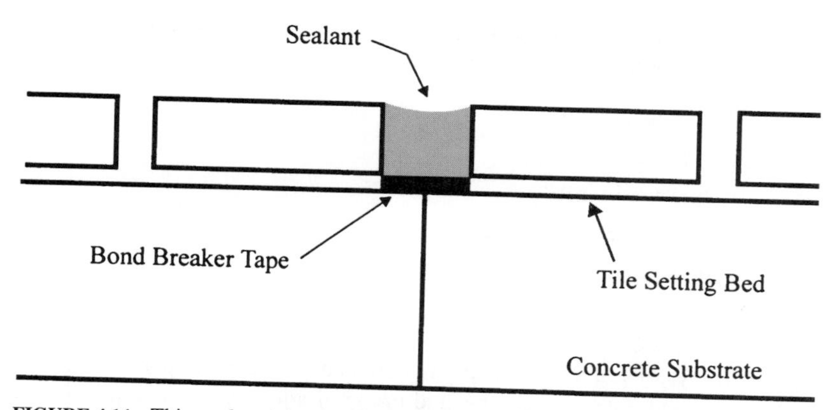

FIGURE 4.14 Thin sealant joint details require bond breaker tape for proper installation.

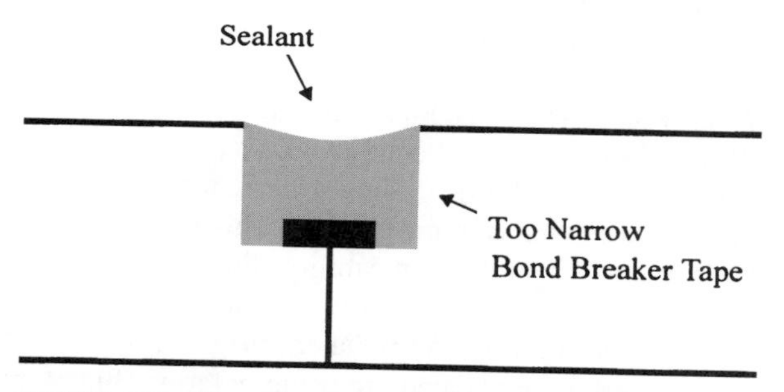

FIGURE 4.15 If bond breaker tape is not as wide as joint, three-sided adhesion will occur.

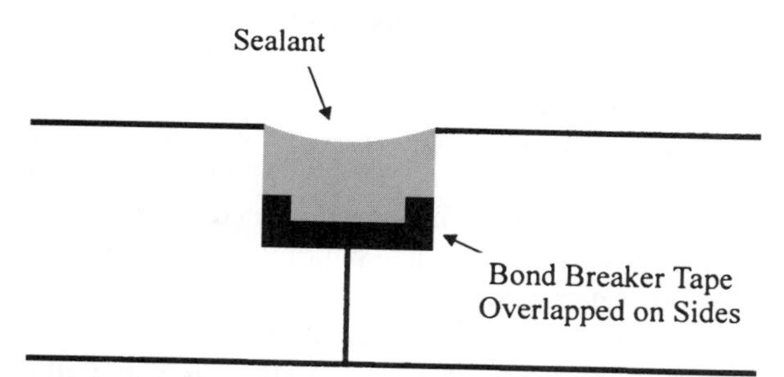

FIGURE 4.16 If bond breaker tape overlaps sides of joints, improper adhesion will occur.

problems associated with joint sealing. All too often, designers depend on single joint application to act as a barrier system against water infiltration. While sealant products have the necessary performance capabilities to act successfully in a special joint design, the conditions encountered in field applications often prevent sealants from performing as intended.

In most cases, considering the relative costs involved and the protection provided, joints should be detailed to receive "belt and suspenders" protection through some form of dual sealing or protection against water infiltration. This can be accomplished by designing a dual-seal joint system or providing diversion protection in the substrate in addition to the sealant joint detail. Double protection joint detailing includes

- Double sealing
- Secondary seals
- Binary systems
- Joint protection systems
- Substrate diversions

Double Sealing

Many construction cladding materials used today are permeable to water. Concrete, precast concrete, masonry blocks, and brick all allow water or vapor to enter directly through the building cladding and bypass sealant joints. In addition water enters through substrate cracks, defective mortar joints, and other envelope voids. In many instances, considering that field construction is not a perfect science, sealant joints may simulate two sponges sealed together.

When water bypasses a joint through a substrate it travels transversely on a path of least resistance. Water then collects at backs of substrate breaks or joints (usually where a sealant joint is installed), finding a path into the interior drawn by the difference in air pressure between interior and exterior. This leakage often appears as joint leakage when in fact it is due to substrate permeability (Fig. 4.17).

Therefore, it is often prudent to double-seal exterior joints. The secondary joint effectively seals interior areas from water intrusion, bypassing initial sealant joints. If accessible, second joints are sealed from the exterior (Fig. 4.18), but they can be sealed from inside the structure (Fig. 4.19). In both cases, joint design should include allowances for drainage of moisture that passes the first joint, back to the exterior, by installing flashing and weeps. Double-seal designs should not include materials that are sensitive to negative moisture drive, which is present in these applications.

Double sealing has several advantages beyond those derived from waterproofing. This joint design stabilizes air pressure in the space between sealant beads, thus eliminating positive vapor transmission into a building by air pressure. The interior bead also stops vapor inside a building from entering cladding where it may condense and cause damage, such as spalling or corrosion, to building components. This double-sealant installation also serves as an energy conservation method by effectively eliminating uneven air pressures that cause airflow into or out of a structure.

Design of inner sealant beads is controlled by design of exterior joints. Since both sides of a joint movement will be equal, the same material should be used on both. Using sealants with low movement capability on interior sides leads to ineffectiveness in preventing water and air transmissions.

Secondary Seals

When envelope components are involved that are not permeable, such as glass or metal, sealant joints can be designed using secondary seal systems to act as a backup barrier to the primary joint sealant material. The secondary seal should be designed to act as a pri-

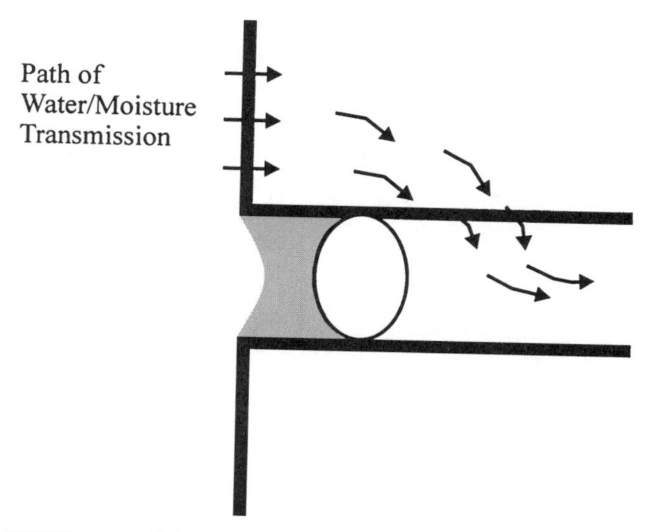

FIGURE 4.17 Water infiltration through pervious surfaces can bypass primary joint seals.

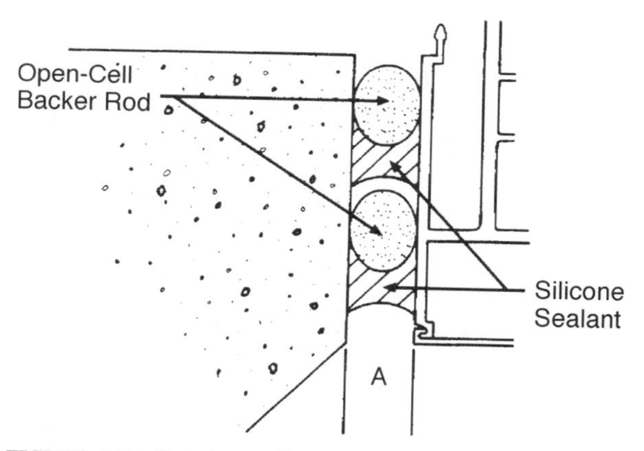

FIGURE 4.18 Exterior sealing of double joint. *(Courtesy of Dow Corning)*

mary barrier itself, preventing water infiltration directly through the joint if the sealant material fails in any manner.

Unlike double sealing, the secondary seal is adjacent to the sealant material and often is used in lieu of backer materials (Fig. 4.20). Figure 4.21 details the difference between secondary seal and doubled-seal joints. The secondary seal should be manufactured to prevent adhesion to the sealant (Fig. 4.22), eliminating three-sided adhesion when the backer material is deleted. Secondary seals can also be used beneath the backer materials, as shown in Fig. 4.23.

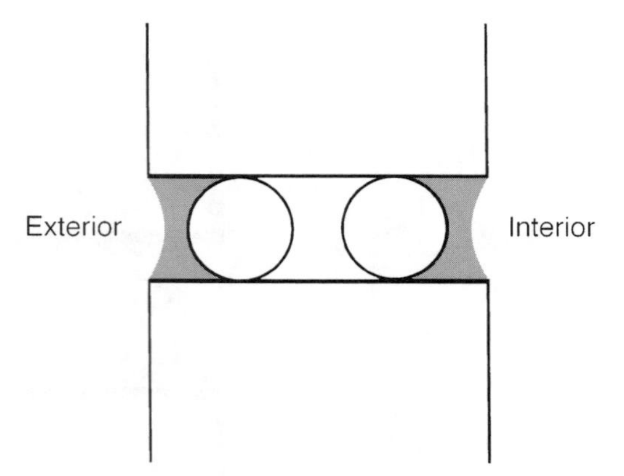

FIGURE 4.19 Interior sealing of double joint.

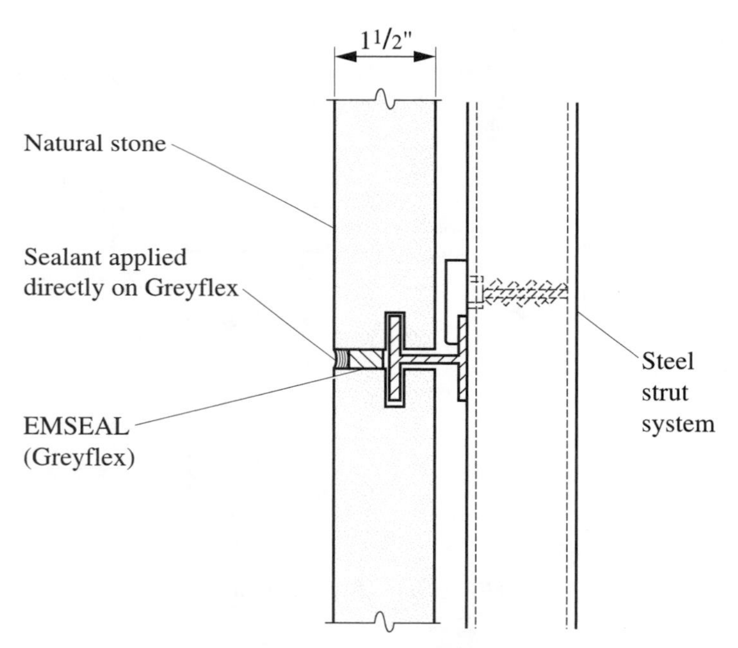

FIGURE 4.20 Typical secondary joint sealant design. (*Courtesy of Emseal)*

The secondary seal is often a precompressed foam sealant, as presented later in this chapter. The secondary sealant should be installed as required by the manufacturer as a primary seal, including proper cure time before the backer rod and sealant is applied. This prevents the secondary seal from not functioning as designed and preventing loss of the dual protection.

Secondary seals are not intended to prevent the problems associated with pervious substrates and water bypassing the primary sealant, as discussed in the previous section. However, the precompressed foam can be used as a double seal in the rear portion of the

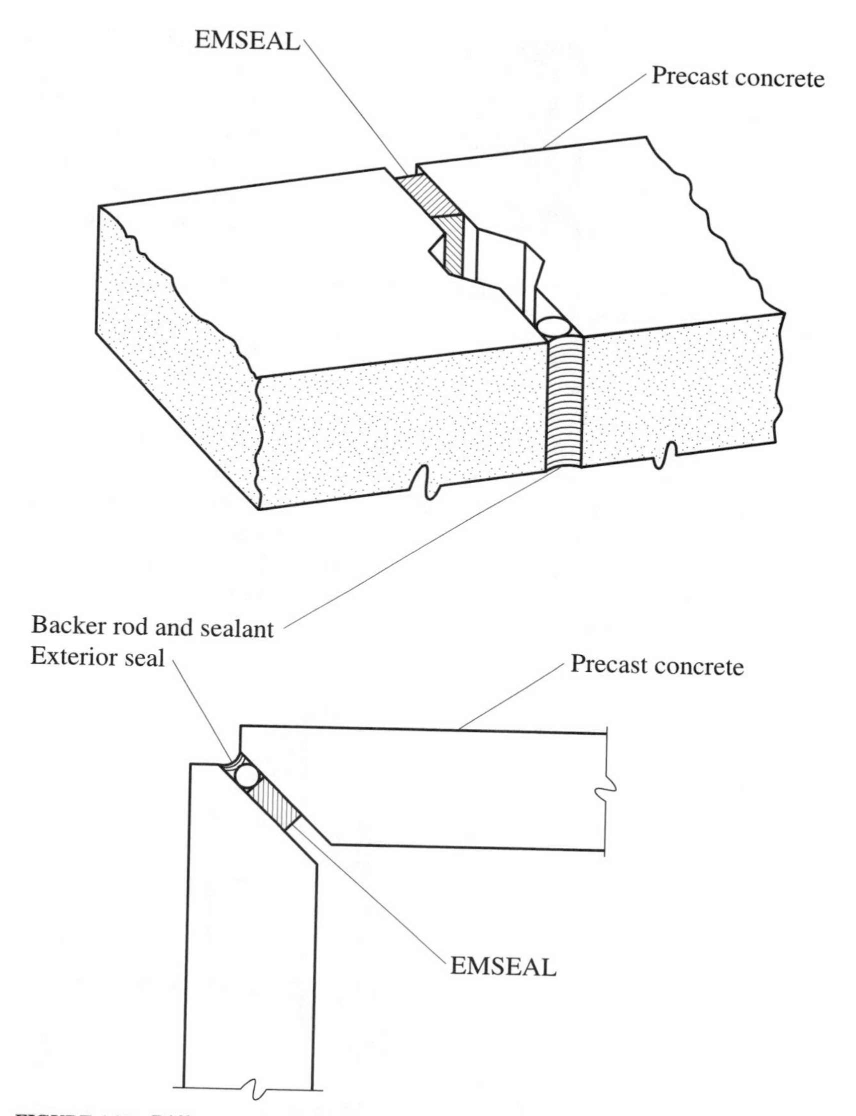

FIGURE 4.21 Differences between double sealing and secondary sealing. (*Courtesy of Emseal)*

joint as well as being part of a secondary seal at the primary barrier sealant. The secondary seal systems should be used whenever an unusual condition exists in the envelope design that requires a belt-and-suspender system for the primary barrier.

Binary Seals

Manufacturers of joint seal products are now combining two (binary) previous stand-alone products into one system, to act as dual primary barrier systems referred to as

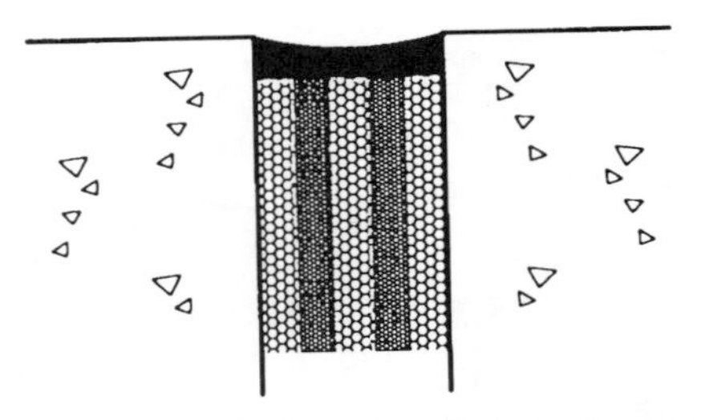

FIGURE 4.22 Primary seal applied directly to secondary seal. (*Courtesy of Emseal*)

binary systems. The field-applied secondary systems (previous section) can be described as a binary system, but for identification purposes, binary systems usually refer to premanufactured systems that consist of two systems capable of providing primary barrier seals if installed individually.

Figure 4.24 presents a typical binary system, composed of precompressed foam with silicone sealant facing. The system eliminates the need for a backing material, but not the requirement for a field-applied edge seal to complete the installation.

Again, these products are designed as primary barriers and will not prevent infiltration that occurs through porous substrates that permit water to bypass the primary barrier. They should be used on impervious envelope components or in conjunction with a waterproofing or sealer applied to the pervious envelope facade substrate. Binary systems are excellent choices when the joint width exceeds the standard one-inch maximum or whenever a joint is expected to experience maximum movement during life-cycling.

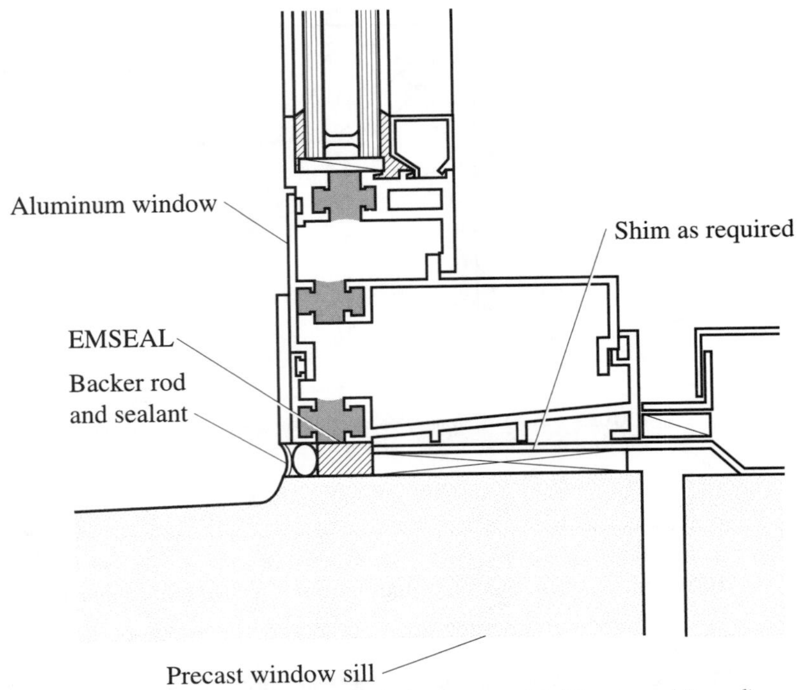

FIGURE 4.23 Secondary seal beneath the backer material. (*Courtesy of Emseal*)

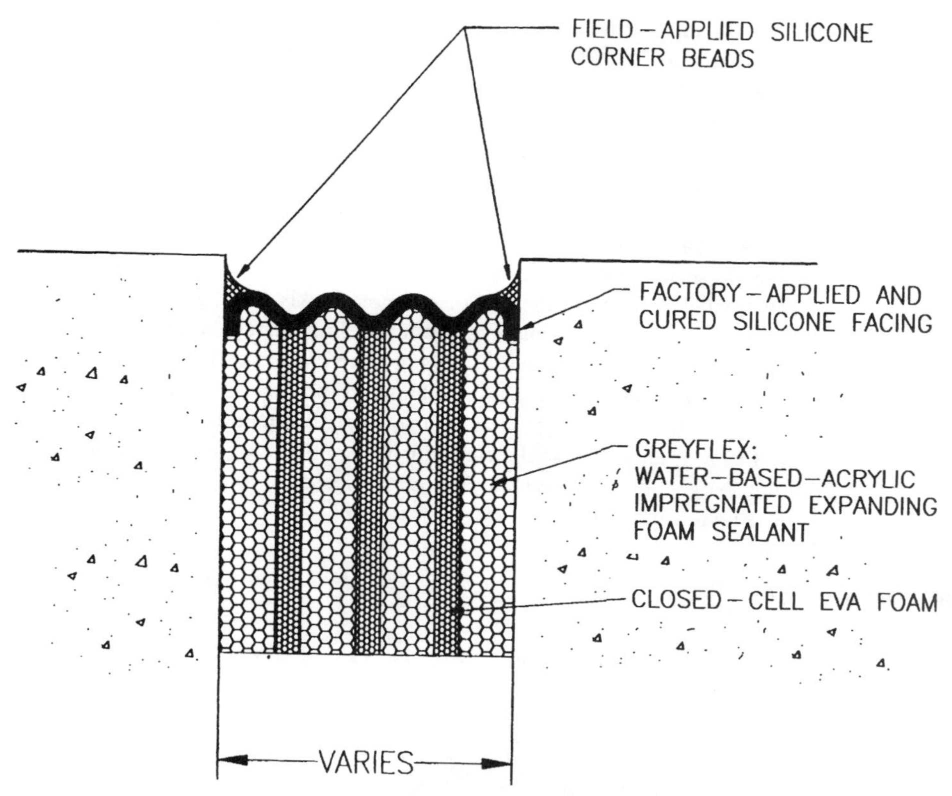

FIGURE 4.24 Premanufactured secondary sealant joint, consisting of closed-cell foam and cured silicone sealant. Fillet sealant beads are field-applied to edges of joint. (*Courtesy of Emseal*)

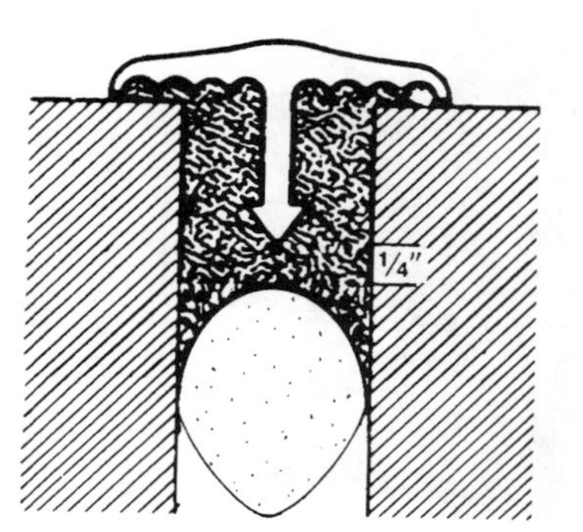

FIGURE 4.25 Preformed joint protector that is applied to joint before sealant is fully cured. (*Courtesy of Weathercap, Inc.*)

Joint Protectors

Available premanufactured and custom-manufactured joint protectors afford additional protection to the primary joint sealant. These products typically are manufactured plastic components in various sizes to cover the entire joint width and have a protruding piece that acts as an anchor in the sealant (Fig. 4.25).

These systems act much like a coping or flashing, adding additional protection to the primary waterproofing system. They make excellent choices whenever the sealant joint might be subject to unusual wear or life-cycle conditions such as exposed and within reach of children at schools, horizontal joints subject to ponding water such as coping stone joints, or joints with numerous changes in direction such as balustrade joints (Fig. 4.26).

The joint protector must be installed before the joint sealant has cured, to ensure that the anchor becomes an integral part of the joint. This prevents the anchor from moving after installation, which can actually damage the joint and permit water infiltration.

Substrate Diversions

If single-sealed joints are to be used on an envelope, substrates that form the joints should be constructed or manufactured to shed water quickly from the joint and envelope. They should also be designed to prevent water from traveling laterally across the joints.

Figure 4.27 shows several primary envelope barrier designs that complement joint effectiveness. These allow joint sealants to be the secondary means of protection against water infiltration. These designs also provide secondary protection against direct water infiltration, should a sealant exhibit minor disbonding or adhesion problems along the joint. Never should an envelope joint be designed that allows water to stand or collect on the sealant material.

MATERIAL SELECTION

In addition to the elongation properties necessary for the expected movement, the most important sealant properties are:

FIGURE 4.26 Preformed joint protector is supplied in rolls and conforms to the joint shape. (*Courtesy of Weathercap, Inc.*)

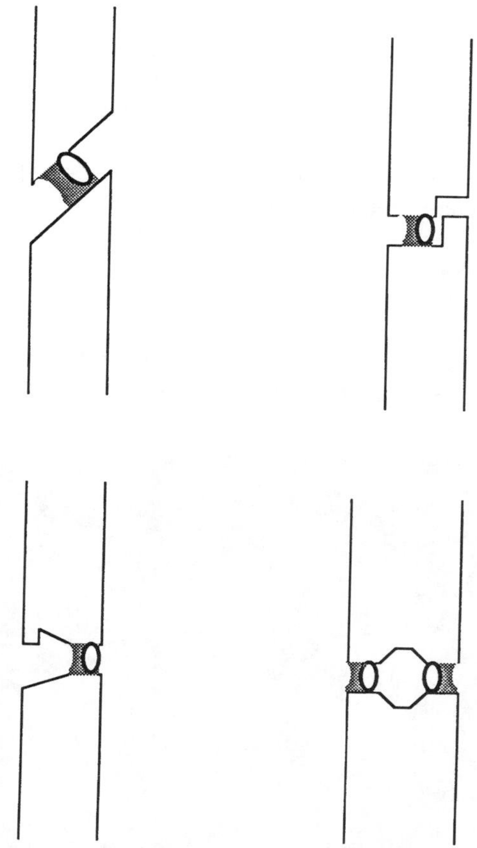

FIGURE 4.27 Envelope-joint construction for effective sealing.

- Adhesion strength
- Cohesion strength
- Elasticity
- Expected life
- Modulus

Additional desirable features or characteristics include color, availability, paintability, substrate compatibility, ultraviolet resistance, and presence of one or two component materials.

Elongation

Elongation is the ability of sealants to increase in length, then return to their original size. Limits of elongation are expressed as a percentage of original size. A material with a 200 percent elongation ability is, therefore, capable of stretching to double its original size without splitting or tearing. Since this is an ultimate measure of failure, joints are not designed to perform to this limit of elongation, rather to a portion of this capability including a safety factor consideration. A joint stretched to its limit will not return to its original shape or size. Thus, a joint will cease to function properly if elongated to its maximum elongation capability.

Modulus of elasticity

The modulus of elasticity is the ratio of stress to strain and is measured as tensile strength, expressed as a given percentage of elongation in pounds per square inch. Modulus has a direct effect on elongation or movement capability. Low-modulus (tensile strength under 60 lb/in^2) materials have a higher ability to stretch than high-modulus sealants. High tensile strength results in lower movement capability. More simply, soft materials are more easily stretched than harder materials. Low-modulus sealants with high-elongation factors are required in high-movement joints.

Elasticity

Elasticity and recovery properties are measures of a sealant's ability to return to its original shape and size after being compressed or elongated. As with elongation, elasticity is measured as a percentage of its original length. In high-movement joints, a sealant with sufficient recovery ability is mandatory. A sealant that does not continually return to its original shape after movement will eventually fail.

Adhesive strength

Adhesive strength is the ability of sealants to bond to a particular substrate, including adhesion during substrate movement. Since materials differ substantially in their adhesive strength to a particular substrate, manufacturers should be consulted for adhesion test samples on proposed substrates.

Adhesive failures occur when the sealant loses its bond with the substrate (Fig. 4.28). Adhesion failures typically occur due the inadequate, improper, or complete lack of primers, not preparing the substrate properly (such as removing curing or form-release agents), or incompatibility between the sealant and substrate.

Cohesive strength

Cohesive strength is the ability of a material's molecular structure to stay together internally during movement. Cohesive strength has a direct bearing on elongation ability.

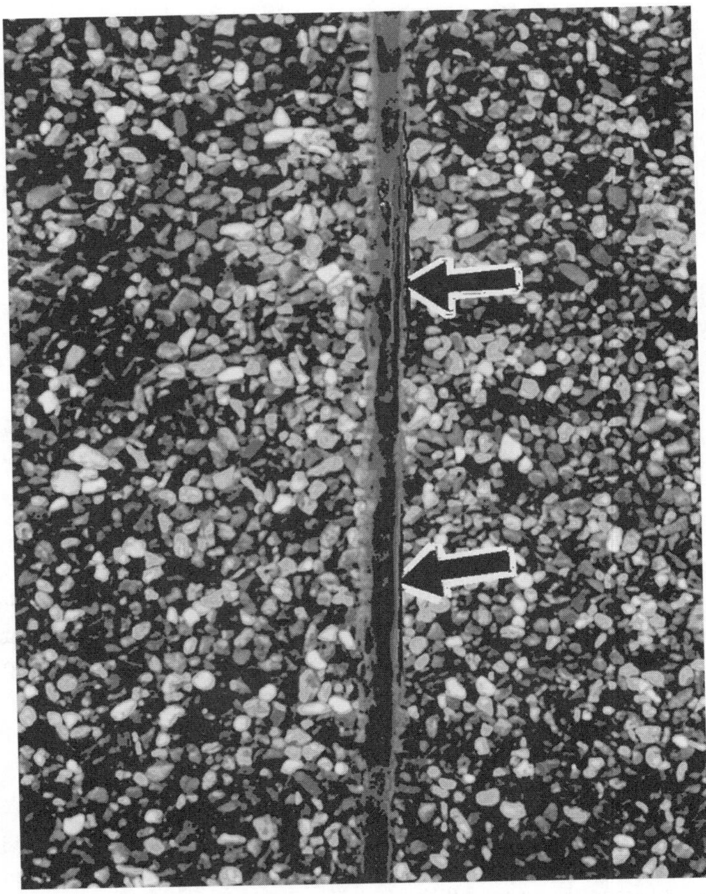

FIGURE 4.28 Adhesive failure of sealant; note disbonding from edges of joint. (*Courtesy of Coastal Construction Products*)

Cohesive failure is when a sealant tears or splits apart due to excessive joint movement or improper installation (Fig. 4.29). Excessive movement failure occurs whenever the material selected does not have the movement capability of expected life-cycle movement or if the joint is designed too small to handle the actual movement that occurs.

Installation problems that facilitate cohesive failure include the sealant being applied too thickly to permit proper movement at the joint. Also, not properly installing the sealant in the "hourglass" shape will often result in cohesive failures.

Shore hardness

Shore hardness is resistance to impact, measured by a durometer gage. This property becomes important in choosing sealants subject to punctures or traffic, such as horizontal paver joints. A hardness of 25 is similar to a soft eraser; a hardness of 90 is equivalent to a rubber mallet.

FIGURE 4.29 Cohesive failure of sealant; exceeds three inches. (*Courtesy of Coastal Construction Products*)

MATERIAL TESTING

All of the properties discussed must perform in unison for materials to function as necessary during joint life cycling. Weathering, ultraviolet resistance, amount of movement, and temperature change all affect sealant durability.

Many tests are available for comparison of different materials as well as different manufacturers. Unbiased testing is completed by the National Bureau of Standards (federal specifications) and the American Society for Testing and Materials. Tests presently used as standards include:

- ASTMC-920 for elastomeric joint sealants
- Federal specification TT-S-227 for two-component sealants
- Federal specification TT-S-00230C for one-component sealants
- Federal specification TT-S-001543 for silicone sealants

The ASTM C-920 involves a series of tests including adhesion-in-peel, effects of accelerated weathering, indentation hardness, and adhesion and cohesion testing under life-cycling movement. Whenever this ASTM specification is referred to, the specific test the manufacturer actually included should be detailed in the product literature.

While these standard tests provide a basis to compare the properties of different manufactured products, there are serious flaws that are created when using test results only for selection of sealants for any specific project. Reference to ASTM C-920, while now widely referred to in the industry, can be easily abused by manufacturers. This ASTM test is made sufficiently basic to include a wide range of generic sealant types including single- and multicomponent products, traffic- and nontraffic-bearing sealants and 25 or 12½ percent joint movement capability. This generality leaves a considerable amount of space for manufacturers to only test a product to a specific test it knows the sealant will pass, and omit the nonpassing grades from its product literature.

It is also important to recognize that these tests are conducted in a very controlled environment of pristine laboratory conditions that are rarely if ever duplicated in actual field application conditions. For example, ASTM C-920 requires that the testing be completed in the ideal conditions of 73.4°F, with a plus or minus of only 3.6°F. In addition the humidity must remain at 50% plus or minus only 5% during the entire test and curing stage. These conditions would rarely be duplicated during the application of any sealant in the field.

The test allows sufficient latitude in how the sealant can be considered to pass. Primers can be used or omitted at the descretion of the manufacturer, and such details are often not referred to in the manufacturer product literature. The test also permits a loss of complete adhesion in a limited area, yet still be considered to pass. In addition, the specification also permits manufacturers to request specific waivers or exclusions (e.g., longer curing time before adhesion test is conducted) and not clarify this fact in their product literature.

Therefore it is imperative that whenever considering sealant for a project, a sufficient safety factor should be incorporated into the design rather than depending solely on the performance during these pristine and perfect conditions of a laboratory test. For instance, a sealant should have a minimum of 100 percent expected joint movement capability rather than the 22½ or 25 percent tested for in C-920 (this results in a safety factor of at least 4).

Fortunately, most sealants produced today are far superior in performance to the minimum standards produced by ASTM testing. However, proper selection of sealants for a required installation still requires comparison of products under some form of standard basis. Appropriately, industry associations including the Sealant, Waterproofing and Restoration Institute (SWRI) are providing a means of comparison and standardization for the industry.

SEALANT VALIDATION PROGRAM

The SWRI Sealant Validation Program eliminates the confusion created when trying to compare the product literature of different sealant manufacturers. The program requires that the manufacturers perform specific ASTM testing and reveal the product's actual test results. This standardization permits industry members to review compare similar products using the same standardized tests.

Rather than having to rely on product literature that merely implies passing test results for the ASTM tests, the SWRI Validation Program provides specific test results. The program details actual test results (not just pass or fail), if primers were used or not, and any waivers granted the manufacturer from the basic test procedures. The sealant must pass three specific tests: C 719 (adhesion and cohesion under movement), C 794 (adhesion), and C 661 (shore hardness).

Once a sealant has passed all three tests and provides supporting and detailed test results, SWRI then provides the manufacturer with a Certificate of Validation that is current for three years before tests must be redone (or whenever a product is reformulated). The Program establishes a baseline for comparison of products produced by different manufacturers and removes the guesswork of deciphering manufacturers' specifications and product literature.

Products meeting the Validation Program receive a documentation seal for their product literature as shown in Fig. 4.30. More information about the program, and a list of sealants that meet the standards, can be obtained directly from SWRI (contact information is provided in Chapter 14).

MATERIALS

The numerous materials used as sealants exhibit a wide range of properties. In choosing a sealant, properties should be matched to expected conditions of a particular installation. The most common materials available and used for sealing joints in building construction include:

- Acrylic
- Butyl
- Latex
- Polysulfide
- Polyurethane
- Silicones
- Precompressed foam
- Preformed derivatives

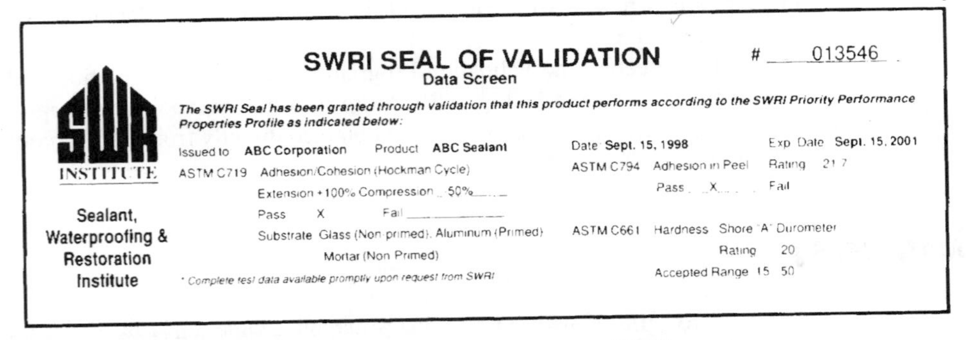

SWR INSTITUTE

Sealant, Waterproofing & Restoration Institute

SWRI SEAL OF VALIDATION # 013546

Data Screen

The SWRI Seal has been granted through validation that this product performs according to the SWRI Priority Performance Properties Profile as indicated below:

Issued to **ABC Corporation** Product **ABC Sealant**

ASTM C719 Adhesion/Cohesion (Hockman Cycle)

Extension +100% Compression -50%

Pass X Fail

Substrate Glass (Non primed), Aluminum (Primed)

Mortar (Non Primed)

Date **Sept. 15, 1998** Exp Date **Sept. 15, 2001**

ASTM C794 Adhesion in Peel Rating 21.7

Pass X Fail

ASTM C661 Hardness Shore 'A' Durometer

Rating 20

Accepted Range 15 50

** Complete test data available promptly upon request from SWRI*

FIGURE 4.30 SWRI seal of validation for sealants. *(Courtesy of SWRI)*

Once the joint design has been completed, a material with required properties must be chosen. Typical properties of each class of sealant are summarized in Table 4.3. Preformed seals are considered in Chap. 5 on expansion joints.

Acrylics

Acrylic-based sealants are factory-mixed, one-component materials polymerized from acrylic acid. These are not used on joints subject to high movement because of their relatively low-movement capability. They are frequently used in remedial applications with acrylic-based waterproof coatings. Acrylic materials are available in brushable or trowel

TABLE 4.3 Comparison of Common Sealant Properties

Property	Acrylic	Butyl	Latex	Poly-sulfide	Poly-urethane	Silicone	Precom-pressed foam
Maximum joint movement capability, %	7	5	7	25	25	50	25
Weathering resistance	Good, excellent	Excellent	Fair	Good	Excellent, good	Excellent	Excellent
Recovery, %	25	Poor	75	80	90	100	100
Adhesion	Good	Excellent	Fair	Good	Good	Excellent	Excellent
Joint design (number of times expected movement)	12	20	12	6	4	4	*
Shrinkage, %	12	18	20	10	5	3	n/a
Tack-free time (hours)	72	24	1	72	72	3	n/a
Water immersion	No	No	No	Yes	Some	No	No
Paintable	Yes	Yes	Yes	No	No	No	No
Primer required	No	No	Some cases	Metal, masonry	Horizontal masonry	Metal, natural stone	No
Ultimate elongation, %	Low	Low	450	1000	700	1600	Very low
Horizontal joints	No	No	No	Yes	Yes	No	Yes
Modulus of elasticity, lb/in^2	40	25	18	30	35	30	25

*Best in compression mode.

grades for use in preparing cracks in substrates before waterproof coating application. They are used in small movement joints such as doors and window perimeters, thresholds, and equipment penetrations.

Acrylic-based sealants do not require primers and have minimal surface preparation. Their general ease of application is offset by low performance characteristics. These materials are not recommended in continually submerged joints or joints subject to vehicular or foot traffic. (See Table 4.4.)

Butyl

Butyl sealants are produced by copolymerization of isobutylene and isoprene rubbers. Butyls are some of the oldest derivatives to be used for sealant materials. However, technological advancements in better-performing sealants have now limited their use to glazing window perimeters or curtain walls with minimal movement.

TABLE 4.4 Acrylic Sealant Properties

Advantages	Disadvantages
No primers required	Long cure stage
Good UV resistance	Low-movement capability
Minimal surface preparation	Poor impact resistance

Although butyls have low-movement and recovery characteristics, they have excellent adhesion performance. They bond tenaciously to most substrates and have excellent weathering characteristics. Butyls should not be used on water-immersed joints or joints subject to traffic.

Butyl sealants are used in metal curtain wall construction because of their ability to function in very thin applications. As long as movement is within the capability of a butyl, materials will function properly in metal wall construction splice joints.

Butyls are relatively easy to install, available in one-component packaging, and easily gunable or workable. They require no priming and are paintable. (See Table 4.5.)

TABLE 4.5 Butyl Sealant Properties

Advantages	Disadvantages
No primers required	Low-movement capability
Excellent weathering	High shrinkage rate
Little surface preparation	Poor recovery

Latex

Latex sealants are typically acrylic emulsions or polyvinyl acetate derivatives. Latex materials have very limited usage for exterior applications. They are typically used for interior applications when a fast cure time is desired for painting. Latex sealants have an initial set of tackfree time of less than 1 hour, fastest of all sealant materials.

Latex materials have very low movement capability, high shrinkage rates, and only fair weathering and adhesion properties. Their exterior use is limited to window or door perimeters where it is desired that the sealant match the frame color opening. Latex materials should not be used in areas subject to water immersion or traffic. (See Table 4.6.)

Polysulfides

Polysulfide materials are produced from synthetic polymers of polysulfide rubbers. Polysulfides make excellent performing sealants for most joint use. However, urethanes and silicones frequently have become specified and used due to their excellent recovery ability and joint movement capability.

TABLE 4.6 Latex Sealant Properties

Advantages	Disadvantages
Past cure stage	Low-movement capability
Paintable	High shrinkage rate
One component	Poor weathering

Polysulfides withstand an average of 16–20 percent joint movement, with a joint design of six times anticipated movement, versus a joint movement of 25 percent for urethanes and joint design of four times anticipated movement.

As with other types of better sealants, polysulfides exceed the movement capabilities of paints and therefore should not be painted. They are, however, manufactured in both one- and two-component packaging in a wide range of colors. With two-component materials, a color additive is blended in during mixing. Color charts are provided by the manufacturer.

Polysulfides are acceptable for a wide range of applications, including curtain wall joints, precast panels, and poured-in-place concrete. Polysulfides require primers on all substrates, and preparation is critical to allow successful adhesion and movement capabilities of installed materials.

Manufacturers usually produce two types of primers—one for masonry, concrete, and stone, and another for glazing, glass, and aluminum work. In a precast panel-to-window-frame perimeter joint, two different types of primer on each side of the joint would be required.

If properly prepared and installed, polysulfides will function in constantly immersed joints. Of all commercially available sealants, polysulfides are best suited for total-immersion joints. This includes swimming pools, water and wastewater treatment structures, fountains, and water containment ponds. Typically, two-component materials are recommended for these types of joint installations.

Polysulfides should not be installed in joints that might have bituminous residue or contamination, such as premold joint filler (e.g., concrete sidewalk joints). Polysulfides should also not be applied over oil- or solvent-based joint sealants. Joint preparation for resealing joints containing asphalt or oil-based products is especially critical if polysulfides are to be used. Sandblasting or grinding of joints to remove all residues is necessary before application of polysulfide materials.

Polysulfides are manufactured in grades for horizontal joints subject to foot or limited vehicular traffic. These materials are self-leveling and ideal for plaza and parking deck joints. (See Table 4.7.)

Polyurethane

Urethane sealants are polymers produced by chemical reactions formed by mixing di-isocynate with a hydroxyl. Many urethanes are moisture-cured materials reacting to moisture in atmospheric conditions to promote curing. Other two-component urethanes are chemically curing mixtures. Their compatibility with most substrates and waterproofing materials has made them a commonly used sealant in waterproofing applications.

Formulations range from one-component, self-leveling materials in a pourable grade for horizontal joints in plaza decks to two-component nonsagging materials used for ver-

TABLE 4.7 Polysulfide Sealant Properties

Advantages	Disadvantages
Immersion applications	Primers required
Good UV resistance	Low-movement capability
Horizontal applications	Low recovery rates

tical expansion joints. Some urethanes are manufactured to meet USDA requirements for use in food-processing plants. As with polysulfides, polyurethanes are available in a wide range of colors. Two-component mixes add coloring to the activator portion that is mixed with base material.

Polyurethanes are available for a wide range of applications, including precast concrete panels, expansion and control joints, horizontal joints, flashing, and coping joints. Urethane sealants are not recommended for continual immersion situations.

Urethane has excellent adhesion to most substrates, including limestone and granite. In most cases, a primer is not required. However, manufacturer's data should be reviewed for uses requiring primers. These include horizontal joints, metals, and extremely smooth substrates such as marble.

Two-component urethanes are low-modulus sealants and have high joint movement capability averaging 25 percent, with joint design limitation of four times the expected movement. Since urethanes exceed the movement capabilities of paint, they should not be painted over because alligatoring of the paint surface will occur. Coloring should be achieved by using standard manufacturer colors.

Urethanes have excellent recovery capability, 90 percent or more, and possess excellent weathering and aging characteristics. Since urethanes are extremely moisture-sensitive during curing, a closed-cell backer rod should be used. However, with one-component urethane sealants an open-cell backing material is acceptable.

Polyurethanes cannot be used in joints containing a polysulfide sealant or residue. These joints must be cleaned by grinding or other mechanical means to remove any trace of sulfides. Urethane sealants also should not be used in glazing applications of high-performance glass, plastics, or acrylics. Joints contaminated with asphalts, tar, or form-release agents must be cleaned before sealing work.

Polyurethane's compatibility with most substrates, excellent movement and recovery capability, and good weathering characteristics have allowed their widespread use in waterproofing applications both above and below grade. Their ability to withstand vehicular traffic and compatibility with urethane deck coatings leads to their extensive use in parking deck applications. (See Table 4.8.)

Silicones

Silicone sealants are derivatives of silicone polymers produced by combining silicon, oxygen, and organic materials. Silicones have extremely high thermal stability and are used as abrasives, lubricants, paints, coatings, and synthetic rubbers. Silicones are available in a wide range of compositions that are extremely effective in high-movement joints, including precast panels and expansion joints. When used properly, silicone sealants provide

TABLE 4.8 Polyurethane Sealant Properties

Advantages	Disadvantages
Good elongation capability	Moisture sensitive
Excellent recovery rates	Unpaintable
Horizontal applications	Require some priming

excellent movement capability, as much as 50 percent, and adhesion and recovery properties after movement.

Silicone sealants cannot be used for below-grade applications, horizontal applications subject to vehicular traffic, and water immersion joints. It is extremely important not to install silicone materials over materials that might bleed through a silicone. This includes oil, solvents, or plasticizers, which will cause staining and possible silicone failure.

Uncured silicone must not encounter nonabradable substrates such as metal, polished granite, or marble. The uncured sealant can leave a residue that stains or changes the substrate appearance. This is also true for primers used with silicone sealants. Masking adjacent surfaces is necessary to protect against damage.

Silicones contaminate all surfaces or substrates they encounter. This makes it virtually impossible to seal over silicone residue with other materials such as urethanes. Abrasive methods are the only acceptable methods for removing silicone from a substrate before resealing it with another product. Most substrates do not require primers for silicone applications; however, natural stone materials such as limestone and marble will require primers.

Most silicone is produced in one-component packaging, although two-component products are available. Silicones have excellent adhesion to almost all building products including wood, ceramic, aluminum, and natural stones. Silicones may be used in curtain wall joints, precast panels of concrete, marble, or limestone, and expansion and control joints.

Silicone materials exceed the movement capability of paints, and as most paints will not adhere to silicones, they should not be painted over. Most silicones are now produced in a wide range of colors; in addition, special color blending is available by the manufacturer. Both open-cell and closed-cell backing materials can be used with silicone joints.

Silicones have excellent recovery capabilities, usually up to 100 percent. They have very little initial cure shrinkage, 3 percent, and a tackfree time of only 1–3 hours. High-tensile-strength silicones with lower movement capabilities are typically used in glazing applications. (See Table 4.9.)

Precompressed foam sealant

Foam sealants are manufactured by impregnating open-cell polyurethane foam with chemical sealant containing neoprene rubbers. An adhesive is applied to one side and covered with a release paper. The foam is then compressed and supplied in rolls with various widths of up to 12 inches.

Foam sealants are applied by unrolling the foam, removing release paper exposing the adhesive, and installing into a joint. The foam then swells and expands to fit tightly against both joint sides, allowing for any irregularities in joint width. Splices in material are prepared by overlapping or butting joint ends. This material eliminates the need for joint backing, primers, and tooling.

TABLE 4.9 Silicone Sealant Properties

Advantages	Disadvantages
High-movement capability	No submersion applications
Excellent adhesion	Possible staining
Excellent recovery rates	No below-grade uses

Joint width determines the size of foam material required. If a joint varies considerably, more than 25 percent in width, different sizes of preformed foam sealant are required in the one joint.

A horizontal grade is available, allowing use in horizontal plaza and deck joints. Some have properties sufficient to withstand vehicular traffic as well. Foam sealant adheres to most clean and prepared building materials, including stone, aluminum, concrete, wood, and glass.

In addition, foam sealant is compatible with other sealant materials and allows elastomeric sealants to be applied over the foam, providing a double barrier in critical waterproofing joints. With these applications, foam sealants are acting as a backing material for elastomeric sealants.

Most foam sealants withstand up to 25 percent movement in either direction, for a total joint movement capability of 50 percent. Foam sealant performs best in compression mode with no long-term compression set, returning to its originally installed size.

Critical to successful foam sealant applications are well-cleaned joints. If the joint is wet or contaminated, the contact adhesive will fail. Materials are usually supplied in black only, but they can be painted to achieve other coloring, although paints will crack during movement. (See Table 4.10.)

SUBSTRATES

Successful sealant installation depends upon ensuring that a substrate is compatible with the material and is in acceptable condition for proper sealant adhesion. Adhesion is essential, without which all other properties are insignificant.

For example, Teflon®-coated materials, Kynar® finishes, or PVC substrates are especially difficult to adhere to. Teflon® is manufactured so that other materials do not adhere to it, thus keeping the surface continually clean. In these cases, butyl rubber sealant might be chosen over a silicone for its better adhesion capability, provided that substrate movement is within the butyl's capability.

Sealant incompatibility with a substrate causes staining or etching of substrates. On the other hand, some substrates, such as oils, asphalt, and coal tar materials, may cause staining or sealant deterioration. To prevent these problems or when compatibility is in question, actual substrate testing with sealants should be done. This can be completed by testing under laboratory procedures such as accelerated weathering, or by preparing mock-up panels with the sealant applied at job sites and allowing sufficient time to determine success or failure.

TABLE 4.10 Precompressed Foam Sealant Properties

Advantages	Disadvantages
No priming or backing	No colors available
Factory-manufactured	Heat and cold affect installation
Nonlabor-intensive	Difficult installation for varying joint widths

Following are descriptions of substrates commonly found in construction and their requirements for proper sealant installation.

Aluminum substrates

A common building component, aluminum substrates present difficulties when choosing sealants. This is due to the architectural coatings now being applied to aluminum to prevent aluminum from oxidizing and to provide color.

By themselves, aluminum surfaces must be cleaned chemically or mechanically to remove any trace of oxidation that will prevent sealant adhesion. With coated aluminum, it is more difficult to choose a sealant.

Baked-on finishes and other coatings contain oils, carbon, and graphite residues, which act as release agents for sealants. Some coatings themselves may have poor adhesion to aluminum, thereby making it impossible to achieve proper adhesion with any sealant application. Other coatings may soften or deteriorate when solvents in sealants or primers come into contact with the finish.

The only positive method to test adhesion with a coated aluminum is actual testing before application. Silicones and butyls have acceptable adhesion to aluminum and are often used in aluminum curtain wall construction. But even these materials should be tested for proper adhesion. Elastomeric sealants such as urethanes are often used around aluminum frame perimeters, and as such should be checked for adhesion especially when coated aluminum products are used.

Cement asbestos panels

Cement asbestos panels are produced with a variety of finishes, including exposed aggregate and tile. The panels are attached to a wood or metal stud frame for support and attachment to a structure.

Often panels without finishes are less than 1/2 inch thick. This composition prevents proper width-to-depth ratios for backing and sealant installation. Cement asbestos panels present difficult problems for sealant applications because of their high moisture absorption properties and thinness of the panel itself. After sealant installation, any water absorbed into a panel can bypass sealant joints and cause damage to interior areas and panel support systems.

Cement asbestos panels have thermal coefficients similar to concrete panels, with movement at joints often exceeding movement capability of sealants. Compounded with absorption rates of panels, long-term performance of any sealant is questionable. In addition form-release agents or sealers used on panels often contaminate joints, prohibiting proper adhesion of sealants. Some panels are manufactured with aggregate exposed in joint sides, which also prevents proper sealing.

These factors all contribute to problems in sealing and keeping panels watertight. To prevent such problems, adequate connections must be incorporated into panel design. Accelerated weathering testing of a panel design with wind and structural loading should be completed to verify the effectiveness of proposed sealant systems. Panel design should include details that make joints acceptable for sealant installation. These include joint sides at least 1 inch thick, panel edges clean of any form-release agents or sealers, and aggregate not exposed on joint sides.

Precast concrete panels

Precast panels, including tilt-up and prestressed ones, are now produced in a variety of sizes, textures, and finishes. These have become a common building facing material for all types of structures. Problems arise not with the panels themselves but with sealers, finishes, or coatings applied to them.

Form-release agents are used in all precast panel fabrications. Since panel edges typically become sides of joints after erection, problems with adhesion occur. Oil- and petroleum-based products used for curing the panels will cause deterioration of silicone and polysulfide sealants. Film-forming curing and release agents can act as bond breakers between sealant and concrete.

Substrate adhesion testing often tests a sealant's ability to adhere to the form-release or curing agents rather than to a panel itself. Therefore, all precast joints should first be abraded or chemically cleaned to remove all residue of these compounds before sealant application.

Often precast panels are designed with exposed aggregate finishes. Although aesthetically pleasing, exposed aggregate often prevents proper joint sealing.

When panels are manufactured with aggregate turned or exposed onto panel sides (that later become joint sides), proper sealing is impossible (Fig. 4.31). Sealants will not adhere properly to exposed aggregate, and the aggregate will prohibit proper movement characteristics of the sealant.

Figure 4.32 shows a typical improperly manufactured panel. If someone attempts to chip the stone out at a project site, pockets are created that must be patched with a cementitious grout. If such a repair is at tempted, grout repairs can actually be pulled away from precast substrates during joint movement and cycling. The only acceptable repair method is to replace panels with precast panels, which are manufactured with no aggregate exposed within joints.

Project specifications often require coating application to panels after installation. These coatings include water repellents, antigraffiti coatings, color stains, and elastomeric coatings. Coatings applied before sealants can create problems. Sealant application should be completed first and protected during coating application; otherwise, sealants would be bonded to coatings rather than to the panel surface.

If a panel finish is porous and water absorption rates are high, water may enter the panel substrate and bypass sealant joints. Water causes bonding problems if faces of joints remain wet or if open-cell backing materials become saturated after sealant installation. This wet backing causes sealants to remulsify.

Wet substrates also cause the release of primers used for sealant installations. Absorptive panels require sealing with a water repellent after sealant installation, to prevent these problems.

Panels should be erected and securely attached to prevent slippage, bowing, or creeping, which causes shearing and ripping of sealants. Panels must be installed so joints are uniform from top to bottom, to prevent joints that are too narrow or too wide for proper sealing. In these situations an applicator may not bother to change the backing material size when the joint width changes, causing performance problems with sealants after installation.

Joints should also be kept uniform from one to the next. For instance, panels meant to have 1/2-inch joints should not be installed with one joint 1 inch wide and an adjacent joint virtually closed. Such variances will considerably shorten the life-cycling of sealants.

FIGURE 4.31 Aggregate exposed on joint sides. (*Courtesy of Coastal Construction Products*)

Tiles

Quarry tile is manufactured with a patina finish, a result of firing tiles for smooth finishes. This finish should be removed by grinding joint sides before sealant application. Sealants should never be applied to grout in place of tile itself. Grout will eventually loosen and cause failures.

If efflorescence has formed on the tile before the sealing of joints, it should be removed chemically before applying sealants. Most manufacturers recommend primers when sealing quarry tile joints.

PVC

Polyvinyl chloride material such as PVC piping does not provide an acceptable substrate for sealant applications. It is necessary to mechanically abrade surfaces of PVC to be sealed.

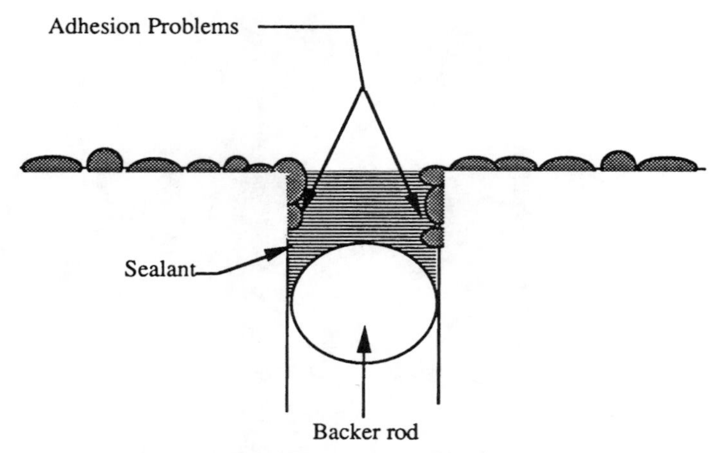

FIGURE 4.32 Exposed aggregate precast joint problems.

This roughens their surface before sealant application. This rough surface may provide an acceptable substrate for sealants such as butyl or silicones, but PVC materials should never be used at high-movement joint areas.

Stonework

Building facades of limestone, marble, or granite generally provide a surface acceptable for sealants. However, adhesion tests should be completed to determine their acceptability, since there are so many finishes of each natural stone type available. Priming is usually required with these types of substrates.

It is important to note that in most cases a primer or uncured sealant may stain stone work. Therefore, precautions including masking joint faces before sealant or primer application will prevent staining. With porous absorptive stone, closed-cell backing material should be used to prevent backing from absorbing water that passes through the stone facade.

Terra cotta

Terra cotta tiles or stones manufactured from natural clay are typically supplied with a baked or glazed surface finish. However, sides of the tile are typically unfinished clay and are very porous and absorptive. Primers are required for adequate sealant bonding.

Should the facing of terra cotta be porous, water absorption may cause adhesion problems or gassing of sealants. Closed-cell backing materials should be used to prevent the backing from absorbing water entering through the terra cotta facade.

SEALANT APPLICATION

Of all factors affecting sealant performance, installation is the most critical and most often causes joint failures. No matter how good a sealant is selected and how well a joint is designed, improper installation will lead to failures. Successful installation depends on several steps, including:

- Joint preparation
- Priming
- Installation of backer rod or backing tape
- Mixing, applying, and tooling of sealant

Of these steps, the most common problem and most widely abused installation step is joint preparation. All remaining installation steps depend on how well this first step is completed. If joints are not properly prepared, regardless of how well joints are primed and sealed, materials will fail.

Joint preparation

The most common joint preparation problems arise when joints are not cleaned or when contaminated and incorrect solvents are used. All joint contaminants must be removed and joints must be dried before sealant and primer application (Fig. 4.33).

To clean a joint, two rags are necessary—one rag to wet a joint with solvent, the other to wipe contaminants from the joint while at the same time drying it. Using a single solvent rag will smear the contaminants in a joint. Continually dipping the same rag in a solvent will contaminate the entire container of solvent.

All loose mortar and aggregates must be removed, since sealant will only pull loose material away from substrates when the joint moves. Other contaminants, such as waterproofing sealers, form-release agents, oils, waxes, and curing agents, must be removed.

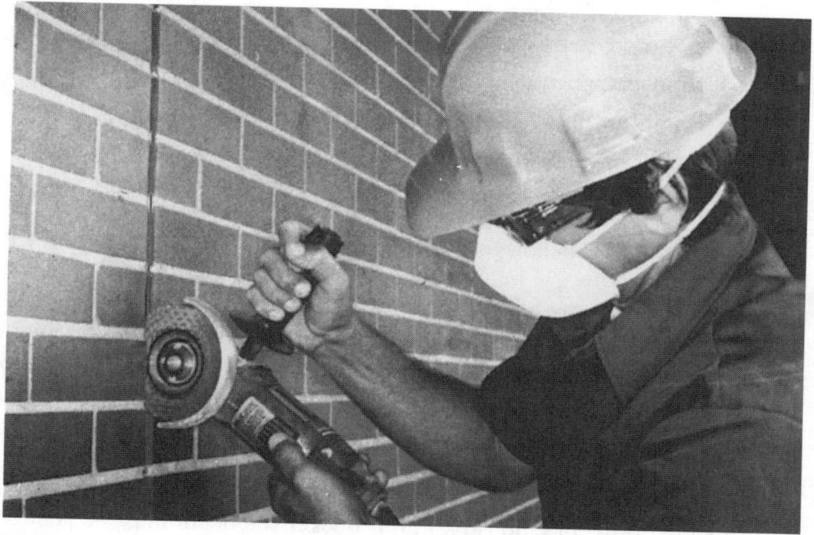

FIGURE 4.33 Preparation of substrate joint prior to applying sealants. (*Courtesy of SWRI*)

This may require mechanical methods such as grinding or sandblasting. It is important to note that after mechanical cleaning, joints must be recleaned to remove dust and residue left behind by mechanical cleaning.

Successful joint preparation steps include:

1. Two-rag method of cleaning:

   Use lint-free rags.
   First rag has solvent poured on it, not dipped in solvent.
   Second rag removes solvent and contaminants.
   Change rags often.

2. Form-release agents, oils, paints, and old waterproofing materials must be removed by mechanical means, followed by recleaning joints, including pressure-washing if necessary.
3. Joint sides must be dry and free of moisture or frost.
4. Loose joint sides must be chipped away and cut smooth. Jagged edges may cause air pockets to develop during sealant installation.

Priming

Primers are used to ensure adhesion between sealants and substrates. If there is any doubt if a primer is required or not, adhesive tests should be completed with and without primers to determine the most successful application methods.

Using too much primer, allowing primer to cure too long before installation, or applying sealants over wet primer will cause sealant failures. Manufacturer's application instructions pertaining to mixing, coverage rates, drying time, and application time vary with different types of sealant and primers. Instructions must be consulted on an individual basis for proper installation, including:

- Use of proper primer
- No overapplication of primer
- Priming within application time recommendations
- Discarding of primers that are contaminated
- Manufacturer's recommendations

Figure 4.34 shows the correct method of installing primer in a joint.

Backing materials

Backer rod and backing tape prevent three-sided adhesion in joint design. Tapes are used where a firm substrate, against which to seal, exists at backs of joints, when joints have insufficient depth for backer-rod installation. Rod is installed in joints where there is no backing substrate. The backer rod or backing tape provides a surface against which to tool material and maintain proper depth ratios.

Failure to install backer rod properly causes cohesive failure due to improper sealant width-to-depth ratio (Fig. 4.35). Backer-rod depth must be kept constant, which requires use of a packing tool (Fig. 4.36). This simple tool, a roller that can be adjusted to various depths, is unfortunately rarely used.

FIGURE 4.34 Proper priming of joint prior to applying sealant. (*Courtesy of SWRI)*

FIGURE 4.35 Improper depth-to-width ratio, by improperly applying backing material, causes failure of sealant joints. (*Courtesy of Coastal Construction Products)*

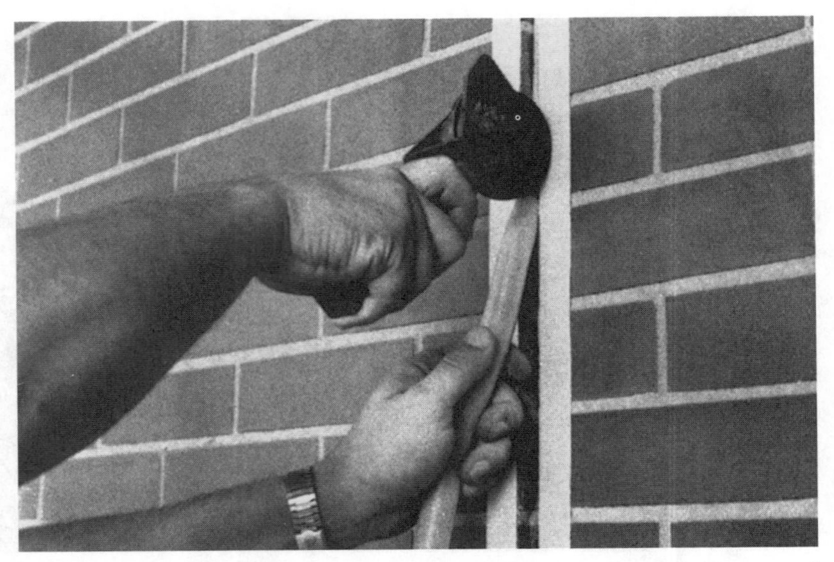

FIGURE 4.36 Proper installation of backing material using packer tool. (*Courtesy of SWRI)*

Applying bond breaker tape incorrectly can cause joint failure both adhesively and cohesively. Tape allowed to turn up sides of joints will not allow sealant to adhere properly to sides, causing adhesion problems. If tape does not completely cover the backs of joints, three-sided adhesion occurs, causing cohesion failure.

Open-cell materials are usually made of polyurethane; closed-cell materials are manufactured from polyethylene. Sealant manufacturers will recommend the appropriate backing to use. Open-cell materials are not recommended for horizontal or submerged joints where water can collect in open cells. Closed-cell materials are inappropriate for moisture-cured sealants, since they prohibit air from reaching the back of joint material.

Proper backer rod and tape installation depends on the following:

- Use of an adjustable packing tool to ensure proper depth (Fig. 4.37)
- A backer rod that is 25 percent larger than joint width
- Backing material that corresponds in width to the varying widths of joints
- Horizontal joints without gaps in the rod that would allow sealant to flow through
- Bond breaker tape that covers the entire back of a joint but is not turned up at the sides
- Installation of tape over existing sealant to prevent three-sided adhesion (in remedial applications, where it is necessary to install new sealant over old)

Mixing, applying, and tooling sealants

Improperly mixed sealants will never completely cure and therefore will never provide the physical properties required. Improperly mixed sealants are evident from their sticky surface or softness of material, which can literally be scooped from a joint. All components of a mate-

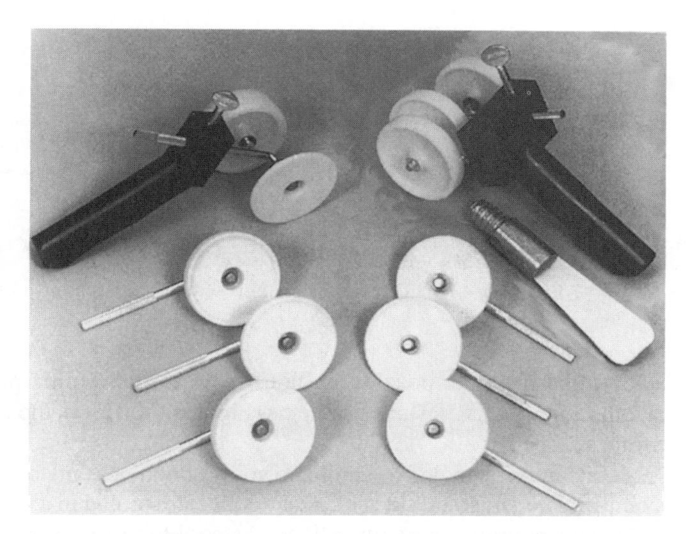

FIGURE 4.37 Packing tools and accessories. (*Courtesy of Albion Engineering Co.*)

TABLE 4.11 Approximate Coverage Rates for Sealant Materials

Depth	Width (in)							
(in)	1/16	1/8	1/4	3/8	1/2	5/8	3/4	1
1/16	4828	2464	1232	821	616	493	411	307
1/8		1232	616	411	307	246	205	154
1/4			307	205	154	123	103	77
3/8				137	103	82	68	51
1/2					77	62	51	39
5/8						49	41	31
3/4	Lineal feet of joint coverage per gallon of material						34	26
1								19

rial must be mixed in provided quantities. Never should less than the manufacturer's prepackaged amounts be used. Table 4.11 summarizes material coverage per gallon of material.

The use of proper mixing paddles and mixing for adequate lengths of time are important. Materials that have passed their shelf life, which is printed or coded on material containers, should never be used.

Proper application tools, including sealant application guns that are available in bulk, cartridge, or air-powered types, are necessary, Figs. 4.38 and 4.39. The cartridge is used for one-component materials supplied in tube form. Bulk guns use two-component sealants. These guns should be filled carefully with the mixed material; air should not be allowed to mix with sealant, or gassing of materials will occur.

Nozzle selection and use is also important. Use a metal nozzle with a 45° angle, and cut the plastic nozzles of tube containers to the same 45° angle. Joints might be overfilled or

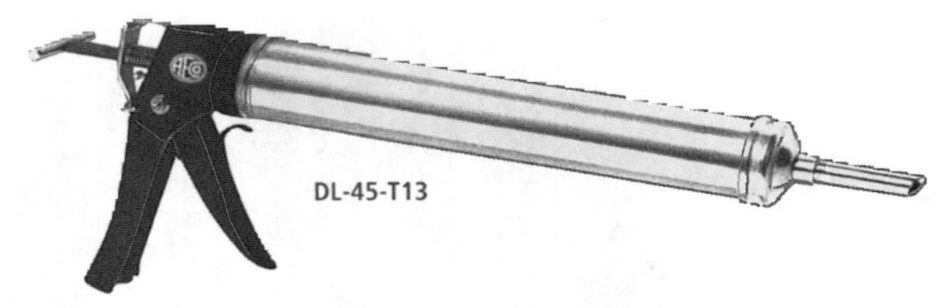

FIGURE 4.38 Typical bulk gun sealant applicator tool. (*Courtesy Albion Engineering Co.)*

underfilled if proper nozzles are not used, causing tooling problems, improper depth-to-width ratios, and adhesions problems. Figure 4.40 shows the proper placement of a nozzle while sealing.

Joints must be tooled to eliminate voids or bubbles and to ensure that the materials press completely against the sides of joints. Joints are tooled in a concave finish as shown in Fig. 4.41. This hourglass structure allows material to move properly and enhances the physical properties of a sealant.

Many types and sizes of tools are available for joint finishing, including those required for recessed joints. Soaps or solvents should never be used in tooling a joint because they will cause improper curing, adhesion failure, or color change (Fig. 4.42).

Proper mixing, application, and tooling of sealants includes:

- Applying only in recommended temperature ranges, typically 50–80°F
- Mixing only complete packages of materials
- Using the appropriate mixing equipment
- Mixing for the proper amount of time
- Keeping air out of sealant during mixing
- Using properly sized nozzles and slopes to fill joints
- Tooling joints by compression, for adequate adhesion
- Avoiding use of soaps or solvents in finishing joints

When the hourglass shape, as described previously, is not properly created, failures often occur because the sealant is either too thick or thin to function as intended and tested. Sealants that are applied too thick, often when the backer rod is installed too deep into prepared joint, will promote cohesion failure.

Cohesion failure results when the sealant is so thick that it can not elongate when the substrate is experiencing expansion movement. The sealant literally rips itself apart, usually in the middle of the joint, when the substrate separates. This is reflected in Fig. 4.43.

Likewise when the sealant is installed in too thin an hourglass shape, again often due to misplacement of the backer material, the joint will likely fail in an adhesive manner. Adhesive failures occur when there is insufficient sealant material adjacent on the sides of the substrate to permit proper movement in the expansion mode. When the substrate moves apart the sealant is ripped off the side or sides of the joint, due to insufficient bonding

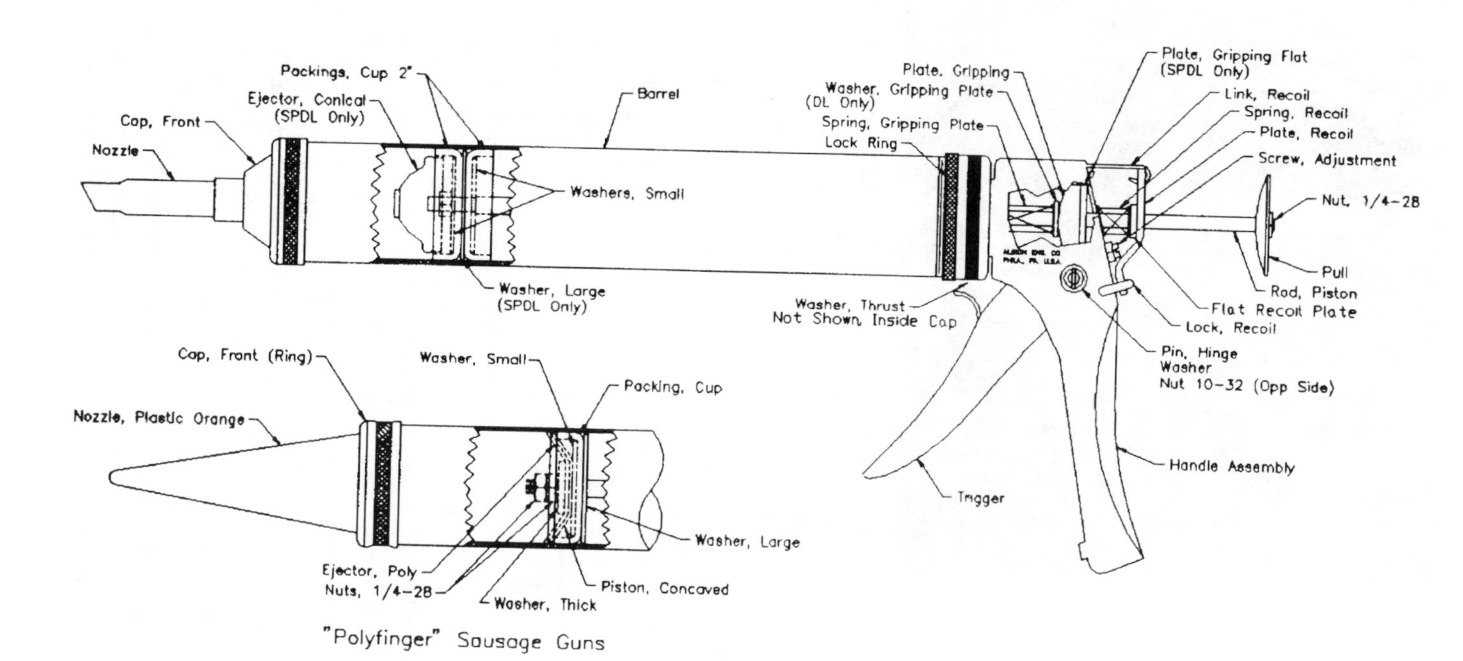

FIGURE 4.39 Detail of bulk gun construction. (*Courtesy of Albion Engineering Co.*)

FIGURE 4.40 Proper positioning of nozzle when applying sealant. (*Courtesy of SWRI*)

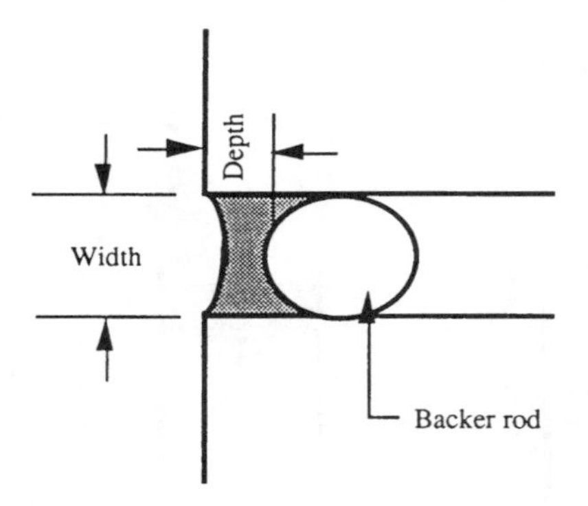

FIGURE 4.41 Typical sealant joint detailing.

capacity caused by the lack of material applied to the sides of the substrate as shown in Fig. 4.44.

These failures both occur in the expansion mode. It is interesting to note that these problems can likely be prevented if the sealant is applied when the joint is completely expanded, or in its widest width stage. This typically occurs when the temperature is the coldest to be experienced over the life cycle of the joint. Sealant applied under this condition will always be in the compression mode, when the substrate is pressing the sealant material together.

Under this situation, poor application techniques are much less likely to produce failures of the sealant material. The joint is constantly compressing the material, and whether it is too thick or thin, the material can usually provide sufficient capabilities to maintain an effective weathertight condition (Fig. 4.45). Often the sealants in the joint under these conditions will bulge outward. This is not a sign of failure, (unless adhesion or cohesion problems are evident), just that the sealant is under a compression mode. This bulging should not cause problems unless at a horizontal condition subject to foot or vehicular traffic that can damage the exposed sealant. In such instances the sealant should be recessed when the horizontal joint is in a contracted state, as shown in Fig. 4.46.

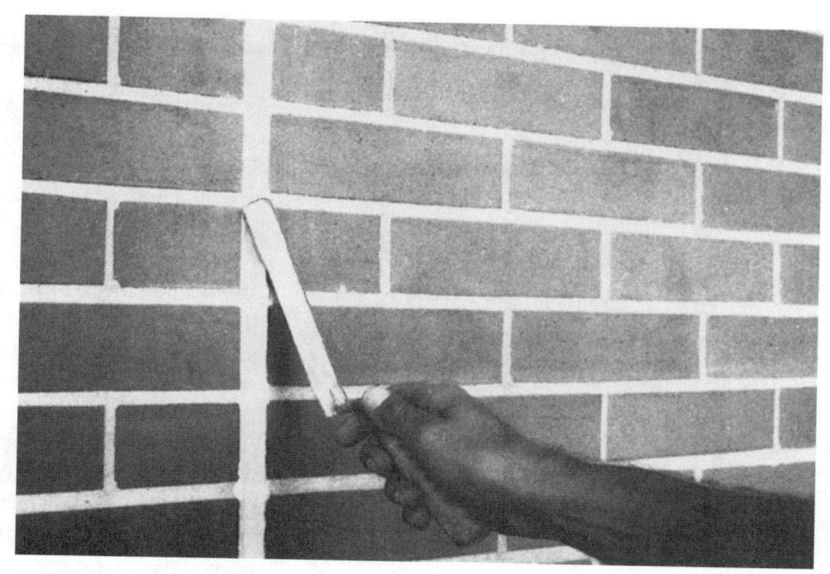

FIGURE 4.42 Proper tooling of joint ensures that material adheres to sides of joint. (*Courtesy of SWRI*)

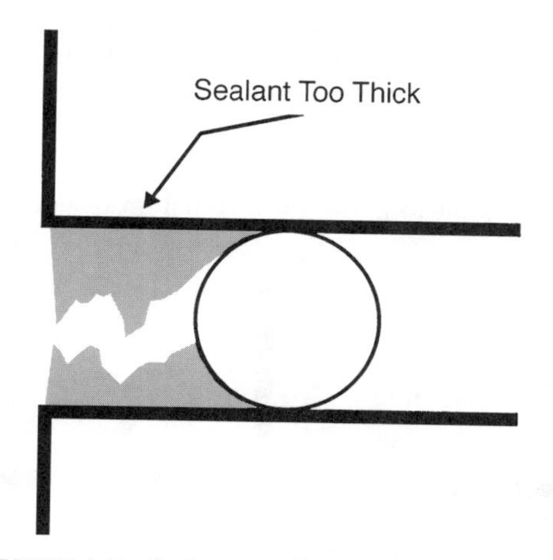

FIGURE 4.43 Sealant material applied too thickly will result in cohesive failure. (Material cannot stretch sufficiently, and splits apart)

Applying sealants in the hottest part of the season in a particular locale (when the joint is completely contracted or in its smallest width stage) is more likely to create material failures. This is because the joint will always be in an expansion mode, constantly pulling at the sealant material as shown in Fig. 4.47.

Cold-weather sealing, therefore, offers advantages over hot or mild weather, and it is advantageous to seal when the joint is in its most contracted stage. Thus, sealing should not be completed when the conditions are likely to put the joint in a complete expanded mode, when the sun and temperature are at their peak. Therefore, sealant applications should keep ahead of the sun around a building, working in the shade as much as possible. Applications should be completely avoided on west elevations at the day's peak temperatures around midafternoon.

Cold-weather sealing

Of the many problems that might occur in sealing joints in temperatures below freezing, the most serious is joint contamination by ice. In freezing temperatures, a joint surface can

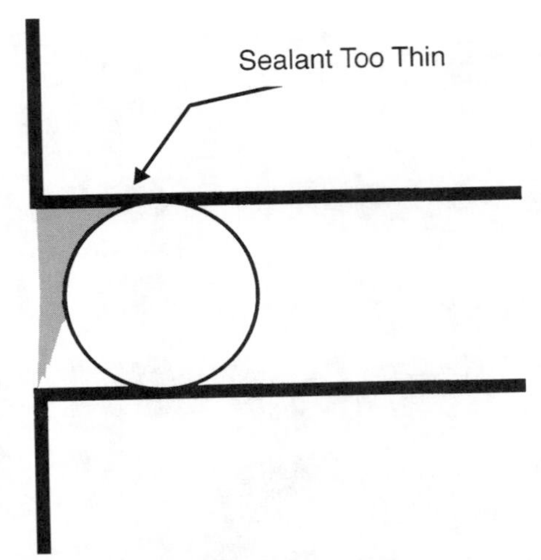

FIGURE 4.44 Sealant material applied too thinly results in adhesive failure (insufficient material bonded to substrate sides to move properly).

be covered with ice that is not visibly noticeable but that will cause the sealant not to bond to the substrate. Even if the sealant is warmed sufficiently to melt this ice, the resulting joint wetness will cause failure. Therefore in freezing temperatures it is critical that joints be heated and dried before sealant application.

Sealants in cold-weather conditions should be stored in heated containers until the actual application. Curing time is slowed considerably, and sealants should be protected from physical abuse during this curing period.

With cold-weather joint applications, joints are installed at their maximum width. These joints will always be in compression mode during movement, and must be designed not to exceed the maximum width limit.

Narrow joints

Sealing thin or narrow joints, such as metal panels of curtain wall construction, presents several problems. The substrate area for sealant bonding is usually minimal, if not totally insufficient. Three-sided adhesion may be necessary if no allowance is available for application of a bond breaker tape.

For proper performance under these circumstances, a splice or backing plate of material should be installed behind the joint to allow for installation of bond breaker tape. In

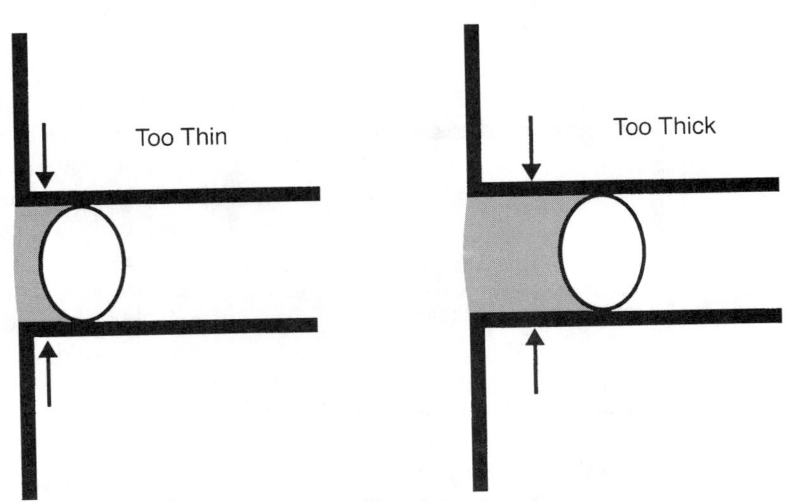

FIGURE 4.45 Compression at the joint can overcome poor installation practices.

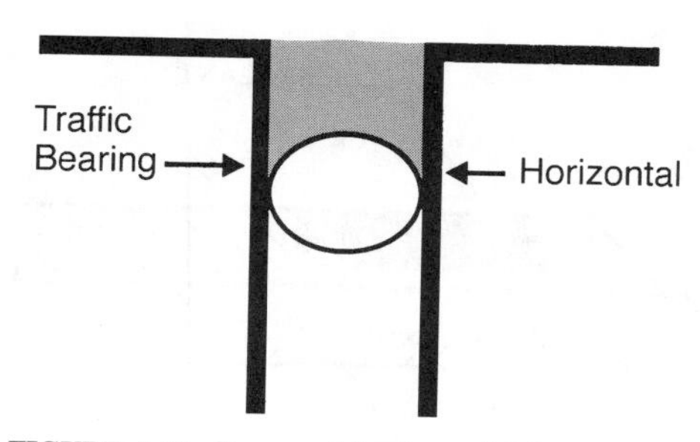

FIGURE 4.46 Recessed horizontal joint that permits compression bulging in the sealant without damage to the material from traffic.

addition, sealants should be tooled flat and flush, not concave, which would leave a narrow section of material in the center. Refer to Fig. 4.48.

Another alternative is to overband the sealant onto sides of metal facing, as shown in Fig. 4.49. Note that in this situation, backing tape is brought up and onto the joint side to prevent three-sided adhesion. The bonding area is determined by movement at a joint but should not be less than 1/4 inch.

Metal-frame perimeters

Sealing of metal-frame perimeters including doors, windows, and storefronts presents problems, since rarely is a proper joint provided on which to apply the sealant. Typically, frames are butted up against surrounding structures including brick, precast, and curtain walls. If frames are smaller than openings, voids are left around the frame perimeters filled only with shims used for frame installation. In such instances either there is no space to install backer rod or tape, or the frame is manufactured without sides against which to compress backer rod. This forces sealant installers to fill joints to incorrect depths, deterring joint effectiveness.

If a frame is butted to substrates, installers will usually place the sealant in a V-shape application by installing the sealant in a cant between the frame and substrate. This three-sided adhesion joint will not function properly. (See Fig. 4.50.)

To seal these situations effectively, frames that allow a joint to be formed between frame and substrate should be manufactured, similar to that shown in Fig. 4.51. If this is not possible—for example, when repairing existing frame perimeters—steps must be taken to allow only two-sided adhesion. Sealant of equal thickness should be bonded to substrate and frame.

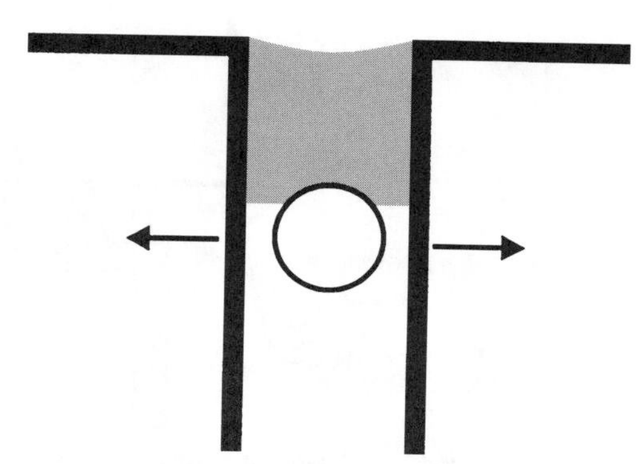

FIGURE 4.47 Expansion mode places considerable stress on sealant materials.

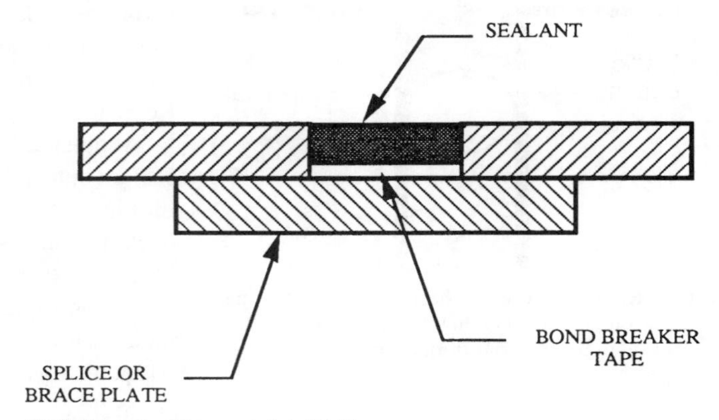

FIGURE 4.48 Narrow-joint detailing.

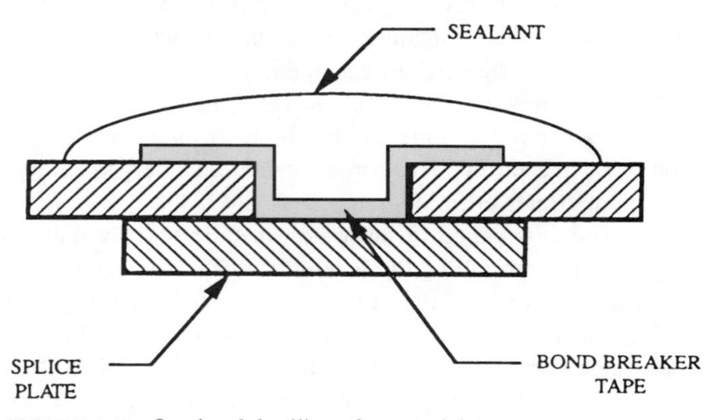

FIGURE 4.49 Overband detailing of narrow joints.

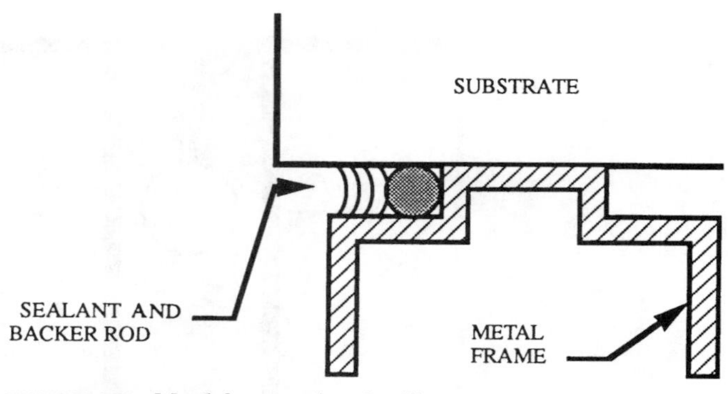

FIGURE 4.50 Metal-frame sealant detailing.

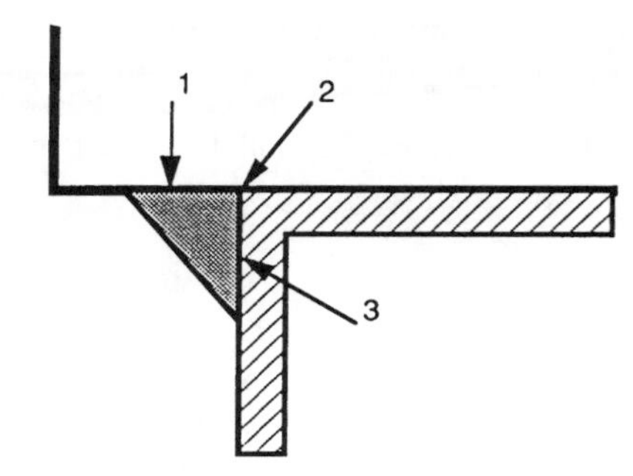

FIGURE 4.51 Incorrect metal-frame sealing.

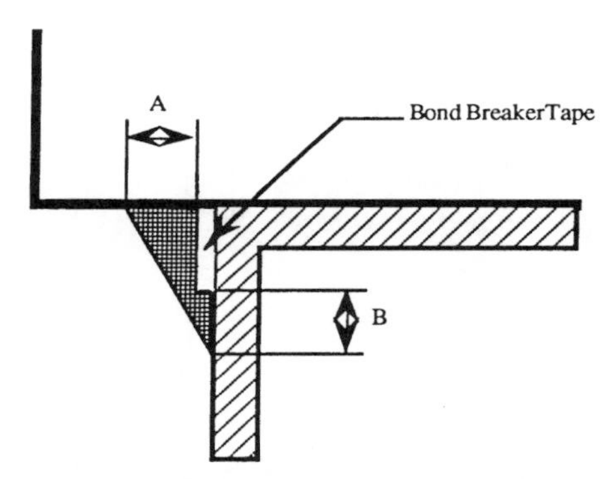

FIGURE 4.52 Correct metal-frame sealing. Distance *A* should equal distance *B*.

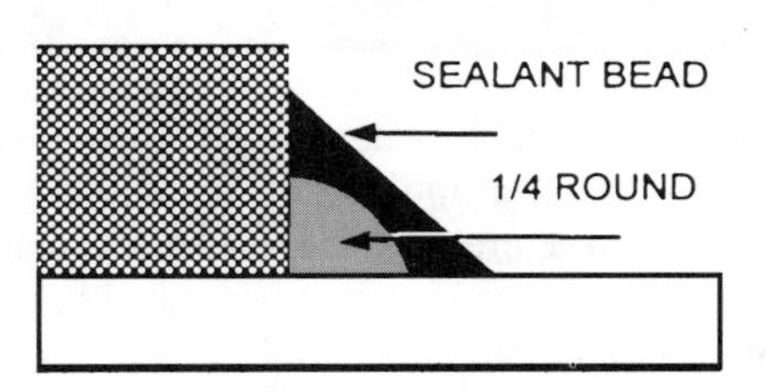

FIGURE 4.53 Proper application of perimeter joint using fillet backer rod. (*Courtesy of Nomaco*)

Bond breaker tape is installed as shown in Fig. 4.52 to allow contact area length to substrate equal to that of frames. Typically, the contact area length should be 1/2 inch. In curtain wall construction or storefront perimeters, based on movement capability of sealant, this length may be increased to allow larger joints. Figure 4.53 shows glass perimeters sealed with the use of "quarter-round" premanufactured backer rod.

Any gap or space between frame and substrate should first be filled with a caulking material. This provides a firm substrate on which to apply bond breaker tape. The sealant should then be tooled flat with a straight edge over the caulking. Table 4.12 summarizes the preferred installations for the major generic sealant materials.

GLAZING

Glazing refers to the use of silicone sealants to adhere glass, metal, tile and other finish materials to the structural components of a building envelope. Silicone materials are used for glazing because of their excellent adhesion qualities, with manufacturers providing warranties for up to twenty years.

Glazing can have structural properties when it acts as the primary attachment mechanism of the facade component to the structure. Glazing can also function as a nonstructural application, in which the glazing is used to seal the glass or other finishes for a weathertight envelope.

While the primary purpose of glazing is adhesive attachment, it also provides a secondary purpose of acting as a waterproofing barrier at the joints in the facade materials. For structural applications, high- or medium-high-modulus materials are used to provide effective adhesion capabilities, since the high movement capabilities of a low modulus material are of secondary importance. In nonstructural glazing, the low modulus materials are typically acceptable.

TABLE 4.12 Generic Sealant Materials and Their Common Uses

Substrate	Acrylic	Butyl	Latex	Poly-sulfide	Poly-urethane	Silicone
Metal frame at interior	X	X	X			
Metal frame at exterior				X	X	X
Precast joints				X	X	X
Glazing and bed joints		X				X
Interior work	X	X	X			
Stucco crack repair	X		X			
Horizontal joints				X	X	
Submerged joints				X		
Wood joints			X			X
Metal curtain walls		X		X	X	X
Stone and masonry joints				X	X	X
Bath fixtures	X		X			X
High movement joints					X	X
Parking deck joints					X	
Marble					X	X
Granite					X	X
Limestone					X	X
Kynar® finish		X				

A typical glazing detail is shown in Fig. 4.54. Note that the silicone structural glazing material is used to attach the glass to the metal mullion of the curtain wall components. The detail also includes the use of a nonstructural silicone material to seal the butt ends of the glass together and provide for a weathertight joint that is capable of expanding or contracting under thermal movement. A low-modulus material is typically used on the nonstructural portion, while higher-modulus and higher-tensile-strength materials are used for the structural glazing attachment.

Typical guidelines for structural glazing joint design (refer to Fig. 4.54) include:

- The "bite" dimension length should be a minimum of 1/4 inch.
- The "glueline" thickness should be a minimum of 1/4 inch.
- The "bite" dimension must be equal to or greater than the "glueline" dimension.

It is recommend that the structural bite dimension be calculated by using the following equation (Courtesy of Dow Coming):

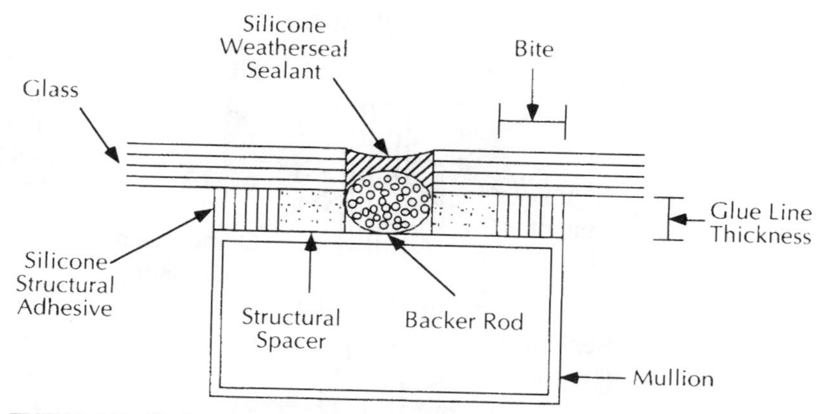

FIGURE 4.54 Design of typical structural glazing joint. (*Courtesy of Dow Corning)*

Bite (inches) = ½ × (smallest leg of the largest single piece of glass, length or width) × expected windloading in lb/sf / Sealant design strength × 12 in/ft

Thus, an 8-foot-by-16-foot piece of glass with a 60 lb/sf, using a sealant with a design strength of 20 lb/square inch, requires a one-inch bite. Whenever a fraction of inch results an appropriate safety factor is added by rounding up the required bite to the next ⅛ inch.

Glazing requires that all proper sealant application techniques described previously in this chapter be used when applying glazing materials in both structural and nonstructural joints. Most silicone glazing materials are moisture- or air-cured and therefore require the use of open- or dual-cell backing materials. In all structural applications, provisions must be made to permit the sealant to cure before applying the nonstructural glazing joints that can prevent exposure to air for curing.

Glazing also is used to seal the glass panes into the curtain wall or window frames to prevent air and water infiltration. Typical recommended detailing for nonstructural glazing is shown in Figs. 4.55, 4.56, and 4.57.

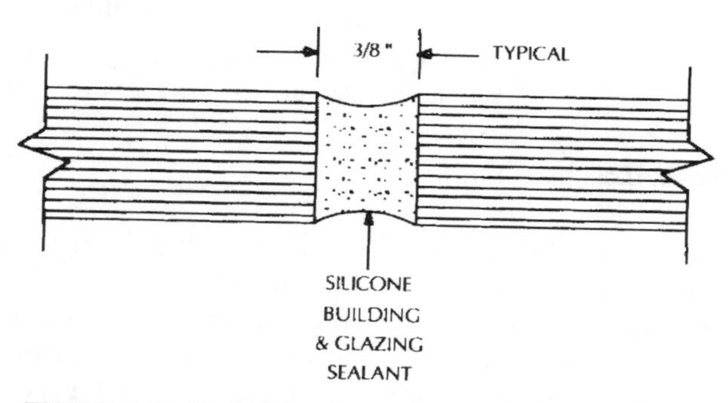

FIGURE 4.55 Typical butt joint glazing. (*Courtesy of Dow Corning)*

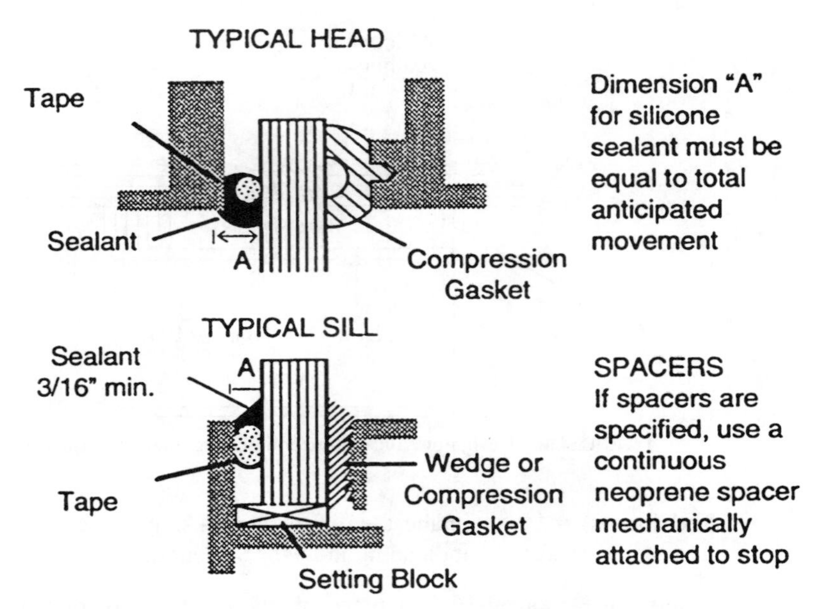

FIGURE 4.56 Nonstructural glazing detail using compression gasket. (*Courtesy of GE Silicone*)

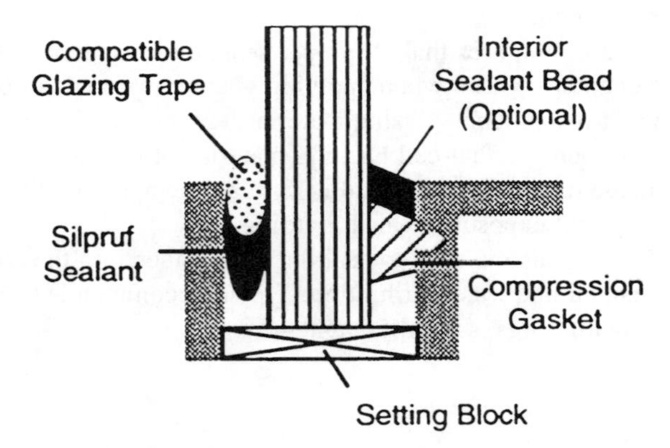

FIGURE 4.57 Nonstructural glazing with compression gasket, alternate detailing. (*Courtesy of GE Silicone*)

GLAZING MATERIALS

Most glazing installations are limited to silicone materials for both structural and nonstructural requirements. Specific silicone structural adhesive products are manufactured for structural applications involving adhering glass or other facade materials to a building envelope's structural components. Unlike silicone sealant materials described previously, the structural glazing materials are typically much higher in tensile, modulus, tear and peel strength, to provide the necessary adhesion qualities during life-cycling.

The higher strengths, including tensile, also result in a material that has much lower elongation capabilities than silicone building sealants. Typically, the structural silicone adhesives have 1/2 to 1/4 the elongation capability of superior silicone sealants. This does not mean the adhesives are poorer performers, only that the required adhesion strengths necessary for optimum performance in structural applications result in a lower elongation capability.

For nonstructural applications and those areas of glazing that are exposed to movement, typical silicone sealants, as described previously, can be used. These materials provide excellent movement capability with the necessary adhesion qualities required in glazing applications. Manufacturers also produce a combination of adhesive and sealing materials that can provide the necessary adhesion performance for light-duty applications, while at the same providing the sealing and weatherproofing qualities necessary. These materials have properties that fall between the high tensile and modulus of adhesion products and the low- modulus, high-elongation proprieties of the building sealants. Manufacturers should be consulted for appropriate products for specific installation requirements.

CHAPTER 5
EXPANSION JOINTS

INTRODUCTION

The variety of expansion joints available is almost as numerous as their failures. Leakage is so common, and failure so expected, that expansion joints are available with integral gutters to channel the water leaking through joints. Manufacturers often recommend installing a gutter system below joints to collect leaking water. One only has to visit a few parking garages and view the numerous attempts at collecting leaking water to confirm this situation. Roof gutters, PVC piping, and metal collection pans are often used in makeshift fashion to collect water leakage.

Leaking water collects salts, efflorescence, lime, sulfites, and other contaminants as it travels through substrates. This contamination causes damage to automobile paint finishes and building structural components. There are numerous causes for expansion joint failure. Among the most prevalent are:

- Selection of one joint for all details
- Improper detailing of joints into other building components
- Improper installation
- Use of too few joints
- Inadequate design
- Joints that are not capable of withstanding existing traffic

EXPANSION JOINT DETAILING

Expectation that one joint design will suffice for all conditions on a single project frequently causes failures. For example, a joint designed for horizontal straight runs is not appropriate for vertical installations, changes in plane, and terminations into walls or columns. Many joints are insufficient for 90° turns and changes in plane and often fail if such installations are attempted. Joint installations at walls or columns that abruptly stop with no provision for detailing joints into other building envelope components will fail. Attempts to install expansion joints continuously throughout a deck, including wall areas, planters, and seating areas, typically fail. Joints at building-to-deck intersections encounter considerable movement forces, including shear and differential movement, that often exceed joint capability.

For expansion joints to function properly over a range of in-place service requirements, they must include manufacturer details, design accessories, and systems components for the following common installations:

- Floor joints
- Wall-to-floor joints
- Building-to-floor joints
- Intersections with curbs
- Intersections with columns
- Joints at ramps
- Ramp-to-floor joints
- Intersections of two or more joints
- Changes in direction
- Joint terminations

Other common problems are connections between joints and substrates. These connections must withstand movement occurring at joints, or they will be ripped away from the substrate. If sufficient protection from traffic conditions is not provided, traffic wear over a joint might eventually break down or damage connections.

Often joints are not designed for the shear or lateral movement occurring in parking decks, especially at ramp areas. When an automobile travels over a joint, live loads induced by the automobile cause one side to lower while the opposite side remains level. Reverse action occurs after the auto passes over the joint. This *shear stress* can be felt by standing directly over a joint when automobiles cross.

Joints must be designed to withstand shear loading in addition to expected expansion and contraction movement. Expansion joints, such as T-joints, with a metal plate beneath the sealant portion, often fail because shear movement forces the plate into the sealant, ripping it apart.

For expansion joints to function when in place, they should also have the following components:

- Connection details for installation to structural components
- Connection details for waterproof coatings or membranes
- Protection against vehicular and snow plows
- Channeling of any water that might collect
- Cleaning provisions to remove accumulated dirt (e.g., leaves)

Expansion joints do not alleviate all movement encountered with deck construction. Concrete may crack at short distances from expansion joints due to shrinkage, settlement, or differential movement. This is common with double-T precast construction incorporating a topping slab. At each panel joint, the topping slab is subjected to differential movement and will crack over each T-joint, regardless of how large an expansion joint is installed. Therefore, adequate allowances must be made for settlement and for differential and structural movement, which expansion joints alone cannot resolve.

For a joint to be successful, it must have the following characteristics:

- The ability to withstand substantially more than the expected movement
- The ability to withstand all weathering conditions (e.g., freeze–thaw)
- The ability to withstand road salts and other atmospheric contaminants
- Facility of installation or training by manufacturer
- A superior connection to deck details
- Be seamless along its length

In addition, adequate allowances for tolerances must be made in deck levelness.

DESIGN OF JOINT

The first step in selecting an effective expansion joint is determining the amount of movement expected at a joint. This can be completed by computing the expected movement of a facade span or deck. This total movement is then divided into a number of strategically placed joints throughout the span.

Actual placement of the required joints is completed by a structural engineer. He or she determines where structural components can be broken to allow for movement, in addition to where this movement is likely to occur.

Besides allowances for substrate movement, it must be determined what movement will occur in such areas as deck-to-building and floor-to-wall intersections. Differential movement and structural movement will occur at these areas, and an expansion joint system that will function under these conditions must be chosen.

For expansion joints at building-to-deck intersections, expansion material should be connected to both building and deck structural components, rather than facade materials. Expansion joints applied to surface conditions become loose and disbonded during weathering and wear cycles. Additionally, structural component movement may exceed movement capabilities of the facade, causing joint failures. If it is necessary to install a surface-mounted joint, a secondary or backup seal should be installed beneath the expansion joint for additional protection.

In considering placement of joints, all design factors should be reviewed to avoid possible problems. For instance it is not practical to place a planter, which is filled with soil and plants and is constantly watered, over an expansion joint. Even with proper protection, failure will occur when dirt contaminates the joint and disrupts movement capability. Furthermore, planter walls placed over a joint may not allow joint movement, causing failure of the joint and wall. Similarly other items, such as equipment placement, column placements, light stanchions, and auto bumpers, should be reviewed.

Although movement amounts expected at joints are calculable, it is difficult to predict all types of movement that will occur. Factors such as wind loading, structural settlement, and distortion of materials impose directional movements that joints are not capable of withstanding. Therefore, selected joints should be capable of taking movement in any direction, a full 180° out of plane in all directions, to prevent failure.

Once a joint has been selected and sized and appropriate accessories selected to cover various details, proper joint terminations are designed. Simply stopping a joint at a wall, column, or termination of a deck will usually cause leakage. Attempting to apply a sealant over terminations is not sufficient. The sealant will not withstand movement that is likely to occur, especially if shear and other forces are encountered.

Manufacturers should provide specific termination detailing for complete weathertightness and movement capability at terminations. Any joints that channel water must incorporate allowances into the joint design to collect and dispose of the water at terminations. Most waterproofing systems are not manufactured to span expansion joints and are not capable of withstanding the movement that occurs there. Therefore, specific details must be designed for successful juncture of expansion joints into other waterproofing and building envelope components.

CHOOSING A JOINT SYSTEM

In choosing joint systems, examine all possibilities and choose a system for each specific need. Although convenient, it is not practical to choose one joint design for all conditions. Accordingly, manufacturers will have several types of systems and designs within each generic type to fulfill given project requirements. This prevents the dividing of responsibility among several manufacturers. Likewise, manufacturers of other building envelope components should approve the use of selected joint systems to ensure compatibility and complete envelope weatherproofing.

Generically, several systems are manufactured for use as expansion joints, including:

- Sealant systems
- T-joint systems
- Expanding foam
- Hydrophobic expansion seals
- Sheet systems
- Bellows systems
- Preformed rubber systems
- Combination rubber and metal systems
- Vertical systems
- Heavy-duty metal systems
- Below-grade applications

Sealants

Sealants are often used as expansion joint materials, but they are successful only for joints with minimal movement. Sealants are not recommended for joint widths greater than 1 inch. Joints larger than 1 inch should be backed by other material such as expanding foam sealants or be used as part of the T-joint system. In designing sealant expansion joints, manufacturers recommend joint widths of four times expected movement when the material is capable of 25 percent movement in one direction.

Chapter 4 discusses the various sealant materials and their properties and uses. For horizontal expansion joint applications, polyurethane sealants are commonly used. Urethanes are capable of withstanding both pedestrian and vehicular traffic. They are compatible with deck coatings, sealers, and protected membrane applications.

Exposed sealant joints will not be effective when subjected to harsh traffic such as snowplows and vandalism. In such instances, it is advisable to protect sealants from abuse by installing a metal plate or other protection over joints. If protection is installed, it is attached to only one side of a joint to allow for movement. Refer to Fig. 5.1 for typical sealant expansion joint installation.

Sealant systems can be installed in a contiguous application with no seams. Terminations and junctures to other building envelope components are easily detailed and installed. Sealant systems are used extensively as expansion joint material in deck coatings and membrane waterproofing applications in which waterproofing systems are carried directly over the sealant (Figs. 5.2 and 5.3). This installation type is effective as long as movement at joints does not exceed the capability of the waterproofing material. (See Table 5.1.)

For vertical applications, joint width also should not exceed 1 inch, unless specifically approved by the manufacturer. In joints exceeding 1 in, a backup material, such as expanding foam sealants, is suggested. In larger joints, it is advisable to cover sealants with a metal plate to protect against vandalism and excessive weathering.

Sealant materials are manufactured specifically for use in horizontal deck joints. They have properties making them resistant to traffic wear and contaminants such as oil, grease, and road salts. These materials are available for specific conditions such as airport runways. Typically, the material is a coal-tar derivative for resistance to gasoline and jet fuels.

In using sealant systems for remedial applications, all traces of previous material should be removed. If asphalt products were used, abrasive cleaning must remove all traces of contaminants. This will prevent problems associated with bonding of new sealants.

In all applications, joints must be protected during the curing stage, with no traffic allowed during this period, which may be as long as 72 hours. Refer to Chapter 4 for specific application details for sealant joints.

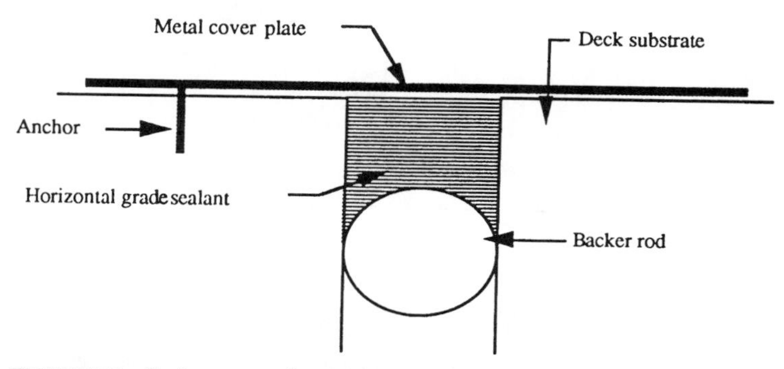

FIGURE 5.1 Sealant expansion-joint detailing.

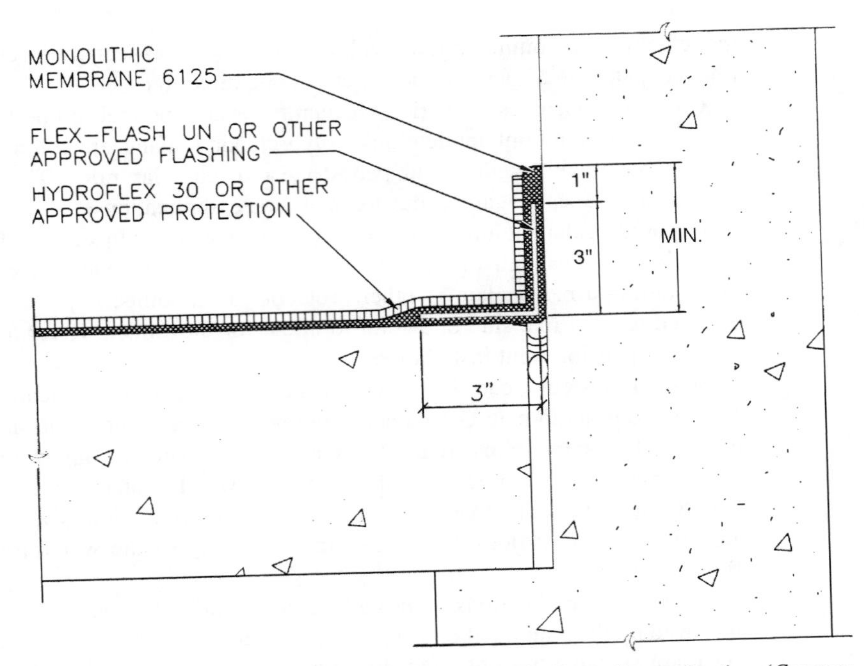

FIGURE 5.2 Sealant at wall-to-floor juncture prior to waterproofing application. (*Courtesy of American Hydrotech, Inc.*)

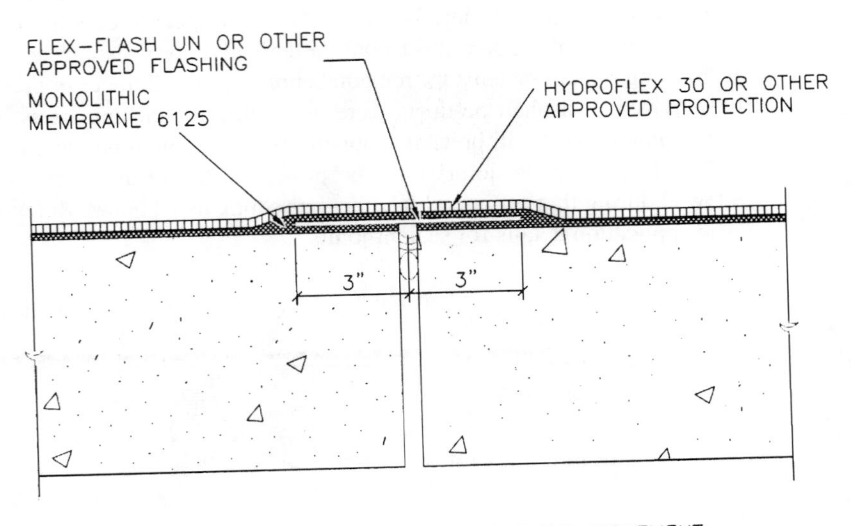

NOTES: DETAIL TO ACCOMODATE MAX. 50% TOTAL MOVEMENT.
BACKER ROD AND SEALANT ARE RECOMMENDED IN JOINT PRIOR TO MM6125/FLEX–FLASH UN INSTALLATION.

FIGURE 5.3 Sealant expansion joint covered by waterproofing application. (*Courtesy of American Hydrotech, Inc.*)

TABLE 5.1 Sealant Expansion Joint Properties

Advantages	Disadvantages
Compatibility with waterproofing systems	Minimal movement capability
Seamless application	Protection recommended
Ease of terminations	Not for excessive wear areas

T-Joint systems

A T-joint system is a sealant system reinforced with metal or plastic plates and polymer concrete nosing on each side of the sealant. This system derives its name from a cross section of the joint, which is in the shape of a T. Figure 5.4 shows a typical T-joint configuration.

The basic T-joint consists of several components, including:

- The sealant, usually urethane
- The reinforcement plate, 1/8-in aluminum or plastic reinforcement
- The bond breaker tape, which prevents three-sided adhesion to reinforcement
- The epoxy or polymeric nosing for attaching to substrates

Whereas sealant joints are recommended only for up to 1 inch widths, T-joint design allows for greater widths by adding reinforcement at both bottom and sides. With T-applications, joints as wide as 12 inches (excluding nosing) are used. Design width of sealant in T-joints is recommended at five times anticipated movement, versus four times with regular sealant joints. This provides an additional safety factor for these size joints. Manufacturers require that joint width be not less than 3–4 inches.

The T-system modifies a regular sealant joint to withstand the abuse and wear encountered in traffic-bearing horizontal joints. The metal plate reinforces the soft sealant during loading by automobiles. Nosings provide impact resistance and additional adhesion properties. It is recommended that this nosing extend approximately 1/8 inch above the sealant material and be sloped toward the sealant, to prevent damage from automobiles and other heavy equipment.

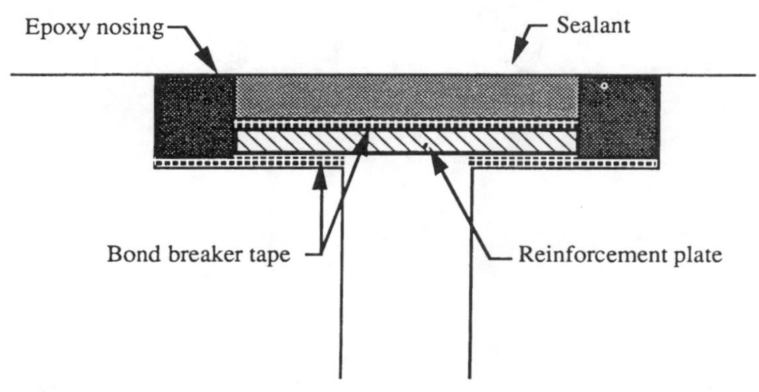

FIGURE 5.4 T-expansion-joint detailing.

The two installation methods for T-joints are fluid-applied and preformed. Both use sealant materials, but a preformed system uses a sealant that has been formed and precured (see Fig. 5.5). This cured material is then placed into a joint at the job site. Fluid-applied systems are placed directly into a joint at sites after mixing and before curing (see Fig. 5.6). Both systems have distinct advantages and disadvantages.

Preformed systems allow for uniform sealant thickness and curing under controlled conditions. This prevents possible abuse that may occur during curing of fluid-applied systems. Preformed systems are not seamless applications and require a site filling of seams with compatible sealant.

Preformed material is usually formed in 8 feet lengths, requiring seams every 8 feet. Preformed systems do not make allowances for irregularities with levelness of a substrate. The nosing is applied after the preformed sealant placement, to alleviate any irregularities in joint width and levelness.

Fluid systems are vulnerable to damage and weathering during the curing stage. Colder temperatures may extend the length of typical curing time from 48 to 72 hours. Sealants used in expansion joints are typically a self-leveling grade. This causes fluid-applied sealants to flow to low ends of a joint, resulting in uneven joint thickness. Fluid systems may shrink somewhat in the long joint, and possibly pull away from the nosing during curing.

Bond breaker tape is required between a concrete deck and a reinforcement plate and between this plate and sealant. If bond breaker tape is installed improperly and turns up joint sides, improper adhesion will occur. Refer to Chapter 4 for further discussion of sealant tape installation.

FIGURE 5.5 Sealing preformed T-joint seams. (*Courtesy of Coastal Construction Products*)

FIGURE 5.6 Finishing of T-expansion joint. (*Courtesy of Coastal Construction Products*)

Epoxy nosings are installed level with the edge of the concrete deck and are installed by troweling to cover minor irregularities within the deck. With preformed joints, sealant is beveled along edges at approximately 45°, upon which nosing material is placed. This provides adequate bonding to secure the preformed sealant to a substrate. Nosing material should not flow or be troweled onto horizontal portions of the sealant surface, as this prevents sealant movement capabilities.

These joints may be applied over existing expansion joints by ramping the polymeric nosing upward to provide the required sealant depth. However, this exposes a joint to abuse from vehicular traffic and nosing will eventually wear, exposing sealant to damage.

T-systems are labor intensive, providing opportunities for job-site misapplications as compared to factory-manufactured systems that require minimal field labor. (See Table 5.2.)

TABLE 5.2 T-Expansion Joint Properties

Advantages	Disadvantages
Reinforced for better wearing	Labor intensive
Seamless application	Sealant portion exposed to wear
Ease of terminations	Not for excessive wear applications

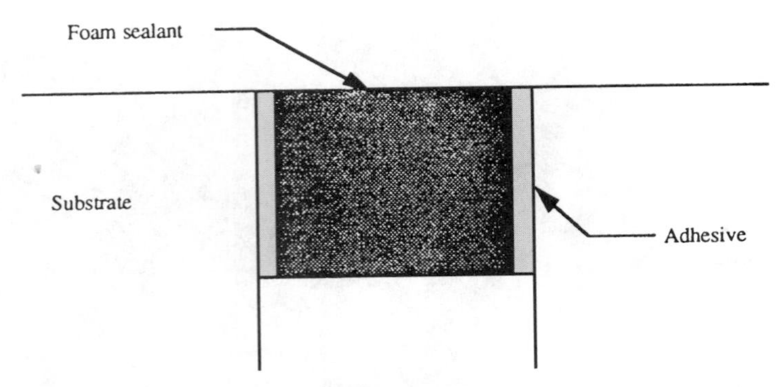

FIGURE 5.7 Foam expansion-joint detailing.

Expanding foam sealant

Foam sealants should not be confused with generic sealants. Expanding foam sealants are composed of open-cell polyurethane foam, fully impregnated with a manufacturer's proprietary product formulation; these include neoprene rubbers, modified asphalts, and acrylic materials. Foam sealants are covered in detail in Chapter 4. A typical foam expansion joint is detailed in Fig. 5.7.

Foam materials are supplied in a compressed state, in rolls of various widths and lengths. For large widths, straight pieces 8–10 feet long are manufactured. A release paper over the adhesive on foam sealant facilitates installation.

These materials have considerably fewer elongation properties than better sealants, (150 versus >500 percent for sealants). They also have lower tensile strengths than sealants (20 versus 200 lb/in^2).

With limited elongation properties, these joints should be designed to be in a continuous compression rather than an elongation mode. Therefore, materials are provided in widths of two to five times the actual joint width, allowing materials to be in compression always.

Foam systems are particularly easy to install. The material is completely premanufactured and requires only that the joint be cleaned, contact paper removed, and the materials adhered to one side of the joint. Foam sealants then expand to fill a joint completely. Timing of this expansion is dependent on weather conditions, being slower in colder weather. These materials expand laterally and will not expand vertically out of a joint if properly installed.

Foam materials are extremely durable considering their low tensile strength. Once installed, foam is difficult to remove and is resistant to traffic and vandalism. Depending on the impregnating chemicals used, they can also be resistant to gasoline and oils.

Manufacturers produce several grades and compositions of materials designed for specific types of installations. These include below-grade and above-grade joints, vertical or horizontal applications, and high-traffic grade for bridges and highways.

For vertical expansion joints, foam is often used as backup for a fluid-applied sealant. Horizontal installations do not require a cover plate or other protection. Foam sealants are also used as secondary protection in T-joints and are installed in place of standard backing material in a joint beneath the support plate.

Due to adhesion characteristics, foam material adheres to itself, providing seamless joint applications. It is recommended that joining ends of material be mitered for additional adhesion. These materials allow for 90° turns with changes in plane, intersections, and terminations easily and effectively detailed. They are compatible with most building materials comprising the building envelope. (See Table 5.3.)

Expanding foam sealant systems also make excellent choices for remedial applications. Existing joints that have failed using generic sealants can usually be easily prepared to receive a new foam sealant. The joint is prepared by removing the failed sealant, including any grinding or solvent wiping necessary to remove traces of old material that remains on the sides of the joint.

The new foam sealant is then applied into the existing joint per manufacturer recommendations. Applying a sealant over the foam sealant, as described in the Chapter 4 secondary sealant section, can provide additional protection at the joint.

Figure 5.8 details the use of an expanding foam system involving the addition of a new wall adjacent to an existing structure. Such locations require allowance for differential as well as thermal movement. In this detail, a cover is provided as the primary barrier system with the foam joint acting as a secondary barrier or double-seal protection.

Hydrophobic expansion systems

Combining hydrophobic resins with synthetic rubber produces hydrophobic expansion seals. *Hydrophobic* refers to materials that swell in the presence of water. Thus, these materials require active water pressure to become effective water barriers. They are similar to below-grade clay waterproofing systems and therefore are limited to below-grade applications. As with foam sealants, materials are provided in rolls in preexpanded form. Due to their reactivity with water, materials must not encounter water until after installation.

The use of hydrophobic expansion systems in expansion joints is extremely limited. Typically, they are used in conjunction with waterproofing membranes to fill expansion, control, or cold joints in below-grade construction. They are also used as waterstop materials in concrete substrates.

These materials swell from 2 to 10 times their initial volume. They have low tensile strength, but their elongation is similar to fluid-applied sealants, with some materials exceeding 500 percent elongation. As with foam, they should only be used in a compression mode. (See Table 5.4.)

Sheet systems

Sheet materials are manufactured from neoprene or hypalon rubber goods. They range from 40 to 60 mil in thickness, and width ranges from 4 to 12 inches. Joint expansion

TABLE 5.3 Foam Expansion Joint Properties

Advantages	Disadvantages
Factory manufactured	Cost
Seamless application	Poor elongation
Ease of terminations	Low tensile strength

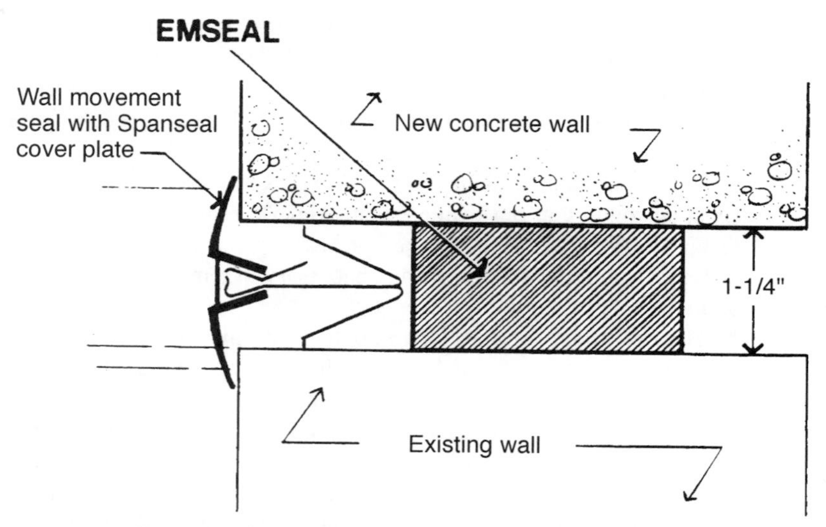

FIGURE 5.8 Expanding foam sealant. (*Courtesy of Emseal Joint Systems*)

and contraction are made weathertight by installing these materials in a bellows or loop fashion. This provides sufficient sheet material for stretching during contraction of a substrate. Material provided to form the bellows should be at least two to four times the expected joint movement. Figure 5.9 is representative of a typical sheet installation.

Materials are supplied in rolls 10–25 feet long. Seams are fused together by vulcanizing the rubber with a manufacturer's supplied solvent. Solvents are applied at seams that are lapped over each other, completely fusing the two pieces of material.

Materials may be applied to a substrate surface or recessed into a joint by installing a cutout along each side of the joint, Fig. 5.10. Sheet systems are typically perforated along their edges for complete embedding of the sheet in an epoxy or polymer mix used to adhere the material to a substrate. This provides an effective mechanical bond allowing for installation over substrates other than concrete, including wood, metal, masonry, and glass.

Sheet installations allow materials to be applied at floor-to-wall joints (Fig. 5.11), besides straight horizontal applications (Fig. 5.12). Sheet systems are, however, difficult to install at transitions between a horizontal floor joint and vertical wall joints. This is because of the bellows that forms in the material when making changes in direction or plane. It is not effective to turn the material in a 90° bend, as the bellows distorts and deters the system's effectiveness and bonding to a substrate.

TABLE 5.4 Hydrophobic Expansion Joint Properties

Advantages	Disadvantages
Chemical resistant	Below-grade applications only
Follows contours of joint	Requires positive waterproofing systems
Good elongation	Used only with waterproofing systems

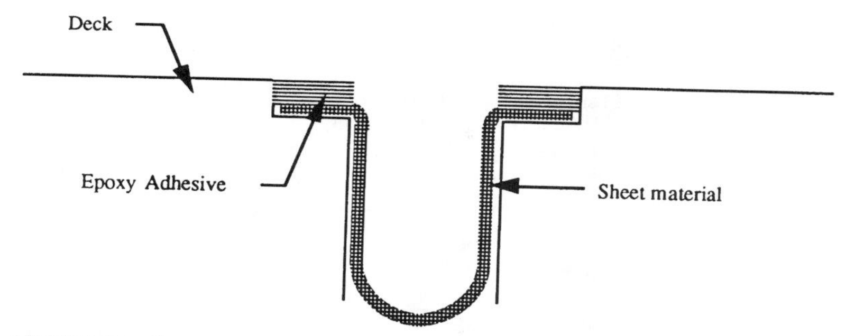

FIGURE 5.9 Sheet expansion-joint detailing.

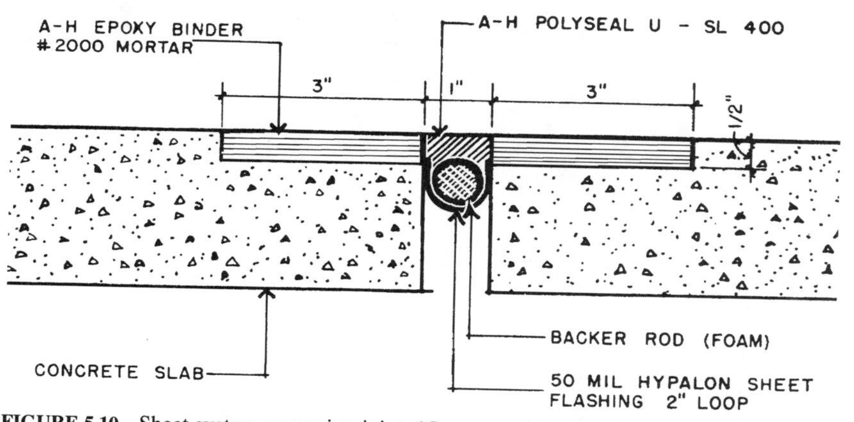

FIGURE 5.10 Sheet system expansion joint. (*Courtesy of Anti-Hydro*)

Sheet systems are also difficult to terminate into structural components such as columns or ramp walls. At such details, material is formed into a box or dam and fused together. This design allows for collection of dirt and debris in the bellows that eventually prevents a joint from functioning. Further, water collecting in a bellows acts as a gutter, with no drainage for water.

Because of these problems, joints should be covered with metal plates to prevent the accumulations. The cover plate also prevents possible safety hazards to pedestrians, who might trip on an exposed joint.

TABLE 5.5 Sheet Expansion Joint Properties

Advantages	Disadvantages
Vertical and horizontal applications	Collects debris in bellows
Good shear and deflection movement	Difficult to terminate
Metal, glass, and wood applications	Changes in-plane detailing

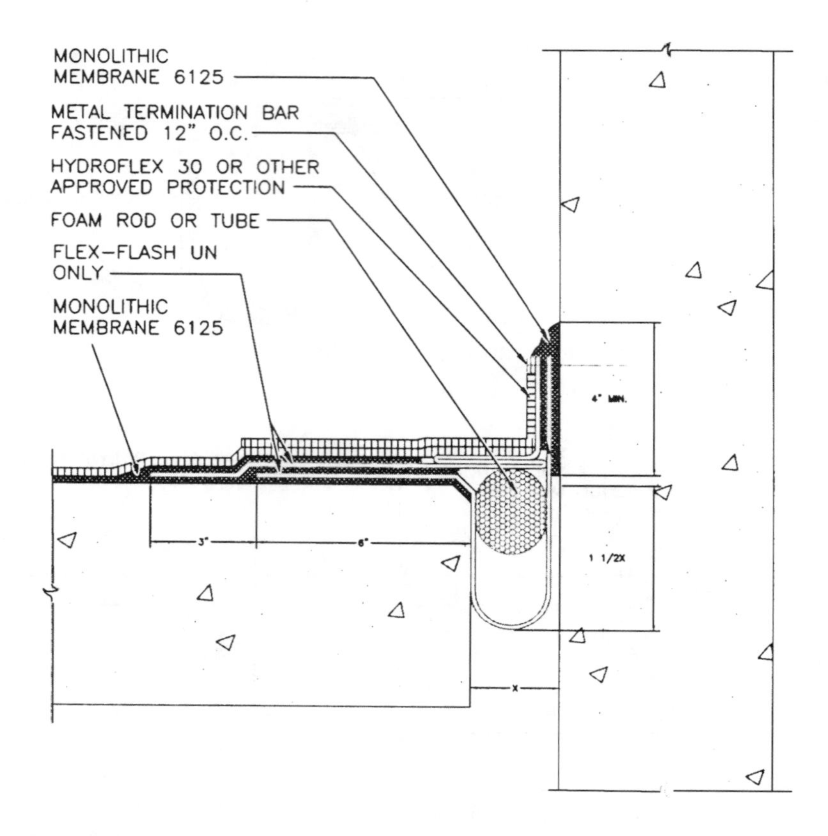

NOTES: DIAMETER OF FOAM ROD OR TUBE SHOULD BE 1 INCH LARGER THAN MAXIMUM JOINT OPENING.

INCORPORATE A FOLD IN THE SHEET OF THE FLEX–FLASH UN TO ACCOMMODATE MOVEMENT OF THE JOINT. A BOND BREAKER MAY BE REQUIRED BETWEEN THE FOLDS.

DETAIL DESIGNED TO ACCOMODATE MAX. 50% TOTAL MOVEMENT.

WHENEVER POSSIBLE THE DETAIL SHOULD BE CURBED OR OTHERWISE CONSTRUCTED TO SHED WATER AWAY.

FIGURE 5.11 Sheet expansion joint system used at wall-to-floor joint. (*Courtesy of American Hydrotech, Inc.*)

Sheet materials are effective choices in remedial applications. They can be surface-mounted to an existing substrate, without requiring the substrate to be grooved or trenched for installation. Epoxy or polymer adhesives are applied in a ramp or slanting fashion, to prevent blunt ends that might be damaged by vehicular traffic.

This type of installation allows the joint to be installed in applications where two different substrate materials, such as brick masonry, must be sealed. Refer to Fig. 5.13 for remedial detailing of a sheet system. Besides expansion and contraction movement, these systems also withstand shear and deflection. (See Table 5.5).

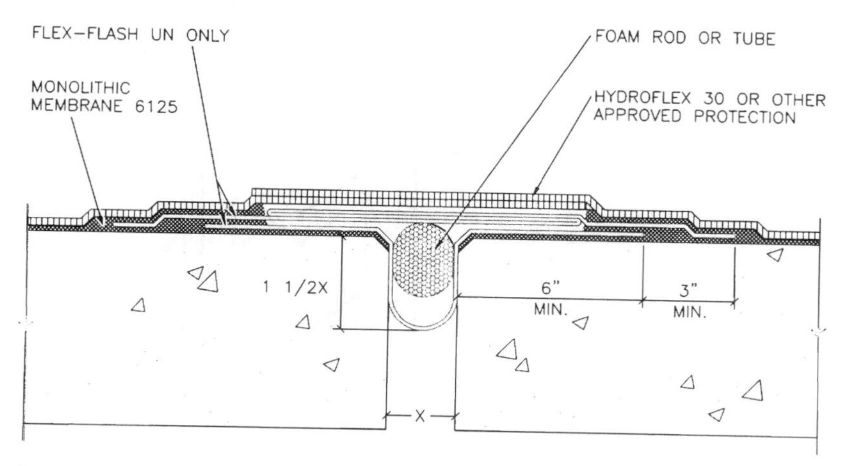

FIGURE 5.12 Sheet expansion joint system. (*Courtesy of American Hydrotech, Inc.*)

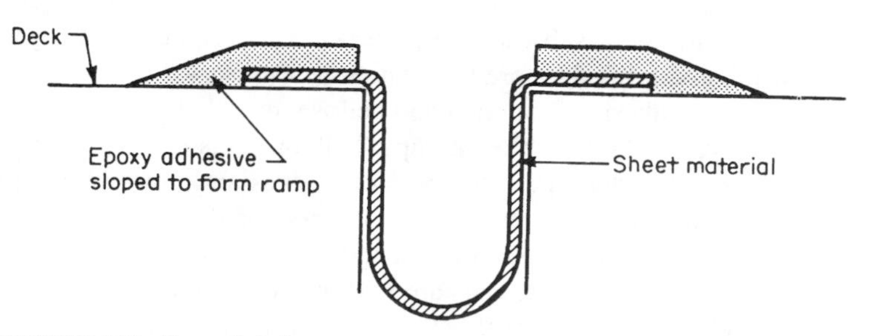

FIGURE 5.13 Remedial sheet expansion-joint detailing.

Bellows systems

Bellows systems are manufactured from vulcanized rubber into preformed joint sections. They are installed by pressurizing the joint cross section during adhesive curing that promotes complete bonding to joint sides. A typical installation is shown in Fig. 5.14. These systems are similar to preformed rubber systems, but use air pressure for installation. Their cross sections are not stiffly reinforced by ribs manufactured in the material, as are other preformed systems. Epoxy or polymeric nosing can also be installed, to provide for better wearing at edges.

Bellows are available in sizes up to 3 inches but are normally applied in joints 1 inch wide. Joint material depth is approximately twice joint width. These systems function under 50 percent compression movement and 50 percent expansion movement.

Since bellows systems use preformed material, traffic can be applied immediately upon adhesive cure. Unlike sheet systems, adhesive is applied to interior sides of a

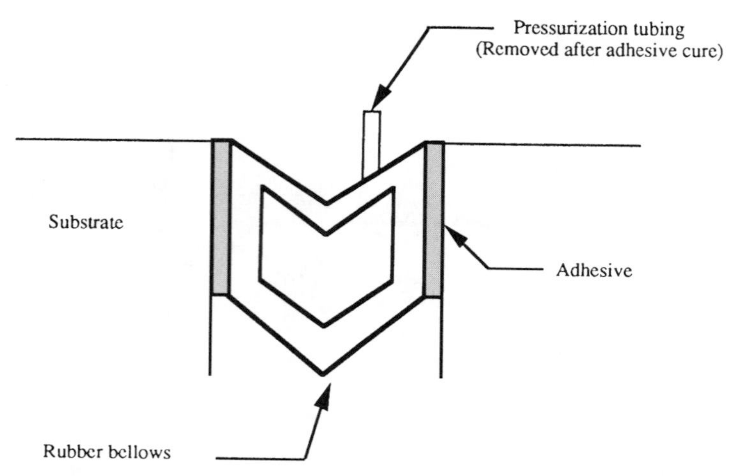

FIGURE 5.14 Bellow expansion-joint detailing.

joint, thereby protecting them from traffic wear. Additionally, the bellows is closed, which prevents accumulation of debris and water and, therefore, does not require a cover plate for protection.

These systems do not allow for major irregularities in joint width. This would prevent materials from performing in expansion or contraction modes. They also cannot take up irregularities in substrate unlevelness. This requires that a joint be saw-cut to uniform width and leveled before installation if necessary.

After adhesive is applied and bellows installed, air is injected to expand the joint cross section, similarly to blowing up a balloon. This pressure is maintained until the adhesive is cured, at which time the pressure valves are removed and pressure holes sealed. This joint functions under movement in stress and deflection.

Material is supplied in roll lengths usually sufficient for seamless application. Should seaming be necessary, ends are vulcanized together with solvents.

Bellows systems are effective for surface-mounted floor-to-wall building joints. It is not, however, possible to turn the material 90° for changes in plane. Additionally, it is difficult to terminate these systems, and manufacturers should be consulted for recommendations.

For remedial installations, any existing joint material must be removed completely and interior joint sides must be cleaned for adhesive bonding. If existing joints are irregular in width or shape, they should be cut to uniform size for proper installations. (See Table 5.6.)

TABLE 5.6 Bellow Expansion Joint Properties

Advantages	Disadvantages
No cover plates required	Maximum 3-inch width
Factory-manufactured	Difficult to terminate
No debris or water collection	Nonconforming to irregularities

Preformed rubber systems

There are numerous preformed rubber systems available. These are manufactured from extruded synthetic rubbers such as neoprene and hypalon. They are available in countless cross sections and sizes. Unlike the bellows systems, they require a blockout or ledge in the substrate on which to place joint material. Many systems require an epoxy or polymeric concrete nosing at joint edges to prevent damage.

Many preformed systems have flanges attached to a compression seal that is perforated to allow for embedding into nosing material for mechanical bonding of joints to substrates. Other systems use metal frames for attachment to concrete substrates before concrete placement. Still others rely on chemical bonding to a substrate with adhesives.

Preformed rubber systems are available in widths ranging from 1/4 to 6 inches. Movement capability varies, but it is usually 50 percent compression and expansion movement. Rubber systems are very resistant to weathering and chemical attack from gasoline, oils, and grease.

These systems are typically used for straight horizontal runs only, but some are designed for use at vertical-to-horizontal junctures. These joints do not allow for 90° changes in plane. Some of the numerous cross sections of preformed expansion joints are shown in Fig. 5.15.

Preformed systems have high impact (tensile) strength, usually more than 1000 lb/in^2. This strength reduces movement capability, and materials should be sized accordingly. The high tensile strength allows for excellent wear resistance on areas subject to large amounts of vehicular traffic. Preformed systems do have limitations on the amount of shear and deflection movement they are able to withstand.

Flange Systems

Adhesive Systems

FIGURE 5.15 Preformed rubber expansion joints.

Only one portion of a preformed joint, the nosing, is job-site-manufactured. This nosing anchors a joint to a substrate. Size of the nosing and adhesive contact area must be installed properly to ensure that the joint does not rip from the substrate during weathering or movement (Fig. 5.16).

Mechanical attachment of preformed systems is completed by using metal anchor bolts installed through holes in the rubber flange joint section (Fig. 5.17). These systems require a blockout to allow a joint to be flush with a substrate. Anchoring should be checked by maintenance crews on a regular basis, since bolts may work themselves loose during joint movement. Horizontal seams are sealed by mitering ends of the rubber portion and fusing them with a solvent.

TABLE 5.7 Preformed Rubber Expansion Joint Properties

Advantages	Disadvantages
Factory manufactured	Cost
High-impact strength	Difficult transitions
Chemical resistant	Limited remedial applications

Premanufactured enclosures are available, to terminate preformed joints into other building envelope systems. Factory-manufactured accessories are available for changes in plane, direction, and intersections with other joints (Figs. 5.18 and 5.19). Premolded joints are used in remedial applications but require concrete substrates to be cut, to provide a ledge to place flanges as shown in Fig. 5.20. (See Table 5.7.)

Combination rubber and metal systems

Combination rubber and metal systems are manufactured with a basic rubber extrusion seal and metal flanges for casting directly into concrete placements (Fig. 5.21). These systems are designed for new construction installations and are placed by the concrete finishers. They are not used in remedial applications unless major reworking of a deck is involved. Joints are manufactured in large sizes, with no joints being less than 1 inch wide.

The metal flanges on each side have reinforcement bars or studs for bonding with the concrete. Some combination joints include intermediate metal strands between the rubber, for additional reinforcement and wear. Others include metal cover plates for additional protection against traffic wear.

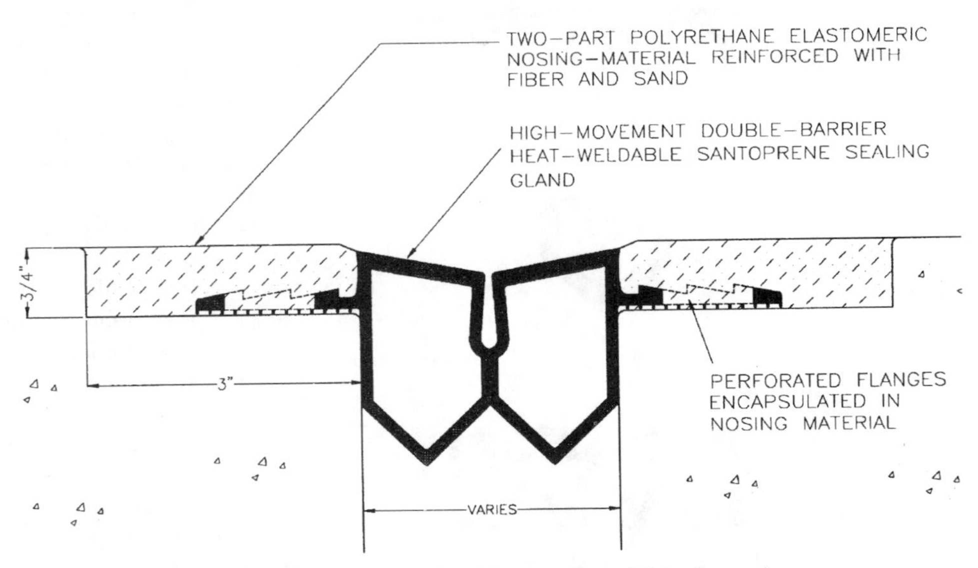

FIGURE 5.16 Preformed rubber expansion joint. (*Courtesy Emseal Joint Systems*)

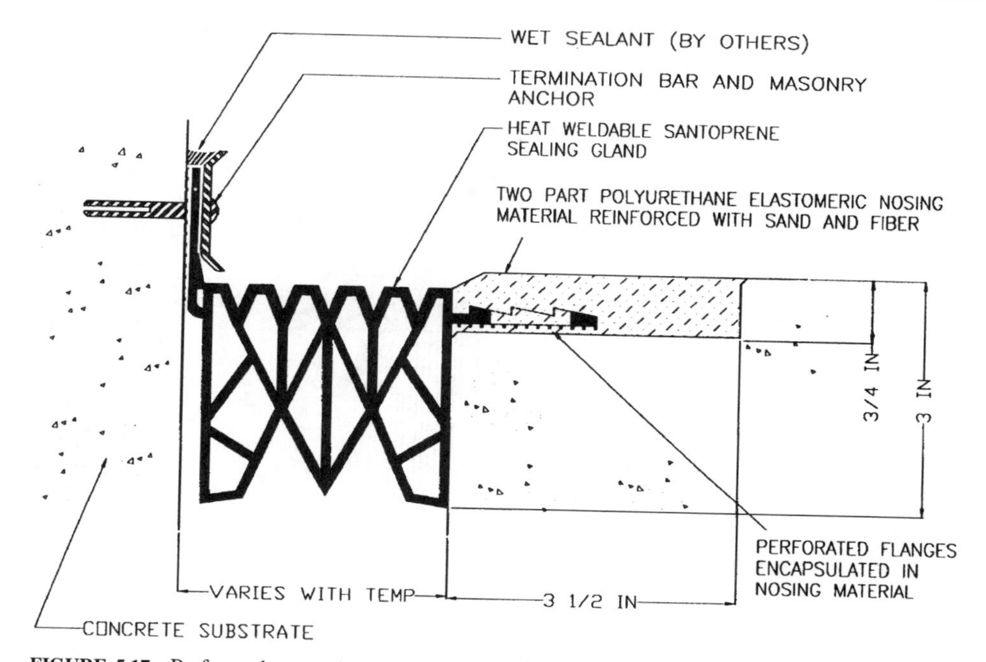

FIGURE 5.17 Preformed expansion system using metal anchor bolts for attachment. (*Courtesy of Emseal Joint Systems*)

Combination metal–rubber expansion materials may be interlocked to cover a joint width of more than 2 feet. These systems are costly, and are used primarily in heavy service areas such as bridges and tunnels.

Combination systems require accessories for terminations, changes in plane or direction, and intersections with other joints. They must be carefully positioned before concrete placements and protected so as not to allow concrete to contaminate the joint. Concrete must be reinforced at the metal flange intersection to prevent substrate cracking and water infiltration that can bypass a joint.

In some instances it may be advantageous to block out a joint location during concrete placement, and install the joint with polymer concrete mix after concrete curing. A typical combination system are shown detailed in Fig. 5.22. (See Table 5.8.)

Combination joint systems are available for exclusive use with topping or sandwich-slab construction, or tile and paver finishes over sandwiched membranes. Figure 5.23 details a combination system for use in paver construction over a structural slab with membrane waterproofing.

TABLE 5.8 Combination Expansion Joint Properties

Advantages	Disadvantages
Available for large joints	New construction only
Factory manufactured	Cost
Extremely durable	For joints wider than 1 in

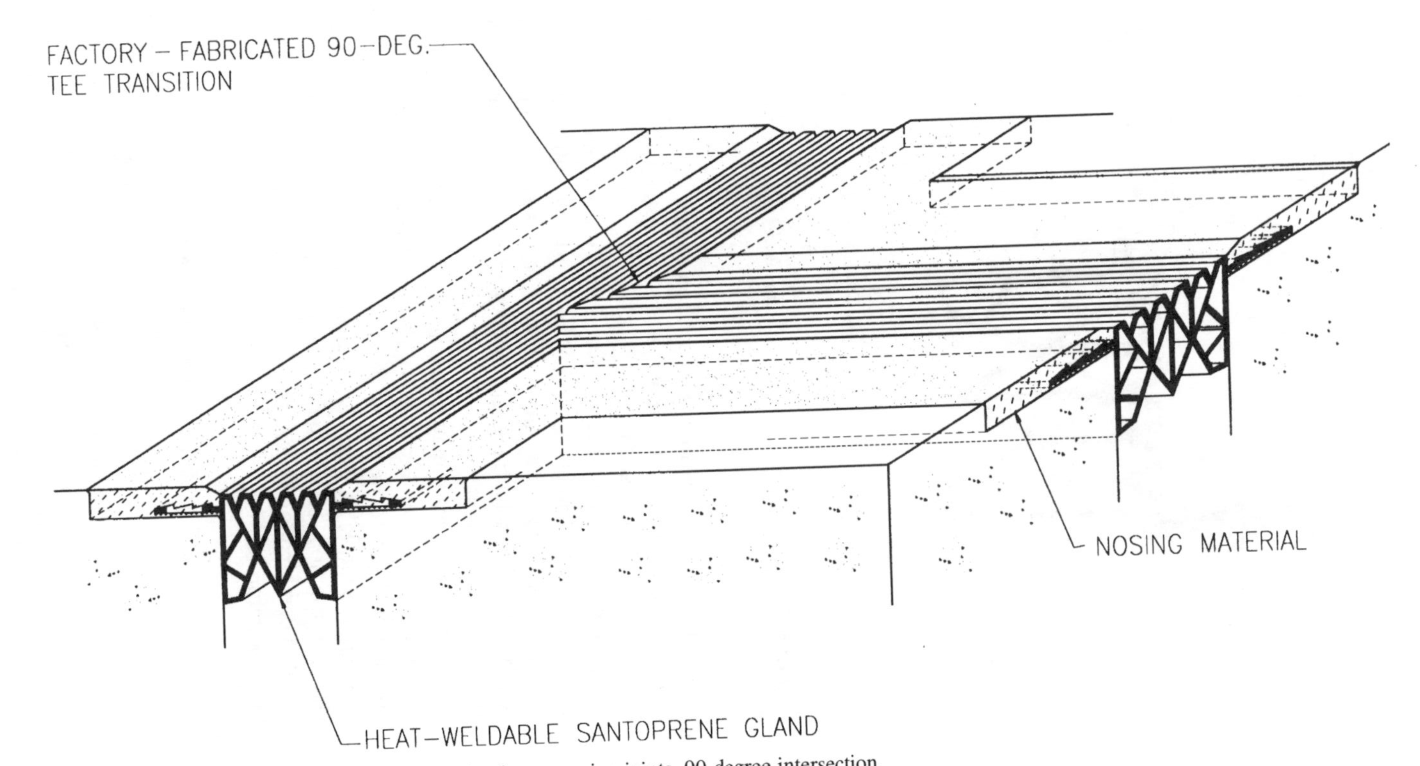

FIGURE 5.18 Factory-manufactured accessories for expansion joints, 90-degree intersection.

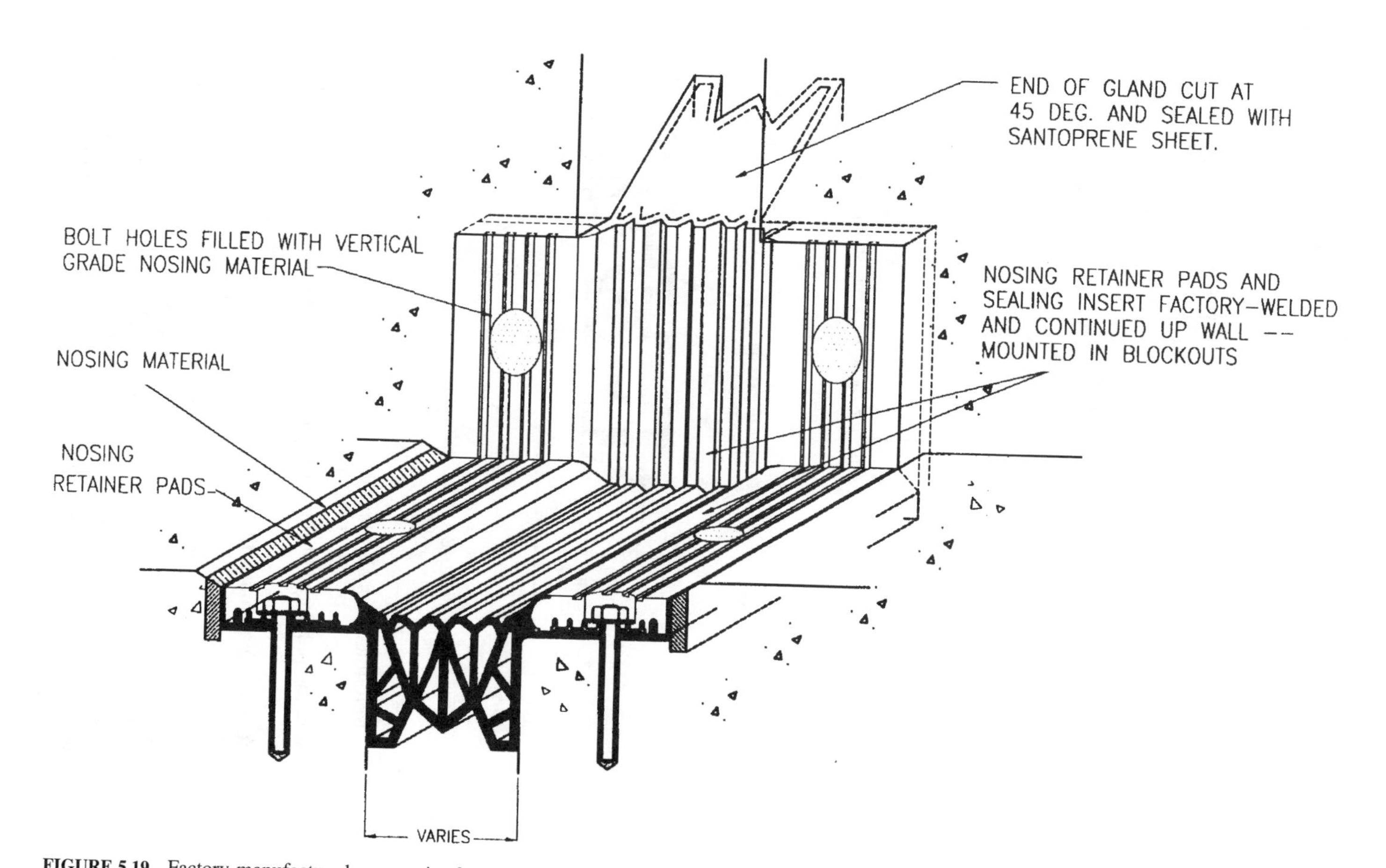

FIGURE 5.19 Factory-manufactured accessories for expansion joints, wall-to-floor transition. (*Courtesy Emseal Joint Systems*)

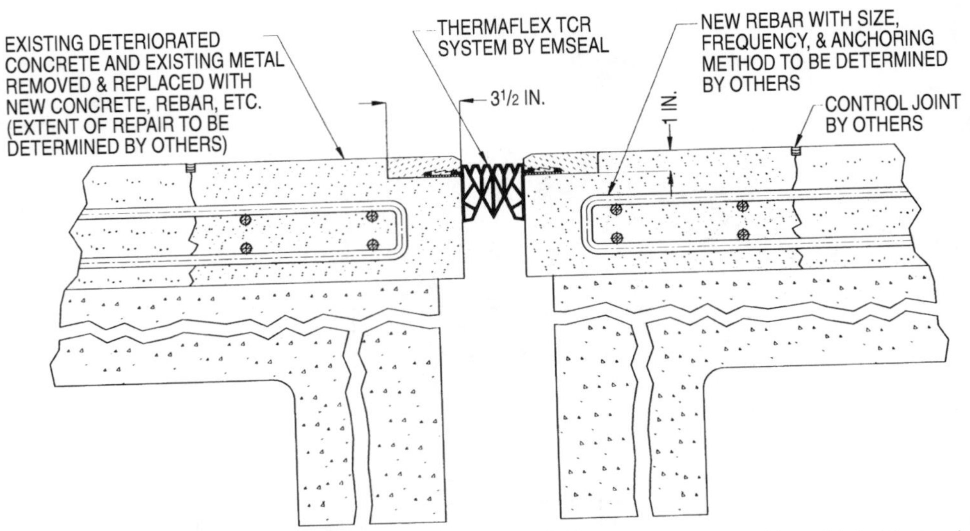

FIGURE 5.20 Ledge cut into substrate for anchoring expansion joints. (*Courtesy Emseal Joint Systems*)

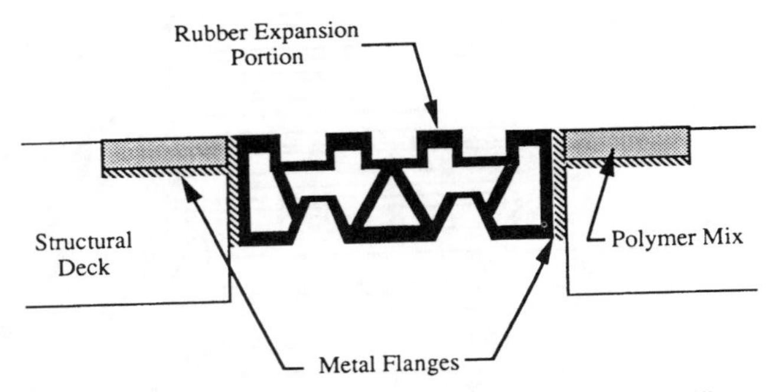

FIGURE 5.21 Combination rubber-metal expansion-joint detailing.

Manufacturers also provide appropriate termination and transition accessories for their joint systems. Figure 5.24 details a combination system at a floor-to-wall intersection. Note the requirement for additional counter flashing to be installed.

Vertical expansion joints

Most systems discussed thus far are also manufactured for horizontal uses. These include premanufactured metal and plastic expansion and control joints. Typical details of ribbed and smooth vertical expansion joints are shown in Figs. 5.25 and 5.26.

Installations of stucco wall systems usually allow no dimension greater than 10 feet in any direction and no area larger than 100 ft^2 without control joints being installed for thermal movement. Joints also are installed where there are breaks in structural components behind stucco facades.

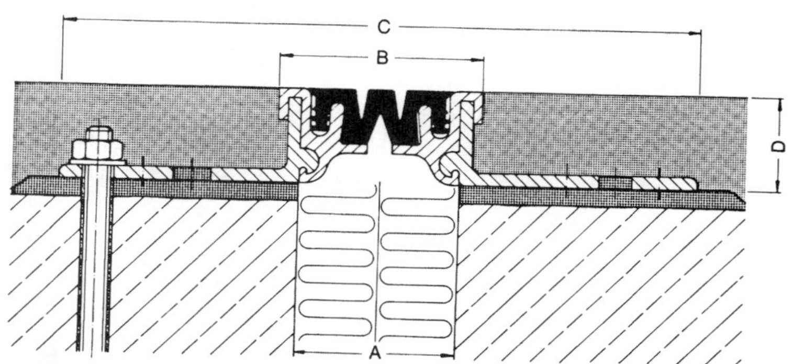

FIGURE 5.22 Substrate block-out installed to facilitate installation of expansion joint. (*Courtesy of Emseal Joint Systems*)

Preformed metal joints for stucco are available in a variety of designs and metals. The most durable metal is zinc, which does not corrode like galvanized metal. Zinc materials withstand greater substrate movement than plastic or PVC materials.

A typical cross section of a stucco control joint is shown in Fig. 5.27. The metal lath flanges are used to attach joints to substrates and are secured in place when stucco is applied over flanges. Flange sides should be secured to separate and structurally break the sides of a substrate to allow structural movement, Fig. 5.28. Applying both flanges to the same structural portion will defeat the expansion joint purpose.

When vertical joints intersect horizontal joints in a facade, they should not be broken, Fig. 5.29. Breaking the horizontal joint instead will prevent the water running down the facade from entering at joint intersections. These intersections should be monitored during installation, as this is the most likely area of infiltration.

Stucco substrates often separate from preformed joints due to differential movement between the materials. This results in cracking along joint faces, allowing access for water infiltration into a structure and its components.

Unfortunately, to repair what might be perceived as a leaking joint, metal joints are often filled with sealant. This restricts joint movement capability and does not address the immediate problem.

In all types of vertical envelope surfaces, expansion and control joints should be placed at changes in plane or direction, at intersections of dissimilar materials, around substrate openings, and where allowances are made for thermal movement or structural movement.

Heavy-duty metal systems

Manufacturers also provide horizontal expansion joints that are designed specifically for heavy-duty wear or use. These typically include such installations as parking garages, especially at loading dock areas, and interior decks subject to forklift or other heavy equipment.

These systems are usually complete metal fabrications, since the systems described in the previous sections could not withstand this type of abusive life-cycle conditions. These joints require blockouts to be installed in the concrete slab for installation. The joints expand and contract by the use of a flange that permits the top of the joint to slide back and forth as required, as shown in Figs. 5.30 and 5.31.

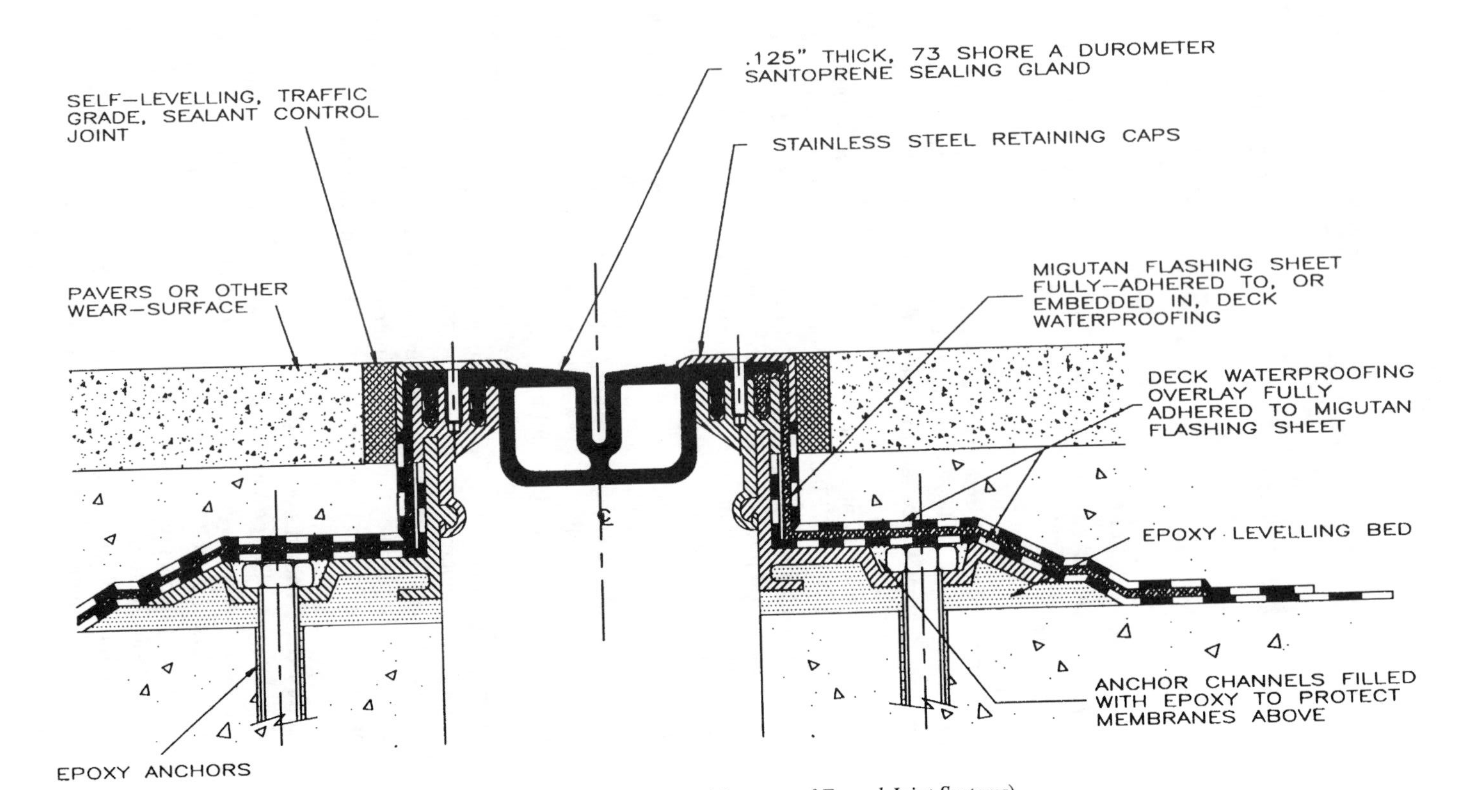

FIGURE 5.23 Expansion joint designed for sandwich-slab construction. (*Courtesy of Emseal Joint Systems*)

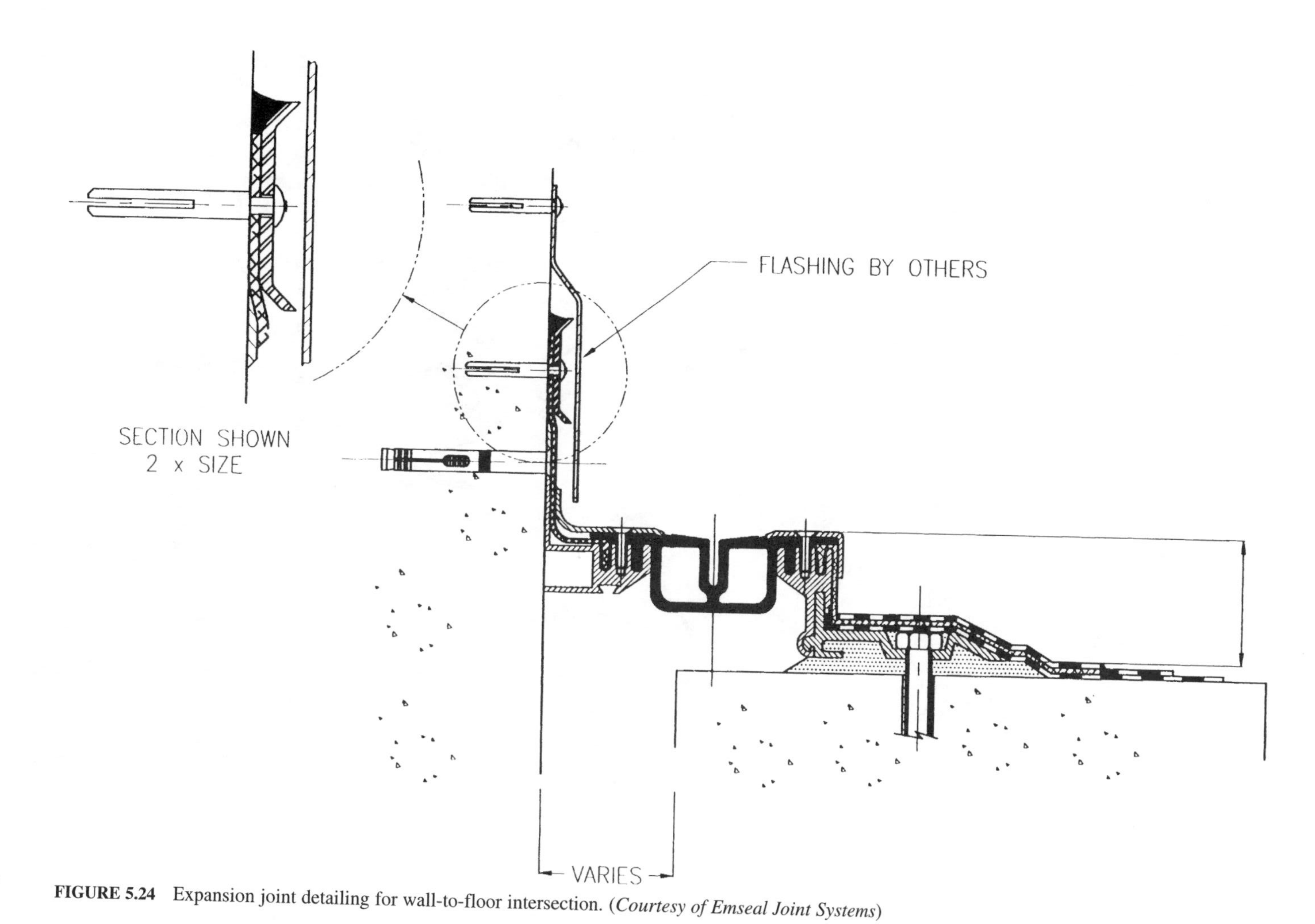

FIGURE 5.24 Expansion joint detailing for wall-to-floor intersection. (*Courtesy of Emseal Joint Systems*)

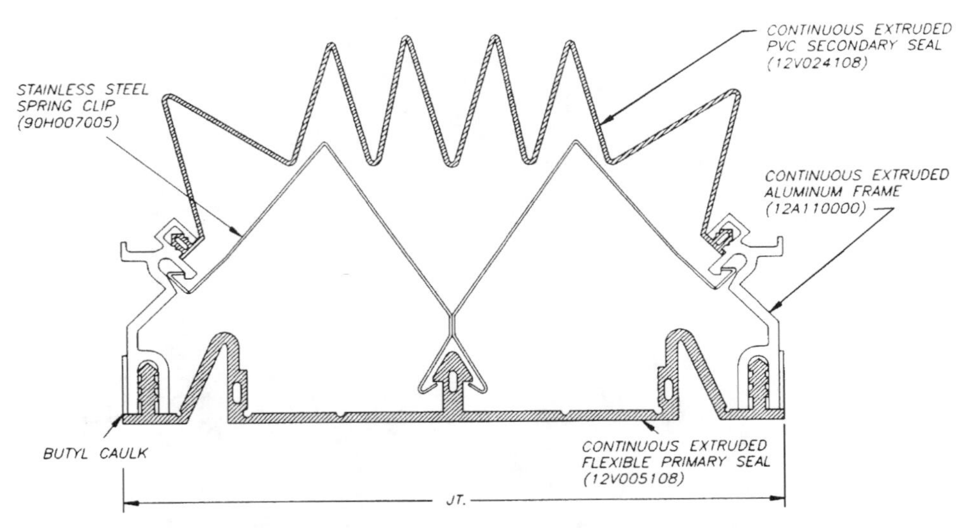

FIGURE 5.25 Premanufactured vertical metal expansion joint. (*Courtesy of the C/S Group*)

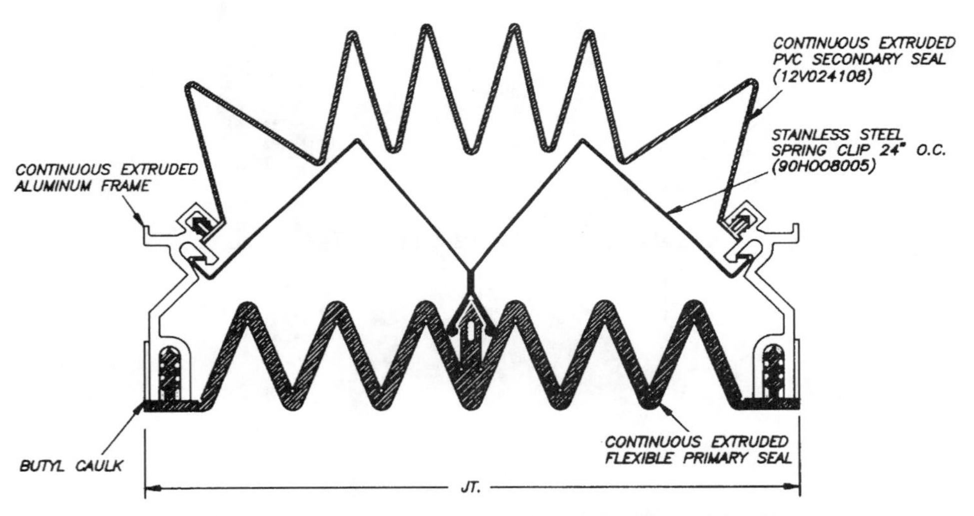

FIGURE 5.26 Premanufactured vertical ribbed expansion joint. (*Courtesy of the C/S Group*)

Obviously, the system cannot be field-adapted for any conditions, with the possible exception of shortening by field cuts. This requires that any termination or transitions be covered by using premanufactured pieces provided by the manufacturer. A typical floor-to-wall transition metal joint is shown in Fig. 5.32.

Below-grade expansion systems

Various treatments are used for below-grade expansion joints. As discussed in Chapter 2, these joints typically are not subject to thermal movement, but only to movement from structural

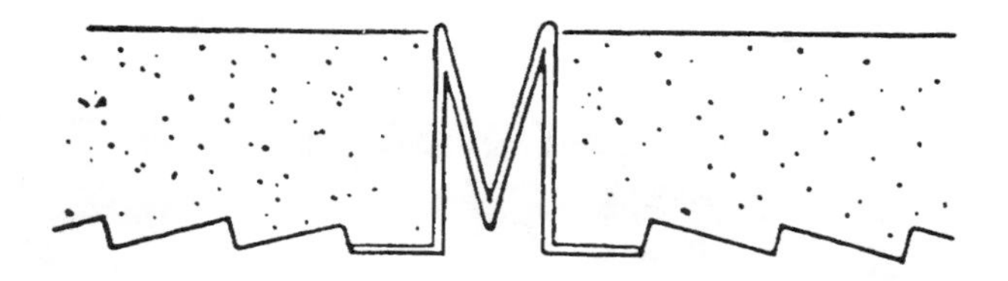

FIGURE 5.27 Stucco expansion joint detail. (*Courtesy Keene Products by Metalex*)

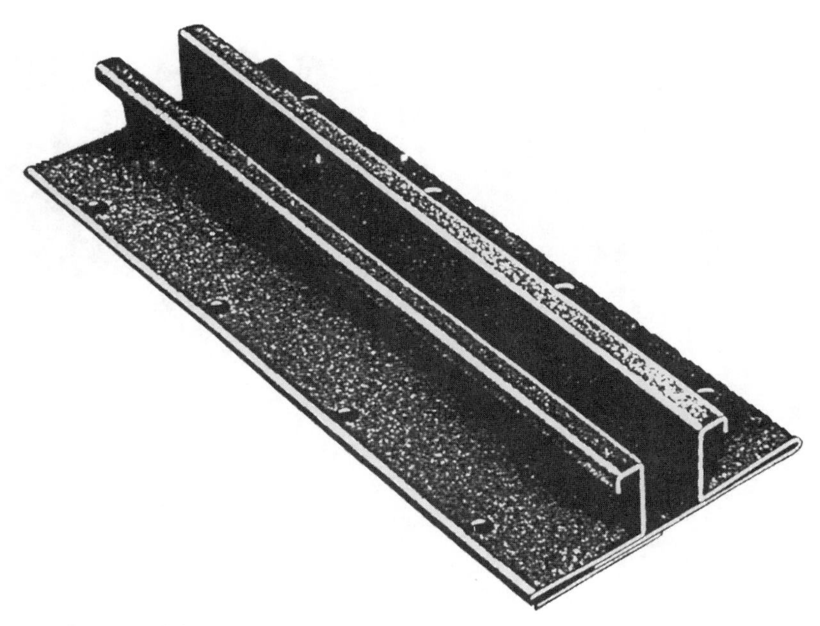

FIGURE 5.28 Schematic view of flanged stucco expansion joint. (*Courtesy Keene Products by Metalex*)

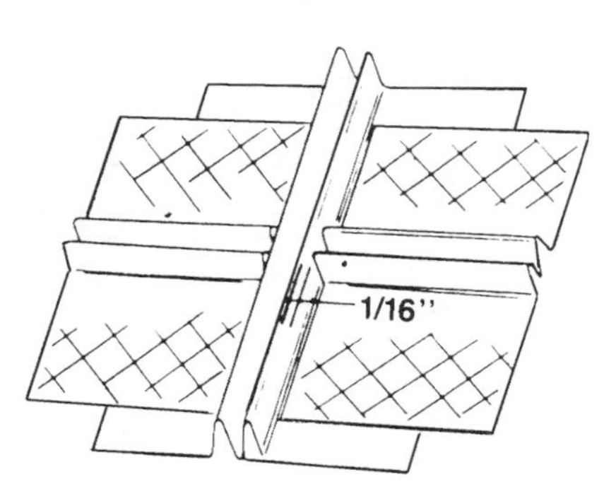

FIGURE 5.29 Premanufactured intersection for stucco expansion joints. (*Courtesy Keene Products by Metalex*)

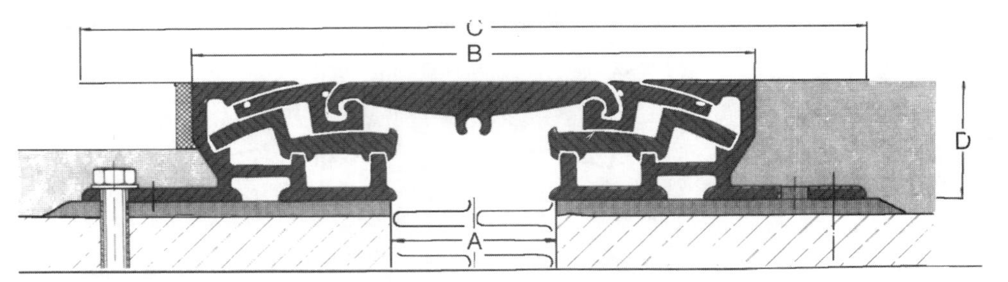

FIGURE 5.30 Metal expansion joint. (*Courtesy of Emseal Joint Systems*)

FIGURE 5.31 Metal expansion joint detailing. (*Courtesy of Emseal Joint Systems*)

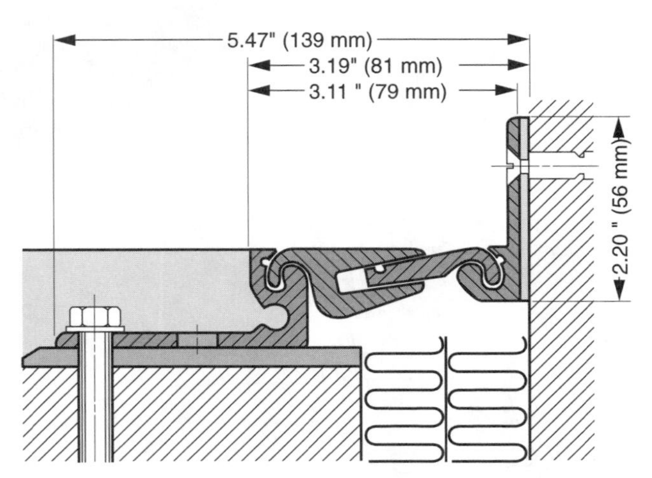

FIGURE 5.32 Metal expansion joint for wall-to-floor transition. (*Courtesy of Emseal Joint Systems*)

settlement during or immediately after construction. As such, these joints do not require the movement capability of those described in the previous sections.

Often the joint is used in combination with a waterstop system as described in Chapter 2. Figure 5.33 details the use of a foam sealant used in combination with waterstop for primary and secondary protection. The foam sealant system can also be used in lieu of regular sealant systems, when used in combination with waterproofing membranes as shown in Fig. 5.34.

Sheet systems are also used, including products that are designed specifically for use over yet-uncured concrete to facilitate the construction schedule. Figure 5.35 details a typical sheet system application for below-grade applications. The sheet is some type of vinyl or synthetic rubber and is applied directly to the concrete substrate using a cemetitious adhesive. This system can be used where waterstop has been inadvertently left out of a below-grade expansion joint area. Figure 5.36 shows a typical sheet system being applied.

EXPANSION JOINT APPLICATION

All joint systems require that substrates be free of all dirt, oil, curing compounds, and other containments. Joints should be smooth, level, and straight to allow functioning and movement of expansion materials. With remedial applications, irregular areas should be sawn out, leveled, and chipped where required. If a ledge is necessary for installation, it must be free of all fins, sharp edges, and honeycombing.

Most joint systems, with the exception of those placed into concrete, require that the substrate be cured and dry. The various expansion systems require unique installation procedures as recommended by the manufacturer. Sealant and T-joints are applied as described in Chapter 4.

Other expansion systems are factory manufactured and require only installation of adhesive or polymeric nosing.

Primers are generally required for horizontal sealant and T-joint applications. Polymer adhesives may require a solvent wipe or solvent primer before application. Rubber, combination, and metal systems require the installation of a blockout in the concrete substrate. A typical blockout detail is shown in Fig. 5.37. Note that the total width is the expansion joint system width plus 2 inches for tolerance and space to adhere the joint to the substrate.

Expansion joint systems have movement limitations. If deflection or shear movement is expected, use only materials expressly approved for this type of movement.

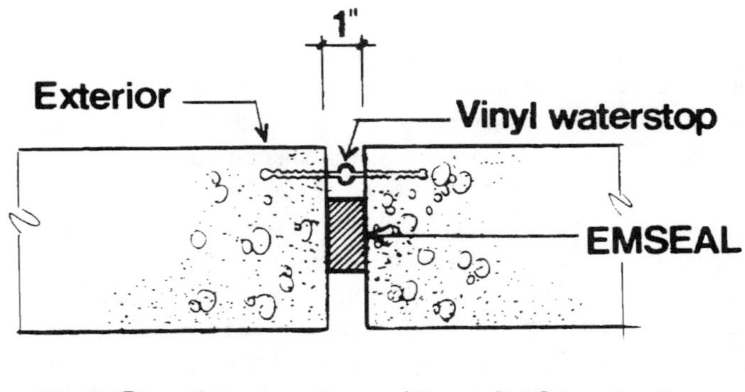

FIGURE 5.33 Below-grade expansion joint detailing using waterstop and foam sealant. (*Courtesy of Emseal Joint Systems*)

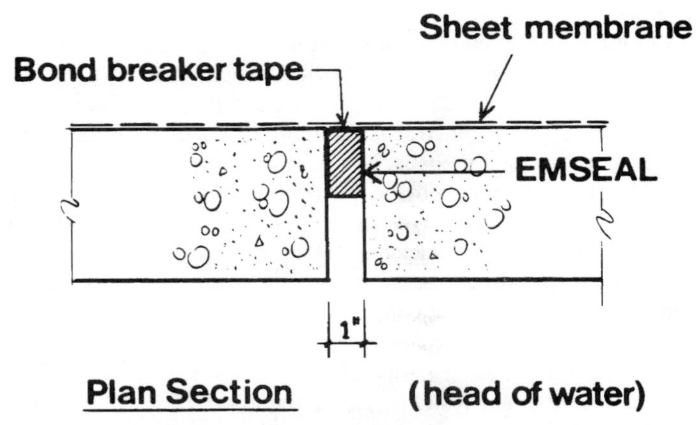

FIGURE 5.34 Below-grade expansion joint detailing without waterstop. (*Courtesy of Emseal Joint Systems*)

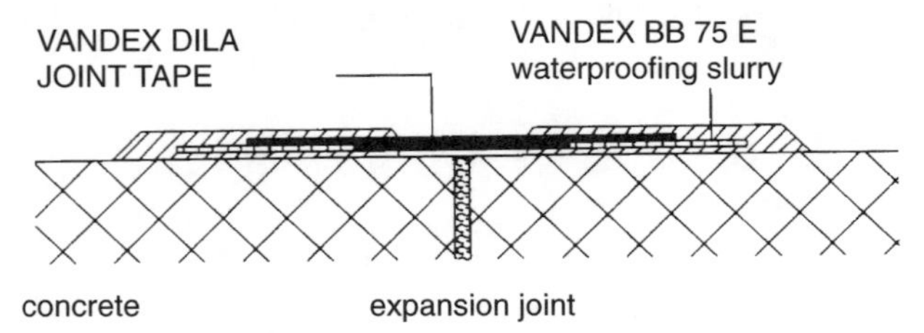

FIGURE 5.35 Below-grade sheet expansion joint system. (*Courtesy Vandex*)

FIGURE 5.36 Below-grade application of sheet expansion joint. (*Courtesy Vandex*)

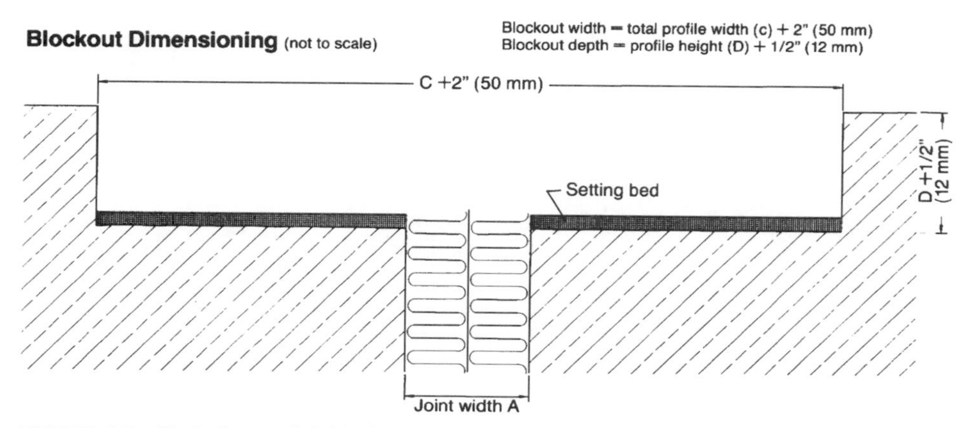

FIGURE 5.37 Typical remedial block-out detail for expansion joints. (*Courtesy of Emseal Joint Systems*)

CHAPTER 6
ADMIXTURES

INTRODUCTION

Admixtures are used with masonry and concrete materials to enhance and improve in-place cementitious product performance. Admixtures are additions, other than normal ingredients of aggregate, sand, water, and cement, which impart desirable qualities to in-place concrete or masonry. These qualities might include:

- Color
- Workability
- Shrinkage reduction
- Improved hydration
- Reduction of porosity
- Faster setting times
- Faster curing
- Waterproofing

Admixtures added during mixing of the concrete or masonry slurries add qualities throughout the in-place product. Surface-applied admixtures only disperse additional qualities to the substrate surface and to the depth to which it penetrates. Admixtures are available in many forms, including:

- Dry form
- Liquid additive
- Premixed cementitious form
- Dry shake or troweled-on (added at finishing stage)
- Liquid mixtures (applied during curing stages)

HYDRATION

Water added to cement, sand, and aggregate forms a paste that cures, hardens, and shrinks to create the finished concrete or masonry product. During curing, water leaves this paste through a process called hydration, which causes formation of microscopic voids and cracks in concrete. Once formed, these voids allow water absorption through the material.

Only controlled conditions of perfectly portioning, mixing, placing, and curing the concrete slurry will produce materials with minimum voids and absorption. Since field construction is never completed perfectly, however, concrete and masonry products often leak through the voids and cracks formed by the curing process.

The purpose of waterproofing admixtures is to provide complete hydration, which in turn promotes internal curing. This allows a reduction in shrinkage, providing a denser, higher-strength, and more water-resistant product by reducing the water absorption rates of a concrete or masonry material.

Admixtures available for concrete and masonry products that impart waterproofing or water-repelling characteristics include:

- Dry shake
- Concrete admixtures
- Masonry admixtures
- Stucco admixtures
- Agents
- Polymer concrete

DRY SHAKE

The dry shake, power troweled, or shake-on methods use materials similar to cementitious membranes for below- and above-grade waterproofing. The difference is that unlike materials for cementitious membranes, dry-shake admixture is applied during initial concrete finishing and curing (green concrete) rather than after curing. Shake-on admixtures consist of a cementitious base with proprietary chemicals that provide water-repellent properties.

These products are broadcast in powder form at $^3/_4$ to 1 lb/ft^2 of substrate area before initial concrete is set. Power troweling then activates proprietary chemicals with the moisture present in concrete.

With this method, the cementitious admixture becomes an integral part of a concrete substrate. These products do not merely add water repellency; they waterproof concrete against water-head pressure. They are effective admixtures, used when waterproofing of concrete substrates is required. These admixtures add compression strength to concrete substrates, and abrasion resistance to withstand heavy traffic and wear. As with all cementitious systems, these products do not withstand cracking or movement in substrates by structural, thermal movement, or differential movement. (See Table 6.1.)

DRY-SHAKE APPLICATION

Dry shake-on surface preparation requires only that concrete be in its initial setting stage, before power troweling. Areas being placed or finished should be no larger than a work crew can adequately cover by broadcasting material during this precured stage. Should concrete begin setting and curing, this method becomes ineffective for substrate waterproofing.

TABLE 6.1 Properties of Dry-Shake Admixtures

Advantages	Disadvantages
Simple installation	No movement capabilities
Above- and below-grade installations	Not completely waterproof
Becomes integral part of substrate	Must be applied during concrete finishing

Use materials only in dry powder form as supplied by the manufacturer. Immediately after broadcasting, admixtures should be power-troweled into the concrete. Dry-shake products, like all cementitious products, are not used when moving cracks or joints are expected.

MASONRY, MORTAR, PLASTER, AND STUCCO ADMIXTURES

Masonry, mortar, plaster, and stucco admixtures are added directly to the water, cement, aggregate, and sand paste, and are available in a liquid or dry-powder form. They consist of organic chemicals, usually stearates, and proprietary chemicals, which impart integral water repellency. These admixtures lower the amount of water required for a paste mix, increase internal curing by increasing hydration, and reduce shrinkage. This results in a high-density material, with high compressive strength, which absorbs substantially less water.

Specific additives, including chlorides, gypsum, metals, and other chemicals, might adversely affect concrete finishes or reinforcing in the substrate. For example metallic additives bleed through finish substrates, causing staining, and increase chlorides, often leading to reinforcing steel deterioration. Therefore, product literature for each type of additive should be reviewed for specific installation procedures.

Admixtures typically reduce water absorption from 30 to 70 percent of regular mixes under laboratory conditions of controlled mixing and curing. Actual reductions, considering field construction inaccuracies, will substantially lower the results of water reduction.

Even with the high reductions of water absorption achieved, these products are not adequate for complete waterproofing of building envelope components. Also, their inability to resist cracking and movement further restricts their waterproofing characteristics. (See Table 6.2.)

Therefore, admixtures should be considered only as support or secondary measures in providing a watertight envelope. This includes admixtures added to mortar for laying brick veneer walls, which assists the primary waterproofing properties of the brick facade. Flashing, dampproofing, weeps, admixtures, and water-repellent sealers all become integral parts of the building envelope.

The admixtures' compatibility with primary waterproofing materials should be confirmed. Admixtures of this type may adversely affect bonding capabilities of waterproofing sealers or sealants. If in doubt, testing is recommended before actual installations are made.

MASONRY AND STUCCO ADMIXTURE APPLICATION

Masonry and stucco admixtures require no specific surface preparation since they are added to the concrete, mortar, or stucco paste during mixing. Admixtures in quantities and

TABLE 6.2 Properties of Masonry, Mortar, and Stucco Admixtures

Advantages	Disadvantages
Simple installation	No movement capabilities
Above- and below-grade installations	Can stain or damage substrate
Becomes integral part of substrate	Not completely waterproof

mixing times recommended by the manufacturer should be monitored for complete dispersal throughout the paste. Water added to the paste must be measured properly so as not to dilute the admixture's capabilities and properties.

These materials are not waterproofing but water-repelling products. They will not function if cracking, settlement, or substrate movement occurs.

CAPILLARY AGENTS

Hydration of concrete or masonry materials leaves behind microscopic pores, fissures, and cracks from water that is initially added to make the paste mixture. This hydration allows in-place concrete and masonry materials to absorb moisture through these voids by capillary action. Capillary admixtures prevent this natural action and limit moisture absorption and water infiltration into a substrate.

Capillary admixtures are available in liquid or dry-powder form that is mixed into the concrete paste, applied by the shake-on method, or rolled and sprayed in liquid form to finished concrete. Capillary admixtures react with the free lime and alkaline in a concrete or masonry substrate to form microscopic crystalline growth in the capillaries left by hydration.

A substrate should be totally damp, to ensure complete penetration of capillary admixtures and provide the filling of all voids. This crystalline growth fills the capillaries, resulting in a substrate impervious to further capillary action. This chemical reaction requires moisture, either contained in a substrate or added if necessary.

As with other admixtures, these systems are not effective when cracks form in the substrate. Nor are they capable of withstanding thermal, structural, or differential movement. Capillary admixtures are further limited by their reliance on a chemical reaction necessary to form an impervious substrate. This reaction varies greatly depending on the following:

- Moisture present
- Alkali and lime available
- Admixture penetration depth
- Number and size of voids present
- Cracks and fissures present in a substrate

In the imperfect world of construction field practices, it is unrealistic to depend on so many variables to ensure the substrate watertightness that is essential to the building envelope. (See Table 6.3.)

TABLE 6.3 Properties of Capillary Admixtures

Advantages	Disadvantages
Simple installation	No movement capability
Above- and below-grade installations	Not completely waterproof
Fills minor fissures in substrate	Relies on chemical reaction

CAPILLARY ADMIXTURE APPLICATION

Dry-shake or capillary admixtures are applied before the initial set and finishing of concrete. Admixtures added to concrete or mortar paste require no additional surface preparation. Liquid-applied materials require that concrete or masonry substrates be free and clean of all laitance, oil, curing, and form-release agents.

Capillary admixtures chemically react with a substrate and require water for complete chemical reaction. Therefore, fully wet the substrate before application. At best, consider capillary admixtures as dampproofing materials, not complete waterproofing systems.

POLYMER CONCRETE

Polymer concrete is a modified concrete mixture, formulated by adding natural and synthetic chemical compounds referred to as *polymers*. These polymers are provided separately to be added to a concrete paste or as a premixed dry form.

Although the proprietary chemical compounds (polymers) vary, the purpose of these admixtures is the same: to provide a denser, higher-strength, lower-shrinkage, more chemically and water-resistant concrete substrate. A comparison of typical concrete mixes versus polymer mixes is shown in Table 6.4.

Admixtures include chemicals to promote the bonding of polymer concrete to existing substrates. This allows polymer overlaying to existing concrete decks after proper surface preparation. These overlays are applicable as thin as 1/8 inch thick, compared to at least 2 inches thick for conventional concrete. This allows slopping of the polymer mix during installation to facilitate drainage and fill birdbaths or water ponding on existing decks.

TABLE 6.4 Comparison of Regular and Polymer Concrete

Property	Regular mix	Polymer mix
Compressive strength	3000 lb/in^2	4000-8000 lb/in^2
Adhesive bonding	Poor	Excellent
Minimum thickness	2–4 in	1/8–1/2 in
Water absorption	High–10%	Low–0.1%
Chemical resistance	Poor	Good
Initial set time	72 hours	4 hours

Additives also promote initial set and cure time, allowing substrates to withstand traffic in as little as 4 hours after placement. This can be very desirable in remedial or restoration work on parking decks.

Whereas capillary admixtures produce chemical reactions that fill the microscopic pores left by hydration, polymer admixtures produce reactions that eliminate or reduce these microscopic pores. Polymer mixes also reduce the shrinkage that leads to cracking and fissures in a substrate, allowing water penetration. These features provide the characteristics of low absorption that makes polymer concrete highly resistant to chloride attack. Polymer concrete products do not completely waterproof a structure. They are subject to cracking should structural, thermal, or differential movement occur.

Due to the high costs of polymer concrete, these materials are often used as overlays, not as complete substitutes for conventional concrete. These materials are used in renovations of existing concrete walks, bridges, and parking garage decks. They are also used for warehouse and manufacturing plant floors, where high-impact strength and chemical resistance are necessary.

Polymer mixes are also chosen for installations not over occupied spaces, such as bridges, tunnels, and decks, where additional structural properties such as high compressive strength are necessary. In remedial installations such as parking decks, where reinforcing steel too close to the surface has caused concrete spalling, polymers provide an overlay to restore structural integrity. (See Table 6.5.)

POLYMER ADMIXTURE APPLICATION

Polymer admixtures that are added to cement paste do not require any specific surface preparation. Polymer concrete applied as an overlay requires that existing substrates be thoroughly cleaned to remove all dirt, oil, grease, and other contaminants. Exposed reinforcing steel is sandblasted and coated with a primer or epoxy coating before overlay application. Existing decks should be thoroughly checked for delaminated areas by the chain-drag method and repaired before overlay application.

Polymer products added to concrete mixes require that proper mixing and preparations be used. Applications including overlays require proper proportioning and mixing according to the manufacturer's recommendations. Materials are placed and finished as conventional concrete. Working times with polymer concrete are substantially less than with conventional concrete.

In preparation for overlays to existing substrates, concrete should be sufficiently damp to prevent moisture from being absorbed from the polymer overlay mix. Polymers require

TABLE 6.5 Properties of Polymer Concrete

Advantages	Disadvantages
Thin applications	Cost
High strength	Not completely waterproof
Chemical resistance	No movement capability

no special primers. If a stiff mixture is required for application to sloped areas such as ramps or walls, a thin slurry coat of material is brush-applied before the final application of the overlay (see Fig. 6.1).

These products are not applicable in freezing temperatures, and are not designed as primary waterproofing materials of a building envelope.

WATERPROOFING SYSTEM COMPATIBILITY WITH ADMIXTURES

Admixtures are intended to directly impact the hydration process of curing concrete. Since most waterproofing systems are designed for installation over regular concrete mixes, it is mandatory that any intended use of admixtures with a waterproofing system be investigated for compatibility.

Most admixtures significantly reduce the size of capillary void sizes within the cured concrete product. Since many of the available concrete waterproofing systems, used both above and below-grade, require the capillary voids in order to deposit or grow chemical formulation to repel water, the reduction of capillary size can adversely effect water repellency or waterproofing of the concrete.

FIGURE 6.1 Polymer concrete overlay application to existing parking deck. (*Courtesy of Coastal Construction Products*)

For example, clear the penetrating water repellents discussed in Chapter 3, use the capillary voids to grow crystalline particles that actually repel the water. Reducing the size of these capillary voids reduces the available space for the crystalline growth and therefore the effectiveness of the water repellency. While the admixture use itself may have increased the concrete's ability to repel water, it has also reduced the repellent's ability to function properly. Therefore in this situation, rather than supplying "belt and suspenders" protection, the admixture provides no additional protection for the additional costs of construction.

The same situation can also affect the cementitious coatings discussed in both Chapters 2 and 3. Since coatings use the capillary voids for waterproofing reactions or bonding to the substrate, the admixture can also detrimentally effect their performance. Therefore, anytime admixtures are specified for use in the concrete portions of the envelope, including precast units, their compatibility with the proposed waterproofing systems should be reviewed with the manufacturer of the waterproofing materials.

Due to the number of different types of admixtures available, it would be appropriate to complete a field test to verify compatibility before the actual installation is completed. Not only can the admixture affect the waterproofing characteristics of the products, it also can deter the bonding capability of the waterproofing system to the concrete.

Use of fluid-applied membranes, cementitious systems, penetrating sealers, and even sealants over concrete supplemented with admixtures should be tested, and manufacturers consulted, to ensure compatibility.

CHAPTER 7
REMEDIAL WATERPROOFING

INTRODUCTION

Thus far, the waterproofing systems discussed have included applications for new construction or preventative waterproofing. Often, however, waterproofing applications are not completed until water has already infiltrated a building. Waterproofing applied to existing buildings or structures is referred to as *remedial treatments* or *remedial waterproofing*.

Leakage into structural components can damage structural portions and facades of a building envelope (Fig. 7.1). In these cases actual repairs to a structure or its components is required before application of remedial materials. This type of repair is referred to as *restoration*. Restoration is the process of returning a building or its components to the original or near-original condition after wear or damage has occurred.

With historic restoration, new waterproofing materials or systems may not be allowable. Weathertightness then depends solely on a building's facade to resist nature's forces. Such facades are typically walls of stone or masonry. Unfortunately, this type of dependence may not completely protect a building from water damage, especially after repeated weathering cycles such as freeze–thaw cycles.

REMEDIAL APPLICATIONS

Many systems already discussed for preventative waterproofing may be used for remedial applications. In addition to these products, special materials are available that are intended entirely for restoration applications. If existing substrates are properly prepared, most products manufactured for new installations can also be used in remedial applications following the manufacturer's recommendations as necessary.

As with preventative waterproofing, in remedial situations no one product is available to solve all problems that arise. The availability of products used specifically for restoration is somewhat limited compared to that of the frequently used products of new construction applications. Applications and use requirements for preventative products are covered in Chapters 2 through 6.

Remedial application needs are determined by some direct cause (e.g., leakage into interior areas). Restoration application needs are usually determined after leakage occurs or maintenance inspections reveal structural or building damage. In both cases a detailed inspection report must determine the causes of leakage or damage, and the repairs that must be made to a substrate or structure before waterproofing. Leak detection is presented in Chapter 11.

FIGURE 7.1 Deteriorated leaking concrete substrate. (*Courtesy of Webac*)

Once an inspection has been completed, causes determined, extent of damage reviewed, and systems or materials chosen, a complete and thorough cleaning of the structure or substrate is done. This cleaning may reveal additional problems inherent in a substrate.

Before waterproofing, application repairs to substrates must be completed, since waterproofing materials should not be applied over unsound or damaged substrates. After this preparatory work is complete, remedial systems should be applied by trained and experienced personnel (Fig. 7.2).

To reiterate, the sequence of events in remedial or restoration applications (as they differ from new applications), the following actions are necessary:

1. Inspection of damage and leakage
2. Determination of cause
3. Choice of systems for repair
4. Substrate cleaning and preparation
5. Restoration work
6. Waterproofing system application

INSPECTION

Once an inspection is determined necessary, through either routine maintenance or direct leakage reports, a thorough analysis of a building's envelope should be completed. This analysis includes an inspection of all envelope components and their termination or connections to other components. This inspection determines causes of water infiltration and the extent of damage to building components (e.g., shelf angles).

Before the inspection, all available existing information should be assembled to assist in analyzing current problems. This information includes as-built drawings, specifications, shop drawings, maintenance schedules, and documentation of any previous treatments applied. Inspection of existing structures typically includes:

- Visual inspection
- Nondestructive testing
- Destructive and laboratory testing

Visual inspection

Visual inspection may be done at a distance from envelope components (e.g., ground level), but preferably a close-up inspection is completed which, if necessary, includes scaffolding a building. Scaffolding may also be necessary for actual testing of facade or structural components.

FIGURE 7.2 Surface preparation and repairs completed before application of remedial waterproofing treatment. (*Courtesy of Coastal Construction Products*)

During visual analysis, documentation of all unusual or differing site conditions should be addressed. Visual inspections should locate potential problems, including:

- Cracks or separations (Fig. 7.3)
- Unlevel or bulging areas
- Presence of different colors in substrate material
- Efflorescence
- Staining
- Spalled surfaces
- Missing elements

In addition to documentation of these areas, inspection should be completed on functioning areas of an envelope, including roof drains, scuppers and downspouts, flashings, and sealant joints.

Accessories available to complete visual inspections include cameras, video cameras, binoculars, magnifying glasses, handheld microscopes, plumbs, levels, and measuring tapes. The better the documentation, the better the information available for making appropriate decisions concerning repair procedures.

Either during visual analysis or after collection of data, further testing may be required to formulate repair procedures and document the extent of substrate and structure damage. Preferably nondestructive testing, which does no harm to existing materials, will suffice. However, in some situations destructive testing is required to ensure that adequate restoration procedures are completed.

Nondestructive testing

Nondestructive testing is completed with no damage to existing substrates and typically requires no removal of any envelope components. Available testing ranges from simple methods, such as use of a knife, to advanced methods of X-ray and nuclear testing. The most

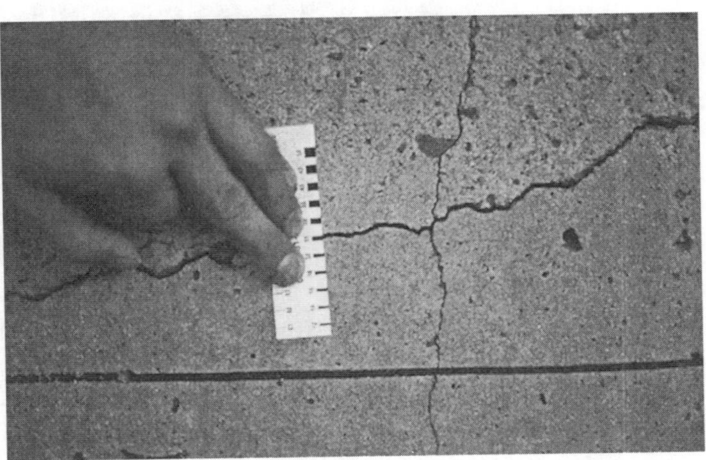

FIGURE 7.3 Visual inspection of substrate cracking. (*Courtesy of Webac*)

prevalent nondestructive testing is water testing. In this analysis, water is applied by some means to a structure to determine areas of infiltration. Water testing is also used to measure moisture absorption rates of the various substrates that comprise a building envelope.

In conducting water tests, water is first applied at the base or bottom of areas being tested. Succeeding applications of water then begin upward. This prevents water from running down onto as-yet untested areas. Water should be applied in sufficient quantities and time in one location to determine if an area is or is not contributing to leakage or absorption. For more information on leak detection, refer to Chapter 11.

Once such a determination is made, testing moves to the next higher location. This testing requires someone to remain inside to determine when water leakage begins to occur. Water testing is limited in that it does not determine specific leakage causes or if leakage is created by damaged envelope systems within a structure such as cavity wall flashing.

Sounding is an effective means of determining areas of disbonding on spalling masonry materials. Such testing uses a rubber mallet to lightly tap substrates to discern differing sounds. For example, hollow sounds usually signify spalled or disbonded areas.

Another sounding method uses chains on horizontal concrete, masonry, or tile surfaces. This test is referred to as *chain dragging*. By pulling a short length of chain along a substrate, testers listen for changes in sounds, carefully documenting hollow sounds. The extent of those areas to be repaired is marked by painting or chalking an outline of their location.

Often, using a simple pocketknife to probe into substrates, without causing any permanent damage, can substantially supplement the information learned from visual inspection. A knife can be used to scrape along mortar joints to determine their condition. Should excessive mortar be removed, it is an indication of an underlying soft porous and poor-strength mortar, which will require attention during remedial repairs. Knives can also be used for testing sealant joints by inserting the knife along joint sides to allow analysis of sealants and to determine if they are properly bonded to a substrate.

Water absorption testing is similar to water testing, but only measured amounts of water are applied to a specific, premeasured substrate area over a specific length of time. This test can accurately determine absorption ratios of substrates. These results are compared to permeability ratios of similar substrates to determine if excessive absorption is occurring.

Modified laboratory testing at project sites can also be completed. This involves constructing a test chamber over an appropriate envelope portion. Static pressure testing, as described in Chapter 10, can then be completed. Such testing requires an experienced firm that has the appropriate equipment to complete testing and the personnel to interpret test results.

Dynamic pressure testing of an envelope at project sites is also possible by using portable equipment that can introduce high air and water pressures. This allows conditions that simulate wind loading and severe rainfall to be applied against an envelope. Chapter 10 reviews job-site testing and mock-up laboratory testing in detail.

Other testing devices include moisture meters, which give accurate moisture content of wood or masonry substrates, and X-ray equipment, which is used to locate and document metal reinforcement. Reinforcement can also be somewhat less accurately located by metal detectors and magnets.

More sophisticated equipment is available to determine existing moisture and its content in various substrates. This equipment includes infrared photographic equipment and nuclear moisture tests completed by trained and licensed professionals.

Destructive testing

Destructive testing involves actual coring (Fig. 7.4) or removal of substrate portions for testing and inspection. Examples include removing portions of a window wall to inspect flashings and surrounding structural damage, and removing small mortar sections from a joint to test for compressive strength. Destructive testing is required when the extent of damage is not visually determinable or when deterioration causes are inconclusive from visual or nondestructive testing.

The most frequently used testing includes laboratory analysis of a removed envelope portion. Testing can consist of chemical analysis to determine if materials meet industry standards or project specifications. Testing can also determine tensile and compressive strength and extent of contamination by chemical or pollutant attack (e.g., by sulfites or chlorides).

Destructive testing includes probing of substrates by removing portions of building components to inspect damage to anchoring systems or structural components. Any removed envelope components should be reinstalled immediately upon completion of analysis, to protect against further damage by exposing components to direct weathering.

Probing is also completed using a borescope. This equipment allows an operator to view conditions behind facade materials through a borehole only $^1/_2$ inch in diameter. The borescope comes equipped with its own light source, allowing close-up inspection without removal of surrounding components or facing materials. (See Fig. 7.5.)

In-place testing is also used frequently, especially in stress analysis. Stress gages are installed at cracked or spalled areas, after which a wall portion adjacent to the gage is removed. Stress readings are taken before and after wall removal, to determine amounts of strain or compression stress that were relieved in a wall after removal.

This test is helpful in such areas as building corners, to determine if shelf angles are continuous around corners. These are areas in which stress buildup is likely to occur, resulting in settlement or stress cracking due to excessive loading.

CAUSE DETERMINATION AND METHODS OF REPAIR

Analysis of compiled information from inspection results and any related data or documents are usually sufficient for a professional to determine leakage causes and the extent

FIGURE 7.4 Destructive coring of concrete substrate for investigation. (*Courtesy of Webac*)

FIGURE 7.5 Removal of pollutants, before visual inspection, is necessary. (*Courtesy of American Building Restoration Products*)

of damage. This analysis includes a review of all pertinent construction documents and maintenance records. Proper repair methods and materials can then be chosen to complete remedial or restoration treatments. For example, if silicone sealants have previously been used in building joints, new sealants must be compatible with silicone, or complete removal, including joint grinding is required.

In reviewing test and inspection results, selecting repair procedures also depends on existing substrate conditions. For instance, if testing has revealed that mortar joints are allowing more water infiltration than existing dampproofing and flashing systems can adequately handle, a sealer application to masonry and mortar to prevent excess water infiltration may be required. However, if mortar joints are cracked or disbonded or have low strength, as determined by testing, a sealer application will not be successful. Additional repairs such as tuckpointing would be required before sealer application, to restore the envelope.

Determining water infiltration causes and choosing repair systems and materials should be done by trained and experienced professionals. Prepared recommendations should be supplied to contractors for bid preparation. This ensures that all bids are prepared on the same basis of information, procedures, materials, and systems.

If recommended procedures are complicated or consist of several required methods to complete restoration, preparation of project specifications may be required. Specifications detail the types of products, materials, or systems to be used, and the manner and location in which they are to be applied.

In addition to specifications, drawings may be required to show repair locations and their relationship to adjacent building envelope components. This enables contractors to prepare adequate bids for completion of remedial repairs and the restoration project. Any additional information that is useful to bidding contractors, such as as-built drawings, original job specifications, and access to site for reviewing existing job-site conditions, should also be provided.

All completed repairs or restoration work should be carefully documented. This documentation should be maintained by the building owner for future reference, should future repair or warranty work be required. Once these remedial procedures are determined and specifications are prepared, the next step in the restoration process is building cleaning.

CLEANING

Existing structures have surface accumulations of airborne pollutants that must first be removed to allow bonding of remedial waterproofing materials (Fig. 7.6). If surface pollutants and dirt accumulations are considerable, it will be necessary to require building cleaning before inspections. This provides adequate conditions to review present conditions and make appropriate decisions (Fig. 7.7).

Besides aesthetic purposes, cleaning is a necessary part of building maintenance. Maintenance cleaning ensures proper life-cycling and protection of building envelope components against pollutant contamination.

Pollutants affect envelopes in two distinct manners. The first is by direct substrate deterioration by pollutants, including salts, sulfites, and carbons that, when mixed with water, form corrosive acids including sulfuric acid. These acids are carried into an envelope system in liquid state. Acids attack not only in-place waterproofing systems but also structural components, such as reinforcing steel or shelf angles.

The second pollutant deterioration is the slowing or halting of the natural breathing process that allows entrapped moisture to escape. Pollutants carried deep into substrates by water and moisture fill microscopic pores of envelope facades. If left unchecked, this collection will stop natural moisture escape that is necessary in substrates and will lead to damage from freeze–thaw cycles, disbonding of coatings, and structural component deterioration.

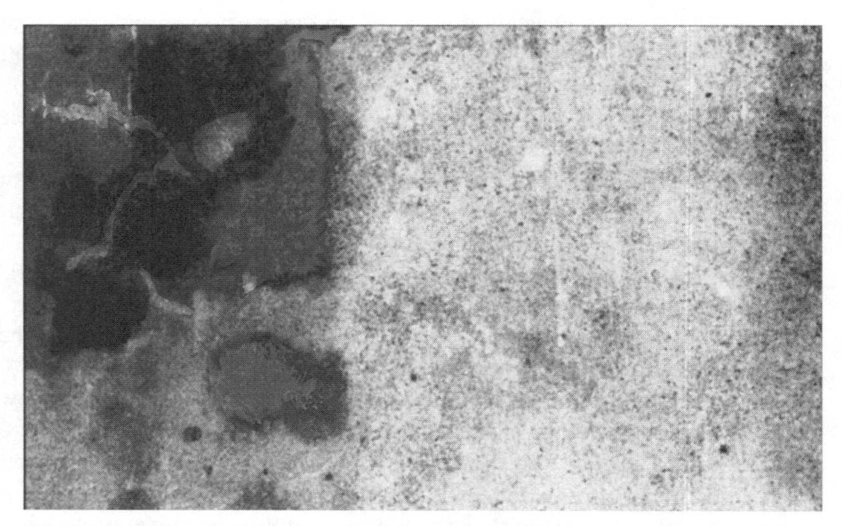

FIGURE 7.6 Chemical cleaning of substrate to adequately access substrate problems. (*Courtesy of American Building Restoration Products*)

FIGURE 7.7 Steam/pressure cleaning of substrate. (*Courtesy of Devcon*)

Building cleaning is therefore a necessary part of building preservation and proper maintenance. Cleaning should not be put off until remedial and restoration applications are necessary.

Exterior cleaning is completed by one or a combination of these methods:

- Water cleaning
- Abrasive cleaning
- Chemical cleaning
- Poultice cleaning

Before cleaning by any of these methods, testing of the proposed method is imperative. Testing ensures that cleaning systems are adequate for the degree necessary and that damage to the existing substrate's adjacent materials and existing waterproofing systems will not occur. Sample-testing a lower building portion in areas away from high traffic and visibility is desirable.

Cleaning is normally completed before starting remedial repairs. In some situations, however, a building can be in such a deteriorated state that introduction of chemicals or water under pressure will further damage interior areas and structural components. In these situations, sufficient remedial repairs may be required before cleaning, to prevent further damage.

Water cleaning

Building cleaning done later includes pressure washing, water soaking, and steam cleaning (Fig. 7.7). Pressure washing is the most common procedure, especially when it is used in conjunction with preparation of substrates for waterproofing applications. It is also used in combination with other cleaning methods such as aggregate and chemical cleaning.

Pressure cleaners are manufactured to produce pressures varying from 300 to more than 25,000 lb/in^2. The lower-pressure cleaners are used for rinsing minor residue accumulation, whereas the higher-pressure machines remove not only pollutant collection but also paints and other coatings. The wide range of pressures available require that testing be completed to determine the pressures required to remove contaminants without damaging substrates.

Equipment spray tips and operators directly control pressure-cleaning results. Fog-type spray nozzles are desirable and impart little harm to a substrate, whereas O-tip nozzles greatly concentrate the stream of water and can cause substrate damage.

Applicators must be experienced in this type of work, especially with higher-pressure equipment. Unskilled mechanics can damage a building by blowing sealants out of joints, damaging masonry or stone finishes, leaving streaks and performing insufficient cleaning in the process. (See Fig. 7.8.)

Water soaking is a cleaning method preferred by preservations and historical restoration professionals due to the minimal amount of damage possible. Water soaking is especially successful on limestone structures, where chemical or pressure washing is unacceptable.

Specially prepared soaker hoses or sprayers are installed on upper building portions and provide a continuous curtain of water flowing down the building facade. After an initial period of soaking, determined by actual project testing, this method loosens dirt and pollutant accumulations. These pollutants are then removed by low-pressure spray cleaning. In highly contaminated areas, a repeat soaking process may be necessary to clean a building to acceptable aesthetic and project requirements.

A major disadvantage of water soaking is the amount of water introduced onto the exterior envelope. If deterioration or leakage is occurring, this system will cause further

FIGURE 7.8 Pressure cleaning operation. (*Courtesy of ProSoCo*)

damage. Soaking will also deepen penetration of salts and other contamination into pores of a substrate, which follow-up pressure washing may not remove.

Available water supplies often contain minerals that stain or streak existing substrates. Water purification equipment is necessary to treat water before application.

Surfaces being prepared for waterproofing applications by using water soaking require a long drying period. As long as one month may be necessary for substrates to dry sufficiently. Also, to prevent damage, preliminary remedial waterproofing, such as tuck-pointing, will be required before start of soaking.

Steam cleaning, although used extensively in the past, is now almost obsolete due to expanding technology in pressure-cleaning equipment. Steam equipment rapidly heats water in a self-contained boiler; then it spray-applies water under low pressure. The heated water swells and loosens collected pollutants, enabling them to be rinsed off a substrate.

Results achieved by steam cleaning are now reachable by pressure equipment that is much lower in cost than steam equipment. However, particular substrates can be so contaminated that too high a pressure may be required to achieve results obtained by steam cleaning. (See Table 7.1.)

Abrasive cleaning

Abrasive cleaning uses an abrasive material to remove mechanically accumulated dirt and pollutants (Fig. 7.9). Abrasive cleaning methods include:

- Sandblasting
- Wet aggregate blasting
- Sanding belts
- Wheel grinders

Abrasive cleaning systems remove not only surface accumulations of dirt but also some portion or layer of a substrate itself. This surface damage is often severe, and substrate restoration may be necessary. These systems are not preferred for substrate preparation, waterproofing applications, or general building cleaning, and are not acceptable to most restoration and preservation professionals.

Abrasive cleaning is now typically limited to paint removal on metal substrates, although this procedure is now economically possible with advanced technology in water-blasting equipment and chemical paint removers.

Wet aggregate cleaning is the mildest abrasive cleaning process. This method uses a vortex attachment on a pressure cleaner that suctions sand to mix with water at the spray

TABLE 7.1 Water-Cleaning Properties

Advantages	Disadvantages
Several methods available, including pressure and soaking	Introduces water to envelope components
Chemicals can be added if required	Improper cleaning can damage substrate or cause streaking
Variable pressures	Environmentally safe

FIGURE 7.9 Abrasive (shot-blast) cleaning process. (*Courtesy of Devcon*)

tip. Water pressure then directs the aggregate against a substrate. This method operates under lower pressures than compressed air-blasting equipment and also wets the aggregate, keeping airborne contaminants to a minimum.

With all abrasive cleaning, some portion of a substrate surface will be removed. Careful testing should be completed to analyze the process before complete substrate cleaning. Additionally, because of potential damage and safety concerns, only highly experienced mechanics should be employed in these cleaning processes.

By using extremely small aggregates, substrate damage is lessened but still produces desired cleaning results. A very fine powdered sand referred to as *flour sand* because it has the consistency of baking flour, is available. By using this sand with low pressures, satisfactory results with insignificant substrate damage are possible.

Sanding and mechanical wheel grinders are used to remove paint and corrosion from metal substrates. Grinders also have limited usage on concrete substrates for removing small contaminated areas of oil, grease, and other accumulations, which pressure cleaning will not remove. This cleaning also removes portions of the substrate and should be used only when other alternatives are not available. (See Table 7.2.)

TABLE 7.2 Abrasive Cleaning Properties

Advantages	Disadvantages
Removes paint layers easily	Can damage substrate excessively
Flour sand reduces damage	Safety concerns
Variable pressures	Equipment required

Chemical cleaning

As with water pressure cleaning, chemical solutions are available in a wide range of strengths for cleaning. Substrates infected with special stains not removable with plain water can be chemically cleaned. These cleaners include mild detergents for mildew removal and strong organic cleaners for paint removal (Figs. 7.10 and 7.11). Chemical cleaning formulations include three types, all of which include a manufacturer's proprietary cleanser:

FIGURE 7.10 Chemical cleaning process. (*Courtesy of ProSoCo*)

FIGURE 7.11 Paint stripping using chemical cleaners. (*Courtesy of ProSoCo*)

- Acidic
- Organic
- Alkaline

Cleaners are toxic, and should be used by trained personnel wearing protective clothing. Cleaners should be applied in sample areas, so that damage such as etching of stonework does not occur. Adjacent envelope components, including glass, metals, and vegetation, should be completely protected before cleaning.

OSHA, EPA, state, and local regulations control chemical cleaner usage, including their collection and disposal. Most municipalities will not allow chemicals to reach city drainage, surrounding soil, or groundwater. Some cleaners have formulations that are neutralized after rinsing with water; others do not. It is important to investigate manufacturer's recommendations and local codes to prevent unlawful use or disposal of chemical cleaners. Refer to Chapter 9 for additional hazardous waste use and disposal regulations.

Chemical cleaners are necessary when water cleaning does not suffice and abrasive cleaners cause too much substrate damage. Removing paint with chemical cleaners only requires rinsing to remove paint residue after cleaner application. It is often necessary to repeat applications several times when previously painted layers are excessive or several different paints have been applied. With lead-based paints, waste from cleaning must be treated as hazardous waste, properly collected, removed, and disposed according to government regulations.

Cleaners are also available for stain and pollutant removal from substrates. These substances include asphalt, tar, and metallic and efflorescence stains. The cleaners remove specific areas of stains on a substrate in conjunction with general pressure cleaning or soaking (Table 7.3).

Poultice cleaning

When existing stains or pollutants have penetrated a masonry surface, water and chemical cleaning are often not sufficient to remove staining. If abrasive cleaning is not acceptable, poulticing may be an alternative method. With poultice cleaning, an absorbent material such as talc, fuller's earth, or a manufacturer's proprietary product is applied to a substrate. This material acts to draw stains out by absorbing pollutants into itself. The poultice is then removed from the substrate by pressure cleaning.

The length of time a poultice must be left on a substrate to absorb pollutants varies with the stain type, pollutant penetration depth, substrate porosity, and general cleaner effectiveness. This cleaning is especially effective on natural stone substrates such as limestone, marble, and granite. Poultice-type cleaners are effective on a wide range of stains, including oil, tar, primer, solvents, paint, and metallic stains from hard water (Table 7.4).

Substrate testing with various types of available cleaning systems should be completed to determine the most effective system that does no harm to facade and adjacent materials.

TABLE 7.3 Chemical Cleaning Properties

Advantages	Disadvantages
Little damage to substrate	Environmental and safety concerns
Ease of paint removal	Clean-up and disposal requirements
Various formulations and strengths available	Damage to surrounding substrates and vegetation

TABLE 7.4 Poultice Cleaning Properties

Advantages	Disadvantages
Removes deeply penetrated pollutants	Requires extensive technological knowledge
No damage to substrate	May force pollutants deeper
Excellent for natural stone substrate	Extensive testing required before application

Complete and thorough cleaning of substrates is necessary before proceeding with the restoration phase.

RESTORATION WORK

Just as remedial waterproofing systems must be applied over clean substrates, so too they must be applied over sound substrates. After cleaning, all restoration work must be completed before waterproofing material is applied. Any substrate deterioration that has occurred, including spalled concrete, damaged structural components, and oxidized reinforcement steel, should be prepared.

Restoration work often requires removal of building envelope portions to repair structural deterioration. This includes anchoring devices, pinning, and shelf angles used for attaching facing materials to structural building components.

This repair work is necessary after years of water infiltration high in chloride content, which corrodes metal components. Other required repairs, including control or expansion joint installation and cleaning of weep holes, also are completed at this time.

After completing all necessary repairs and substrate preparation, remedial waterproofing systems installation can begin (Fig. 7.12). Preventative waterproofing materials, discussed in Chapters 2 and 3, can be applied as remedial treatments if the surfaces are properly prepared. Remedial treatments also include installation of flashing, sealants, and other envelope transitional materials found inadequate in the original construction.

In addition to preventative waterproofing systems, several waterproofing materials and systems are manufactured specifically for remedial and restoration projects. In some cases even before a building is completed remedial products are required, to repair damage occurring during construction.

Remedial waterproofing systems now available include:

- Tuck-pointing
- Face grouting
- Joint striking
- Mass grouting
- Grout injection
- Epoxy injection
- Cementitious patching
- Shotcrete and gunite

FIGURE 7.12 Cleaning completed before restoration work commencing. (*Courtesy of ProSoCo*)

TUCK-POINTING

In most masonry structures, unless the masonry was handmade and is excessively porous, any leakage is usually attributable to mortar joints. The water, moisture, or vapor that passes through the masonry itself is usually repelled by dampproofing or flashing or weep systems.

Through the aging process, all mortar joints eventually begin to deteriorate, caused by a multitude of weathering factors. These include swelling of masonry, which when wetted places pressure on mortar joints from all sides. This causes fractures and cracks along the masonry and mortar junctures. Other factors contributing to mortar deterioration include freeze–thaw cycling, thermal movement, and chemical deterioration from sulfites and chlorides in atmospheric pollutants.

During life-cycling, weathering begins to allow significant amounts of water and moisture through a masonry wall. Eventually this water may exceed the capabilities of existing dampproofing systems, allowing water to penetrate interior areas. Entering water also begins structural deterioration behind masonry facades.

If building maintenance inspections reveal that mortar deterioration is contributing to excess water infiltration, tuck-pointing of mortar joints will be necessary. Tuck-pointing is a restoration treatment used to restore the structural integrity of mortar joints. Tuck-pointing procedures include removing existing deteriorated mortar and replacing it with new mortar (Table 7.5).

Inspections may reveal that only certain wall joints require tuck-pointing, or an entire wall area may require complete tuck-pointing to restore the building envelope. If miscel-

TABLE 7.5 Tuck-Pointing Properties

Advantages	Disadvantages
No aesthetic changes to substrate	Labor-intensive
Environmentally safe	Cost
Repairs can be limited to a specific area	Mortar removal may damage surrounding masonry

laneous tuck-pointing is required, specifications or bid documents should be explicit as to what constitutes sufficient deterioration to require removal and replacement. The tuck-pointing type of repair requires inspection to ensure that deteriorated joints are being repaired as per the contract.

For complete tuck-pointing projects, all joints will be restored, but inspection procedures should also be structured to ensure that all joints are actually tuck-pointed. Economics of complete tuck-pointing often lead to considering alternate repair methods, including face grouting or complete regrouting.

TUCK-POINTING APPLICATION

Masonry walls should be thoroughly checked for contaminants before tuck-pointing. Existing mortar should be removed to a minimum depth of $^3/_8$ inch, preferably $^1/_2$ inch. Up to 1-in removal of severely deteriorated joints is required. These depths allow bonding between existing and newly placed mortar and the masonry units.

FIGURE 7.13 Tuck-pointing application. (*Courtesy of Western Group*)

Joint removal is completed by hand or with power tools such as hand grinders (see Fig. 7.13). On historic structures or soft masonry work, power tools damage existing masonry too extensively. Power tools often cause irregular joint lines, or actual portions of masonry be removed. Sample areas on older masonry structures should be analyzed for acceptability of power tool usage.

Once defective mortar is removed, joint cavities must be cleaned to remove dust and mortar residue. This residue, if left, will deter the effective bonding of new mortar. A preferred method of residue removal is spraying joints with compressed air.

Once preparatory work is completed, existing mortar cavities should be wetted just before tuck-pointing application. This prevents premature drying and curing, which results in structurally weak joints.

Only premixed materials specifically manufactured for tuck-pointing should be used. These dry mixed cement and sand-based products contain proprietary additives for effective bonding and waterproofing, and are nonshrinking. Materials higher in compressive strength than the masonry units are not recommended. If joints are stronger than the masonry, spalling of masonry units during movement in the wall system will occur.

Materials should be mixed using only clean water in amounts specified by the manufacturer. Pointing materials are available in premixed colors, or manufacturers will custom-match existing mortar. Field mixing for color match should be prohibited, as this results in inadequate design strength and performance characteristics.

Pointing mortar must be applied using a convex jointer that compresses and compacts material tightly into joints and against sides of masonry units. This creates an effective waterproof mortar joint. The tooler or jointer should be slightly larger than joint width, and enough mortar should be placed in joints so that after jointing, excess material is pushed from joints. This ensures that joints are properly filled to capacity. Figure 7.14 summarizes the steps involved in tuck-pointing.

After initial mortar set, joints should be brushed or scraped to remove fins formed by applying this material. Finished joint design should be concave or weathered for longevity and weathertightness. Refer to joint design in Chapter 8.

Priming of joints and bonding agents is not required. Dry mixes supplied by manufacturers contain all necessary components. Pointing should not be applied in conditions under 40°F or over extremely wet surfaces.

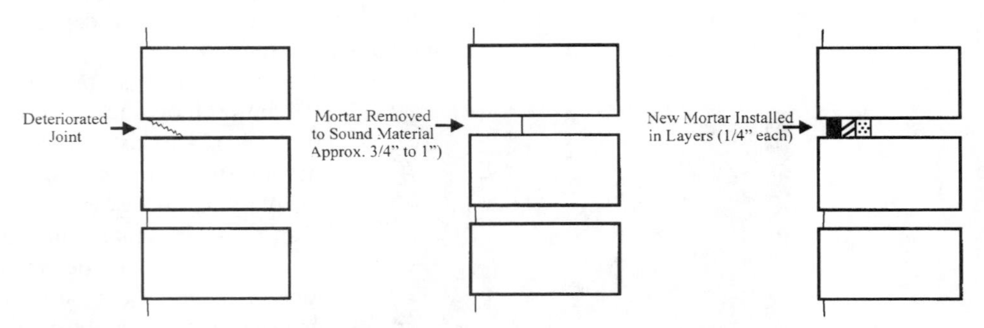

FIGURE 7.14 Tuck-pointing processes.

FACE GROUTING

Certain restoration projects include deteriorated masonry units requiring remedial procedures for both joints and masonry units. In a process referred to as *face grouting* or *bag grouting*, a cementitious waterproofing material is brushed and scrubbed into mortar joints and masonry faces. This grout is then brushed off just before complete curing of grout.

Grout materials are cement- and sand-based products with proprietary waterproofing chemicals and bonding agents. Some materials contain metallic additives that may change the color of a substrate when metallic materials begin oxidizing. Manufacturer's data should be reviewed, to judge product suitability for a particular installation.

Bag grouting refers to a technique using burlap bags to remove grout after application to wall areas. Grout is used to fill pores, cracks, and fissures in both the joints and masonry, waterproofing an entire wall facade. Face grouting does not change the color or aesthetics of wall surfaces nor the breathability of facing materials. Face grouting will, however, impart a uniform color or shading to walls; the effects depend on the grout color chosen. Testing of sample areas should be completed to analyze application effectiveness and acceptability of the finished appearance.

Grouting is a highly labor-intensive system, and the mechanics doing it should be trained and experienced in system application. Should grout be brushed off too quickly, material will be removed from masonry pores and will not sufficiently waterproof. If grout is allowed to cure completely, it will be virtually impossible to remove, and the entire substrate aesthetics will be changed.

Application timing and removal varies greatly, and is affected by weather (dry, humid, sunny, or overcast), substrate conditions (smooth, glazed, or porous), and material composition. Mechanics must be experienced to know when the removal process should begin, as this may change daily depending on specific project conditions, including weather (Table 7.6).

FACE GROUTING APPLICATION

Masonry walls should be cleaned completely to ensure that grout will bond to both existing masonry and mortar. All contaminants, including previously applied sealants, must be removed. Walls should be checked for residue of previous waterproof coatings or sealer applications that hinder out-bonding. All seriously deteriorated mortar joints should be tuck-pointed before grout application.

Grout materials are supplied in dry mix form with acrylic or integral bonding agents. Dry bag mix products are mixed with clean water in specified portions for existing conditions.

Grout should be brushed and scrubbed in circular motions to an entire wall area. The wall surface must be kept continually and uniformly damp to prevent grout from drying before removal. Grout should be applied uniformly and completely, to fill all voids, pores, and cracks (Fig. 7.15).

At the proper time, determined by job conditions, removal should begin. Grout is removed using stiff bristle brushes, burlap bags, or other effective methods. Proper removal will leave masonry free of grout deposits with no change in color or streaking.

No priming is required, although surfaces should be kept properly damp. Materials should not be applied to unsound or defective substrates or joints. Temperature must be above freezing during application.

TABLE 7.6 Face Grouting Properties

Advantages	Disadvantages
Repairs both masonry and joints	Cost
Environmentally safe	Labor-intensive
Low water absorption after installation	Difficult installation

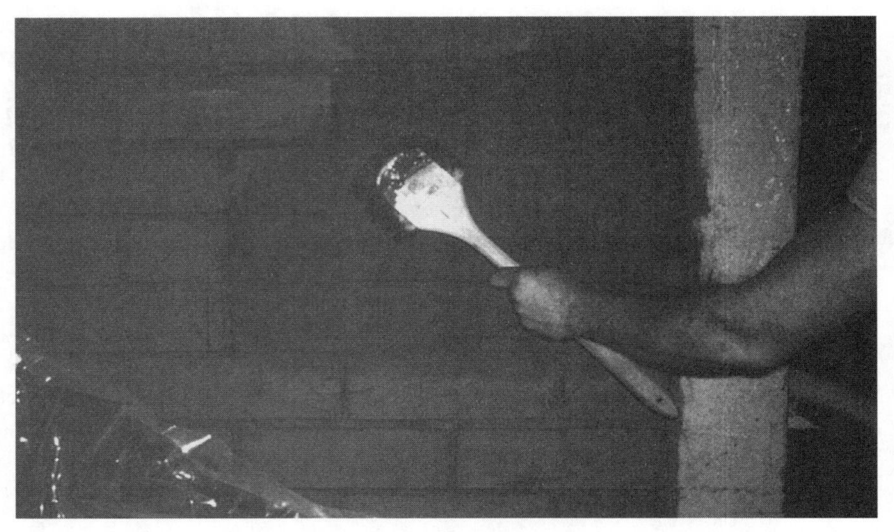

FIGURE 7.15 Face or bag grouting application. (*Courtesy of Western Group*)

JOINT GROUTING

Joint grouting is an application of cementitious grout to all surfaces of existing mortar joints. This application is sometimes referred to as mask grouting, which is grouting walls that have all masonry units masked (taped or otherwise covered). This protects them from grout application on masonry unit faces. Materials used and surface preparation are the same as that for face grouting; only applications are different.

Cementitious grout material is brushed onto joint surfaces to fill voids and cracks, while keeping material off masonry facing. In restoration projects where joints have been tool-recessed, grout application should fill joint recesses completely.

This application effectively points joints without requiring joint cutout. However, as with all joint grouting systems, severely deteriorated joints should be removed and properly tuck-pointed before grout application. (See Table 7.7.)

JOINT GROUTING APPLICATION

If joints exist with a minor recess, 1/8 inch or less, masonry units are masked and grout is applied to fill joints flush with the masonry facade. Masking is removed before complete

TABLE 7.7 Joint Grouting Properties

Advantages	Disadvantages
No aesthetic changes to substrate	Repairs only masonry joints
Less labor intensive than other methods	Adjacent surfaces should be masked
No damage to surrounding substrates	Joint removal required may be overlooked

curing of grout, so that any fins formed may be removed before final grout is set without affecting waterproofing integrity (Fig. 7.16).

This system is not designed to replace tuck-pointing seriously deteriorated joints. As with other systems, sample test areas should be completed to analyze effectiveness under specific job conditions.

Materials should be mixed according to the manufacturer's recommendations. Materials are brushed on existing mortar cracks and voids, or applied by jointers to fill joint recesses completely. Grout materials are available in standard colors or are manufactured to match existing colors.

No priming is required, but joints should be kept damp during application. Materials should not be applied to frozen substrates or in freezing temperatures.

EPOXY INJECTION

During original construction or structure life-cycling, cracks often develop that allow water and pollutants to enter a substrate (Fig. 7.17). If this cracking is nonmoving but

FIGURE 7.16 Mass grouting application. (*Courtesy of Western Group*)

structural, it is repaired through injection of a low-viscosity epoxy. The epoxy seals the cracks and restores the monolithic structural nature of a substrate (Fig. 7.18).

High-strength epoxies can return a substrate to its original design strength but do not increase load-bearing capability. Epoxy used for injection has compressive strengths more than 5000 lb/in^2 when tested according to ASTM D-695.

Injection epoxies are two-component, low-viscosity materials requiring mixing before application. Low viscosity allows materials to flow freely and penetrate completely into a crack area. Epoxy used for injection applications has no movement capabilities, and will crack again if original cracking or movement causes are not alleviated. Expansion and control joints must be installed if it is determined that cracks may continue to move. Otherwise, cracks should be treated with a material that allows for movement.

FIGURE 7.17 Cracking of concrete substrate. (*Courtesy of Webac*)

FIGURE 7.18 Repair of substrate using epoxy injection. (*Courtesy of Webac*)

Epoxy injection is a restoration system as well as a waterproofing system. Injection can restore substrates to a sound condition before waterproofing application, or be used as waterproofing itself by stopping leakage through a crack.

Epoxy injection has been used on concrete, masonry, wood, metal, and natural stone substrates. Large wood timber trusses in historic structures have been restored structurally with epoxy injection systems. Typically, epoxy injection is used to restore concrete and masonry substrates to sound condition.

Cracks to be injected must be large enough to allow entrance of epoxy, approximately 5 mil thick, and not so large that material flows out, 35–40 mil. Cracks that meet these size limitations can be injected through any of the above listed substrate materials.

Application is completed by the pressure injection method using surface-mounted or drilled ports through which to apply epoxy. In some cases, e.g., horizontal surfaces such as parking decks, epoxy is installed by the gravity method, in which epoxy simply penetrates by gravity. In all cases a low-viscosity material is used, to allow for better epoxy penetration into a substrate.

Surface-mounted ports are applied directly over a crack surface (Fig. 7.19). Drilled ports require a hole to be drilled at the crack location and a mechanical packer placed into these holes for injection (Fig. 7.20). Drilled ports are required for large, deep cracks, to allow complete saturation of cracks with epoxy.

In both port applications, cracks are sealed with a brushed-on epoxy to prevent epoxy from coming out of the crack face during injection. The port surround is also sealed and adhered completely to substrates, preventing them from blowing off during injection. If cracks penetrate completely through a substrate, the backside must also be sealed before injection.

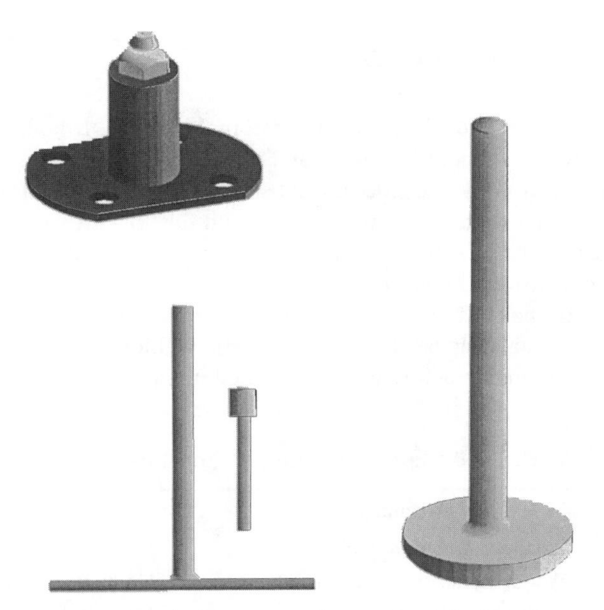

FIGURE 7.19 Surface-mounted ports for injecting epoxy under low pressure. (*Courtesy of Webac*)

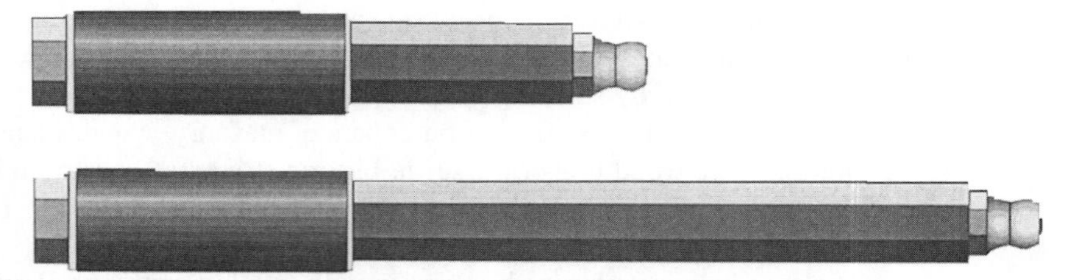

FIGURE 7.20 Mechanical packer for injecting epoxy under high pressures. (*Courtesy of Webac*)

The premise of injection work is to allow for maximum epoxy penetration so as to ensure complete joint sealing. Ports are placed approximately the same distance apart as crack depth, but not exceeding 6 inches. Epoxy is then injected into the lowest port, and injection is continued until epoxy flows out the next highest port. The lower port is then sealed off, and injection is continued on the next highest port. After epoxy curing, ports and surface-applied sealers are removed.

Epoxy crack sealing will cause staining of a substrate or possible damage to it during its removal. In restoration procedures where such damage is not acceptable, such as glazed terra cotta, hot-applied beeswax may be applied as a sealer in place of epoxy. This wax is then removed after injection without damage or staining to substrates.

Epoxy injection requires technical knowledge and experience of an installer. Proper mixing of the two-component materials, proper injection pressures, and knowledge of the injection process are mandatory for successful installations. (See Table 7.8.)

EPOXY INJECTION APPLICATION

Substrates must be cleaned and completely dry. If both sides of a substrate are accessible, they should both be sealed and injected to ensure complete crack filling. Cracks accessible from only one side of a substrate lose a quantity of material out to the unsealed side.

Cracks that are contaminated with dust or dirt cannot be properly injected. All cracks should be blown with compressed air to remove dirt accumulation (Fig. 7.21). Steel substrates should be free of oxidation.

If epoxy is to be installed by the gravity method, horizontal cracks should be grooved to form a V-shape. The groove should be blown out, to remove all concrete dust and other contaminants before epoxy is placed in the groove.

TABLE 7.8 Epoxy Injection Properties

Advantages	Disadvantages
Restores structural integrity	Extensive installation requirements
Wood, metal, and concrete substrate	No movement capability
High-strength	May stain surrounding substrate

FIGURE 7.21 Cracks must be cleaned and prepped before injection. (*Courtesy of Abatron, Inc.*)

Cracks must be sealed with brushable epoxy gel or wax, to prevent epoxy runout during injection (Fig. 7.22). Ports should be installed using either surface-mounted or drilled ports as recommended by the manufacturer.

Injection should begin at the lowest port and be injected until epoxy is visible at the next higher port (Fig. 7.23). After sealing and capping of the injected port, the injection is then moved to the next port. Upon completion of port injection, substrate sealers and ports are removed.

Most epoxy used for injection is two-component, and must be properly mixed before application. Working life or pot life is extremely limited, and epoxy must be installed before it begins to cure in the applicator equipment. Epoxy injection equipment that mixes and injects epoxy under constant uniform pressure, approximately 100–300 lb/in^2, is available (Fig. 7.24). Low-viscosity epoxy thickens in cool weather and may not flow sufficiently to fill the crack. Hand pump injectors are adequate as long as enough ports are used (Fig. 7.25).

Upon injection completion, a core sample of substrate and installed epoxy should be taken. This allows for inspecting penetration depth and testing strength of cured epoxy and repaired substrate. Epoxy materials are extremely hazardous and flammable. Care should be taken during their use as well as their disposal. Equipment must be checked frequently to ensure that proper mixing ratios are being maintained.

CHEMICAL GROUT INJECTION

Epoxy materials are used for restoring substrates to sound structural strength, with waterproofing of cracks a secondary benefit. Epoxy joints do not allow for movement. If movement should occur again, leakage can resume. Chemical injection grouts, on the other

FIGURE 7.22 Sealing large cracks with gel before injection. (*Courtesy of Abatron, Inc.*)

FIGURE 7.23 Epoxy injection begins at the lowest-elevation ports. (*Courtesy of Abatron, Inc.*)

hand, are used primarily for waterproofing a substrate and are not intended for structural repair. Chemical grouts also allow for future movement at joint locations.

Injection grouts are hydrophobic liquid polymer resins, such as polyurethane formulations. They react with water present in a crack and substrate, creating a chemical reaction. This reaction causes a liquid grout to expand and form a gel or foam material that fills voids and cracks. Expansion of materials forms a tight impervious seal against substrate sides, stopping water access through a joint (Fig. 7.26).

Grout material is supplied in low-viscosity formulations to enhance its penetrating capabilities. However, unlike epoxy, substrates do not need to be dry for grout application. In fact substrates should be wetted before application and, if necessary, grout can be applied directly into actively leaking joints.

Grouts are typically used for concrete or masonry substrates, although they will bond to metals, wood, and polyvinyl chlorides such as PVC piping. Some grouts are also available in gel form, which is used to stabilize soils in areas of bulkheads or soil banks of

FIGURE 7.24 Two-component injection mixer and injection equipment. (*Courtesy of Webac*)

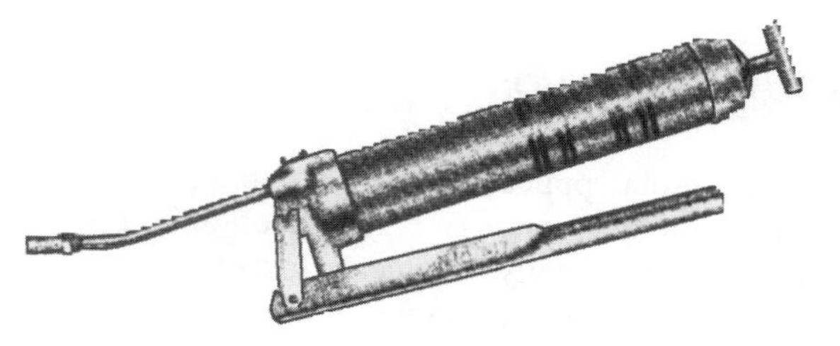

FIGURE 7.25 Hand-held grout injector. (*Courtesy of Webac*)

FIGURE 7.26 Chemical grout injection. (*Courtesy of de Neef Construction Chemicals*)

retention ponds. In these applications, materials react with groundwater present, binding together soil particles.

Grout injected into wetted substrates fills fissures and pores along a crack surface (Fig. 7.27). Once-cured grouts, similar to sealants, have excellent movement capability with elongation as much as 750 percent. This flexibility allows material to withstand thermal movement or structural movement at a joint without deterring its waterproofing capabilities. These grouts have been used successfully in remedial below-grade applications where leakage is occurring directly through a crack itself rather than through entire substrates (Fig. 7.28).

Since water moves through a path of least resistance, during remedial repairs injecting cracks can redirect water and start leakage in other areas of least resistance. Therefore, complete remedial waterproofing treatments can require grout injection of cracks and application of a waterproofing system. (See Table 7.9.)

CHEMICAL GROUT APPLICATION

Chemical grout applications are very similar to epoxy injection, with similar equipment and injection tubes necessary. The major difference is that grouts require water and epoxies do not. Additionally, chemical grouts are supplied in one-component rather than two-component epoxy formulations.

Substrate preparation is almost unnecessary with chemical grouts. Surfaces do not need to be dry, but should be cleaned of mineral deposits or other contaminants along a crack area. If a waterproofing system is to be applied after grout injection, the entire substrate should be cleaned and prepared as necessary.

On minor cracks, or substrate of 6-inches thick or less, holes for ports are drilled directly over cracks. In thicker substrates and large cracks, ports are drilled approximately 4–6 inch away from cracks at an angle that intersects the crack itself. Test holes should be com-

FIGURE 7.27 Chemical grout injection into large joint. (*Courtesy of Webac*)

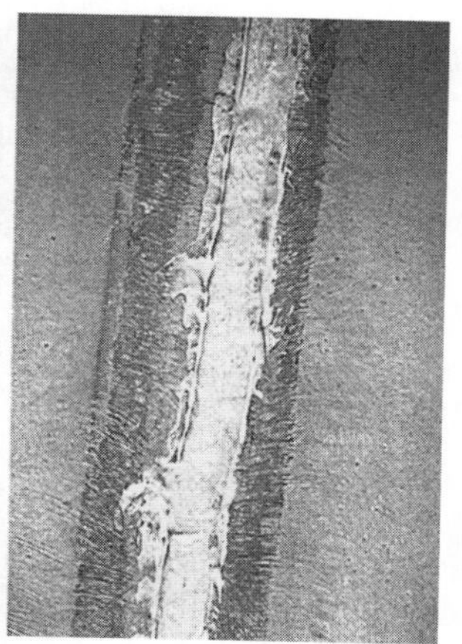

FIGURE 7.28 Repair of existing joint using chemical grout to allow for future movement. (*Courtesy of Webac*)

TABLE 7.9 Chemical Grout Properties

Advantages	Disadvantages
Excellent movement capability	Cost
Several formulations available	Requires specialized installation methods
Concrete, masonry, wood, and soil substrates	Toxic chemical used

pleted with water injected for testing to ensure that grout will penetrate properly. Port spacing along cracks varies, depending on crack size and manufacturer recommendations. Spacing varies from 6 to more than 24 inches (Fig. 7.29).

In smaller cracks it is not necessary to surface-seal crack faces. However, on large cracks, temporary surface sealing with a hydraulic cement-patching compound is necessary to prevent unnecessary grout material waste. As with epoxy applications, begin at the lowest port; grout is injected until it becomes present at the next higher port. This process is then moved to the next higher port (Fig. 7.30).

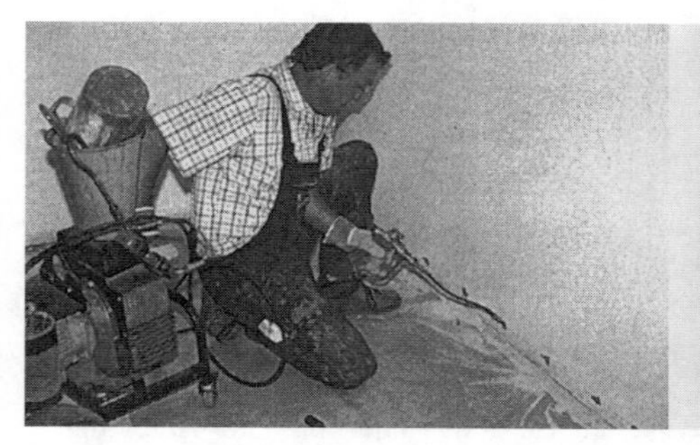

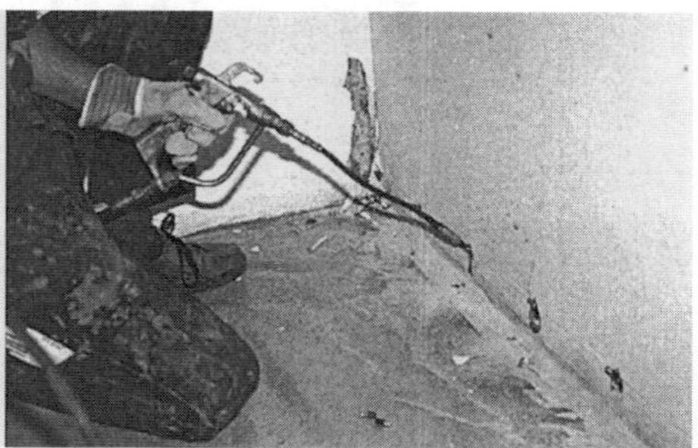

FIGURE 7.29 Chemical grout injection. (*Courtesy of Webac*)

FIGURE 7.30 Injection of grout starting at lower elevations. (*Courtesy of de Neef Construction Chemicals*)

Pressures required to inject materials are generally 300–500 lb/in^2 (Fig. 7.31). After injection, material should completely cure before injection ports are removed, approximately 24 hours. Portholes should then be patched with quick-set hydraulic patching material. Excess grout is then removed from the repair area (Fig. 7.32).

Chemical grouts should not be used in temperatures below 40°F nor on frozen substrates. Chemical grout materials are flammable and hazardous. Extreme care should be taken during its use and storage, as well as in the disposal of the chemical waste. In confined spaces, ventilation and respirators are required for safe working conditions.

CEMENTITIOUS PATCHING COMPOUNDS

For restoration of concrete and masonry substrates, a host of cementitious-based products is available to restore substrates to a sound condition before remedial waterproofing applications. These cement-based products are high-strength, dry mixes, with integrally mixed

FIGURE 7.31 Mechanical grout injector. (*Courtesy of Webac*)

FIGURE 7.32 Excess grout is removed from substrate after injection repairs. (*Courtesy of de Neef Construction Chemicals*)

bonding agents or a bonding agent to be added during mixing. The premixed products have similar properties: they are high strength (compressive strength usually exceeds 5000 lb/in^2), fast setting (initial set in less than 30 minutes), applicable to damp substrates, and nonshrinkable.

Some products are used as negative-side waterproofing systems, whereas others are used to patch a specific area of leakage. Cementitious materials are used in a variety of restoration procedures. A typical use is repairing spalled concrete surfaces after repair and preparation of exposed reinforcing steel.

Cementitious patching systems are also used for concrete overlays to add substrate strength, patching of honeycomb and other voids, and patching to stop direct water infiltration. Cementitious patching systems include:

- High-strength patching compounds
- Hydraulic or hot-patch systems
- Shotcrete or gunite systems
- Overlays

High-strength patching

High-strength patching products are supplied in premixed formulations with integral bonding agents. They are used to patch spalled areas or voids in concrete or masonry to prepare for waterproofing application, Fig. 7.33. These products contain a variety of proprietary chemicals and additives to enhance cure time, strength, and shrinkage. High-strength patching systems require only water for mixing and can be applied in a dry or stiff mix that allows for vertical patching applications.

Properties of cementitious high-strength mixes vary, and product literature should be reviewed to make selections meet specific repair conditions. Most important, installation procedure with all products is to use maximum single-application thickness.

Most products require maximum 1-inch layers due to the chemical process that creates extreme temperatures during curing. If an application is too thick, patches will disbond or blow off during curing.

Hydraulic cement products

Hydraulic products are frequently referred to as *hot patches* because of heat generated during the curing process. *Hydraulic* refers to running water, such as water leakage through a

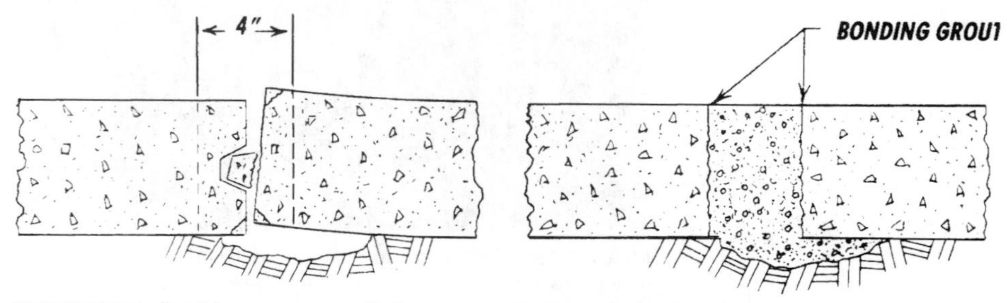

FIGURE 7.33 Patching existing spalled concrete with high-strength grout. (*Courtesy of Anti-Hydro*)

crack. These materials set in an extremely short time due to this internal chemical curing, which dries the material rapidly. This property enables these materials to patch cracks that exhibit running water in concrete or masonry.

To complete repairs to substrates, leaking cracks are sawn out, approximately 1 × 1 inch (Fig. 7.34). This groove is then packed with hydraulic cement, regardless of any running water present (Fig. 7.35). The material sets in approximately 5 minutes, in which time leakage is effectively sealed (Fig. 7.36). Should water pressure be great enough to force patch material out before initial set, relief holes must be drilled along the crack to redirect water (Fig. 7.37). These holes will continue to relieve water pressure until the crack patching has cured, after which relief holes themselves are patched to stop water infiltration completely.

These products are also used to seal portholes in epoxy and chemical grout injections. Hydraulic materials are often used in conjunction with negative cementitious waterproofing applications to complete substrate patching before waterproofing material application.

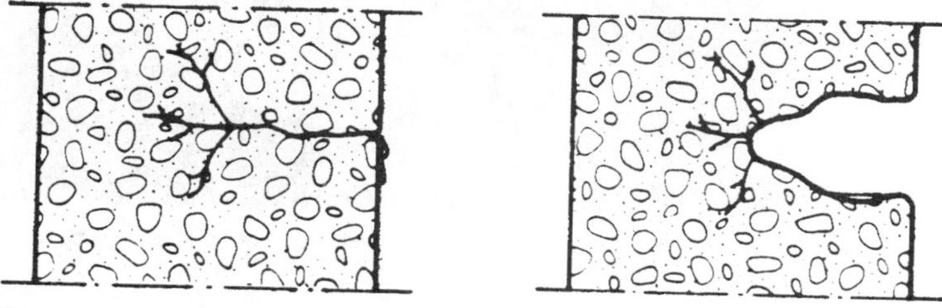

FIGURE 7.34 Crack repair preparation of existing substrate. (*Courtesy of Vandex*)

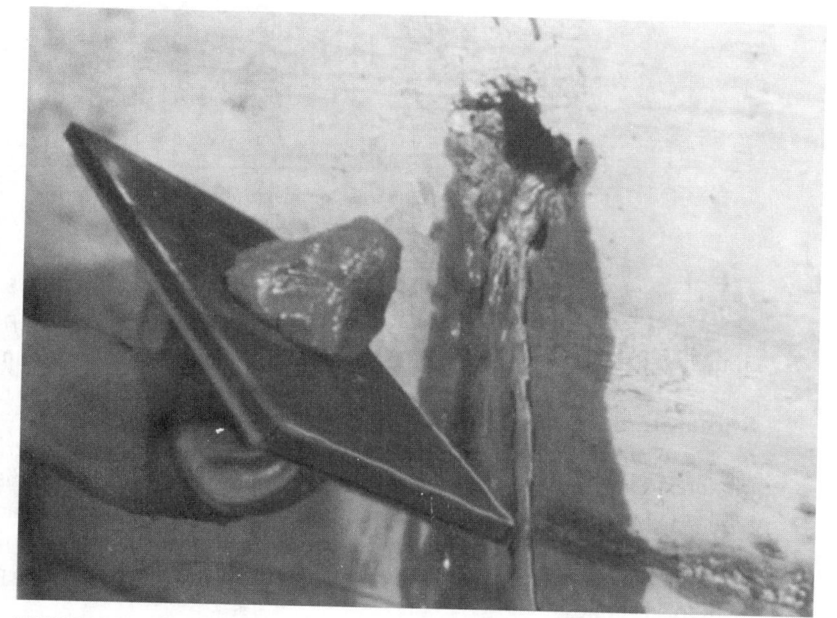

FIGURE 7.35 Application of hydraulic cement to stop leakage. (*Courtesy of Vandex*)

FIGURE 7.36 After a short cure period, leak is effectively stopped. (*Courtesy of Vandex*)

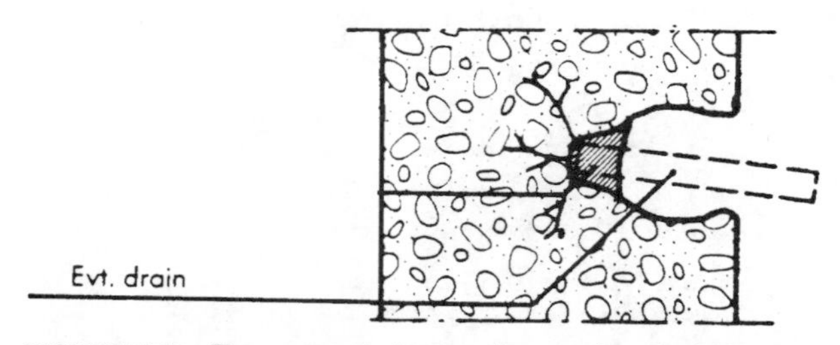

FIGURE 7.37 Temporary drainage tubes are installed if necessary, to alleviate water pressure during curing of grout. (*Courtesy of Vandex*)

The extremely fast initial set prohibits their use over large wall or floor areas. They are limited to patching cracks or spalled areas that can be completed within their short pot life.

Shotcrete or gunite

Shotcrete and gunite are pneumatically applied small aggregate concrete or sand–cement mixtures. These are used to restore existing masonry or concrete substrates to a sound structural condition, waterproofing preparation, or both. These methods are used when areas requiring restoration are sufficiently large, making hand application inefficient. Pumping and spray equipment used in shotcrete can automatically mix materials, then pneumatically apply them to substrates.

Gunite or shotcrete mixtures used with this equipment vary from field ratio mixes to premixed manufactured dry materials requiring only the addition of water. Materials are applied as a dry mix for vertical applications. After initial application, materials are troweled or finished in place as necessary.

Surface preparation requires chipping and removal of all unsound substrate areas and repairing of existing reinforcing steel as necessary. Additional reinforcing steel may also be installed if necessary before gunite operations.

Overlays

Cementitious overlays are used for restoring deteriorated horizontal concrete substrates. Overlays are available in a wide range of mixtures containing various admixtures that add strength and shorten curing time. They are used in a variety of applications including bridge repairs and parking deck restorations.

Often they are sufficiently watertight to eliminate a need for waterproofing coatings. Others are designed specifically for use as an underlay for deck coating applications. If additional structural strength is necessary, qualified engineers should be consulted for selection and use of such products.

These materials are usually self-leveling, conforming to existing deck contours to which they are applied. They are used also to fill ponding or low areas of existing decks before deck-coating application, Fig. 7.38. Stiffer mixes are available for ramp areas and inclined areas. (See Table 7.10.)

ELECTRO-OSMOSIS

Electro-osmosis is a process that introduces electric current into a substrate to control the flow of water and humidity, a process originally identified in the early 1800s. The electro-osmosis process is now available commercially and used for a variety of construction techniques including removal of hazardous contamination from groundwater and facilitating dewatering of soils.

Electro-osmosis is also used today as an effective remedial waterproofing and humidity control method, but only for concrete structures at or below-grade. The process creates a pulsating direct current (DC) electrical current that causes cations within the concrete to move from the dry or interior side of a structure to the wet or exterior side. The movement of cations attracts the available water and moisture to follow, against the flow induced by the hydrostatic pressure, eliminating water infiltration into the structure.

A commercial electro-osmotic pulse system (EOP) consists of a control unit that plugs directly into a common 110-volt outlet that delivers electric pulses to an anode (positive

TABLE 7.10 Cementitious Systems Properties

Advantages	Disadvantages
Variety of systems available	No movement capability
Negative or positive systems	Only masonry and concrete substrates
Large or small repair areas	Mixing controlled at site

FIGURE 7.38 Application of overlay prior to waterproofing application. (*Courtesy of Devcon*)

electrodes) installed into the concrete substrate. A negative electrode (cathode) is installed into the adjacent surrounding soil. The flow of current causes the water to flow away form the interior of the structure back towards the exterior. A schematic diagram of the system is shown in Fig. 7.39.

The process works for both reinforced and nonreinforced concrete, and is very efficient in improving air quality by substantially reducing humidity in basement areas. The system can be an effective remedial means to eliminate water infiltration into basement or below-grade structures without having to expend money for excavating and repairing or replacing existing waterproofing membranes. The operating costs are relatively low since it operates on 110-volt power.

Installation

All obvious cracking and free flowing leaks should be repaired using chemical grouts. Any structural repairs necessary might require epoxy injection. A cathode (copper-clad steel

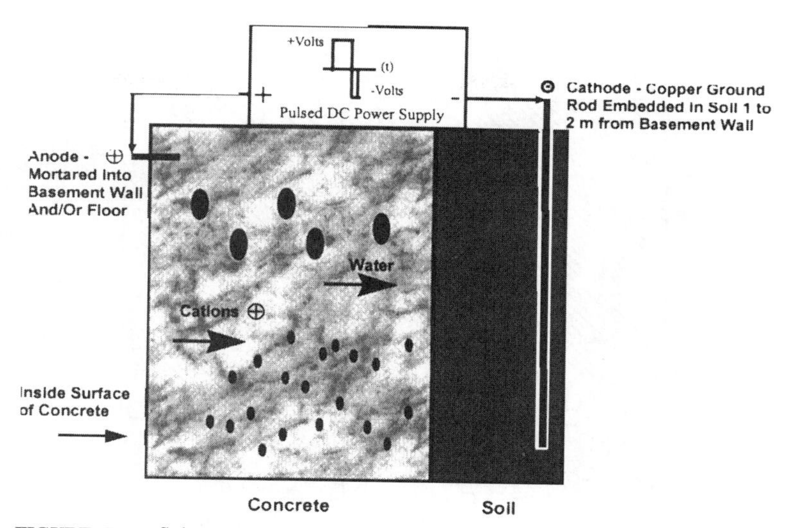

FIGURE 7.39 Schematic diagram of electro-osmosis equipment. (*Courtesy of Dytronic, Inc.*)

rod) is installed into the adjacent soil from the exterior. If not accessible, an access hole can be cored into the concrete structure floor or wall, installing the cathode from the interior side. The core hole is then sealed with a high-strength grout as shown in Fig. 7.40.

Anodes (ceramic-coated titanium) are installed in holes drilled in the concrete in a pattern and amount recommended by the manufacturer. Once the anodes are in-place, the holes should be patched using a high-strength grout as shown in Fig. 7.41. Both the anodes and cathodes are wired in accordance with local electrical codes and connected to the control panel.

The control panel usually includes a warning system that alerts of malfunctions. Otherwise, the system operates itself with no oversight necessary.

DIVERTOR SYSTEM REPAIRS

Divertor systems that fail to function are as difficult to access as exterior or positive sides of below-grade building envelope areas. In many situations, it is more cost-effective to add additional protection to reduce the water infiltrating to a level manageable by the divertor systems. If the system has failed completely, a barrier system is the only alternative.

The first step in the process is to determine if the infiltration is due to negligent maintenance of the divertor system, such as clogged weeps or insufficient drainage away from the structure (this can be determined by the process described in Chapter 11). If it is determined that the water flow entering the envelope is beyond the capability of the divertor system, then remedial waterproofing systems should be considered.

Often when masonry envelope areas are involved, the water infiltration is due to excess water entering through deteriorated mortar joints. The joints can be repaired using the tuckpointing methods described earlier and the area retested to determine if the leakage is

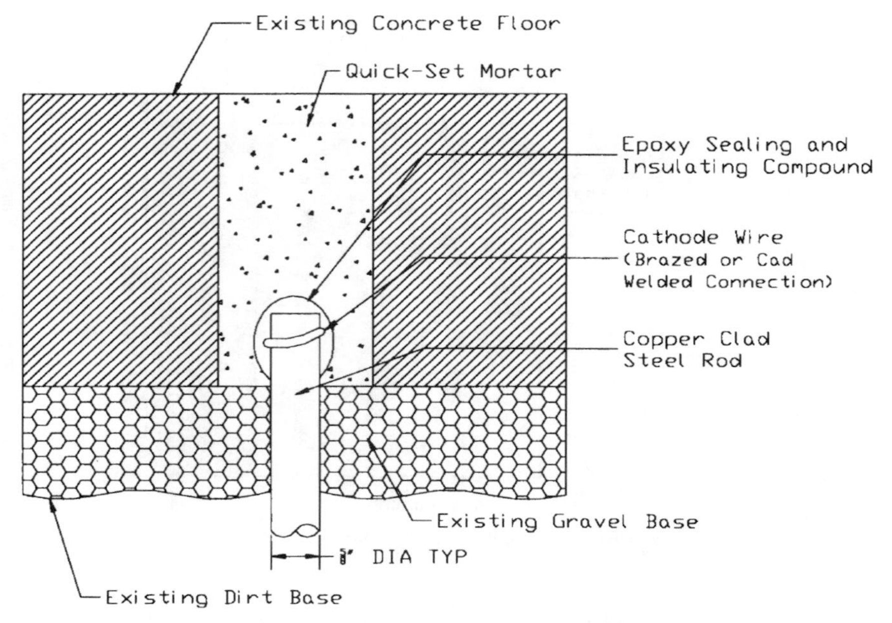

FIGURE 7.40 Cathode installation detailing. (*Courtesy of Dytronic, Inc.*)

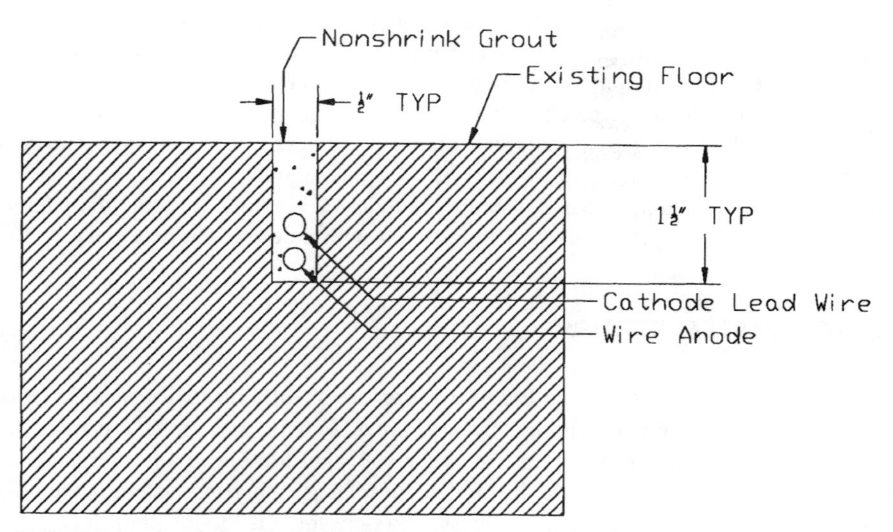

FIGURE 7.41 Anode installation detailing. (*Courtesy of Dytronic, Inc.*)

corrected. If leakage persists, the problem may require a clear repellent application to the entire masonry envelope facade. The materials and systems presented in Chapter 3 are applicable to remedial installation if the existing substrate is cleaned and repaired prior to the sealer application. The sealer should reduce the infiltration into the masonry substrate sufficiently to eliminate leakage.

If the divertor system has failed completely, the cladding must be treated with a barrier waterproofing system that is likely to completely change the appearance of the facade. Barrier systems that are applied to masonry walls include the cementitious systems described in previous sections. In addition, an elastomeric coating can be used, providing that deteriorated mortar joints are treated first. Both these repair methods with change the aesthetics of the existing cladding.

In certain situations, testing and investigation might reveal that infiltrating water is caused by water entering an adjacent envelope component and traveling to the divertor area, exceeding the capability of the system to adequately control the flow of water. For example, precast panels above a window or curtain wall might permit water to travel through the precast and down into the integral window head flashing. While it might appear the window is contributing to the leakage, the leak investigation should confirm that the precast is causing the problem. In this situation, the precast should be repaired and sealed to properly eliminate the water infiltration.

When retrofits or additions are involved to existing structures, care must be taken to preserve the integrity of the existing envelope. Manufacturers should be consulted to determine appropriate transition details between new envelope components (including waterproofing systems) and existing systems. In addition, building owners are to verify any current warranties that might be voided by renovation work adjacent to or on the existing systems. Chapter 8 presents in detail appropriate termination and transition details that can be used in retrofits and remedial applications.

Anywhere that a new envelope component meets an existing component, the applications methods presented in the previous chapters should be closely followed. For example, installation of a new concrete wall adjacent to an existing wall should not prevent the installation of waterstop to adequately seal the envelope. There are remedial waterstop systems to properly tie the concrete placement together as shown in Fig. 7.42.

It is important to remember that all proper waterproofing practices presented in Chapters 1 through 6 should be implemented on remedial applications. Since many remedial repairs are necessitated by these principles, superior installation methods for remedial work can prevent the 90%/1% and 99% principles from reoccurring.

RESIDENTIAL BASEMENT RETROFITS

Residential basements often experience leakage, often because of the 90%/1% and 99% principles, but also due to residential contractors not taking necessary precautions in the original construction to protect the interior areas. Since most basement walls are constructed of concrete or concrete block, all the products covered in Chapter 2 are applicable in residential construction. This includes both positive- and negative-side applications.

For remedial treatments, if the interior side of the basement is accessible (including through removal of paneling or drywall), negative cementitious applications make excellent choices for repairs. Often the critical floor-to-wall juncture is the point of least resistance to leakage, and installing a cove or cant with metallic or high-strength cementitious grout will correct the leakage problem. However, it is important to recognize that the entire basement floor-to-wall joint must be repaired, otherwise the leakage will just move to the next path of least resistance.

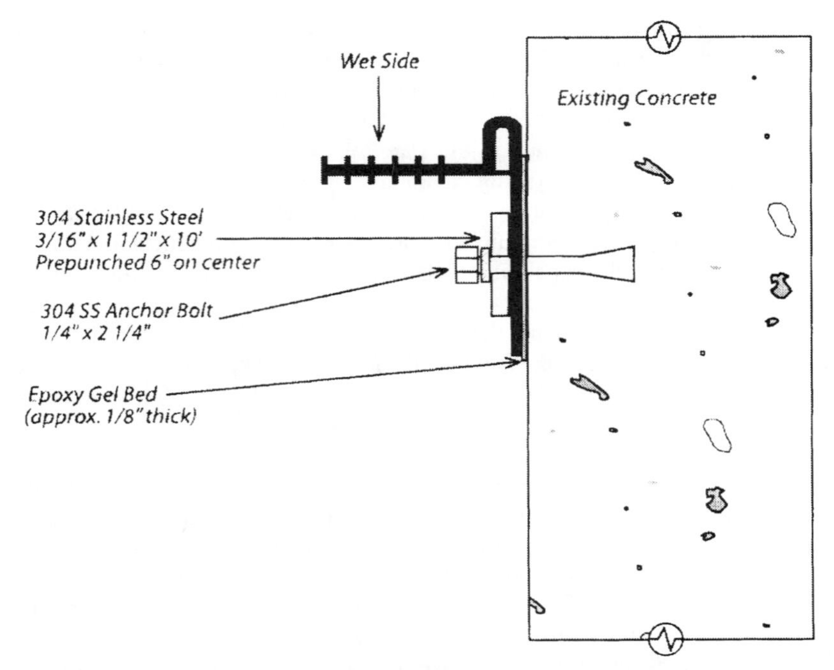

FIGURE 7.42 Mechanically fastened retrofit waterstop. (*Courtesy of Earth Shield*)

Many alternative remedial residential applications are available. Often recommended is the installation of a sump pump that can drain water away from the exterior envelope areas into a drainage area where it can be pumped to appropriate drains. The application of a sump pump and the cove treatment described above are usually sufficient to correct all but the most serious leakage problems. See Fig. 4.43.

Manufacturers have created wall panels that create drainage paths for infiltrating water. The water runs down the panel and is collected at the base of the wall and into a drainage area where a sump pump is used to drain the water. The panels in themselves can be used as a finish surface for basement areas, as shown in Figs. 7.44 and 7.45.

If the basement slab is contributing to the problem, typically fixing cracks in the slab, and remedial treatment of the floor-to-wall juncture, will correct the infiltration. Slab cracks can be filled with chemical grouts as described in previous sections, or routed and filled with a sealant or nonshrink grout. If the problem persists with seepage directly through the concrete, a sump pump might reduce the hydrostatic pressure sufficiently to alleviate the problem. In the most serious situations, prefabricated drainage panels can be used to drain the water to the sump pump area.

PREFABRICATED DRAINAGE PANEL REMEDIAL APPLICATIONS

The same prefabricated drainage systems described in detail in Chapter 2 can be used for remedial treatments, usually for below-grade or slab-on-grade negative-side appli-

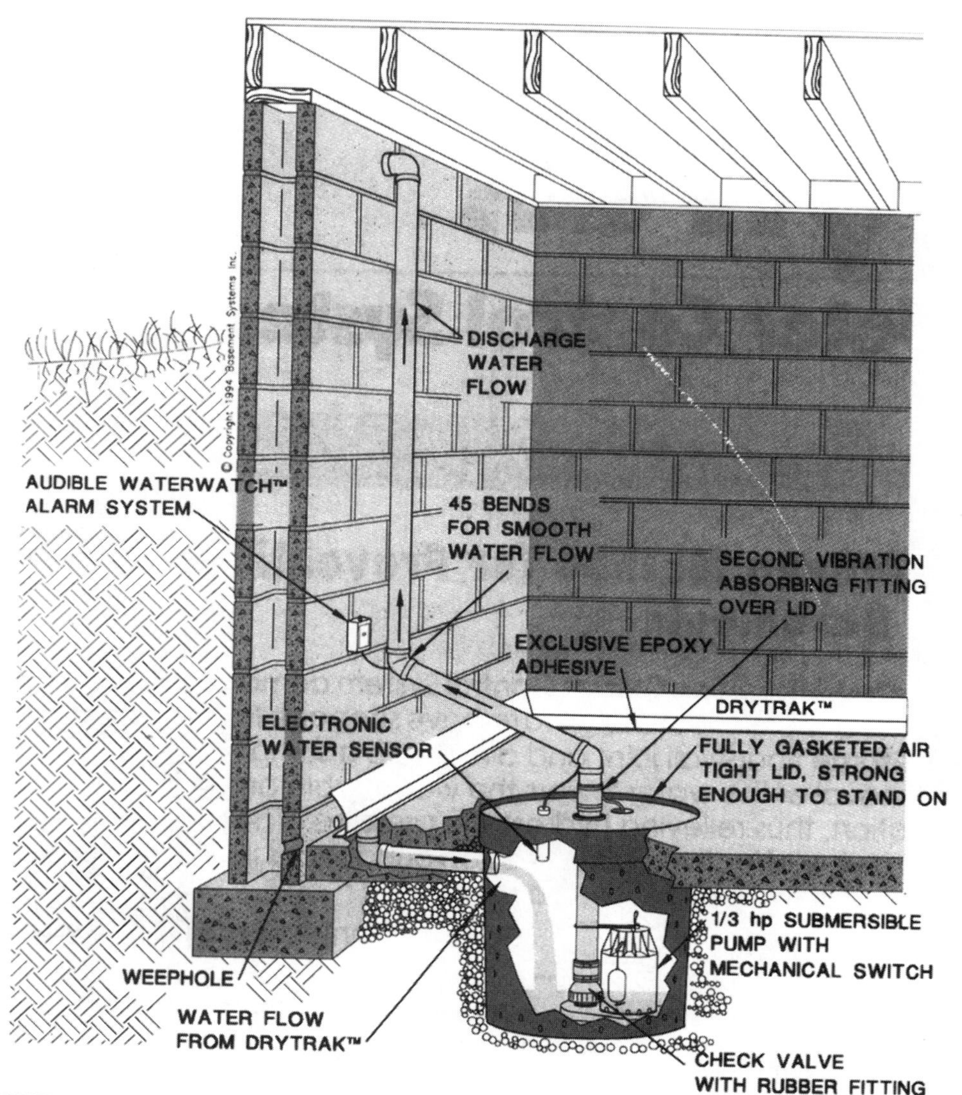

FIGURE 7.43 Remedial basement waterproofing systems. (*Courtesy of Basement Systems, Inc.*)

cations. The prefabricated drainage is simply laid on the floor slab, with the drainage grooves installed correctly so as lead the flow of water to new or existing floor drains (Fig. 7.46).

The plastic material is nonpermeable, so it does not transmit moisture or water vapor into the interior spaces. The drainage board used must be of sufficient compressive strength to withstand the loading to be placed on it as a finished product. A space is left at the floor-to-wall juncture to allow for adequate edge drainage as shown in (Fig. 7.47).

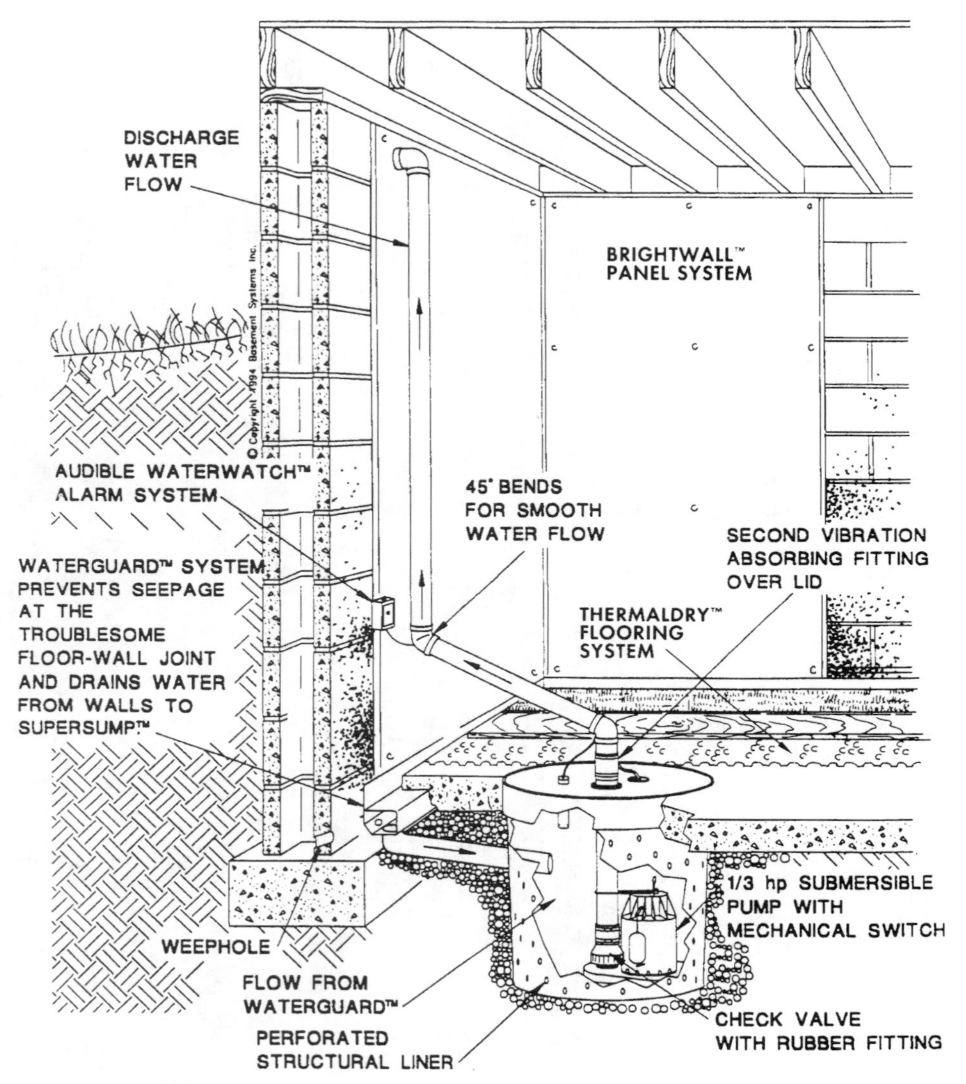

FIGURE 7.44 Wall panels used to divert water to drainage systems in remedial basement treatments. (*Courtesy of Basement Systems, Inc.*)

The drainage mat is then covered with a remedial subfloor such as plywood, and finished flooring such as carpet or vinyl as shown in Fig. 7.48. The edge can be finished as shown in Fig. 7.49. If necessary, the walls can be treated with a drainage panel system (similar to the system described in the above residential section) that drains water to the edge treatment for removal through the prefabricated drainage system on the floor.

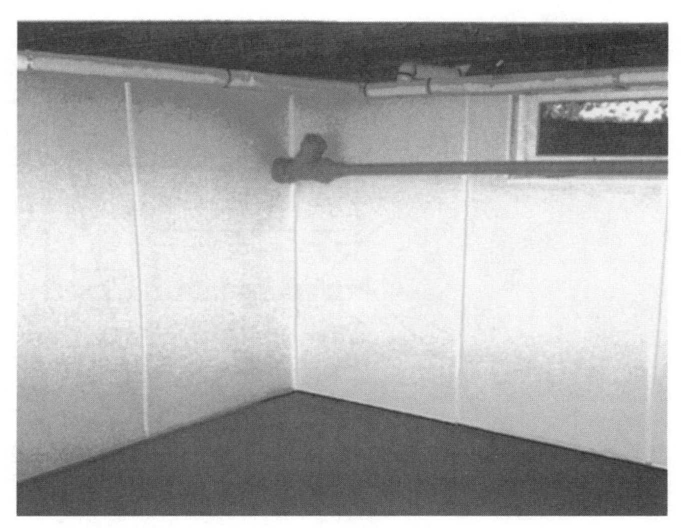

FIGURE 7.45 Repair effectiveness using the remedial wall divertor system. (*Courtesy of Basement Systems, Inc.*)

REMEDIAL SEALANT APPLICATIONS

Most failed sealant joints can be repaired by removing the existing sealant and properly preparing the joint and applying new sealant as described in Chapter 4. However, before replacing failed sealants it should be determined if there are reasons for the joint failure besides material failure.

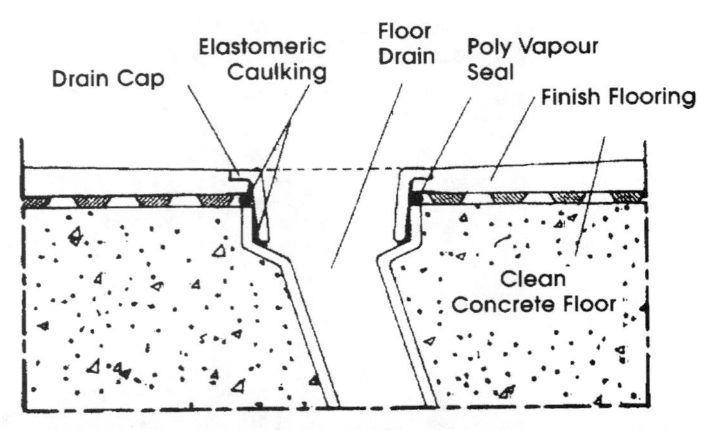

FIGURE 7.46 Application of remedial floor drainage systems to alleviate leakage through slab. (*Courtesy of Cosella-Dorken*)

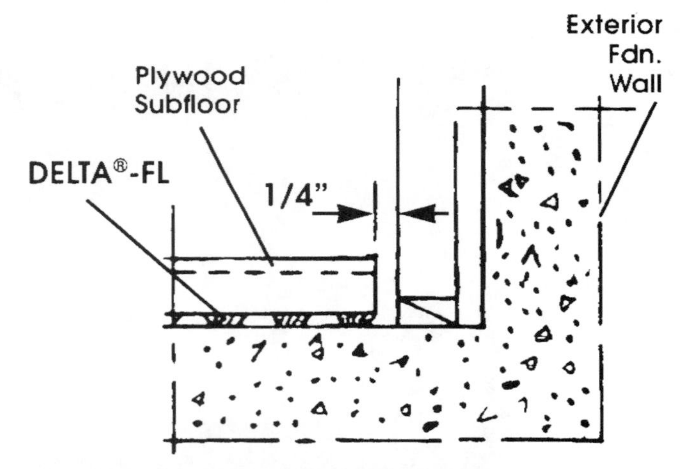

FIGURE 7.47 Floor-to-wall detailing for prefabricated drainage systems in remedial applications. (*Courtesy of Cosella-Dorken*)

Movement at the joint might exceed the capability of the sealant material originally selected. The joint design width may be undersized and insufficient to permit the sealant material to function properly. Actual leakage attributed to the joint might be caused by water infiltration bypassing the primary seal, requiring a secondary seal or joint to be installed to correct the leakage problem. Investigation practices discussed previously in this Chapter 11 should be implemented to determine the exact cause of failure before proceeding with repairs. If it is determined that the sealant has failed due to installation problems only, then the practices described in Chapter 4 should be used to replace the sealant.

If it is determined that the joint movement is exceeding the capability of the sealant material or that the joint has been undersized, there are remedial applications specifically

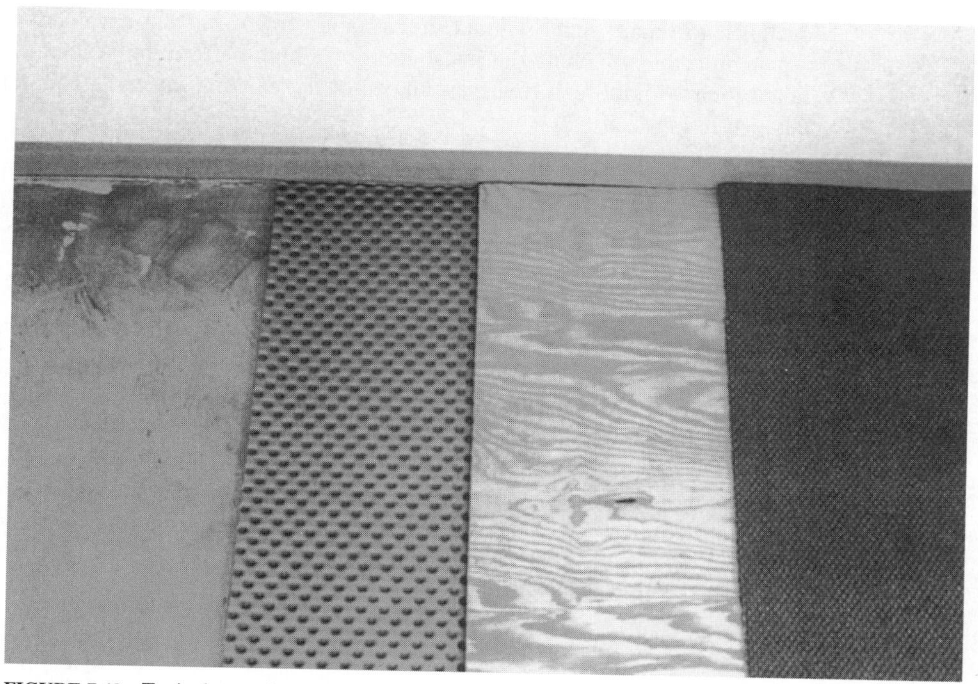

FIGURE 7.48 Typical remedial floor drainage system. (*Courtesy of Basement Systems, Inc.*)

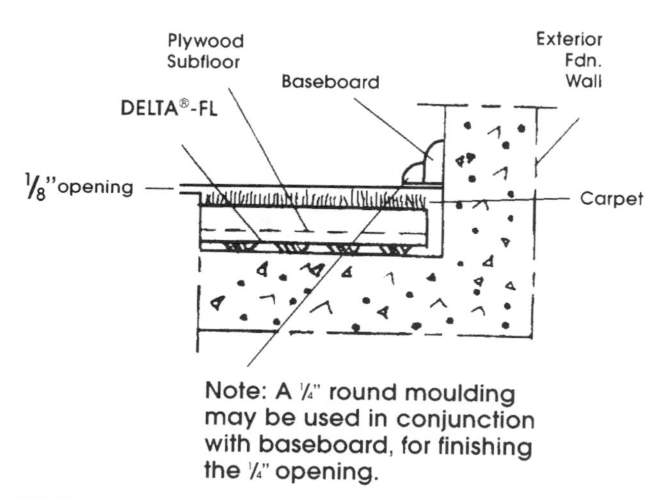

FIGURE 7.49 Finishing detail for remedial floor drainage systems. (*Courtesy of Cosella-Dorken*)

designed to repair joints without necessitating rework of the envelope substrate. These repair procedures include the installation of a high-performing sealant directly over the failed joint without requiring the removal of the existing material. This repair method is detailed in Fig. 7.50.

If the substrate joint width originally was undersized, a remedial overband joint design can be installed to permit for sufficient space at the joint without having to enlarge the substrate joint width or remove the existing failed sealant. Figure 7.51 details such a remedial application method, and Fig. 7.52 shows the application method.

EXTERIOR INSULATED FINISH SYSTEMS RESTORATION

EIFS or synthetic stucco systems, as presented in Chapter 3, have a history of problems with terminations and transitions leakage that fits naturally the description of the 90%/1% and 99% principles. In particular when the EIFS system has been installed as a barrier system in lieu of a water-managed, drainage, or divertor system, extensive damage can occur to the structural components of the building envelope. Figure 7.53 shows the damage found under a residential EIFS system that was posted on a Web site relating to the problems of EIFS in North Carolina.

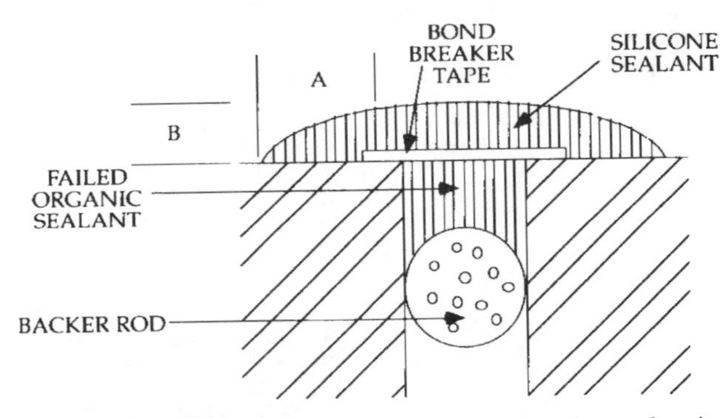

FIGURE 7.50 Remedial sealant repair leaving existing sealant in-place. *(Courtesy of Dow Corning)*

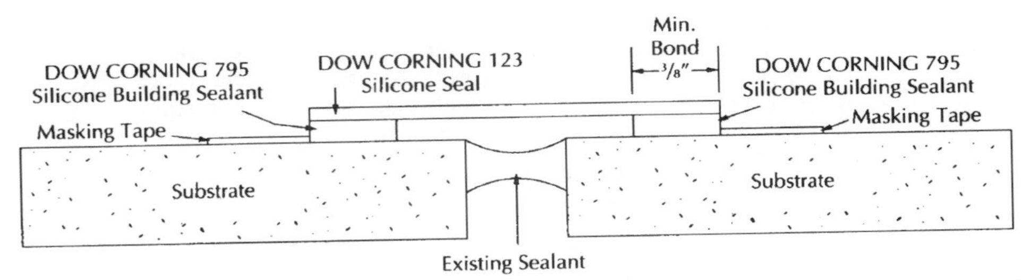

FIGURE 7.51 Remedial treatment for repairing undersized joints with excess movement. *(Courtesy of Dow Corning)*

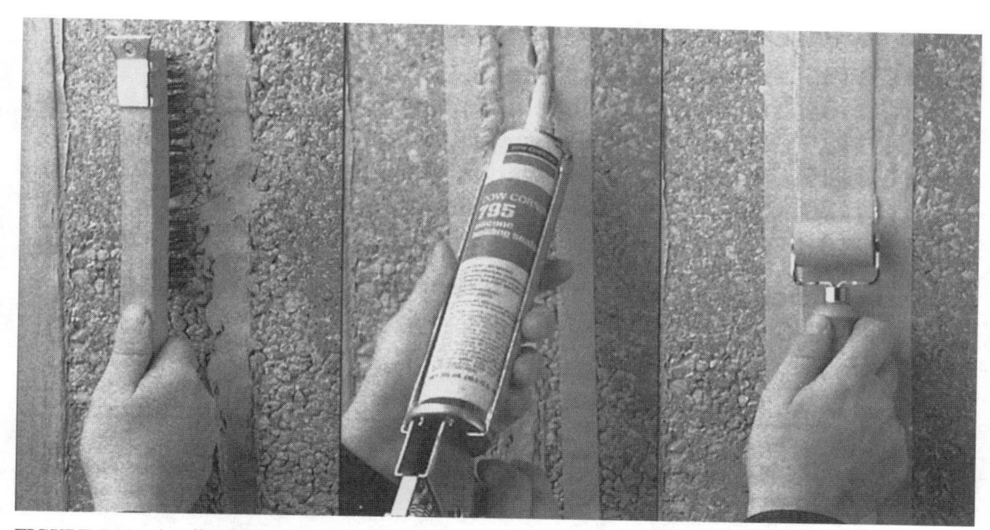

FIGURE 7.52 Application procedures for remedial joint treatments. (*Courtesy of Dow Corning*)

Barrier EIFS systems are more difficult for correcting leakage problems than the now more frequently used EIFS divertor systems that include drainage systems behind the synthetic finish. Repairing barrier systems can involve the complete removal and replacement of the system due to the structural repairs necessitated, as shown in Fig. 7.54. The new systems should be replaced in accordance with manufacturer's recommendations and the suggested waterproofing guidelines for the EIFS systems presented in Chapter 3.

Remedial repairs that do not require complete replacement of the system should first begin with any required structural repairs. Then the new EIFS system should be applied as a patch to the envelope areas that are removed to gain access for the structural repairs.

The existing EIFS system is then repaired, generally using elastomeric coatings to create a barrier waterproofing system on the entire envelope. If repairing a divertor EIFS system, the elastomeric coating must be permeable to ensure that moisture vapor can escape back out to the exterior. All existing cracks in the EIFS system must be repaired using the same techniques described for elastomeric coatings, including the overbanding of minor cracking and the sealing of larger cracks with a coating-compatible, low-modulus sealant.

Since the majority of transition and termination details in EIFS systems are based on sealant joints, all existing details must be carefully examined and repaired or replaced as necessary. If expansion and control joints have failed because of adhesion problems, they should be completely removed and replaced with a low-modulus urethane or silicone sealant. If the joint was originally recessed, it is possible to perform repairs by overbanding the sealant using a silicone sealant with appropriate bond breaker tape as shown in Fig. 7.55.

Sealant overbanding can also repair existing joints that are flush with the surface, as detailed in Fig. 7.56, which is similar to the previously described overbanding joint repair. This technique is also applicable in situations where the existing joint is

FIGURE 7.53 Damage done to substrate envelope components by insufficient EIFS application. (Posted at www.new-hanover.nc.us\ins\)

undersized and failure has occurred due to movement exceeding the capability of the sealant.

In other transition details where removing the existing sealant might damage the envelope further, overbanding can be used as a remedial treatment. Figure 7.57 details the use of overbanding at a window perimeter. Similar repairs can be made at protrusions and penetrations if appropriate.

FIGURE 7.54 Repairs being made to failed EIFS residential project.

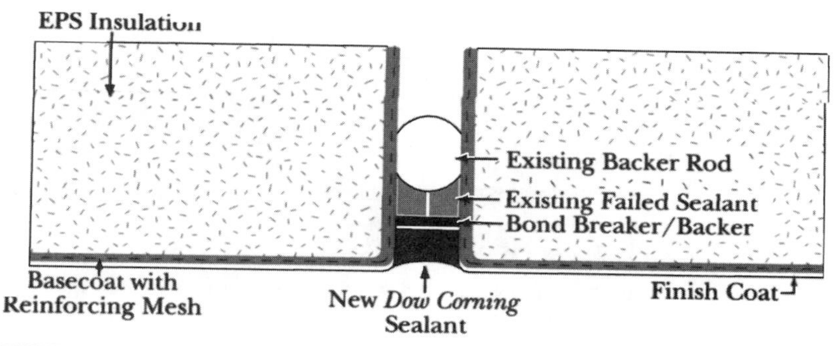

FIGURE 7.55 Using silicone sealant and bond breaker tape to repair sealant joint in an EIFS system. (*Courtesy of Dow Corning*)

The remedial repairs of an existing EIFS barrier system must facilitate restoration of the envelope barrier to prevent all water and moisture transmission through the installed system, since there is no method to redirect entering water back to the exterior.

Repairs to water-managed EIFS systems also should have a goal to completely eliminate water infiltration using the techniques described for barrier systems. However, the first step should be to verify that the existing drainage systems are functioning, and that repairs are made wherever necessary before commencing other repairs. These systems can divert small amounts of water infiltration, as described in Chapter 3. However, failed or improperly installed termination and transition detailing will contribute leakage that exceeds the capability of the divertor systems or completely bypasses these systems and enters interior space.

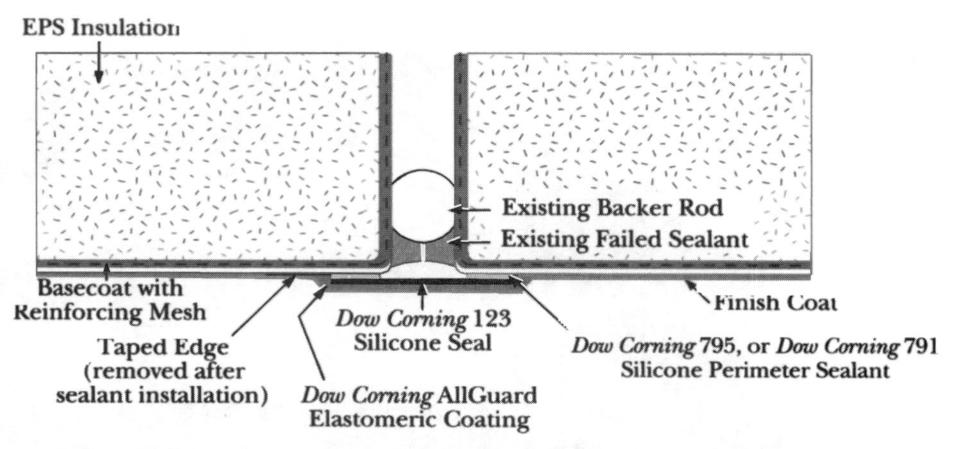

FIGURE 7.56 Remedial treatment for undersized joint in EIFS system. (*Courtesy of Dow Corning*)

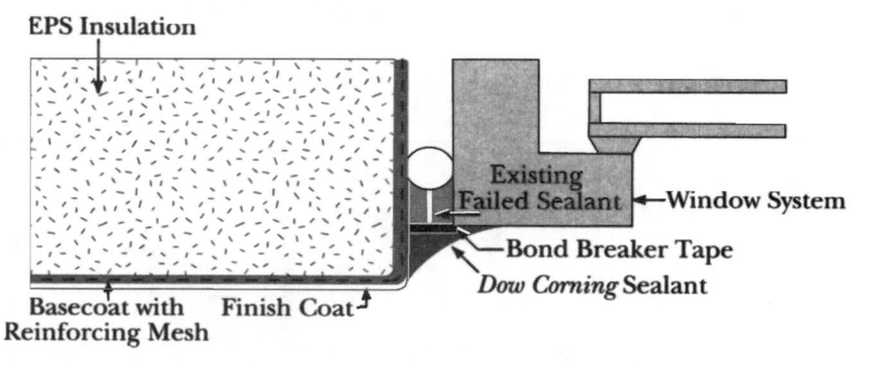

FIGURE 7.57 Remedial treatment for window perimeter seals in EIFS system. (*Courtesy of Dow Corning*)

SUMMARY

This chapter has presented the various materials and methods available to correct or repair existing envelope waterproofing problems. In most situations, other than normal replacements due to age or life-cycle limitations of the materials, the remedial treatments are necessary because of a failure to address the 90%/1% and 99% principles.

Chapter 8 addresses the termination and transition detailing necessary to prevent the need for remedial treatments, and Chapter 9 presents the maintenance necessary to ensure the maximum life-cycling of existing systems. However, should leakage occur, Chap. 11 addresses in more detail the practices available to investigate and detect leakage and make the proper selection of remedial treatments.

CHAPTER 8
THE BUILDING ENVELOPE: PUTTING IT ALL TOGETHER

INTRODUCTION

The entire exterior facade or building skin must be completely watertight to protect interior areas, maintain environmental conditioning, prevent structural damage, and provide economical life-cycling. It is only with exacting attention to details by designers, installers, and maintenance personnel alike that building envelopes maintain their effectiveness.

Too often, items are specified or installed without adequate thought as to how they will affect envelope performance and whether they will act cohesively with other envelope components. For example, rooftop mechanical equipment must by itself be waterproof, but connections attaching the equipment to the building envelope must also be waterproof.

Rooftop mechanical equipment relies on sheet metal flashing, gasketed closure pieces, and drip pans for transitions and terminations into other envelope components. Once installed, these details must act cohesively to remain watertight.

Little thought is given to the performance of these transitions during weathering, movement, and life-cycling. This results in leakage and damage, which could be prevented by proper design, installation, and maintenance. Envelope portions that are often neglected and not made watertight include lightning equipment, building signs, vents, louvers, screens, heating/ventilating/air conditioning (HVAC) and electrical equipment, lighting fixtures, doorways, and thresholds.

ENVELOPE WATERPROOFING

The first step in designing new envelopes or reviewing existing envelopes is to ensure that all major facade components are waterproof, acting as first-line barriers against water infiltration. Brick, glass, metal, roofing, and concrete must be waterproof; otherwise, allowances must be incorporated to redirect water that bypasses these components back out to the exterior.

Considering masonry wall construction, it is typically not the brick that allows water to enter but the mortar joints between the brick. One square foot of common brick wall area contains more than 7 linear feet of mortar joints. These field-constructed joints are subject to installation inconsistencies that occur with site construction.

It takes but a $^{1}/_{100}$-in crack or mortar disbonding for water infiltration to occur. This cracking occurs as a result of shrinkage of the mortar, settlement, differential movement, wind loading, and freeze–thaw cycles. Multiply these 7 linear feet by the total brick area, and the magnitude of problems that might occur becomes evident.

To offset this situation, brick joints should be expertly crafted, including properly mixing mortar, using full bed joints and proper tooling of joints, Joint toolers compress mortar against both sides of attached brick, compacting the material, which assists in preventing water from passing directly through joints.

Water that passes through joints carries salts extracted from cement content in mortar. The whitish film often occurring on exposed masonry walls is referred to as *efflorescence*. It is formed by salt crystallization after water carrying the salts is drawn by the sun to the surface. This water then evaporates, leaving behind the salt film.

When salts crystallize within masonry pores, the process is called *cryptoflorescence*. Formation of these crystals can entrap moisture into masonry pores that cause spalling during freeze–thaw cycles. Additionally, if cryptoflorescence is severe enough, it will prevent the natural breathing properties of a masonry wall. Both forms of salt can attack and corrode reinforcing and supporting steel, including shelf angles. This corrosion often leads to structural damage.

An effective joint tooling method is a weathered joint finishing. In this tooling installation, a diagonal is formed with mortar, with the recess at top, allowing it to shed water. Recessed joints, including struck and raked joints, can accumulate water on horizontal portions of the recess and exposed brick lip. This water may find its way into a structure through mortar cracks and voids. Commonly used joint types appear in Fig. 8.1. All masonry mortar joints as well as all building envelope components should be designed to shed water as quickly as possible. Figures 8.2, 8.3, and 8.4 provide proper sloping details of envelope components.

Once major envelope components are selected and designed, transition systems are chosen to detail junctures and terminations of the major components. Transition materials and systems ensure the watertight integrity of an envelope where changes in facade components occur, or at terminations of these components.

TRANSITION MATERIALS

There are two basic cladding details for above-grade envelope facades used to prevent water infiltration. The first type is a solid, single-barrier wall system with no backups or secondary means of protection (e.g., single-wythe block walls, stucco over metal lath, and exterior insulated finish systems). The second type is a multicomponent system, or divertor system, providing at least two waterproofing methods. These include the cladding itself, and backup systems consisting of flashing and weeps that redirect water passing through first-line barriers (e.g., brick cavity wall, metal and glass curtain walls with integral gutters, flashing, and weeps).

Multicomponent systems often provide better resistance to water infiltration by providing systems that redirect water bypassing the initial barrier back out to the exterior (e.g., EIFS systems with drainage capability). These systems are most effective when this redirecting is channeled immediately out to the exterior and not allowed to drain into other interior systems. The latter allows an envelope to become susceptible to leakage into interior areas.

As previously emphasized, however, it is typically not the primary waterproofing barriers themselves that directly cause water leakage, but rather the transitions between these

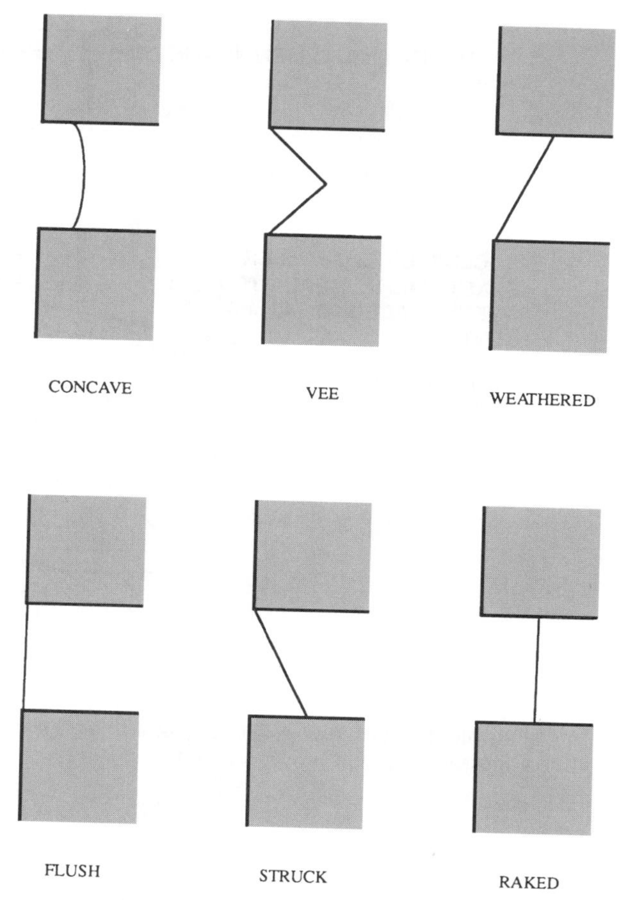

FIGURE 8.1 Common mortar joint detailing.

envelope components. Transitions and detailing between major components create 90 percent of leakage problems, although they represent only 1 percent of an envelope area (90%/1% principle presented in Chapter 1). Besides permitting leaking water into interior areas, this leakage creates damage to structural components.

Manufacturers provide recommended installation details for their products and systems, and these should be followed without exception. If a particular installation presents specific detailing problems, a manufacturer's representative will review the proposed transitions and actual installations as necessary. This inspection requirement is one of the major advantages of the joint manufacturer and contractor warranties discussed in Chapter 9.

In addition to manufactured system components for transitions, several frequently used transition systems and materials are used in building construction. These systems provide watertight transitions between various primary envelope components when installed properly. Standard available systems include:

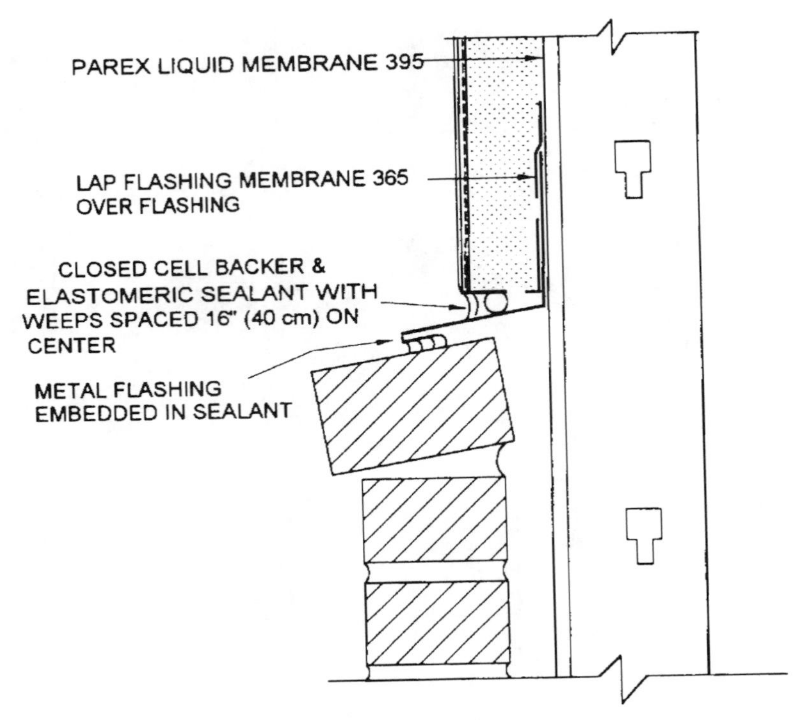

FIGURE 8.2 Detail showing sloping of masonry sill and flashing to facilitate water drainage away from envelope. (*Courtesy of Parex*)

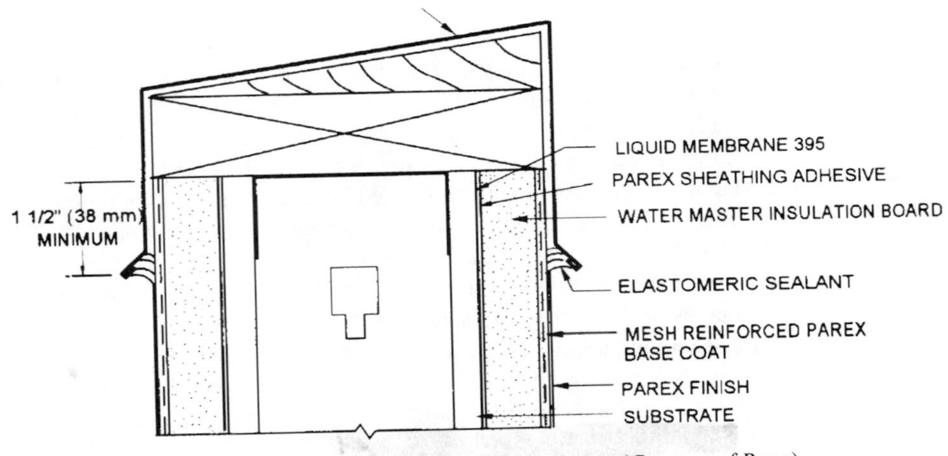

FIGURE 8.3 Detail with sloped coping cap to facilitate drainage. (*Courtesy of Parex*)

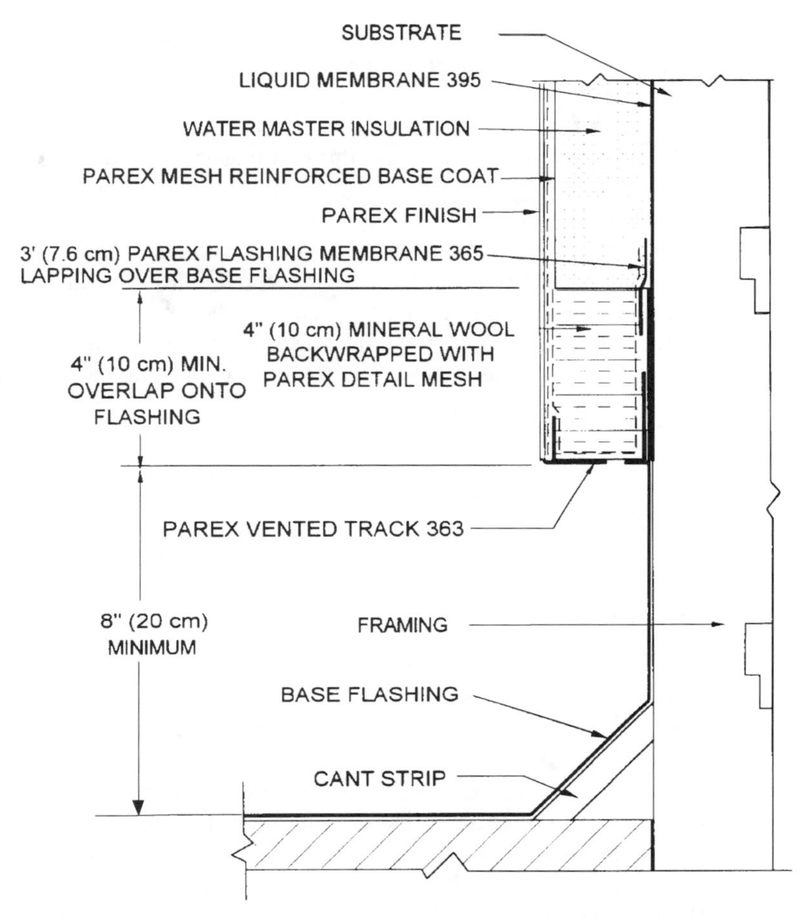

FIGURE 8.4 Cant used at roof-to-wall transition to drain water away from envelope, including water that drains from the cladding system. (*Courtesy of Parex*)

- Flashings
- Dampproofing
- Sealant joints
- Reglets
- Waterstops
- Pitchpans
- Thresholds
- Expansion joints
- Cants

Any of these standard systems proposed to be used as part of a waterproofing or envelope system should first be reviewed and approved by the waterproofing system manufacturer. This eliminates unnecessary problems that prohibit envelope components from acting cohesively together and preventing water infiltration.

Flashings

Because water is likely to pass through masonry facades, cavity wall construction with dampproofing and flashing systems are necessary to redirect entering water. Dampproofing materials, usually asphaltic or cementitious compounds, are applied to the outer faces of interior wythes. This prevents minor amounts of water or moisture vapor from entering interior spaces. Dampproofing requires flashing to divert accumulated water and vapor back to the exterior through weep holes.

Envelopes often depend on flashing to maintain watertight integrity. Flashings are used not only in brick masonry veneer structures but also in the following:

- Poured-in-place concrete
- Precast concrete panel construction
- Stucco or plaster veneer walls
- EIFS systems
- Stone veneers
- Curtain and window wall systems

Flashings are manufactured from a variety of materials, including noncorrosive metals and synthetic rubber sheet goods. Metal flashings include copper, aluminum, stainless steel, galvanized steel, zinc, and lead. Sheet-good flashings are usually a neoprene rubber or a rubber derivative.

Thermal expansion and contraction that occurs in a facade also introduce movement and stress into flashing systems. If installed flashing has no movement capability, it will rupture or split, allowing water infiltration. Adequate provisions for thermal movement, structural settlement, shear movement, and differential movement are provided with all flashing systems (Fig. 8.5).

Flashing installed in cavity walls typically is the responsibility of masonry contractors, who often do not realize the importance of properly installing flashings to ensure envelope effectiveness. Common flashing installation problems include

- Seams not properly spliced or sealed (Fig. 8.6)
- Inside or outside corners not properly molded
- Flashing not meeting at building corners
- Flashing improperly adhered to substrates
- Flashing not properly shedding water

Besides these problems, masons often fill cavities with mortar droppings or allow mortar on flashing surfaces. Additionally, weep holes are filled with mortar, damming water from exiting.

These examples, and all problems associated with site construction, make it necessary for all subcontractors to be made aware of their responsibility and of the interaction of all

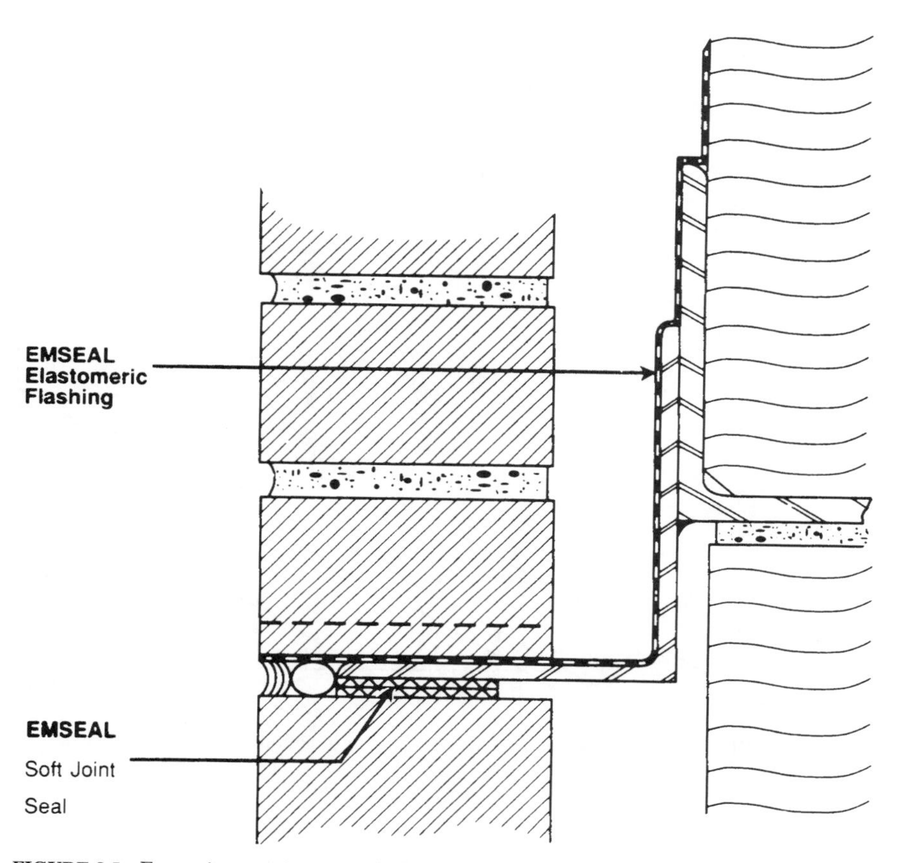

FIGURE 8.5 Expansion provisions at flashing joint. (*Courtesy of Emseal*)

building envelope components. Making frequent inspections during all envelope work ensures quality, watertightness, and cohesiveness. As with all envelope components, attention to flashing details, only 1 percent of building area, ensures the success of waterproofing envelopes.

Flashings should be installed at vulnerable areas in envelopes and where necessary to redirect entering water. Typical flashing locations and their basic functions are summarized in Table 8.1. (See Figs. 8.7-8.12)

Flashing Installation

Flashing applied to concrete substrates requires that the concrete be clean, cured, and free of all honeycomb, fins, and protrusions that can puncture flashing materials. If applied by mastic, substrates should be clean and dried. Mechanically attached systems require that substrates be sound to allow anchoring of attachments. Flashings should extend up vertically at least 12 in from the horizontal installation point.

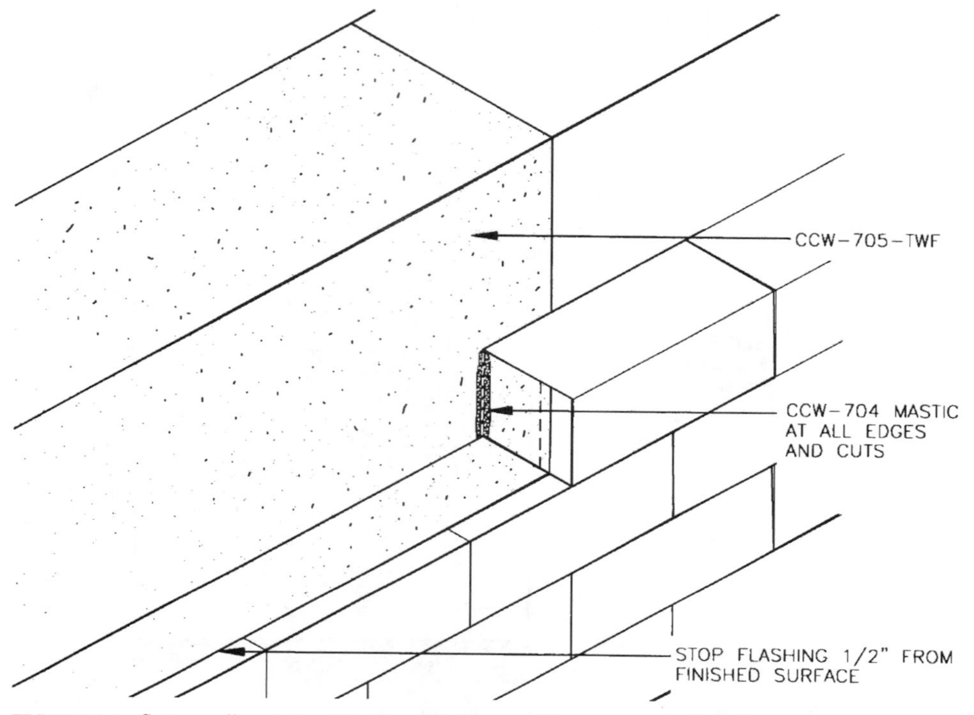

FIGURE 8.6 Seam sealing at sheet flashing termination. (*Courtesy of Carlisle Corporation*)

Flashings set on the top of shelf angles should be manufactured only of materials that do not cause galvanic reaction with the steel angles. Often the horizontal joints along shelf angles function as control joints. If sealant is installed in this joint, it is applied below the shelf angle and flashing to prevent interference with water exiting the envelope.

Exposed flashing such as roof, cap, or coping is installed to provide a transition between dissimilar materials or protection of termination details. These flashings should be securely attached to structural elements to provide resistance against wind loading. Only seams that allow for structural movement, usually as a slip joint design, should be used.

All seams in both metal and sheet materials must be properly lapped and sealed. Flashing systems allow water infiltration because of improper attention to seams, bends, and turn details. Often flashings and shelf angles are inadvertently eliminated at building corners, allowing a continuity break and a path for water to enter.

Flashing must have adequate provisions to allow for thermal and differential movement as well as for shear or deflection of wall areas, not only for longitudinal movement but for vertical movement and shear action occurring when inner walls remain stationary while outer walls experience movement.

Detailings of flashings and intersecting expansion joints in exterior wall systems are also likely failure areas. There is usually a break in structural framework at these locations, allowing for structural movement. At these locations, flashings are terminated, with their

TABLE 8.1 Commonly Used Flashing Systems and Their Functions

Location	Function
Base flashing	Prevents capillary action of water from wicking upward in a masonry wall, Figure 8.7
Sill flashing	Installed beneath window or curtain wall sills, Figure 8.8
Head flashing	Installed above window head detail, just below adjacent facing material that the window abuts, Fig. 8.8
Floor flashing	Used in conjunction with shelf angles supporting brick or other facade materials
Parapet flashing	Installed at the parapet base, usually at ceiling level; may be used on roof side of parapets as part of roof or counterflashing, Fig. 8.9
Counterflashings	Surface mounted or placed directly into walls with a portion exposed to flash various building elements into the envelope, including roof flashings, waterproofing materials, building protrusions, and mechanical equipment, Figure 8.10
Exposed flashings	Used in a variety of methods and locations; can be an integral part of an envelope system, such as skylight construction, or applied to provide materials protection between two dissimilar ,including cap flashings, coping flashings, gravel stops, edge flashings, and terminations, Figure 8.11
Remedial flashings	Typically surface mounted and applied directly to exposed substrate faces; can include a surface-mounted reglet for attachment; do not provide for redirecting entering water; only by dismantling a wall or portion thereof can remedial through-wall flashings be installed, Figure 8.12

ends dammed and detailed to allow for this movement. Typical flashing installations in common building materials are shown in Fig. 8.13.

Dampproofing

Dampproofing materials are typically used in conjunction with flashing and weep systems as part of secondary or backup systems for primary envelope waterproofing materials such as masonry walls. In addition, dampproofing systems are used at below-grade applications to prevent moisture vapor transmission or capillary action through concrete or masonry walls. Dampproofing can be applied in either negative or positive installations.

Dampproofing prevents damage to envelope components where surface water can collect and drain to below-grade areas. It also protects when improper surface water collection systems are not used in conjunction with below-grade drainage mats.

Dampproofing materials, as defined, are not intended to, nor do they, function as primary envelope waterproofing systems. They function only as additional protection for the

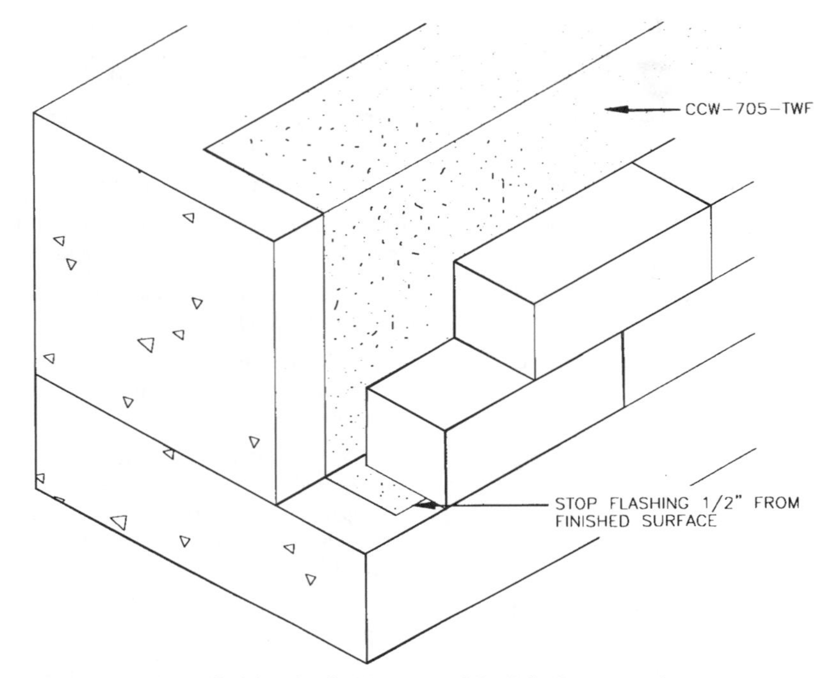

FIGURE 8.7 Base flashing detail. (*Courtesy of Carlisle Corporation*)

primary envelope systems, or are used where water vapor or minor amounts of water are expected to be encountered on an envelope. They are used in below-grade applications when no hydrostatic pressure is expected, or when it will occur in only the severest of expected weathering.

Dampproofing systems are available in a variety of materials and systems, including:

- Cementitious systems
- Sheet vapor barriers
- Bituminous dampproofing

Clear water repellents are also sometimes referred to as dampproofing materials because they are not effective against hydrostatic water pressure. However, clear sealers typically are applied directly to the face of primary envelope waterproofing materials or facades. As such, they do not function as dampproofing materials or backup and secondary systems. Clear water repellents are discussed in detail in Chapter 3.

Cementitious systems

Cementitious dampproofing systems, which are available in a wide range of compositions, are sometimes referred to as *parge coats*. Parging is an application of a cementitious material applied by trowel to a masonry or concrete surface for dampproofing purposes. Parging is also used to provide a smooth surface to substrates before waterproofing material application.

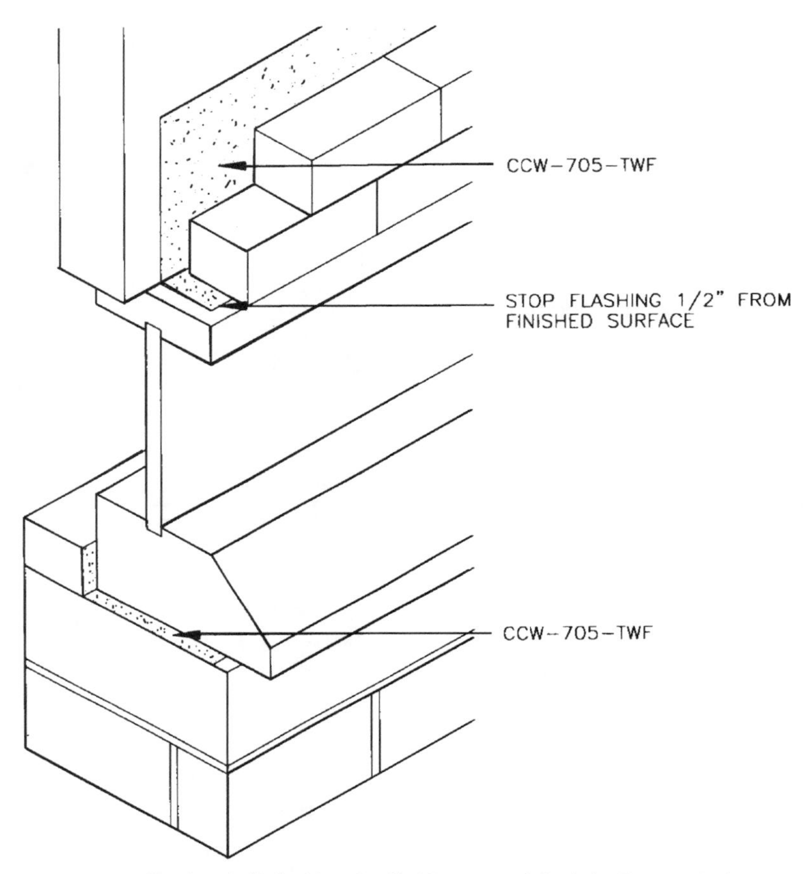

FIGURE 8.8 Head and sill flashing detail. (*Courtesy of Carlisle Corporation*)

Cementitious dampproofing materials are usually supplied in a dry premixed form. They are cementitious in a base containing the manufacturer's proprietary water-repellent admixture. Mixes also include bonding agents. Their bonding capability to masonry or concrete substrates provides an advantage over sheet materials or bituminous materials, since cementitious materials become an integral part of a masonry substrate after curing.

In addition, cementitious systems can be applied to damp substrates in both above- and below-grade positive or negative locations. An example is application to the interior of elevator pits not subject to hydrostatic pressure.

Cementitious applications also prevent capillary action at masonry walls and foundations or floor slabs placed on soils that are subject to capillary action. Dampproofing in these areas prevents water-vapor transmission to upper and interior envelope areas that can cause damage, including deterioration of flooring and wall finishes.

Sheet or roll goods

Sheet dampproofing materials are manufactured from polyvinyl chloride, polyethylene, and combinations of reinforced waterproof paper and polyvinyl chloride. They are available

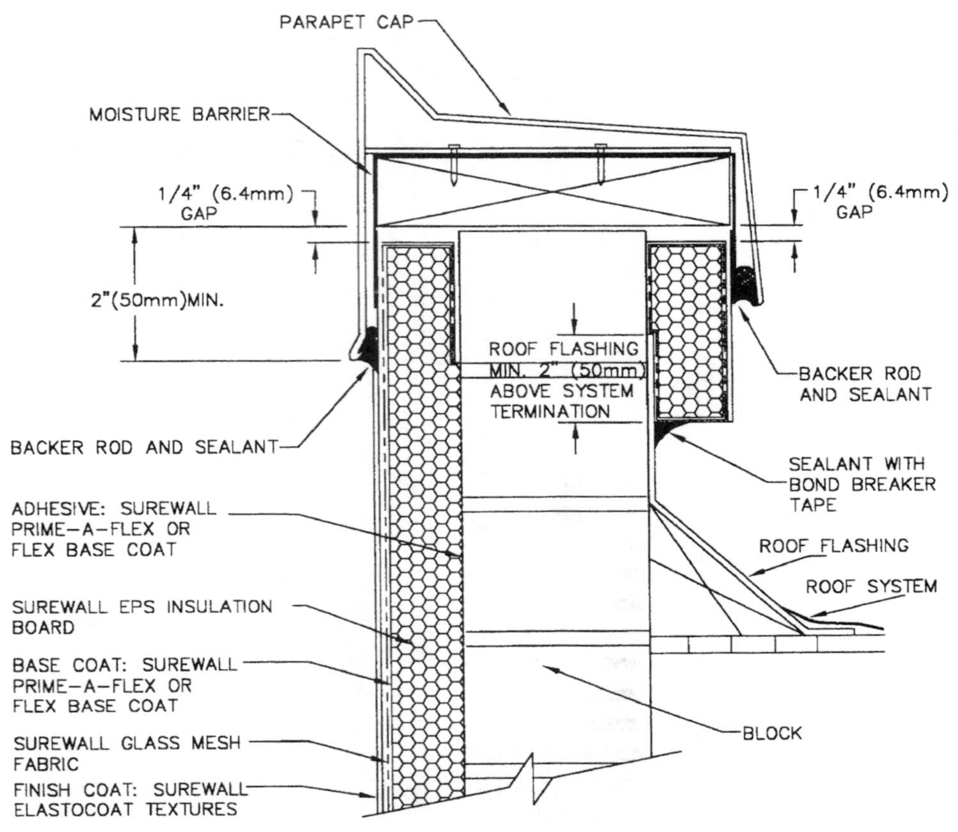

FIGURE 8.9 Transition details between coping and cladding, cladding and roofing. (*Courtesy of Bonsal*)

in thicknesses ranging from 5 to 60 mil. Sheet materials are often used for dampproofing horizontal slab-on-grade applications to prevent capillary action through floor slabs.

Sheet systems have become widely used as divertor systems or dampproofing behind EIFS systems. Often these systems are applied over exterior sheathing that cannot be used as a substrate for cementitious or asphaltic dampproofing compounds. Figure 8.14 shows a typical EFIS system application using a sheet system for dampproofing protection.

Sheet systems are more difficult to transition into other envelope waterproofing systems, particularly at below-grade to above-grade dampproofing transitions (Fig. 8.15). Typically, at these areas a mastic material is used to adhere the sheet material and provide a transition to the above-grade materials. Refer to Fig. 8.16 for a typical detail for above-grade mastic dampproofing material transitioning into slab-on-grade sheet materials.

Bituminous dampproofing

Bituminous dampproofing materials are either asphaltic or coal-tar pitch derivatives. They are available in both hot-applied and cold-applied systems, with or without fiber reinforcing. Coal-tar derivatives are seldom used today due to health risks and safety concerns during installation.

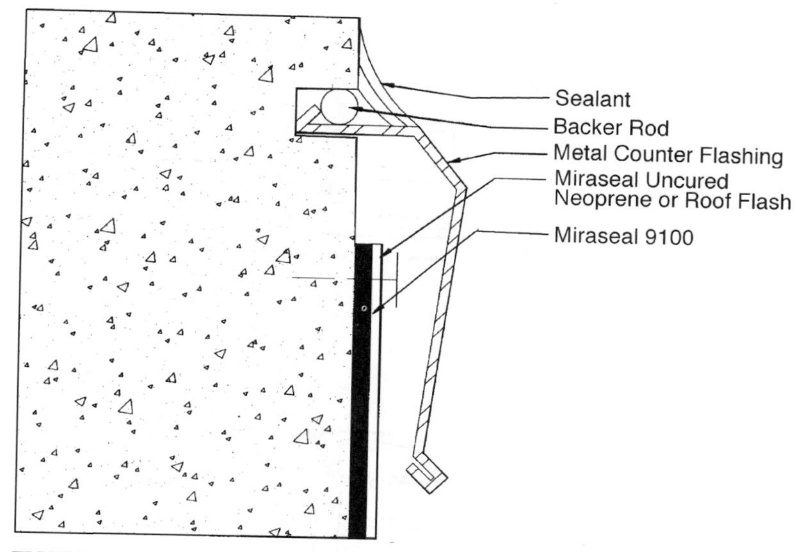

FIGURE 8.10 Typical counterflashing detail. (*Courtesy of TC MiraDRI*)

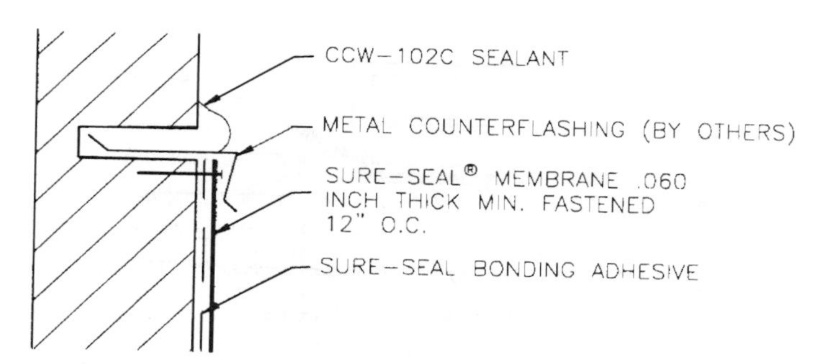

FIGURE 8.11 Termination flashing detail. (*Courtesy of Carlisle Corporation*)

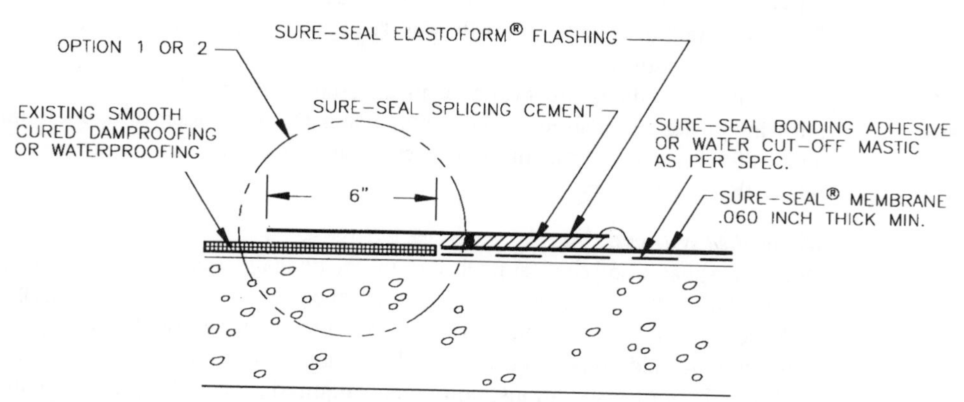

FIGURE 8.12 Remedial flashing detail. (*Courtesy of Carlisle Corporation*)

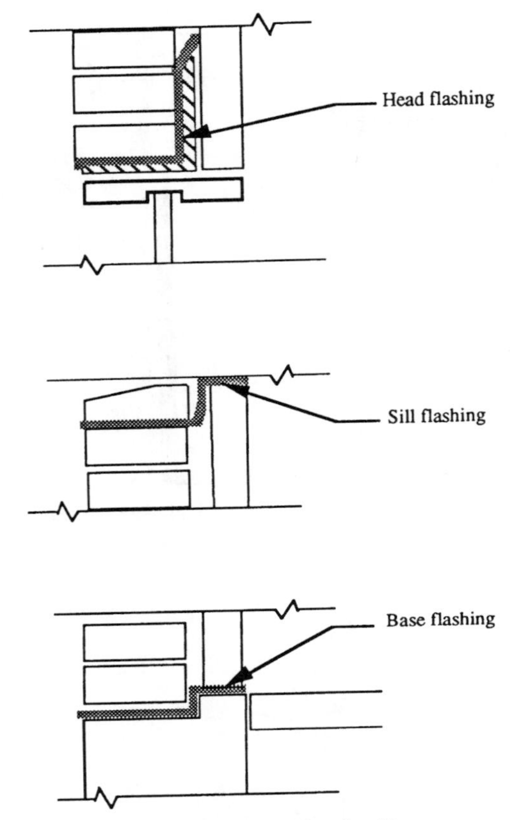

FIGURE 8.13 Envelope flashing detailing.

Glass or fabric fibers are added to dampproofing materials that allow trowel or brushable applications by binding the material together in a thicker consistency. Reinforcement also adds minor durability characteristics to the material, but not water-repellency capabilities.

Asphaltic products are available in an emulsion formulation (water-based). Besides allowing easier applications and cleanup, water-based dampproofing materials are breathable, allowing vapor transmission in envelope areas, such as parapet wall applications, where this is necessary.

Hot-applied systems

Both hot asphaltic and coal-tar pitch systems are typically used for below-grade positive applications. Difficulties involved in installation prohibit most interior (negative) applications.

Materials used in hot-applied systems are typically those used in built-up roofing applications, with the addition of roofing felts. These are usually applied in a one-coat application.

Difficulties in installations, equipment required, and field quality control has greatly limited hot-applied system usage. Cold-applied dampproofing systems that meet or exceed

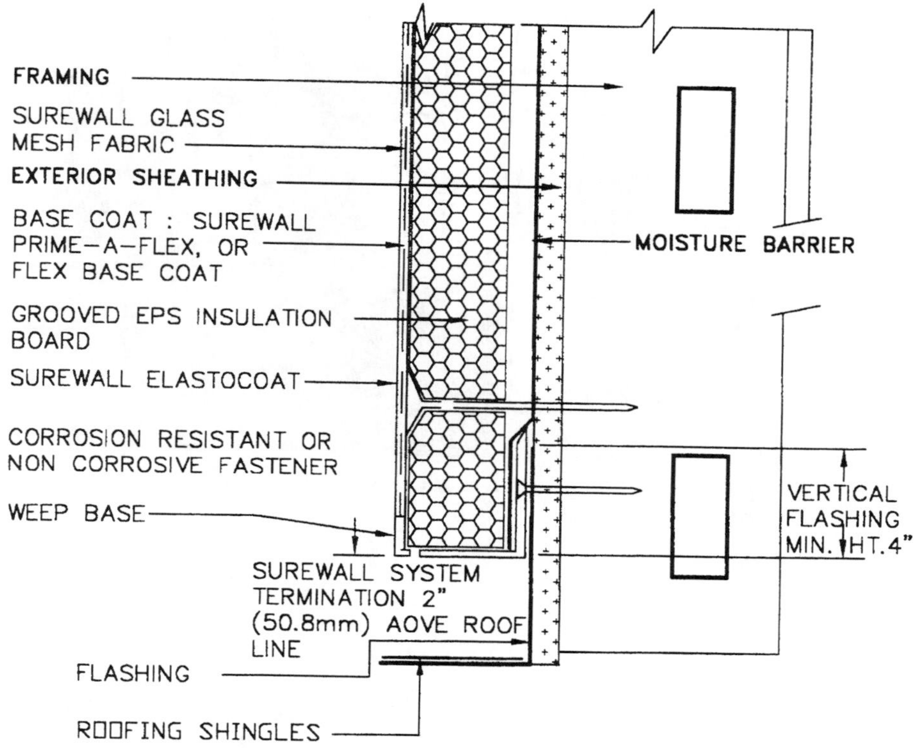

FIGURE 8.14 EIFS using sheet flashing. (*Courtesy of Bonsal*)

the performance of hot-applied systems are available, including those incorporating fibrous reinforcement.

Cold-applied systems

Cold-applied dampproofing systems are available in both coal-tar and asphaltic-based compositions. These systems are solvent-based derivatives, with or without fibrous reinforcement, that cure to form seamless applications after installation. Unfibered or minimal fiber systems are applicable by spraying. Heavily reinforced systems materials are applicable by trowel or brush.

Typically used on concrete or masonry substrates, cold-applied dampproofing materials can also be used on metal, wood, and natural stone substrates. Cold systems are used in both positive and negative applications, both above and below grade. Negative systems are applied to walls that are furred and covered with drywall or lath and plaster.

Negative systems do not allow for the collecting and redirecting of water entering an envelope. Therefore, negative applications are used only when vapor transmission through the primary waterproofing barrier is expected.

Cold-applied emulsion-based asphalt systems are also available. These water-based systems offer easy cleanup and are used where solvent systems can damage adjacent flashings, waterproofing materials, or substrates themselves. Some cold-applied emulsion-based

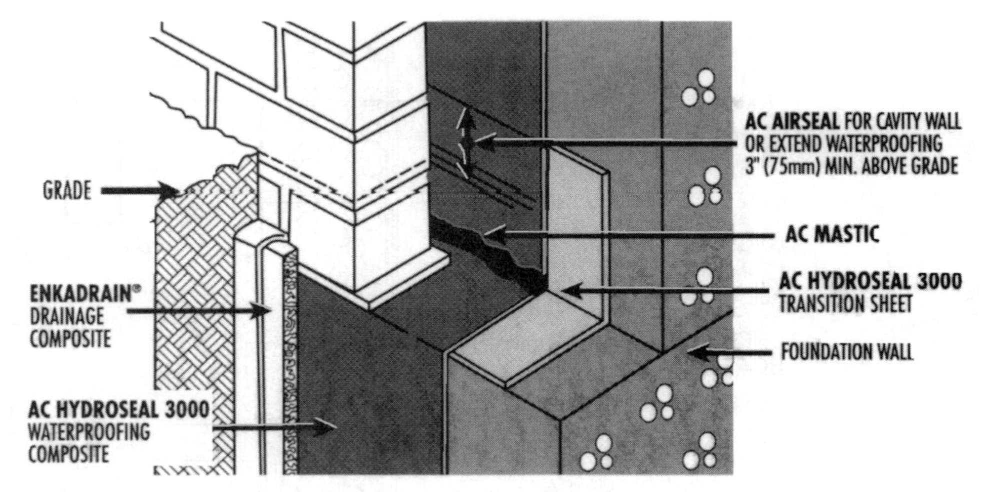

FIGURE 8.15 Waterproofing-to-dampproofing transition detailing. (*Courtesy of NEI Advanced Composite Technology*)

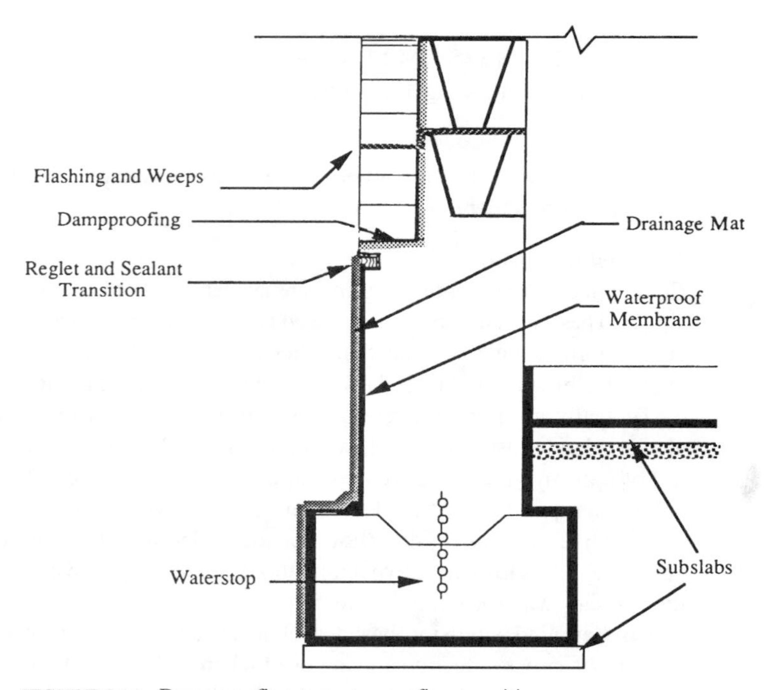

FIGURE 8.16 Dampproofing-to-waterproofing transition.

systems are applicable over slightly damp or uncured concrete, allowing for immediate dampproofing after concrete placement.

Emulsion systems should not be used in any below-grade applications or above-grade locations where sufficient amounts of water are present that can actually wash away the dampproofing material from a substrate. In addition, emulsion-based systems must be protected from rain immediately after installation. This protection must be adequate to keep installations protected until primary envelope materials are in place and backfill operations are complete.

Dampproofing Installation

Dampproofing applied to concrete or masonry substrates requires surfaces to be clean, cured, and free of all honeycomb, fins, and protrusions. Some emulsion systems allow application over uncured or slightly damp concrete. Sheet materials applied directly over soil should be placed on compacted and level granular soils that do not promote capillary action.

Negative applications are used when only vapor transmission is expected through primary envelope waterproofing systems. If water is expected to enter through the primary envelope components, negative systems should not be used. Negative systems provide no means to collect and redirect this water to the exterior.

Positive systems are used in conjunction with flashing and weeps to redirect entering water to the exterior. Water-based systems should not be used where substantial amounts of water are expected to enter and collect on the dampproofing, as this can wash the water-based material off the wall, particularly in below-grade construction.

Mastic applications are applied in thicknesses ranging from 30–35 mil. Sheet systems are generally 10–20 mil thick. Cementitious systems are trowel-applied to thicknesses of approximately $^1/_8$ in. Millage applications should be checked regularly to ensure that proper thicknesses are being applied.

When applying dampproofing to inner wythes of masonry veneer walls incorporating brick ties, a spray application of mastic is most suitable. Spraying allows for a uniform coverage around the ties, which is difficult using a trowel.

Dampproofing used in conjunction with flashing systems should be installed after flashings are adhered to the substrate. Applying dampproofing after the flashing fasteners are in place is preferable to having the dampproofing punctured during flashing application.

The dampproofing used should be compatible with flashing materials. Some solvent materials can damage sheet-flashing systems. Dampproofing should extend over the flashing and attachments to allow adequate transition detailing and ensure proper drainage of water onto the flashing where it can be redirected to the exterior. Refer to Fig. 8.17 for a typical flashing and dampproofing transition.

For negative installation, first, furring strips should be installed; then dampproofing materials should be installed. This prevents damage of the dampproofing continuity by fasteners used for attaching the furring strips. With cementitious negative systems, furring strips can be directly applied with adhesives to the dampproofing to prevent damage.

Sealant Joints

Sealant materials are frequently useful in providing transitions between dissimilar materials or systems in a building envelope. They also provide watertight allowances for thermal

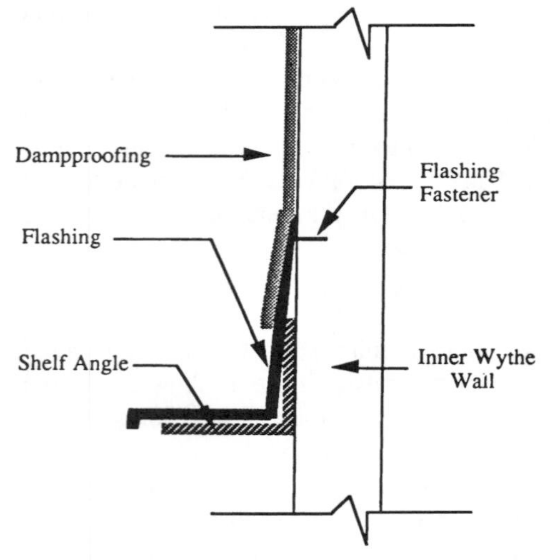

FIGURE 8.17 Dampproofing-to-flashing transition detail.

or dissimilar movement between these components. For instance, joints between metal window frames and wall facades provide a watertight transition between window and wall components and allow thermal and differential movement as shown in Figs. 8.18 and 8.19.

Sealants are often overused, especially in remedial waterproofing repairs. Simply applying sealants over failed areas will not adequately address failure problems. A thorough investigation or study should first be completed to determine why materials or systems originally failed.

For example, sealants are often used in place of tuck-pointing to correct mortar joints. This is a poor approach if existing mortar is not of sufficient strength to maintain envelope integrity. Such applications also allow three-point adhesion and improper sealant depth, and under structural movement sealant installation, repairs will fail.

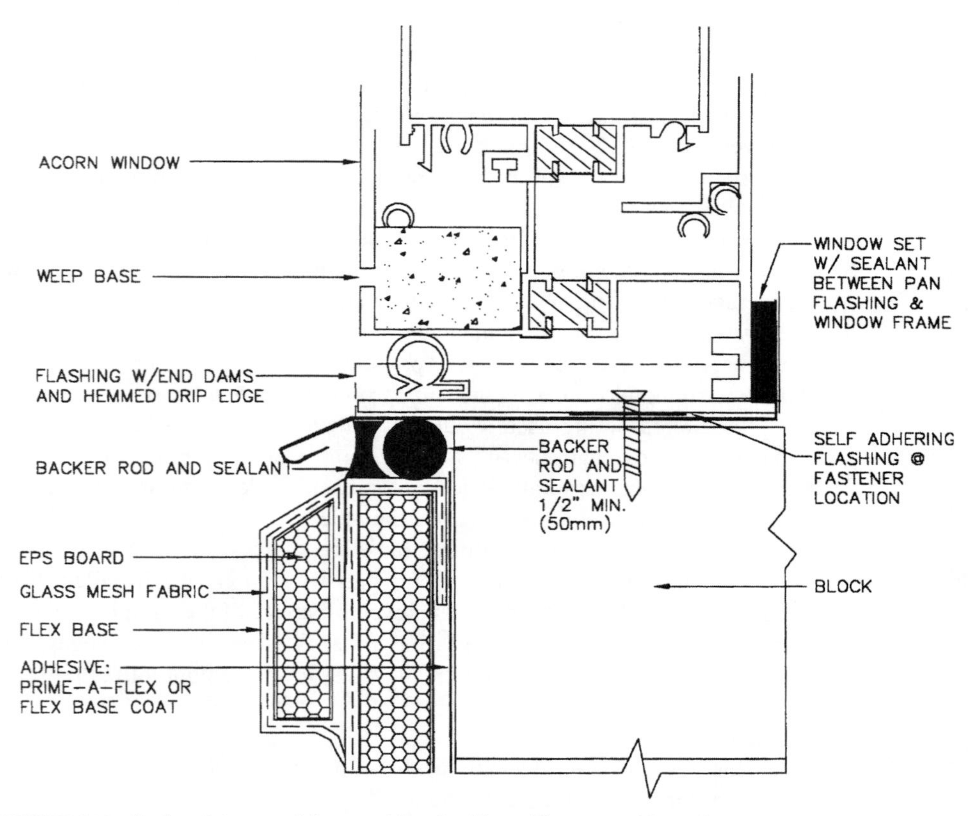

FIGURE 8.18 Sealant joints used for transition detailing. (*Courtesy of Bonsal*)

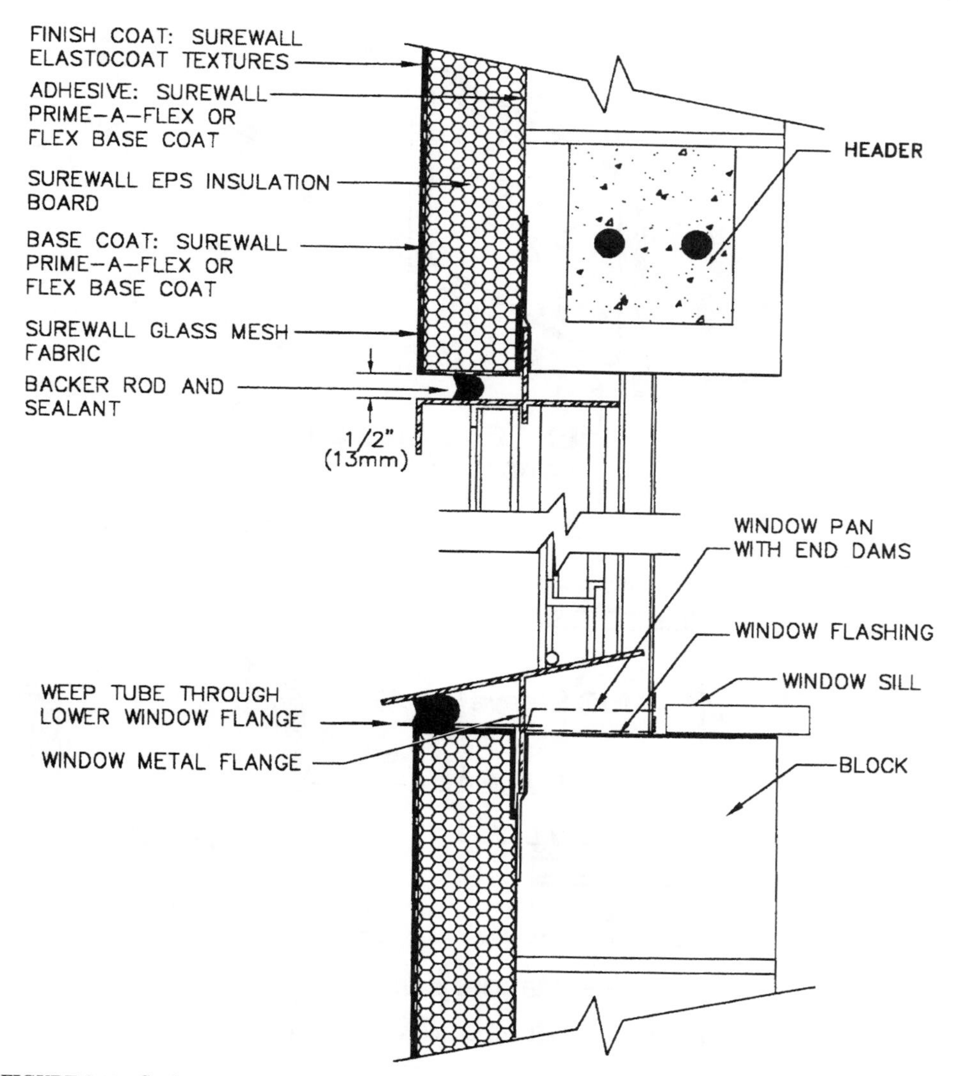

FIGURE 8.19 Sealant joints used for transition detailing. (*Courtesy of Bonsal*)

Control joints provide terminations or transitions that use sealants to seal and waterproof a joint. Common control joints include joints between dissimilar products, transitions between vertical and horizontal junctures, and equipment protrusions such as plumbing, electrical piping, lighting equipment, and sign supports as shown in Figs. 8.20 and 8.21.

Sealants are commonly used as termination detailing for a variety of envelope components. Figure 8.22 shows sealant used as terminating the above-grade masonry at the counterflashing to the below-grade waterproofing. Figure 8.23 details the termination of an EIFS system at grade to the concrete foundation using a sealant cant, with the manufacturer specifying the required bond breaker tape to ensure that the sealant performs prop-

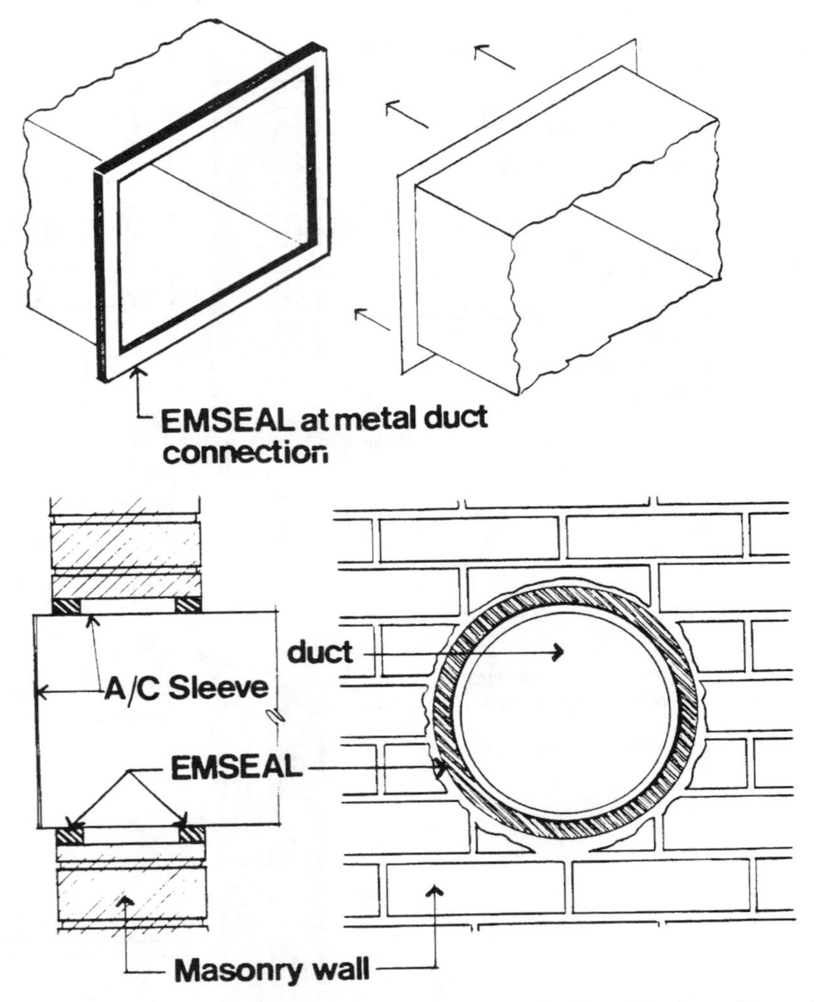

FIGURE 8.20 Sealant used for penetration, transition, detailing. (*Courtesy of Emseal*)

erly as described in Chap. 4. Figure 8.24 shows the correct use of sealant as a termination detail for waterproofing membranes.

Precast panel construction with porous finishes allows water transmission directly through the panels, bypassing joint sealants. Absorption of water through panels is further enhanced by negative air pressure between the exterior and interior areas. This uneven air pressure causes water to be drawn into a structure by a suction process. Likewise, wind-driven rain forces water through minor cracks and fissures in a masonry or concrete structure.

These natural phenomena are addressed by double-sealing joints, with sealant being applied to both exterior and interior sides of panelized construction. This double sealing allows air pressure in wall cavities to remain relatively constant by pressure-equalizing the sealed space between exterior and interior areas. Double sealing also provides additional air-seal protection for buildings, reducing heating and cooling costs.

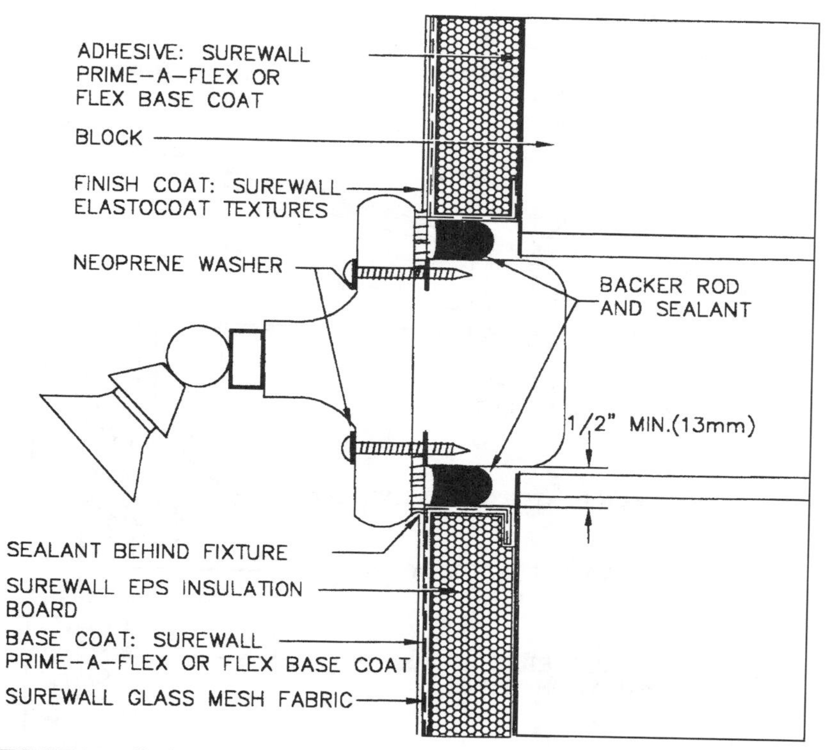

FIGURE 8.21 Sealant used for penetration, transition, detailing. (*Courtesy of Bonsal*)

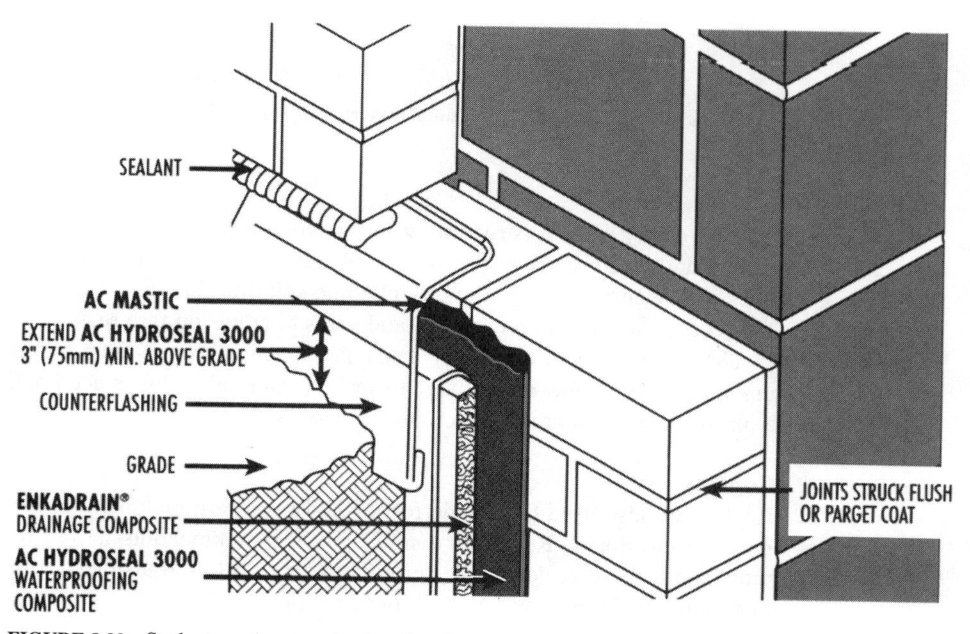

FIGURE 8.22 Sealant used as termination detailing. (*Courtesy of NEI Advanced Composite Technology*)

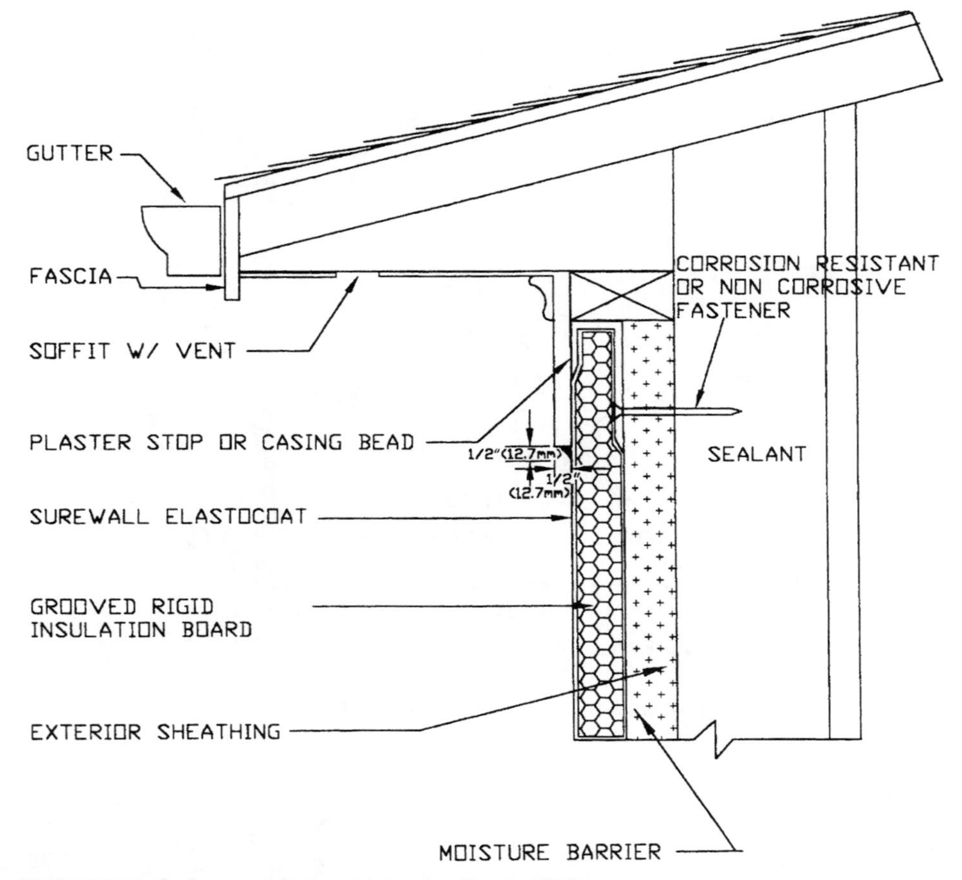

FIGURE 8.23 Sealant used for termination detailing in EIFS system. (*Courtesy of Bonsal*)

With cavity wall construction, water enters through initial weather barriers (e.g., brick facing) and is redirected to the exterior. Water entering in such conditions reacts with alkalines in masonry, causing a highly alkaline solution that deteriorates all types of sealant. This causes sealants to reemulsify and leads to adhesion failure.

Therefore, weeps in masonry walls and at sealant joints must be kept clear and working effectively, to prevent damage to sealant materials. This is accomplished by installing a plastic weep tube at the bottom of each sealant joint.

Reglets

Reglets are also used to provide for transitions or terminations in materials or systems of building envelopes. Reglets are small grooves or blockouts in substrates. Materials are turned into these reglets to be terminated or to allow transitions between two different materials. Reglet uses within building envelopes include:

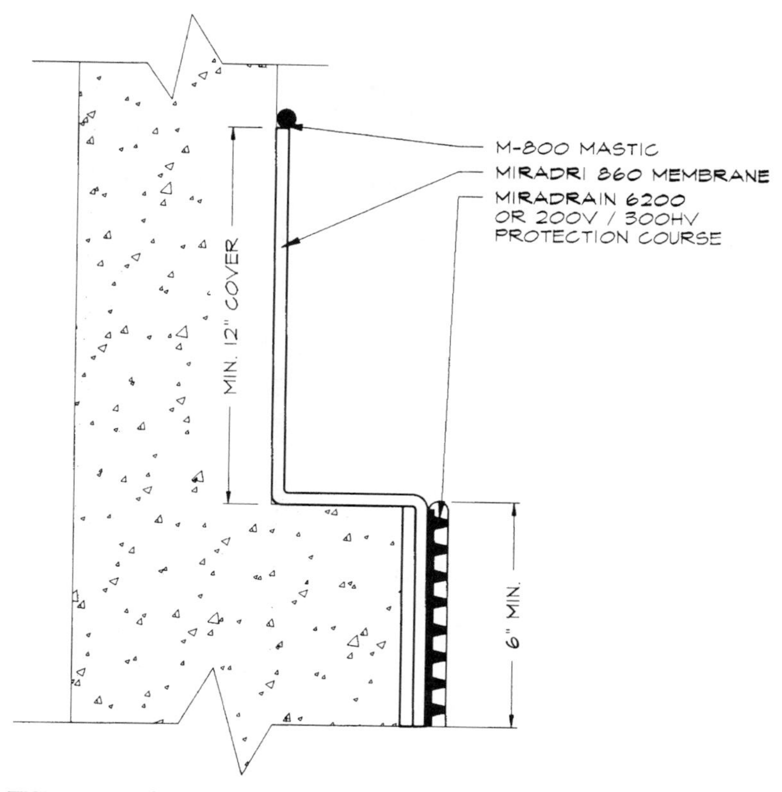

FIGURE 8.24 Sealant used as termination detail for waterproofing membranes. (*Courtesy of TC MiraDRI*)

- Substrate termination
- Waterproofing material to substrate transition
- Waterproofing system to waterproofing system transition
- Waterproofing system termination

Waterproofing materials that run vertically up from a foundation wall often terminate in reglets above-grade (Fig. 8.25). Below-grade waterproofing materials, changing to dampproofing systems above-grade, use reglets for transitions between these two systems.

Reglets are formed in substrates during the placement of concrete by using blockouts. They are also formed by sawing concrete, masonry, or wood substrates to form a reglet recess (Fig. 8.26). After a reglet is in place, it is inspected for cracking, honeycomb, or other problems that can cause leakage.

With certain flashing systems, surface-mounted reglets, mechanically fastened to substrates, are used. These are often used in remedial waterproofing repairs or roofing installations where existing reglets are not functioning or do not exist. Reglets are not recesses

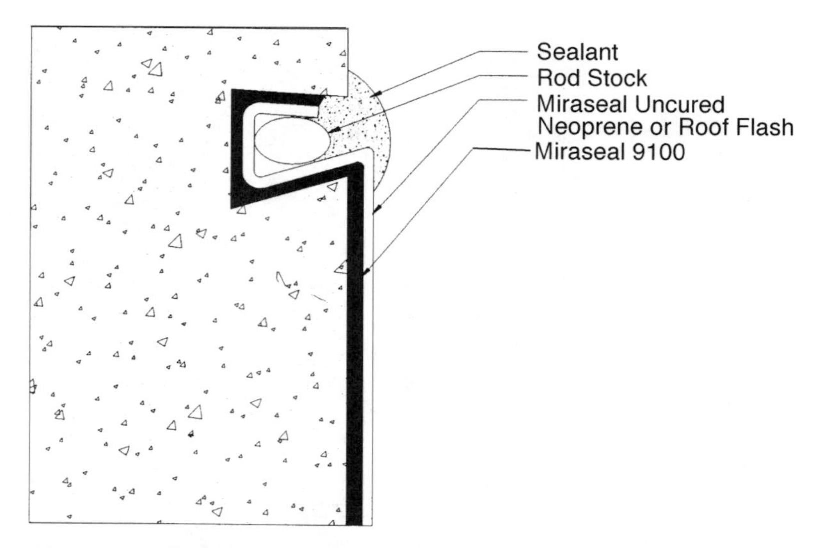

FIGURE 8.25 Sealant and reglet used for terminating waterproof membrane. (*Courtesy of TC MiraDRI*)

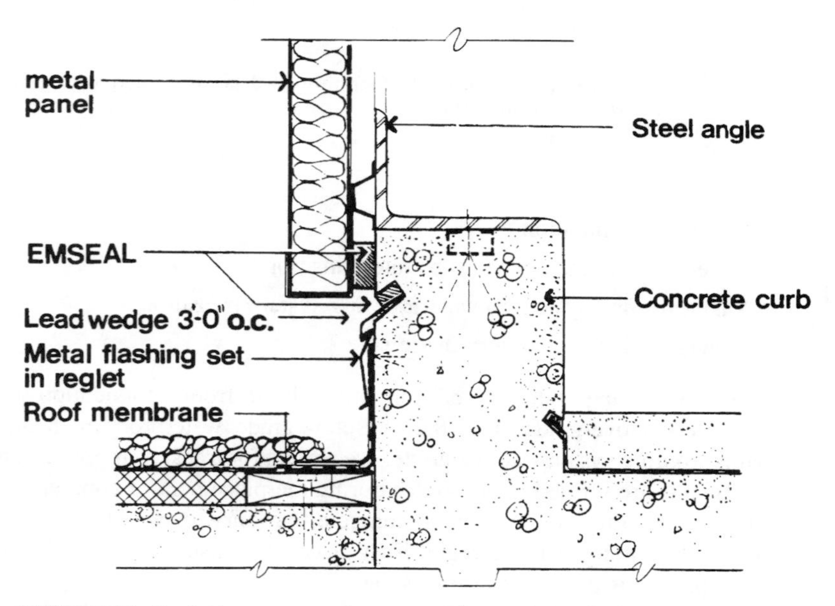

FIGURE 8.26 Reglet in concrete substrate used for transitioning detailing. (*Courtesy of Emseal*)

placed into a substrate for crack control, such as recessed control joints. These control joints are later sealed or provided for aesthetic purposes, as in precast panel construction.

Waterstops

Waterstops, although limited to concrete construction, are highly effective for transitions between separate concrete pours (referred to as *cold joints*) and terminations between vertical and horizontal concrete placements. Waterstops are now produced in a variety of designs and materials, including extruded rubber and hydros clay materials.

Remedial waterstop applications are now available. Cold joints with or without failed waterstops are chipped or sawn out along cracked areas. Manufactured tubing is then placed into this chipped-out area, and the joint is packed with nonshrink grout and cured. Then a urethane grout material is pumped into the tubing and filled, forcing expansion of the tubing and effectively sealing the joint.

Waterstops are effective in preventing lateral movement of water at cold joint areas. These joints are subject to cracking due to concrete shrinkage, and allow water penetration if a barrier such as a waterstop is not installed.

Waterstop materials are manufactured with flanges that allow each side of a joint to be securely anchored to the concrete. A common problem with waterstop usage is that waterstops are frequently installed by concrete finishers who do not understand their importance and effect on a building envelope.

Waterstops often end up bent over, cut, or not lapped properly at seams. They are even completely removed during concrete placement because they get in the way of the concrete finishing process. Waterstop installations must be carefully inspected by construction management personnel during concrete placement operations, to ensure that such activities do not occur. Waterstops are covered in detail in Chapter 2.

Other Transition Systems

A variety of other transition materials and systems frequently are used in construction practices to provide complete waterproofing of the building envelope. Among these are:

- Pitch pans
- Thresholds
- Integral flashings of curtain and window wall systems
- Expansion joint systems
- Cants

All of these systems ensure envelope watertightness by providing a transition or termination between dissimilar materials. They also allow for differential movement between various waterproofing systems, or allow entering water to be redirected to the exterior.

For instance, thresholds placed under doorways prevent rain and wind from entering interior spaces. They also provide a watertight transition among the exterior surface, doorway, and interior areas. Pitch pans provide a watertight transition among mechanical equipment supports, roofing, and structural roofing components.

It is the lack of such systems, or inattention to proper application of these transitory materials, that leads to most leakage in an envelope. It is the 1 percent of detail problems that causes 90 percent of water infiltration problems.

After all the major envelope components and their transitions and termination details are complete, a review of the total envelope is made to ensure it will act cohesively. This review should begin by ensuring that all types of water reaching an envelope drain away quickly. This prevents unnecessary infiltration and weathering of envelope components.

DRAINAGE REVIEW

An important point in reviewing any particular building envelope, existing or in design, is that water should be shed and removed from a building as quickly as possible. From below grade to roof, faces should be sloped whenever possible to shed water quickly.

Drainage must be provided to remove this water from surrounding areas. This not only prevents water leakage but also prevents premature weathering or wear of building envelopes from sources such as acid rain, chloride contamination, algae attack, and standing ponding water. Refer to Fig. 8.27 for recommended drainage requirements.

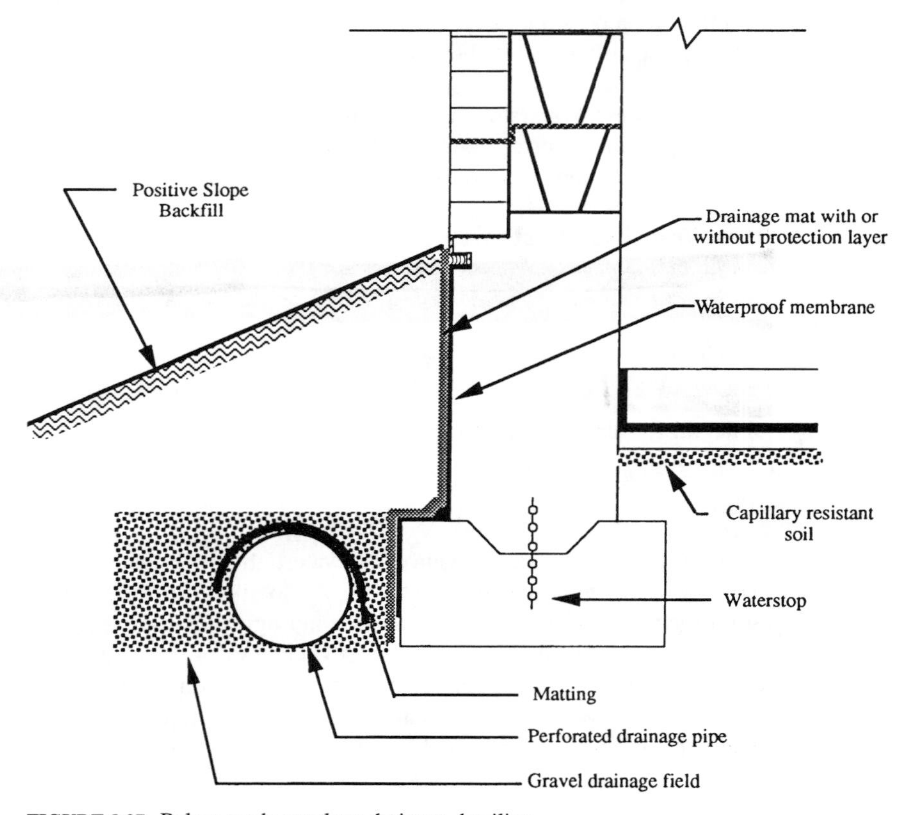

FIGURE 8.27 Below-grade envelope drainage detailing.

Walls and floor areas below grade are waterproofed to form a complete monolithic enclosure. Structures are subjected to water not only from groundwater conditions but also from runoff from surface collected water. Buildings built at or below-grade level are also subject to a head of water. They are also exposed to water from capillary rise from being in contact with the ground and water percolating downward from above-grade sources.

Below-grade surfaces typically have the most severe conditions with which to contend, with the exception of wind-driven rain that presents water pressures higher than a head of water. Particular attention to all of these conditions is necessary to ensure integrity of below-grade envelopes.

Existing soil conditions also require consideration. Below-grade horizontal surfaces should be placed on top of coarse granular soil (e.g., sand, gravel) to allow positive drainage away from foundations. Granular soils also resist capillary water rise that occurs in dense soil materials such as clay or silts. Areas adjacent to wall surfaces below grade should consist of coarse granular material to promote drainage.

Proper drainage of subsurface water is a necessity for adequate protection against water infiltration and longer life-cycling of building components below grade. Foundation drains are installed as shown in Fig. 8.5. Drains are placed at foundation level or slightly above, to prevent washout of soil beneath foundation structures. Drains are usually perforated piping with holes facing downward so as not to fill the pipe with soil, and are set in a coarse granular bed. They should be sloped away from building structures, with water collected at a natural out-face or sump area such as a retention pond.

It is also recommended that drainage mats be installed on vertical surfaces of waterproofing membranes below-grade. Several synthetic compositions are available and compatible with waterproofing membrane systems. These mats promote quick drainage of water off below-grade walls into available drainage systems.

At grade level, grading should be sloped away from structures to provide positive drainage of surface waters away from buildings. Slope ratios differ depending on the type of soil and adjacent exposed finishes at surface level. For planted or grassed areas, slopes should be 5 percent minimum. For paved areas, 1 percent minimum slope is acceptable. Draining water is collected and properly diverted, to prevent excess water from percolating into soil adjacent to structures.

Walls above-grade are subjected to water from rain, snow, and capillary action of soil at grade level. Water conditions can become especially severe when rain is present in high-wind conditions, forcing water through minute cracks and openings in above-grade envelopes. Wall areas must also withstand weathering conditions to which a below-grade envelope is not subjected. These conditions include ultraviolet degradation, air pollutants, acid rain, chloride attack, freeze–thaw cycling, and thermal shock. Therefore to be completely effective, exposed portions of envelopes must not only be watertight but also weather-resistant. This ensures longevity of these systems and protection of interior areas.

Above-grade envelopes must also be provided with provisions to drain water away from a structure adequately, not allowing it to percolate down to below-grade areas. All building horizontal portions should be sloped to shed water. This includes not only roof areas but also coping caps, sills, overhangs, ledges, balconies, decks, and walkways.

To allow areas of standing or ponding water not only makes an envelope subject to water infiltration but also intensifies weathering from such sources as acid rain, chlorides,

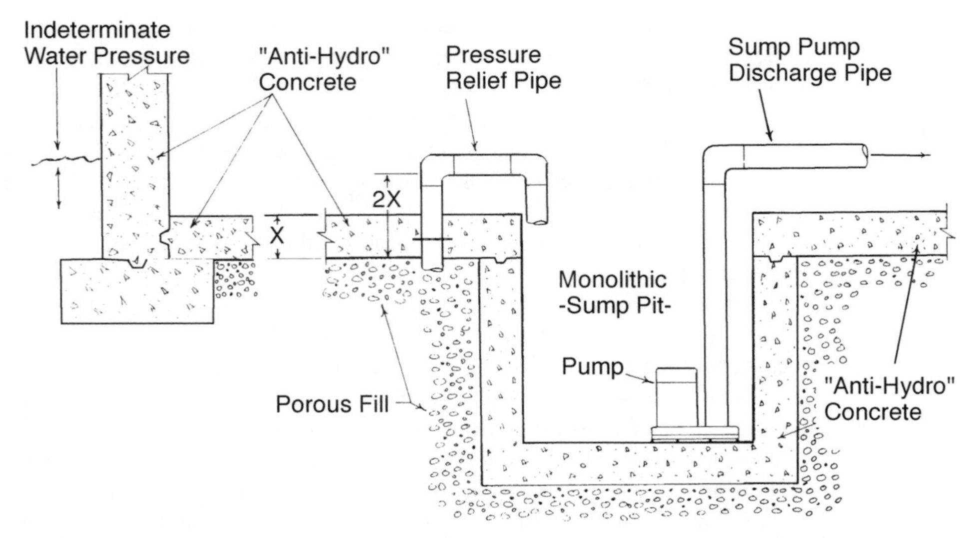

FIGURE 8.28 Pressure relief piping and sump system to reduce hydrostatic pressure on envelope. (*Courtesy of Anti-Hydro International, Inc.*)

and algae. Where applicable, deck drains, roof drains, gutters, and downspouts should be installed to gather collected water and disperse it without subjecting above- and below-grade areas to this water. Figure 8.28 details the use of pressure relief piping and a sump pump system to reduce hydrostatic pressure when site drainage is inadequate.

ENVELOPE REVIEW

Successful building envelopes include most if not all of the following features:

- Few protrusions and penetrations on exposed envelope portions
- Minimal number of different cladding and waterproofing systems to limit termination and transition detailing and trades involved
- Minimal reliance on sealant systems for termination and transition detailing
- Joints designed to shed water
- Minimal reliance on single-barrier systems
- Secondary systems installed where practicable, including:

  Flashing
  Dampproofing
  Weeps
  Drainage tubes

- Proper allowance for thermal expansion, contraction, and weathering cycles
- Absence of level or horizontal envelope areas that would allow ponding water at roofs, balconies, and walkways

- Drainage of water away from an envelope as quickly as possible, both above and below-grade; gutters, drains, slopes, drainage mats are used where appropriate
- Recessing windows and curtain walls at slab edges
- Adequate space provided to detail all termination and transition details properly
- Preconstruction and envelope review meetings with all trades, manufacturers, and supervision that will affect envelope performance involved
- Testing and review of detailing where necessary, to ensure effectiveness before construction begins
- Joint and several warranties for all envelope components
- Quality-control procedures during construction
- No substitution of materials or systems after approvals, testings, and reviews
- Proper envelope maintenance

Successful envelopes require reviewing all building envelope components, including how they interrelate. To understand these requirements, start at below-grade construction and work upward, exploring and reviewing the qualities and designs necessary for an effective building envelope with the longest life-cycling possible. Refer to the typical envelope building detail shown in Fig. 8.29.

Wall-to-floor junctions are areas susceptible to leakage. These areas should receive waterstops for additional protection, with waterproofing materials installed properly along this intersection with necessary cants and reinforcements. Installation details of floor membranes at these locations are continued up to transitions with vertical or wall membranes. If necessary, mud slabs should be used to provide a sound substrate on which to apply horizontal waterproofing membranes.

Similar detailing should continue at any below-grade pits or structures such as elevator or escalator pits. Pile caps and similar structural foundations should be wrapped completely and continually and transitioned into floor or horizontal membranes. Structural design may prohibit membranes from being applied between concrete pours such as floor-to-wall details. This requires wrapping membranes beneath foundations and up sides, to allow for unbroken continuity as shown in Fig. 8.30.

Transitions from below- to above-grade waterproofing should be watertight (Figs. 8.31 and 8.32), but they also must allow for thermal and differential movement that occurs at this intersection. Reglets, flashings, or other means of protection should be installed to provide for this transition (Fig. 8.33).

The above-grade envelope must then be carried completely up vertical surfaces and tied into the horizontal or roofing portion of the envelope (Fig. 8.34). This horizontal portion is in turn tied back into the opposite side of vertical envelopes, back to below-grade, forming a complete envelope on a structure.

All transitions and terminations in a vertical envelope must be completely water- and weather-resistant (Fig. 8.35). Transitions between such features as walls and window frames must be waterproof and allow for thermal and differential movement. In this case, it is typically a well-designed and installed sealant joint. Sealants are frequently used for transitional waterproofing. However, sealants are often overused when better materials or different systems, such as flashings, should be used.

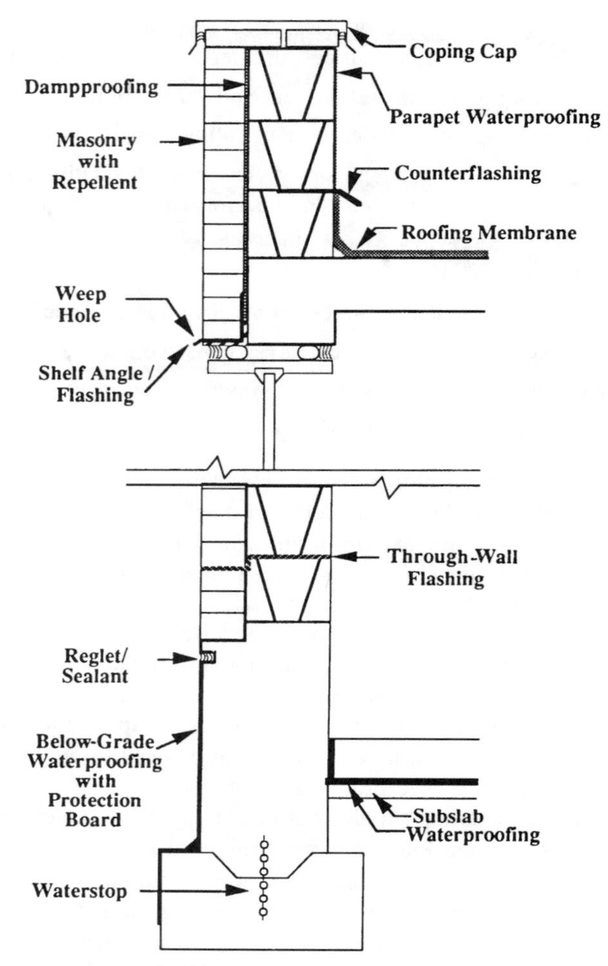

FIGURE 8.29 Building envelope detailing.

Weathering of exposed envelope systems creates movement in all materials, and allowances must provide for this movement. Movement above-grade is created by several phenomena, as summarized in Table 8.2.

To provide for this movement and volume change, control and expansion joints must be designed and placed where such movement is expected. Among the envelope locations for placement of control joints are these:

- Changes in materials
- Changes in plane
- Material volumetric expansion
- Construction joints
- Junction of facade materials to structural components

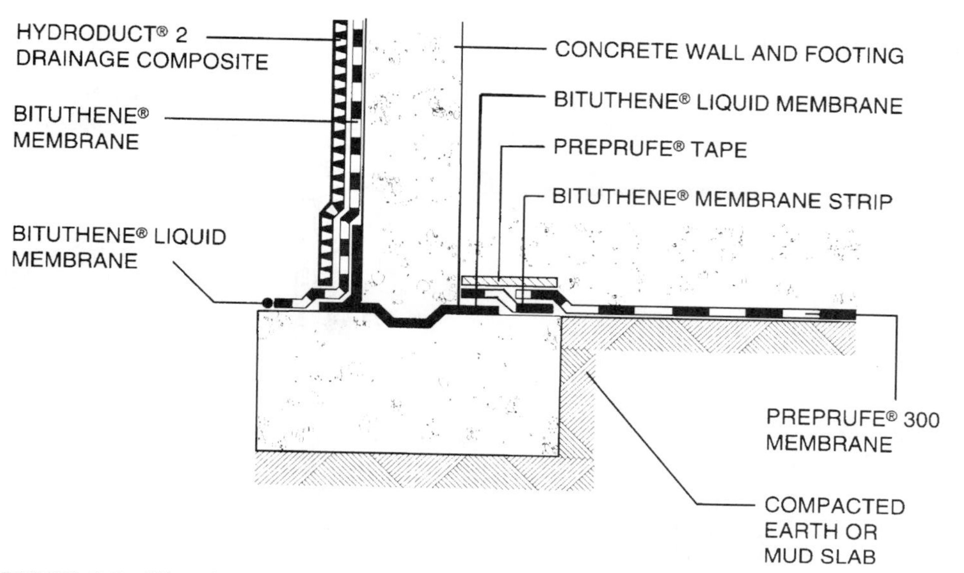

FIGURE 8.30 Wrapping membrane waterproofing to ensure waterproofing. (*Courtesy of Grace Construction Products*)

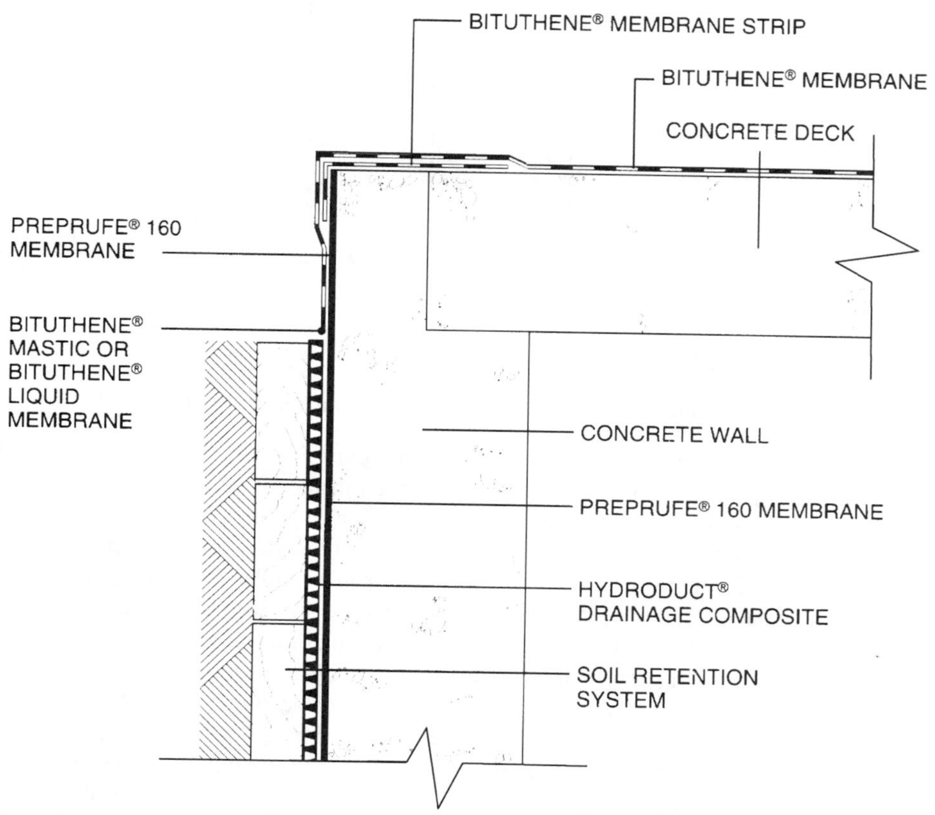

FIGURE 8.31 Below-grade to above-grade waterproofing transitions. (*Courtesy of Grace Construction Products*)

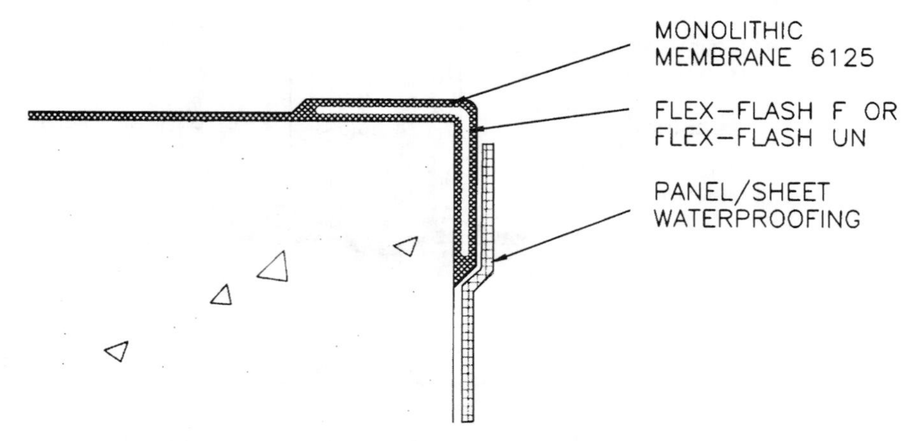

FIGURE 8.32 Below-grade to above-grade waterproofing transitions. (*Courtesy of American Hydrotech*)

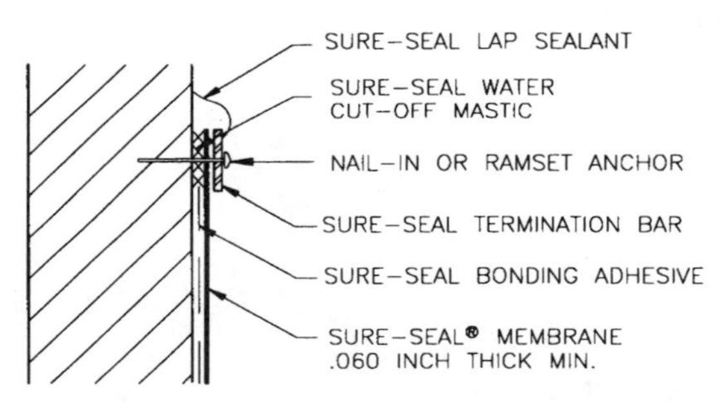

FIGURE 8.33 Flashing used for transition detailing. (*Courtesy of Carlisle Corporation*)

- Changes in direction
- Concentration of stresses (such as openings in a structural wall)

All such construction details in a composite wall area should be reviewed, and, where appropriate, control or expansion joints installed. Sealants installed in these areas should be completed according to the application requirements presented in Chapter 4.

All appurtenances on a building should be checked for watertight integrity. Often-overlooked items in this category include exhaust ventilators, fresh-air louvers, mechanical vents, signage, lightning equipment, pipe bollards, and mechanical and electrical piping. All should be watertight and weather-resistant, including transitions into adjacent materials.

Envelope review then proceeds to roof areas. Roofing systems must be adequately transitioned into the adjacent wall system. This is accomplished either with flashings and counterflashings on a parapet, or edge flashing directly covering adjacent wall facades. As with vertical portions, roof areas must allow for movement with adequate expansion and control joints. Additionally, all surfaces should be sloped so as to shed water as quickly as possible.

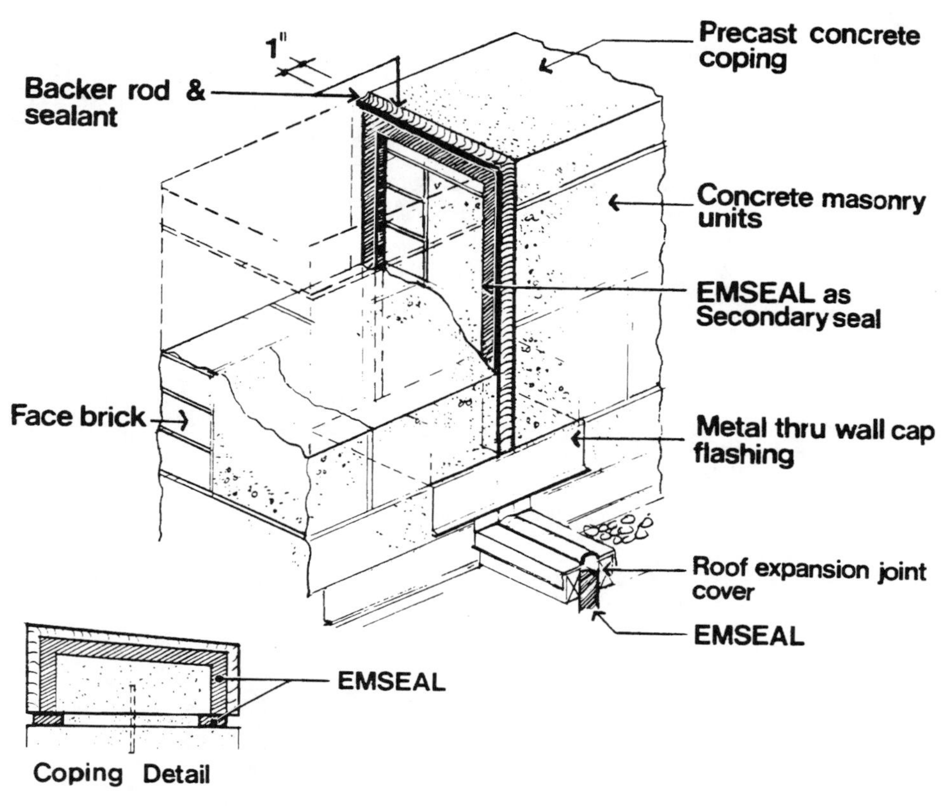

FIGURE 8.34 Transition from vertical envelope areas to roof or horizontal areas. (*Courtesy of Emseal*)

Parapet walls create a particular problem in the life-cycling of a structure. Many unique stresses occur at parapet walls due to the imposed designed loads. Parapet walls move both vertically and horizontally due to thermal movement. In addition, both sides of parapet walls are exposed to weathering, which exemplifies this movement of horizontal and vertical expansion and contraction.

Introduction of an adjacent roof slab, which expands against or contracts away from a parapet, imposes a great amount of additional stress on a parapet wall. These stresses may cause bowing of parapet walls and cracking of facing materials or systems that lead to water intrusion. Once begun, this entering water imposes additional stresses such as swelling, freeze–thaw cycling, and corrosion of reinforcement. Often, parapets eventually fail to function properly.

When unbreathable coatings are applied to a roof side of parapets, such as black asphalts, heat absorption into parapets increases. This type of waterproofing can damage the integrity of a parapet structure. The numerous situations involved with parapet construction require special designs to ensure that sufficient structural components as well as control and expansion joints are included in construction.

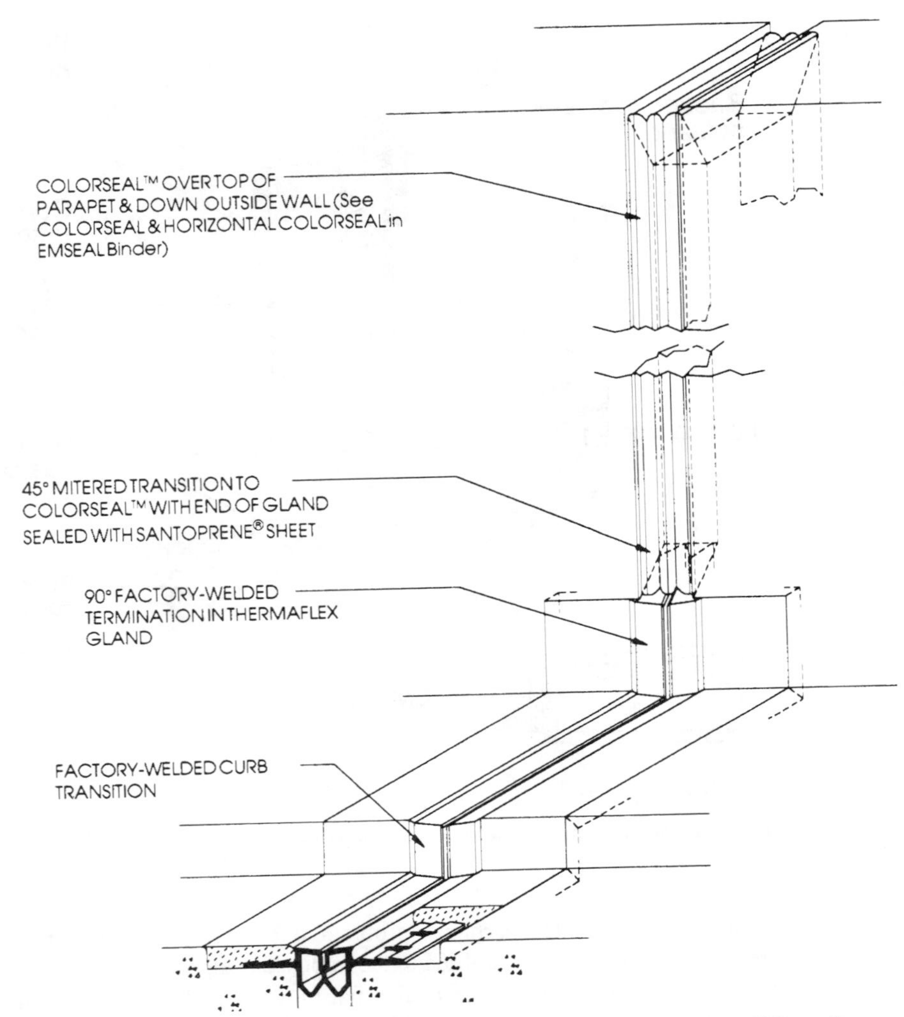

FIGURE 8.35 Vertical-to-horizontal transitions must be watertight. (*Courtesy of Emseal*)

As with waterproofing systems, water leakage through roof areas is typically located within 1 percent of the entire surface area, most often occurring at termination and transition details. This often occurs at equipment supports, equipment pads, plumbing stacks, scuppers, drains, skylights, and lightning and electrical equipment. Detailing transitions properly, and providing differential movement at these areas, ensures watertight transitions to roofing materials. Movement allowance for roofing details includes movement created by vibrations from mechanical equipment. Figures 8.36 and 8.37 detail transitions between multi-ply and single-ply roofing systems into deck-coating or membrane waterproofing systems.

TABLE 8.2 Common Envelope Movement Causes

Movement	Cause
Thermal	Expansion or contraction movement caused by temperature changes
Structural	Caused by the curing process of concrete during settlement or structural loading of the building
Differential	Materials have individual coefficients of movement, which will differ from surrounding materials, causing differences in movement between the materials
Moisture content	Certain materials, particularly masonry, swell when subjected to wetting; this movement or enlarging, when calculated as an aggregate total of the entire facade area, can be considerable

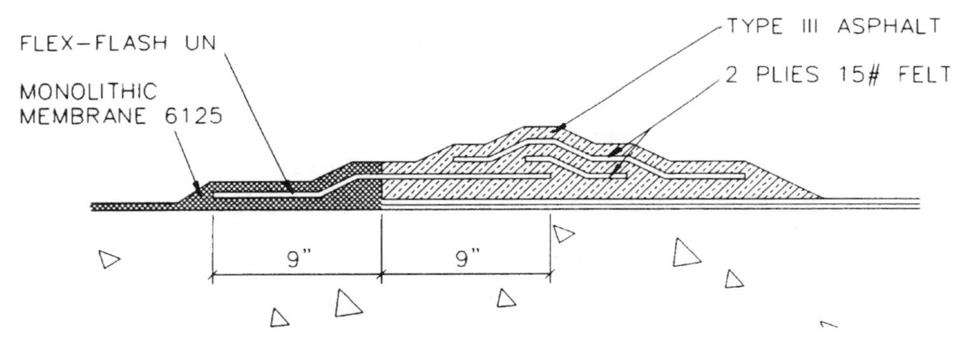

FIGURE 8.36 Transitions between multi-ply roofing and deck coating. (*Courtesy of American Hydrotech*)

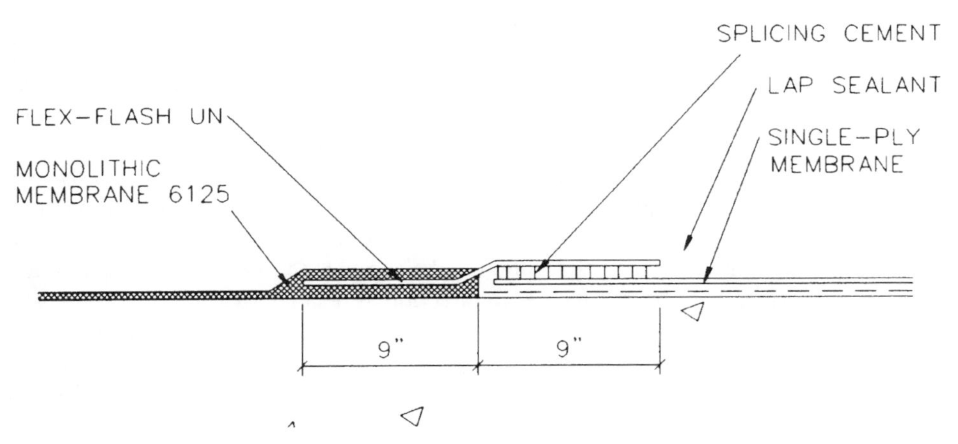

FIGURE 8.37 Transitions between single-ply roofing and deck coating. (*Courtesy of American Hydrotech*)

VERIFYING THE ENVELOPE BARRIER

Water infiltration is a common occurence throughout completed products, despite the high-quality materials and systems available to waterproof structures. Recognizing that leakage is generally attributable to the 90%/1% principle, an envelope barrier meeting should be part of all contractors' quality-control programs.

One of the most important preconstruction practices a contractor can implement is a building envelope meeting attended by all subcontractors having work scopes involving the building facade. The meeting should concentrate on verifying that the entire envelope water barrier is complete by following it through each elevation wall section.

This is the most effective means to ensure that the envelope termination and transition detailing is clearly understood by all subcontractors. Responsibility and coverage for every termination or transition detail should be assigned to a specific subcontractor. The meeting should also be documented by drawing the barrier line through each elevation and assigning contractor responsibility for each envelope transition.

All meeting notes can be documented directly on the plans as shown in Fig. 8.38. Here the previous discussed simplified envelope/wall section is used to shown how the termination and transition detailing can be reviewed. First, the barrier has been followed by drawing a line over the waterproof barrier or divertor envelope components; in this case, starting with the roof membrane, to parapet waterproofing, to coping cap, and so forth.

Then whenever a transition or termination detail is highlighted, a circle is drawn around the detail and assigned to a specific contractor. In Fig. 8.38, note that the termination of the below-grade membrane to the above-grade masonry has been assigned to Top Notch Waterproofing, the coping cap to masonry transition assigned to XYZ Roofing.

After each detail is assigned to a subcontractor, the specifications and plans for that detailing, including any shop drawings by the product manufacturer, should be reviewed. For example, the shelf angle should have a detail associated with it similar to Fig. 8.39, which very clearly shows the interaction of each component and how in this case the flashing is terminated and transitioned into the dampproofing. Similar details should be provided for the coping cap and masonry to below-grade membrane as shown in Figs. 8.40 and 8.41.

In many situations, the transition or termination detail may be supplied by the product manufacturer in appropriate shown drawings in lieu of the original plans and specifications. Figure 8.42 provides shop drawing manufacturer details for an intricate horizontal-to-vertical expansion detail, with Fig. 8.43 providing a plan view of the required layout of the expansion joint. These details are critical to the successful completion of a watertight envelope. Shop drawings are often superior to the original construction documents that might merely show the location of the expansion joint with no details for transitioning onto vertical surfaces or changes in direction as detailed in Figs. 8.42 and 8.43.

The envelope barrier meeting often highlights areas where there are insufficient details provided to properly install components, or where details occur with no corresponding or contractual assignment having been made by the general contractor. The meeting can make field supervisors aware of these voids and ensure that appropriate assignments are made to ensure that the detailing is provided before related construction begins.

Figure 8.44 shows a recommended detail for transitioning from an EIFS system to a masonry facade. Note that the manufacturer has detailed a moisture barrier behind both the

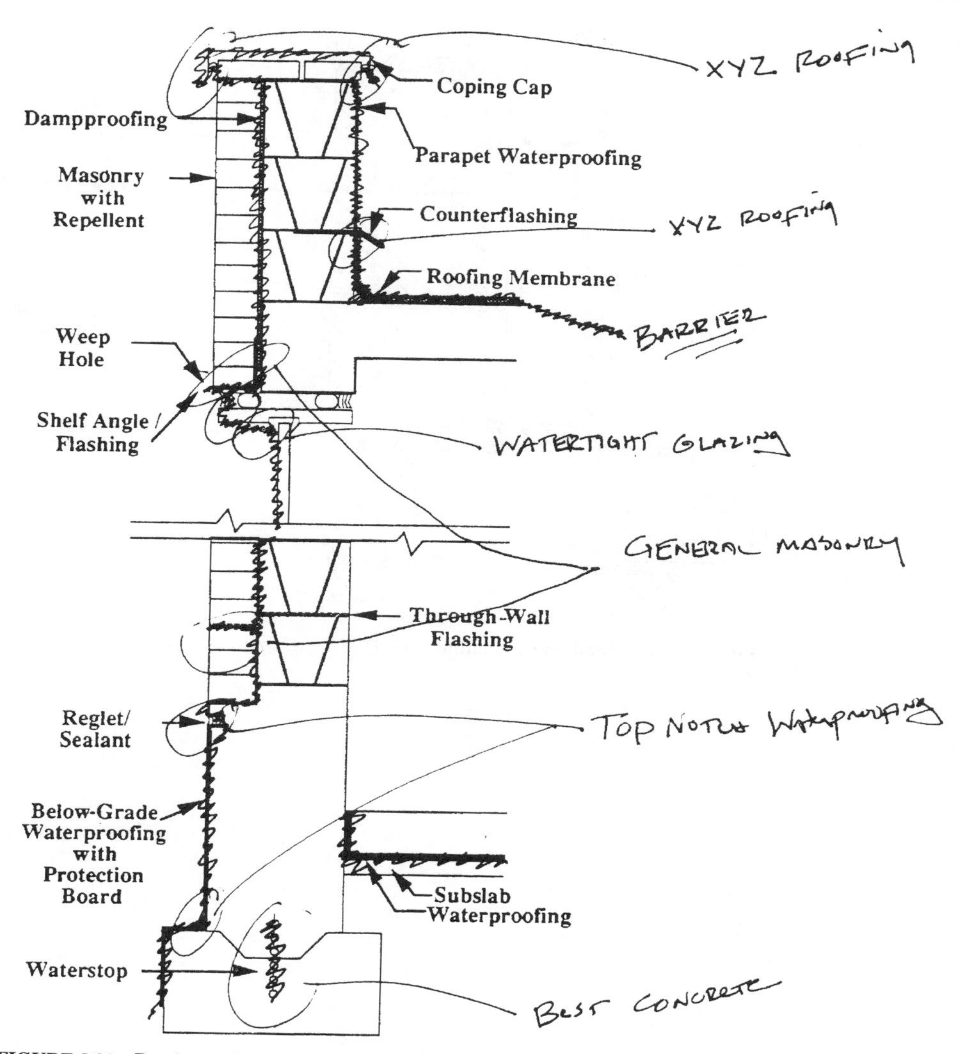

FIGURE 8.38 Barrier review meetings notes, applied directly to documents for future reference.

EIFS and masonry wall that acts as a collection means for divertor system necessary to protect both the masonry and EIFS claddings from leakage. The barrier meeting would uncover the fact that since the manufacturer is correctly recommending, for the best waterproofing protection, that the dampproofing run continuous behind the EIFS and masonry walls, a decision must be made as to the responsibility for the dampproofing installation. Should it be the mason or the EIFS subcontractor? This is a specific case where the barrier meeting can prevent problems during actual construction and later for the building owner.

While each subcontractor involved included the portion of dampproofing directly behind their work scope in their bids and proposals, it would have been unlikely that either would have included or recommended that they be responsible for the dampproofing behind the other

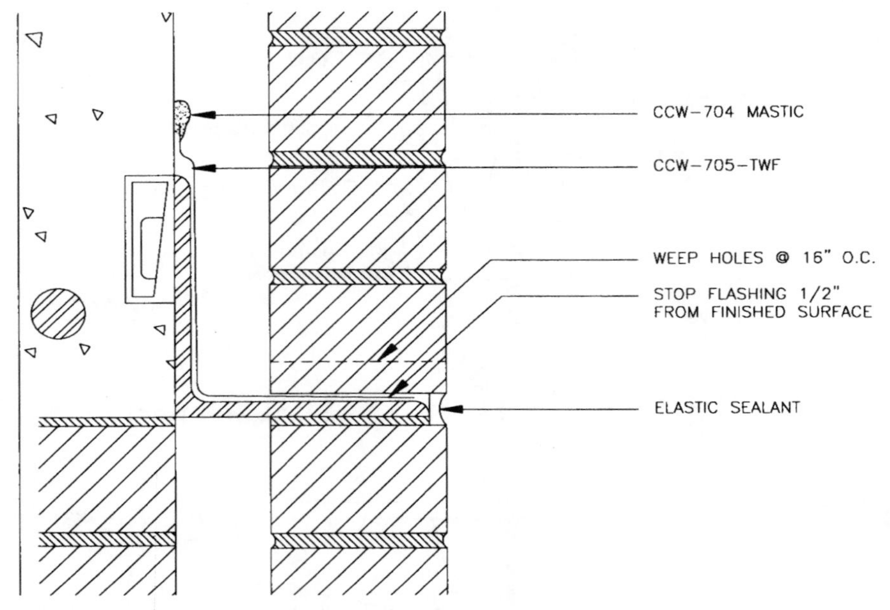

FIGURE 8.39 Shelf angle detailing. (*Courtesy of Carlisle Corporation*)

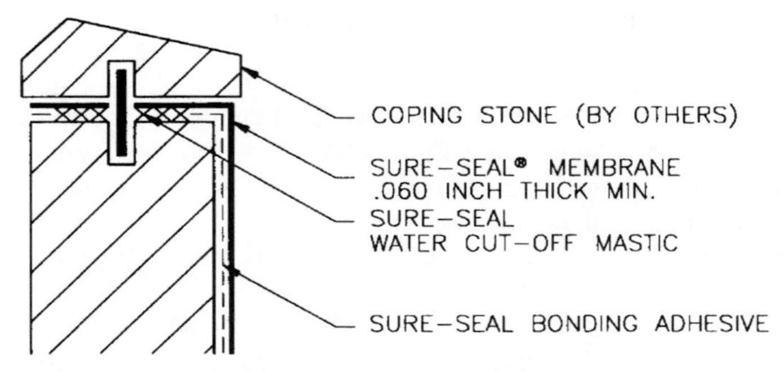

FIGURE 8.40 Coping envelope detailing. (*Courtesy of Carlisle Corporation*)

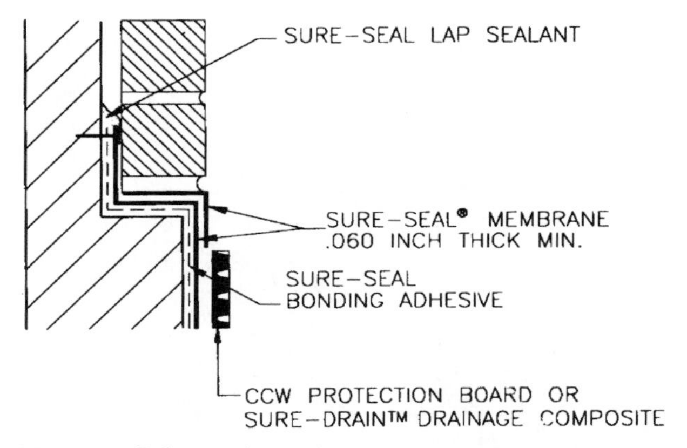

FIGURE 8.41 Below-grade transition detailing. (*Courtesy of Carlisle Corporation*)

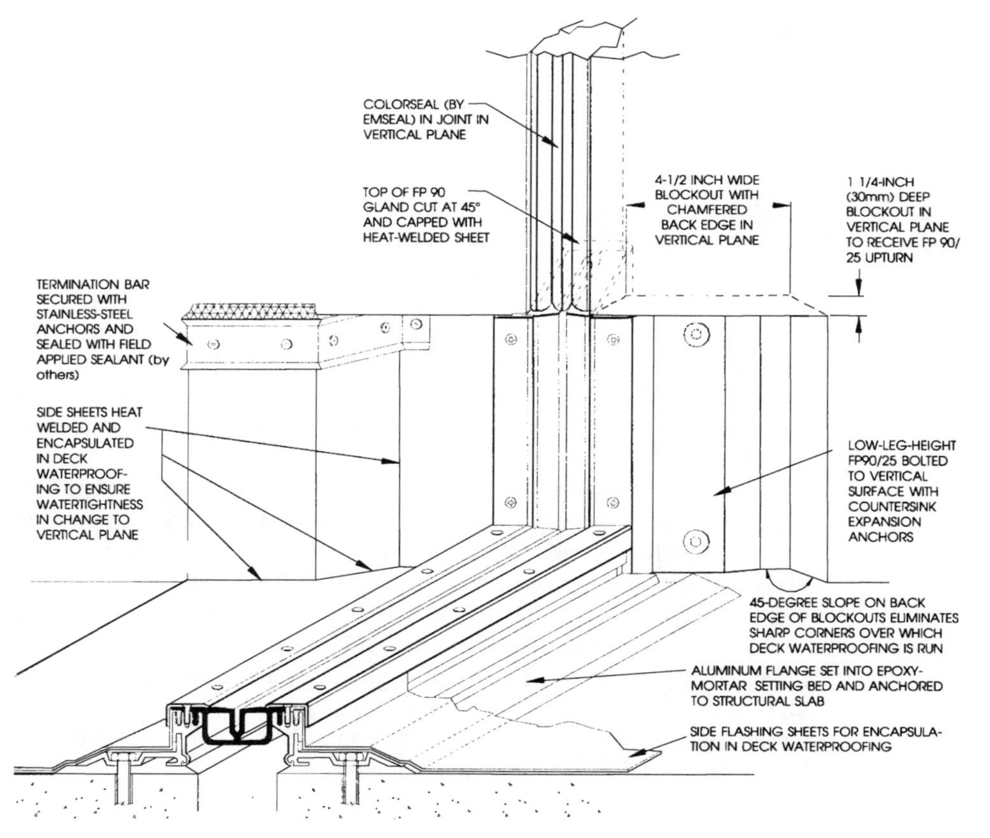

FIGURE 8.42 Typical manufacturer shop drawing for transition or termination details. (*Courtesy of Emseal*)

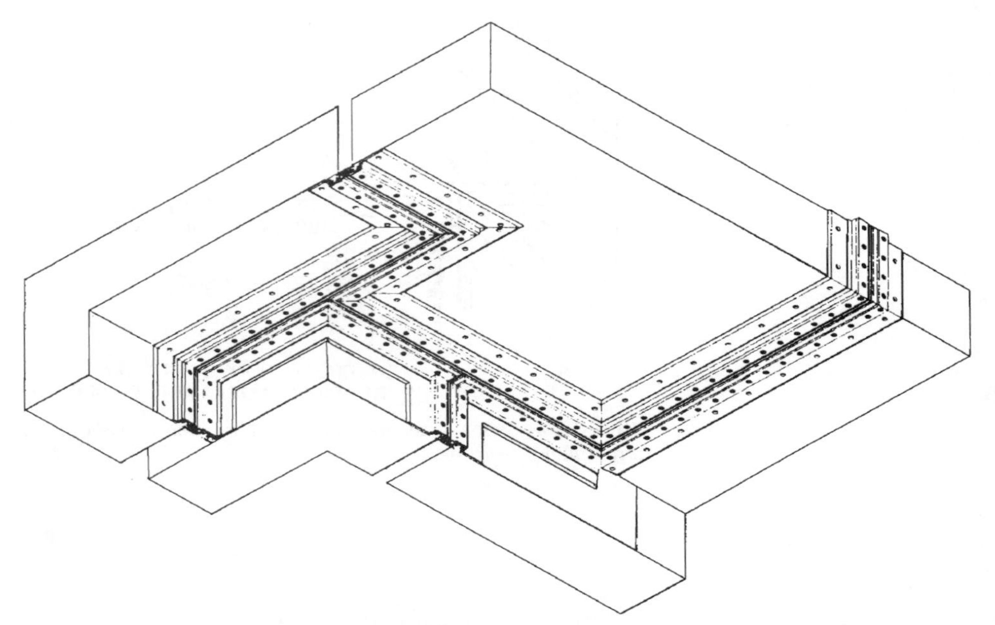

FIGURE 8.43 Typical manufacturer shop drawing for transition or termination details. (*Courtesy of Emseal*)

cladding component. It is also likely that the general contractor accepted the bids as such, and made no note of the importance of this situation until the barrier meeting.

It would not be a recommended solution that the dampproofing stop at the transition and that one subcontractor be given the responsibility for sealing the two installations together. This would likely lead to water or moisture infiltration, exactly the situation that upholds the 90%/1% principle. The general contractor should assign the work to the subcontractor that is likely to complete their work on the envelope first, in this case most likely the mason.

As importantly, the general contractor, to prevent problems of "finger-pointing" if leakage occurs, should require the EIFS subcontractor to inspect and accept responsibility for the dampproofing installed by the mason but directly behind the EIFS system. This eliminates any probability that, if leakage occurs in the EIFS system, the EIFS subcontractor will blame the mason for improperly installing the dampproofing as the cause of leakage.

Figure 8.45 presents a similar problem. The manufacturer has provided a recommended penetration/transition detail that involves numerous subcontractors. Note that this EIFS divertor envelope system has a dampproofing application that is to run continuously around the penetration. To prevent the water traveling along the dampproofing or drainage mat from entering the building at the penetration, the manufacturer has detailed the membrane flashing to run continuously around the pipe penetration. This water also bypasses the sealant joints used around the exterior perimeter of the pipe as a transition system for the EIFS to the pipe. Note that for additional protection the manufacturer has detailed that the dryer vent cover is to extend over this transitioning.

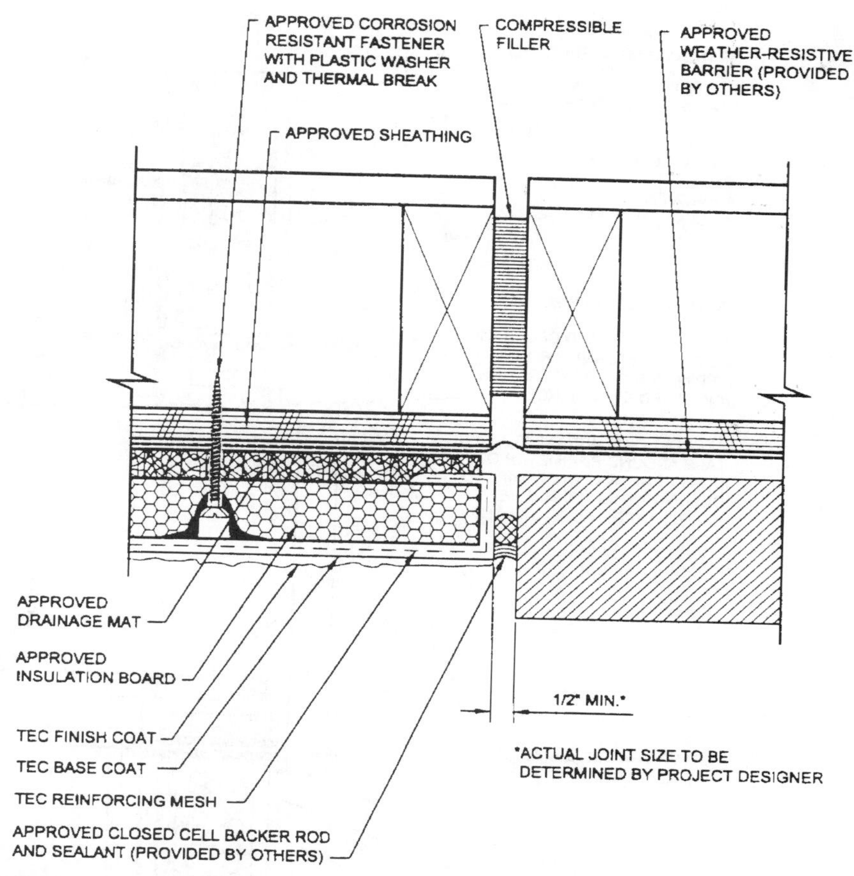

FIGURE 8.44 Transition detailing for cladding materials involving numerous subcontractors. (*Courtesy of TEC Specialty Products, Inc.*)

A barrier meeting should be used to assign responsibility for each of these waterproofing transition details. In this situation, while the manufacturer notes that the membrane flashing around the pipe is not part of the EIFS system, the general contractor should correctly note that this material should be installed by the EIFS subcontractor to ensure they have complete responsibility for the entire water drainage system. The sealant joints around the pipe penetration should also be reviewed, and it is likely that the waterproofing contractor would complete this work. However, should leakage occur in this area, it would be difficult to determine if the water was penetrating directly through the primary transition sealant joint or from the divertor system that includes the membrane flashing that might have lost adherence to the pipe. The contractor in the barrier meeting might require the EIFS subcontractor to take complete responsibility for this entire detail, to prevent problems in assigning cause and responsibility if leakage in this area should occur.

These two details show the importance of barrier meetings and how common 90%/1% installation problems can be prevented, and also how best to provide a quality installation

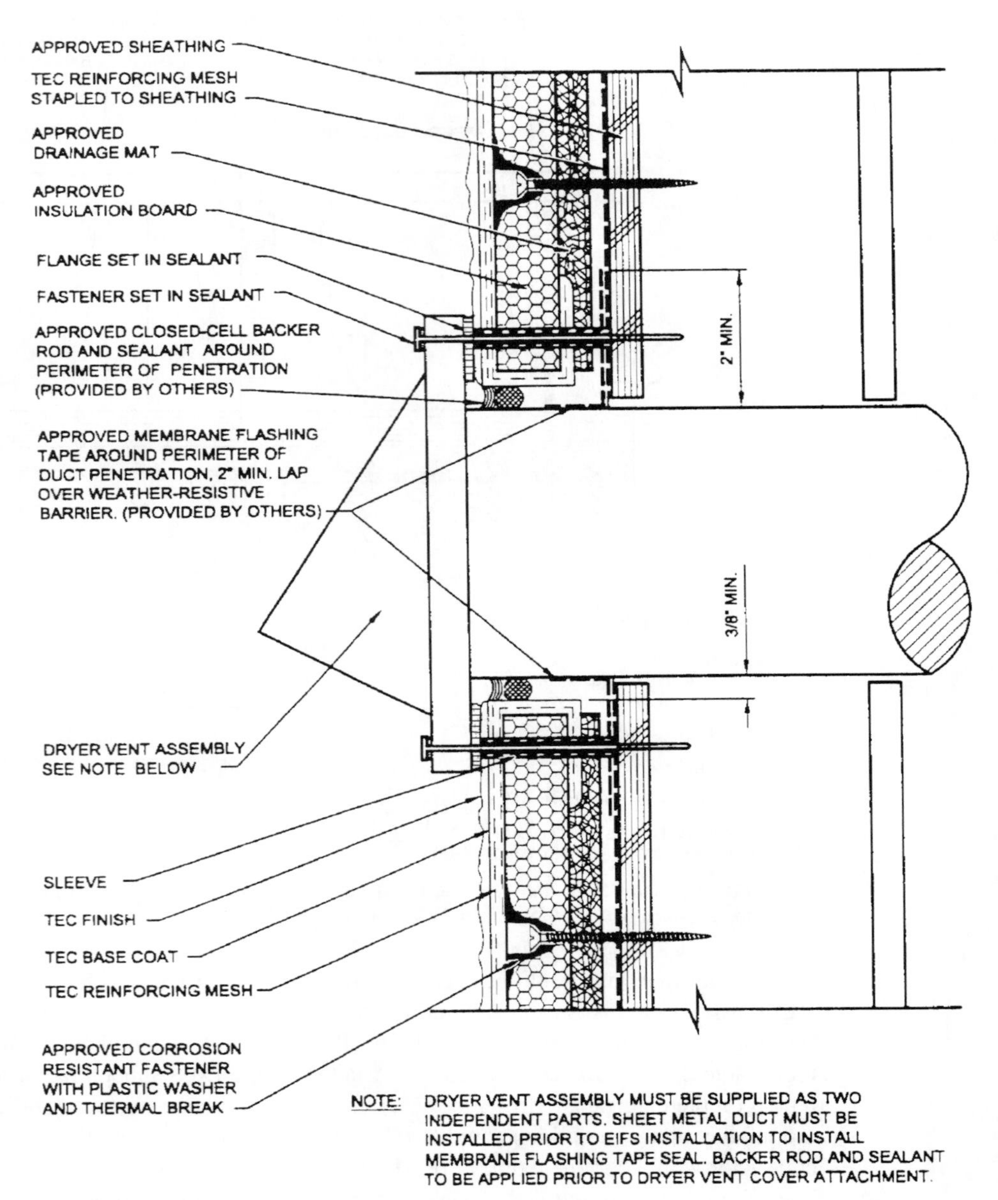

FIGURE 8.45 EIFS penetration detail involving numerous subcontractors. (*Courtesy of TEC Specialty Products, Inc.*)

FIGURE 8.46 Barrier meetings should prevent detailing such as this. (*Courtesy of Coastal Construction Products*)

for the building owner. For example, barrier meetings should alert the project management team to the placement of structural columns in the expansion joint line as shown in Fig. 8.46. This poor transition detailing requiring 90-degree turns in the expansion joint will ultimately support the 90%/1% principle over the course of the building's life cycle. Barrier meetings can also prevent poor transition details that will fail as shown in Fig. 8.47. This transition of expansion joints between a sealant and T-Joint System failed because no transitioning detailing was included in the original documents. Barrier meetings that prevent such mistakes are an effective means to eliminate the 90%/1% principle on every building project.

ROOFING REVIEW

There are frequently more subcontractors and different tradespeople involved in a typical roof installation and related construction than in any other envelope component. Trades often involved in roof construction include:

- Roofing
- Mechanical and HVAC
- Carpentry
- Masonry
- Miscellaneous metalworking
- Waterproofing

FIGURE 8.47 Lack of transitioning and termination detailing in original construction will certainly lead to failure and leakage. (*Courtesy of Coastal Construction Products*)

- Concrete deck installing
- Sheet metaling
- Electrical
- Curtain or window wall contracting

Among the related envelope components involved within a typical roofing envelope are

- Roof drains
- Lightning rods
- Balustrades
- Mechanical equipment
- Electrical equipment
- Signs
- Copings
- Skylights
- Parapets
- Penthouse walls

It is important that each of these envelope components be completely waterproof itself to ensure a roofing envelope's effectiveness. Terminations and transitions necessary to incorporate these trades and systems into an envelope are candidates for errors and resulting water infiltration. This multiple-discipline requirement highlights the importance of requiring a preconstruction meeting that involves all parties who affect roof performance and resulting envelope watertightness.

This conference must include electricians, mechanical contractors, and curtain wall and waterproofing subcontractors, in addition to any of those people listed above who are included in specific project requirements. Each contractor must be made aware of his or her responsibility to interact with all other trades for successful completion of an envelope.

Terminations and transition details are reviewed, and require the manufacturer's preapproval of all project details. This review should include discussion and resolution of the following frequent causes of envelope infiltration related to roofing:

- Inadequate and improper transition and termination details
- Inadequate drainage (no flat roofs), and absence of testing for proper slopes before installation
- Too many separate roof penetrations
- Too much equipment and traffic on roof areas

Manufacturers should be consulted by designers to review proposed detailing, to ensure that the system will adequately function under the proposed job-site conditions. Any unusual conditions expected to be encountered, such as equipment penetration and traffic on roof, should be carefully reviewed to ensure a material's adequacy.

During project bidding stage, manufacturers should preapprove proposed installers and allow only those contractors who are familiar with the procedure and trained to compete for the roof installation contract. This coordination should continue through the actual installation, with reviews and inspections conducted as necessary by the manufacturer. Finally, by requiring joint manufacturer and contractor warranties, the manufacturer continues its involvement throughout the warranty period. Warranties are discussed in detail in Chapter 9.

1/90 PERCENT PRINCIPLE

Throughout this book, emphasis is given to proper selection and installation of envelope waterproofing systems. As this chapter has shown, however, successful installation goes beyond selection and application of a single envelope component. Only if all individual components of a building's envelope have adequate transitions with one another will a building remain watertight and weather-tight.

This is especially true of buildings that use a variety of composite finishes for exterior surfacing such as brick, precast, or curtain wall systems. These designs incorporate a variety of waterproofing methods. Although they might each act independently, as a whole they must act cohesively to prevent water from entering a structure. Sealants, wall flashings, weeps, dampproofing, wall coatings, deck coatings, and the natural weather-tightness of architectural finishes themselves must act together to prevent water intrusion.

As many as 90 percent of all water intrusion problems occur within 1 percent of the total building exterior surface area.

Field construction is predicated on bringing numerous crafts together into the completion of a structure. Too often, these crafts are supervised and inspected independently of one another, without regard for coordinating their work into a solidarity effort.

Quality of field construction must be expanded to monitor and supervise the successful installation of transitions and terminations of envelope components. Project plans and specifications by the architect and engineer must clearly detail the responsibility for this work. Contractors must then take the responsibility for supervising and coordinating proper installations. Building owners must implement maintenance programs required throughout envelope life-cycling.

Any intrusion of water and weather at any envelope detail will create further problems by compounding itself. Leakage promotes deterioration of substrates and structural reinforcement that begins and accelerates throughout the entire process. Each cause feeds the other: further leakage causes further deterioration, further deterioration causes further leakage, resulting in eventual building envelope failure, damaging the structure and interior contents.

Such action results in the lawsuits, wasted energy, increased repair costs, loss of revenue, and inconvenience to tenants so frequent in the construction industry. Applying the 1/90 percent principle prevents these situations.

CHAPTER 9
LIFE CYCLES: QUALITY, MAINTENANCE, AND WARRANTIES

INTRODUCTION

The success of any waterproofing system depends not only on the initial application, but also on the in-place conditions encountered after installation. The successful ability of the envelope to remain watertight over its useful life expectancy is referred to as its *life cycle*. Any waterproofing system or building envelope subjected to life-cycle conditions that exceed the intended use of the product will result in failures and water infiltration; for instance, sealant joints that are subjected to expansion due to thermal movement that exceeds the capability of the material or the parameters of the joint design.

The majority of the *Construction Waterproofing Handbook* addresses the processes involved in successfully applying waterproofing systems and integrating them with other components of the building envelope. However, once successfully installed the systems must be properly maintained for the envelope to operate optimally throughout its life cycle. This chapter address the issues that can affect the performance of waterproofing products and the overall building envelope, including

- A contractor's quality performance
- Product and system manufacturers' quality
- In-place maintenance
- Waterproofing warranties

The last becomes important only if the other three have not prevented a leakage or failure problem. As important as it is to have effective quality processes throughout envelope design and construction, a building owner should recognize the protection, or lack thereof, provided by their warranty.

CONTRACTORS

Regardless of the quality, performance characteristics, and cost of a waterproofing system, the systems are only as effective as the caliber of installation. Even the best systems may prove worthless or ineffective if not installed and transitioned properly into other envelope systems.

Considering that most construction systems are field-manufactured, it is mandatory for properly trained mechanics and competent contractors to complete installation of any envelope component. Ineffective installation can not only destroy the performance of a material itself, but it can also lead to structural and interior contents damage. This results in costly repairs and loss of revenue for a building owner.

Any waterproofing contractor should have the experience and skills necessary to complete the required installation successfully. As importantly, the contractor should also have the required fiscal responsibility to not only complete the work but also honor punch lists and repairs that might become necessary. Industry clients can confirm a contractor's capability to carry out the work satisfactorily by requiring any potential bidder to complete a qualification statement. The client can use a qualification process to prequalify a list of bidders that meet the standards necessary to complete the specific project. A typical qualification form is shown in Fig. 9.1.

By prequalifying and selectively choosing competent and experienced contractors, unnecessary problems are eliminated during life-cycling. Additionally, this process should ensure that the contractor will be available for repairs and will honor warranty items that my occur. There are many qualities a potential contractor should possess, including:

- Experience in the specific type of installation
- Properly trained mechanics to complete work
- Certification by material manufacturer
- Organized and enforced safety policy
- Payment and performance bonds for total contract sum
- Insurance as required by federal, state, and local laws
- Joint manufacturer and contractor warranties
- Availability of maintenance bonds for warranties
- Sufficient equipment to complete installation
- Financial and customer references
- In-place hazardous waste programs
- Structural quality programs

When selecting a contractor, all of the above points should be considered. Reliance solely upon low bids often ends up costing more in maintenance and repairs over-life-cycling of installed systems.

Bonding capacity is a reliable means of allowing only responsible firms to complete work. Bonding and insurance companies run extensive background investigations of contractors before providing either bonds or insurance to a firm. Upon investigation of the contractor's experience, financial, and other capacities, bonding companies will set bonding limits in dollar amounts for the contractor.

Bonds act as insurance policies for the benefit of an owner. Requiring contractors to provide a payment and performance bond in the full contract amount assures owners that the contract will be completed and all materials, suppliers, and employees will be paid. If not, a bonding company will take over the contract, complete the work, and ensure payments of all outstanding invoices.

CONTRACTOR OR SUBCONTRACTOR QUALIFICATIONS

Date of Submittal ______________

Business name: ______________

Address: ______________

____

Phone: ______________

Home office address if different from above: ______________

____

Name of parent organization of above firm or subsidiary: ______________

____

Type of firm:

- Corporation ☐
- Partnership ☐
- Joint Venture ☐
- Sole Proprietor ☐

How long has firm been operating as an architectural/engineering firm? ______________

IF CORPORATION (If joint venture, answer for each firm represented)

Date of incorporation: ______________

State of incorporation: ______________

President/CEO: ______________

Vice president(s): ______________

Secretary: ______________

Treasurer: ______________

IF PARTNERSHIP

Date of Organization: ______________

Type of partnership: ______________

General partners: ______________

IF SOLE PROPRIETORSHIP

Date of establishment: ______________

Name of owner: ______________

Page 1 of 6

FIGURE 9.1 Contractor or subcontractor qualification statement.

LABOR CLASSIFICATION

Union ❑
Merit Shop ❑

Does the firm operate under any job site agreements? Yes ❑ No ❑

List classifications of work completed by own forces:

____
____
____
____

List states / local jurisdictions in which the organization is legally licensed or qualified to operate as a contractor or construction manager. Include license numbers.

____
____
____
____

Has the firm ever failed to complete any projects awarded to it? If so, provide details.

____
____

Has any officer or owner of the organization previously been associated with a firm that has failed to complete a project awarded to it within the last seven years?

____
____

Are there any judgments, claims, arbitrations, or legal suits pending or outstanding against the organization?

____
____

Has the organization filed any lawsuits, arbitration or other alternate dispute resolution hearings in association with any contracts, including subcontracts, in the last seven years?

____
____

Page 2 of 6

FIGURE 9.1 (*Continued*) Contractor or subcontractor qualification statement.

Likewise, maintenance bonds ensure that warranties will be honored even in the event a contractor goes out of business before the warranty expires. All bond premiums will be added to the cost of a project, but they offer protection that otherwise may not be available for performance and warrantability.

Field mechanics installing waterproofing systems have ultimate control over success or failure of the in-place waterproofing system. Field mechanics must be properly trained and

CURRENT WORKLOAD

Annual dollar volume: ____________________

Current year to date: ____________________

Last year: 19____________________

Previous year: 19____________________

EXPERIENCE

List as attachments all related construction and waterproofing projects currently in progress, including name, address, phone numbers of owner and architect. Include contract amount, percent complete, and expected completion date.

Largest singe contract ever managed by the office qualifying: $____________________

FINANCIAL REFERENCES

Include an audited profit and loss statement, including income and balance sheet. Provide name and address of firm preparing the audited statement.

____

____

If the provided audited statement is for any firm other than that listed as the qualifying contractor, explain the fiduciary relationship between the firms.

____

____

If the organization's financial statement is for any firm other than that qualifying, provide evidence that this firm will act as guarantor for the duration of this project.

____

____

CORPORATE QUALITY PROGRAM

Does the organization's main or home office have a structured and active total quality management program? Yes ❑ No ❑ If yes:

Name: ____________________

Date established: ____________________

Name of person managing program: ____________________

Details: ____________________

____

____

FIGURE 9.1 (*Continued*) Contractor or subcontractor qualification statement.

Does the organization have a structured employee training program? If so, describe.

List courses provided last year:

Number of employee - hours given in program:

Name of person managing program:

Does the organization participate in a site-managed, total quality improvement program separate from above? If so, describe.

Names of projects currently participating:

Provide details of programs.

List the recent projects for which the organization has completed structured partnering programs.

Does the firm have or participate in any long-term partnering agreements? Yes ❑ No ❑
If yes, provide details.

Describe in detail the CAD systems, budgeting processes, and constructability review processes used by the firm.

Page 4 of 6

FIGURE 9.1 (*Continued*) Contractor or subcontractor qualification statement.

motivated to complete all installations in a professional manner. Supervision must be provided to ensure that installations act cohesively with other envelope components.

Although owners can require that a mechanic have experience and training to install materials, contractors have ultimate control over job-site and working conditions. These conditions include wages, benefits, and safety conditions, all of which influence installation quality.

SAFETY

Provide details of main-office and site-managed safety programs, including persons responsible for the programs. At minimum, include the following information.

When and how are main office safety programs provided?

____

____

When and how are job-site safety programs provided?

____

____

Provide positions and names of personnel managing these programs.

Provide copies of actual written programs.

Provide the organization's Interstate Experience Modification Rate for the last three years:
19 ____: ____________________________
19 ____: ____________________________
19 ____: ____________________________

From the most recently completed yearly OSHA 200 log, provide the following:
Number of hours worked: ________________________
Number of employees: ________________________
Worker's compensation claims: ________________________
Recordable injuries: ________________________
Lost- time injuries: ________________________
Lost work days: ________________________
Lost-work-day case rate: ________________________
Recordable case rate: ________________________

RELATED EXPERIENCE

By attachment, list a minimum of five projects completed within the last five years that most closely match the scope and dollar value of the project description included in the RFI. For each project listed, provide the following information:

Name of project: ________________________
Location: ________________________
Owner, with contact person, address and phone number: ________________________

____

Architect, with contact person, address and phone number: ________________________

____

Page 5 of 6

FIGURE 9.1 (*Continued*) Contractor or subcontractor qualification statement.

Detailed project description, including:
Original contract amount: ____________________
Final contract amount: ____________________
Dollar amount of change orders: ____________________
Percent of change orders compared to contract amount: If over 2.5%, explain. ________
____
____
Original scheduled substantial completion date: ____________________
Actual completion date: ____________________
Originally scheduled final completion date: ____________________
Actual final completion date: ____________________
Reason for delays or improvements in project schedule: ____________________
____
Details of any site quality management programs and innovations used on the project: ___
____
____

PROPOSED TEAM MEMBERS

For each of the proposed team members, provide the following information:

- Complete resume, including project-specific experience related to the scope of the proposed project
- Proposed position on this project
- Name, address, and phone number of owner's project representative on the employee's last two projects
- Quality, safety, and additional training the employee has received in the last two years
- Partnering programs in which the employee has participated
- Involvement in any site-managed, site-quality programs
- Include letter of management's commitment, to guarantee this person's availability for the duration of the proposed project
- Safety record of the last two projects in which the person participated
- Have the proposed project personnel previously worked together as a team on any previous projects?

Page 6 of 6

FIGURE 9.1 (*Continued*) Contractor or subcontractor qualification statement.

Finally, warranties provided by a contractor should cover both the labor and the materials. Questions to consider when reviewing a proposed warranty include:

- Will it cover material failure?
- Whose responsibility is it to uncover buried or covered systems?
- Are consequential damages (interior contents) covered?

- Is the warranty bondable?
- Is there a dollar limit to the repairs?
- What escape clause does the warranty contain (e.g., structural settlement)?
- Is the warranty issued jointly by the manufacturer and contractor?
- Does the warranty cover transition and termination detailing?

If necessary, proposed warranties should be reviewed by counsel, and necessary changes should be made to protect the owner before a contractor is awarded the contract.

MANUFACTURERS

Equally important to the success of a waterproof system are the experience, assistance, and reputation of the material manufacturer. The quality of the material manufacturing process itself is of high importance, but choosing a manufacturer should review considerations beyond quality, including:

- Manufacturer warranties
- Length of time product has been manufactured
- Sufficient number of previous installations
- Adequate testing of and test information on product
- Detailed installation instructions
- Availability of representatives to review installations
- Certification and training of applicators
- Maintenance instructions
- Material safety data sheets
- Manufacturer's assistance in specification preparation and in detailing terminations and transitions

The manufacturer's representations for a particular product should be reviewed, and test results compared to similar products. Test results based on recognized testing laboratories such as ASTM should be consulted. This testing allows materials to be compared with those of other manufacturers as well as with completely different systems.

Manufacturer warranties should also be carefully reviewed. Often a warranty only covers material failure and provides no relief for building owners in case of leakage. Considering that all waterproof systems require field application or construction, too often it is easy for a manufacturer to pass liability on, citing poor installation procedures.

Therefore, warranties should cover both labor and materials. This places requirements on a manufacturer to ensure that only experienced mechanics and contractors install their products. A labor and materials warranty from a manufacturer and contractor provides protection when one or the other goes out of business. It also prevents attempts to place blame elsewhere when there is a question of liability for repair problems.

MAINTENANCE

No building or structure is maintenance-free. In fact of total costs, 30 percent consist of original construction costs and 70 percent of maintenance costs. Considering the possible damage and costs that might occur, it is just as important to maintain the exterior as the interior of a building. Regular exterior maintenance prevents water intrusion and structural damage that might be associated with water infiltration.

An effective maintenance program involving the building envelope depends on using qualified inspection procedures to determine the required maintenance. A building requires complete inspection from top to bottom, including a review of all exterior elements, at recommended intervals of every 5 years but no longer than every 10 years.

Any building portion inaccessible by ordinary means may require hiring a contractor for scaffolding and inspection. Only competent building trades personnel should make these inspections, be it an architect, engineer, or building contractor.

In view of 90 percent of all leakage being caused by 1 percent of the building envelope, all components of an envelope must be inspected. All details of inspection, including exact locations of damage and wear, that will require attention after an inspection should be documented.

Among envelope components, the following require complete and thorough inspection:

- *Roofing*, with particular attention to terminations, flashings, protrusion, pitch pans, skylights, and copings
- *Above-grade walls*, with attention to expansion and control joints, window perimeters, shelf angles, flashings, weeps, and evidence of pollutant or chemical-rain deterioration
- *Below-grade walls*, checking for proper drainage of groundwater, signs of structural damage, and concrete spalling
- *Decks*, with attention to expansion and control joints, wall-to-floor joints, handrails, and other protrusions

These are only the highlights of maintenance relating to waterproofing materials. Inspection procedures for existing damage and buildings that have not been maintained are discussed in Chapter 7.

During inspections, the effectiveness of a waterproof system should be monitored. This includes water testing if necessary, to check systems already in place. This requires inspection for items such as clogged or damaged weeps, cracks or disbonding of the elastomeric coating, deteriorated sealants, damage to flashings, and wear of deck coatings.

Most waterproof systems require maintenance procedures of some type, and these recommendations should be received from the manufacturer. Certain items will require more maintenance than others, and provisions need to be made to monitor these systems more frequently. For instance, vehicular traffic deck coatings receive large amounts of wear and require yearly inspections. Maintenance for traffic areas includes replacement of top coatings at regular intervals to prevent damage to base-coat waterproofing.

Dampproofing behind a brick veneer wall usually requires inspection to ensure that the weeps continue to function. Other unexposed materials, such as planter waterproofing systems, may require more attention; they should be checked to ensure that water drainage is effective and protection surfacing remains in place during replantings.

Exposures to elements affect waterproofing systems, and required maintenance often depends upon this exposure. Factors affecting waterproofing systems include

- Thermal movement
- Differential movement
- Weathering

  Ultraviolet exposure
  Freeze–thaw cycles
  Rain

- Wind loading
- Chemical attack

  Chloride (road and airborne salt)
  Sulfides (acid rain)

- Settlement

  Structural
  Nearby construction
  Acts of God (hurricanes, earthquakes, tornadoes)

Regular maintenance inspections should monitor any damage that might be caused by these types of wear and weathering, and repairs should be completed promptly according to the manufacturer's recommendations. Manufacturers should make representatives available to assist in the inspection and to make recommendations to the building owner if repairs or maintenance work is necessary.

Should a particular portion of an envelope be under warranty by a manufacturer, contractor, or both, necessary maintenance or repairs should be completed by firms warranting these areas. This prevents nullifying warranties or obligations of a manufacturer or contractor by allowing others to complete the repairs. If extended warranties are available, manufacturers should be consulted for proper maintenance procedures. For example, a contractor or manufacturer can provide a 5-year warranty plus an optional 5-year renewal. This requires that after the initial 5-year period, manufacturer and contractor make a complete inspection, at which time all necessary repairs are documented. Upon an owner's authorization for repair completion and payment for these repairs, the manufacturer or contractor extends the warranty for an additional 5 years.

WARRANTIES

It would require several legal courses to cover warranties, guaranties, and their legal consequences completely. This section, therefore, approaches the subject as summarized in the phrase, "Let the buyer beware." Due to all intangibles involved in field construction, it is rare to find any warranty that simply states, "System is guaranteed to be waterproof." All warranties typically exclude circumstances beyond the manufacturer's or contractor's control. However, owners should review warranties to ensure they are not full of exclusion clauses that negate every possible failure of material or installation, making the warranty in effect worthless.

No warranty is better than the firm that provides it, and should the manufacturer or contractor go out of business, a warranty is useless unless bonded by a licensed bonding company. A manufacturer or contractor that places emphasis on its reputation may be more likely to take care of repairs or warranty items, regardless of the limitations that appear in the warranty or guarantee.

Types of warranties

The terms *warranty* and *guaranty* typically are used interchangeably and have no distinct difference. In preferential order, here is a list of warranty and guaranty types:

- Bonded
- Joint manufacturer and contractor warranty
- Combination of separate contractor and manufacturer warranties
- Manufacturer warranty covering both installation and materials
- Contractor labor warranty only
- Manufacturer material warranty only

Any of these warranties must be specific to be enforceable. Clauses such as, "warrant against leakage" leave open the responsibility of a contractor or manufacturer. Does this mean any leakage into the building, or leakages only through the applied systems? What happens if a juncture between the warrantied system and another system fails? Who covers this failure? Regardless of the warranty type, it should be specific as to what is and is not covered under guarantee terms.

Bonding of a warranty provides complete protection for a building owner and ensures against failure by both contractor and manufacturer. A bonded warranty should be underwritten by a reputable, rated bonding company, licensed to do business in the state in which it is issued.

Some bonded warranties may limit the extent of monies collectible under warranty work. This works as a disadvantage to an owner, should a system require complete replacement. For example, consider an inaccessible system, such as below-grade waterproofing, where costs for obtaining repair access can well exceed the actual cost of repairing the leakage.

Joint warranties, signed by both contractor and manufacturer, offer excellent protection. This warranty makes both firms liable, jointly and severally. This ensures that if one firm is not available, the other is required to complete repairs. These warranties typically have separate agreements by manufacturer and contractor, agreeing to hold each other harmless if repairs clearly are due to defective material or defective workmanship. This separate agreement does not affect the owner, as the document issued makes no mention of this side agreement.

Manufacturers are selective as to whom they choose to become signatories to such agreements. They qualify contractors financially, provide training, and make available manufacturer's representatives to ensure that materials are installed properly. Additionally, most manufacturers thoroughly inspect each project installation and require completion of its own punch list before issuing warranties.

Warranties that cover labor and materials separately have consequences an owner should be aware of. By supplying separate warranties, contractors and manufacturers often attempt to pass the blame to each other rather than correct the problems. Owners may have to contract out work to other parties to complete repairs, and attempt to recover from the original manufacturer and contractor by legal means. In some situations, this method is used to provide warranties of separate length for labor and material (e.g., 5-year labor, 10-year material). These have the same limitations and should be reviewed carefully to combine the two into one agreement.

Other warranties—labor only or materials only—have limited protection and should be judged accordingly. Since most systems are field-installed, labor is most critical. However, materials can fail for many reasons, including being used under the wrong conditions. Therefore, both materials and labor should always be warrantied. By not requiring a material warranty, manufacturers may not be under obligation to ensure that materials are being used for appropriate conditions and with recommended installation procedures.

Warranty clauses

Actual terms and conditions of warranties vary widely, and assistance from legal counsel may be necessary. For common warranty clauses, special attention should be paid to the following:

- Maintenance work required of an owner to keep the warranty in effect
- Alterations to existing waterproofing systems that can void a warranty
- Wear on systems that may void the warranty (e.g., snowplows, road salting)
- Unacceptable weathering (e.g., hurricanes, tornadoes)
- Requirement that prompt notification is given, usually in writing, within a specific time
- Contractor and manufacturer refunding of original cost, to satisfy warranty instead of doing needed repairs
- Specific exclusions of responsibility:

  Structural settlement
  Improper application
  Damage caused by others
  Improper surface preparation

- Complete replacement of a faulty system versus patching existing system

All warranties are limited, and must be reviewed on an individual basis to eliminate any unacceptable clauses before signing the contract, purchasing materials, and installation. Items such as the actual specific location of a waterproofing system should be clearly included in the warranty and not limited to the building address. The terms of what is actually covered should also be addressed (e.g., installation, leakage, materials, or all three). The warranty should be specific, allowing those interpreting a warranty years later to understand the original intended scope.

UNACCEPTABLE WARRANTY CONDITIONS

Many manufacturers' standard warranties contain clauses or requirements that do not afford the proper protection for the purchaser. Any warranty should be reviewed before the project is awarded to a particular contractor or manufacturer, as this is the most appropriate time to make the necessary changes and corrections that will ensure the proper protection for the building owner.

Everything is negotiable, including warranties, particularly during the sales process. A warranty itself can say much about a company and their product. Warranties that are full of escape clauses can be directly related to how difficult a product is to install and to expectations of performance problems by the manufacturer. The following are standard warranty conditions, requirements, or exclusions that should be avoided under most situations. The statements following come from actual manufacturers' warranties, and are typical of what an owner might discover when the coverage is needed.

Maximum obligation limit

"XYZ's obligation for materials and labor combined shall not exceed $ 1.00 per foundation, accumulative for the life of the warranty."

"In no event will XYZ be obligated to pay damages in any amount exceeding the original price of the materials shown to be defective."

Each of these clauses, and similar types, place an upper limit on the damages and or repairs for which a manufacturer can be held liable. Obviously, the costs of repairs will usually exceed the cost of the original application, particularly in situations where the installation is inaccessible (e.g., below-grade, positive-side waterproofing).

These clauses can prevent a building owner from recovering reasonable expenses associated with repairing below-grade waterproofing membranes, when obtaining access to the product far exceeds the cost of actually repairing the membrane. If the warranty is limited to the cost of materials only, there is virtually no protection for the building owner.

Maximum monetary clauses should be completely avoided; they offer no realistic protection if a claim or repair is ever necessitated involving waterproofing systems. The monetary limitations also directly relate to Limits of Liability clauses that further limit a contractor's or manufacturer's liability.

Limitation of liability

"In no event shall will XYZ be liable for special, indirect, incidental or consequential damages (including loss of profits) arising out of or connected to the materials or the system . . . regardless of any strict liability or active or passive negligence of XYZ and regardless of the legal theory (contract, tort, or other) used to make a claim."

This clause attempts to completely disavow any liability, even if required by law. In most states such exclusions are not permitted even if signed by the parties, for no company has the right to change or negate governing laws. However, the manufacturer is attempting to ensure that their liability will be completely limited, in itself not speaking highly of any company.

Obviously such requirements should be avoided, and the manufacturer required to assume the liability they might incur if their product or application fails. There are reasons for limiting the consequential damages, since the manufacturer has no idea what they might be at time of contracting.

For example, consider a hospital, where leakage has caused the failure of a piece of operating equipment that in turn passively causes the death of a patient. The waterproofing manufacturer might possibly be sued for contributory negligence, for which no warranty can negate liability. However, the damages related to the loss profits while the equipment is not functioning are considered consequential damages and often made exempt by the warranty as in the clause quoted above. Such exclusions generally are acceptable, since the manufacturer would not be able to determine the cost of doing business without knowing the exact circumstances of the building uses and potential liability.

These warranty conditions should be carefully reviewed with legal council when appropriate, and changes made or negotiated before the contract is executed. The manufacturer should not be expected to provide coverage for unreasonable situations, but be made to accept liability for damages caused directly by the leaks, such as interior drywall repairs necessitated by leaks. If these incidental damages are excluded, often because the owner maintains building insurance for such events, at minimum the manufacturer should be made to reimburse the insurance deductible amount. In addition to outright limitation of liability, some manufacturers attempt to reduce their remaining liability on a depreciable value.

Prorated or depreciable value

> "Maximum value of warranty is reduced 10 percent each calendar year from the date of this warranty." This clause is used in conjunction with an actual stated maximum value of the warranty or limiting the value to the original purchase price of the materials or product.

By limiting liability in this way, it is implied that the manufacturer expects their product's capability to become increasingly ineffective. The obvious risks the owner assumes in losing practically all the coverage in the out-years of a warranty make such warranties unacceptable.

Access provisions

> "It is the owner's responsibility for all costs associated with moving, removing, restoring, repairing or replacing any of the following, but not limited to, grass, trees, shrubs, landscaping, fences, patios, decks, sidewalks, utility service lines, and structures in order to reach the affected area. In addition, it is the owner's responsibility to remove exterior soils to reach the affected area, backfill, and recompact if necessary."

> "Is it the owner's responsibility to pay the cost to remove interior finishes to reach the affected area and replace with the same or similar materials."

Often the major expense involved in repairs to building envelope waterproofing arises out of gaining access to the leakage area. High-rise and below-grade structures are difficult areas to access positive-side waterproofing repairs. Warranties that pass the access liability to the building owner should be carefully reviewed and rarely accepted, as stated in the above examples.

Escape clauses

> "Warranty is void if material not installed in strict compliance with the specifications and instructions."

> "Warranty void if material not applied with temperature above 40 degrees F and below 60% relative humidity."

Many manufacturers and contractors will attempt to include stipulations that can create sufficient means for them to deny any responsibility for the repairs. Keep in mind the 90%/1% and 99% principles when reviewing warranties. Often manufacturers will negate their warranty coverage if the material was not installed in strict compliance with their instructions.

Actual site conditions encountered at the site are rarely ideal, and the manufacturer can often point to improper application methods regardless of how minor. For instance, temperature and humidity conditions documented by the National Weather Service might not meet the manufacturer's standards although they have no relationship to the actual leakage problem encountered.

After project completion, warranties are typically an owner's only recourse and protection against faulty work and materials. With this in mind, warranties should be given the same close scrutiny and review as the original design and installation procedures to protect the owner's best interests.

Finally, recalling the 90%/1% percent principle, all too often transitions and terminations are not specifically included in each of the envelope component warranties. By making contractors and manufacturers responsible for the 1 percent of a building's area that creates 90 percent of leakage problems, their attention is directed to this most important waterproofing principle. By including these areas in warranties, contractors and manufacturers are prompted to act and to ensure that these details are properly designed and installed. This prevents numerous problems during the life-cycling of a building or structure.

CHAPTER 10
ENVELOPE TESTING

INTRODUCTION

There are several steps, methods, and means to test individual or complete portions of a composite envelope. These tests begin with the manufacturer's testing, which ensures that materials are suitable for specified use, longevity, and weathering. Next an entire composite envelope sampling is tested to ensure that all components, when assembled, will function cohesively to prevent water infiltration.

No project is built or renovated without some testing having been completed. Too frequently, however, the only testing completed, that of material systems by manufacturers, is insufficient to prevent problems that continue to occur at the job site.

Rarely is attention given to testing the 1 percent of a building envelope that creates 90 percent of the water intrusion problems. This 1 percent of a building's area, the terminations and transitions of various independent systems, never is fully incorporated into proper testing.

WHEN TESTING IS REQUIRED

Testing frequently is used to test new designs, materials, or systems. However, envelope designs that incorporate standard materials also require testing under certain circumstances. For example, masonry walls constructed of typical brick composition but having intricate detailed slopes, shapes, and changes in plane should be tested. Testing in these cases will determine whether flashings as detailed will perform adequately in the various detail changes incorporated into the design.

Testing should also be completed when envelope components contain areas such as expansion joints in unusual or previously untested areas; for example, a sealant expansion joint in a sloped area that may pond water.

Specially manufactured products, such as specially colored sealants, brick manufactured in unusual textures, metal extruded in unusual shapes, and joints, are examples of envelope components that should be tested to prevent problems after complete envelope installation.

Any time a new design comprising several different materials is developed for a proposed envelope mock-up, testing is imperative. This is particularly true for high-rise construction.

Cladding materials used in today's designs and construction are lighter-weight and thinner, requiring fewer structural materials and supports. This lowers overall building costs but, in turn, presents numerous problems in envelope effectiveness, particularly in watertightness. This is in comparison to the massive masonry walls of more than 1-foot thickness used in early high-rise construction, where shear magnitude of the envelope eliminated the need for such testing.

TESTING PROBLEMS

Manufacturers, although making recommendations for termination or transition detailing, will not incorporate these areas in their material testing. A manufacturer of sheet-good membranes will test the actual rubber materials for weathering, elastomeric capabilities, and similar properties. The manufacturer will not, however, test the adhesive material used to adhere materials to a termination detail for weathering, movement characteristics, and so forth.

Likewise, transition details, such as between above-grade and below-grade areas, detailed by an architect are not tested by either the waterproofing or dampproofing material manufacturers. Lack of testing in these and similar details of the 1 percent of the building's area reveals another reason for the continuing cause of these areas contributing 90 percent of envelope water infiltration problems.

Mock-up testing of a building envelope portion often eliminates replication of termination and transition details. Testing often is completed on envelope curtain wall portions only and does not include masonry portions, transitions from curtain wall to parapet wall, coping, and roofing transitions.

Ensuring that each material or system is tested independently does not ensure that the composite envelope when completed will be successful. Any material, even if it performs singularly, does not ensure that the composite envelope will be successful.

These are some reasons testing fails in preventing water infiltration. As long as buildings are manufactured from a variety of systems which must be transitioned or terminated into other components, and these details are untested and improperly installed, problems will continue to occur.

STANDARDIZED TESTING

A thorough review of all testing available including manufacturer's testing, independent testing, laboratory testing, and site testing should be made. This will make sure that the entire envelope is properly tested to ensure watertightness and envelope longevity.

Available testing includes:

- Laboratory analysis
- Mock-up testing under simulated site conditions
- Job-site testing
- Long-term weathering testing

Such tests are completed by both government (including state and local municipality) and private agencies. The most commonly referred-to private agency testing standard is the American Society for Testing and Materials (ASTM).

ASTM

ASTM was established in 1898 as an organization for establishing standards in characteristics and performance of materials, including some waterproofing materials, particularly sealants and caulking. These standards are used as a basis of comparison among various products and similar products of different manufacturers.

ASTM standards are specific test requirements described in detail to ensure that individual materials are tested in a uniform manner. This allows for a standard of comparison between different manufacturers or materials. Their characteristics can then be compared for judgments of suitability in such issues as weathering, performance suitability, and maximum installation conditions under which a material will function.

ASTM develops test methods and performance specifications that allow comparisons between different products and systems. Manufacturers document that their materials have been tested according to the ASTM standards. Tests are completed by independent laboratories (refer to Chapter 14), in controlled laboratory conditions. Tests are designed to test a material's maximum capabilities and limits (e.g., elastomeric sealant expansion limits). Also available are accelerated weathering tests to determine if materials are suitable for use in the extremes of weather including freeze–thaw cycling and ultraviolet weathering. These laboratory tests are typically applied to specific materials themselves (e.g., sealants, coatings) but not to the composite envelope systems, transitions, or termination detailing. The frequently referred-to ASTM testing, adaptable to waterproofing materials, is summarized in Table 10.1.

Other testing agencies

Laboratory analysis is completed by the National Bureau of Standards, a federal government agency. The most commonly referred to federal specifications for waterproofing

TABLE 10.1 **ASTM Testing for Waterproofing Products**

ASTM Test number	Test type
D-412	Tensile strength
D-412	Elongation
E-96	Moisture vapor transmission
D-822	Weathering resistance
C-501	Abrasion resistance
E-154	Puncture resistance
D-71	Specific gravity
D-93	Flash point
D-2240	Shore A hardness
E-42	Accelerated aging
E-119	Fire endurance
D-1149	Ozone resistance and weathering
C-67	Water repellency
E-514	Water permanence of masonry
C-109	Compressive strength
C-348	Flexural strength
D-903	Adhesion

materials are TT-S-227 for two-component sealant performance, TT-S-00230C for one-component sealants, and TT-S-001543 for silicone sealants. Underwriter's Laboratory tests (UL) for fire endurance of specific materials include test UL-263.

Other independent laboratory analysis and testing include firms or organizations established to test specific application or installation uses. These include the National Cooperative Highway Research Program (NCHRP), which tests clear deck sealers used in concrete construction in highway and bridgework.

Local government agencies, such as Dade County Building and Zoning Department's Product Approval Group, often establish their own minimum requirements and approve independent laboratories to test materials and approve materials for use in local construction and renovation. This Miami-area public agency tests materials to ensure that products will perform in the harsh environment of south Florida. Tests include ultraviolet and hurricane weathering. Such agencies test individually manufactured materials in laboratory conditions to ensure the adequacy of the material alone. They do not perform tests for complete envelope systems but only for individual components.

This testing allows selection of individual materials that will function under an expected set of conditions, including weathering and wear. For example, a deck-coating material is chosen that will function under extreme ultraviolet weathering and heavy vehicle traffic. Testing does not, however, determine the acceptability of transitions used between the deck coating and deck expansion joints, or whether coatings will be compatible with curing agents used during concrete placement. The tests allow proper selection of individual materials for use in a composite envelope but do not test individual systems joined in the envelope construction. Several private laboratories are available to complete testing of composite envelopes.

MOCK-UP TESTING

Independent laboratories are often used to test mock-ups or composite envelope systems. These tests assume that individual components have been tested and will suffice for jobsite conditions including extended use and weathering. Independent laboratories test the effectiveness of composite materials against water and wind infiltration. They can create conditions that simulate expected weathering extremes at the actual building site.

These tests are limited, however, in that they do not recreate long-term weathering cycles and temperature extremes and are rarely applied to the entire composite envelope. Testing envelopes after weathering and movement cycles, particularly at the transitions and terminations, is mandatory to ascertain the effectiveness of the details.

An envelope or curtain wall mock-up is constructed at the laboratory site using specified exterior envelope components. They are applied to a structural steel framework provided by the laboratory, or framing is constructed specifically for testing, (Figs. 10.1 and 10.2). This framework should include flashing and appropriate transition details if they occur in the selected area to be tested, (Fig. 10.3). Testing is completed on a minimum floor-to-ceiling segment height of the envelope, (Fig. 10.4).

Preferably, testing goes well beyond this minimum to allow testing of the most advantageous and economically feasible portion of the entire envelope, (Fig. 10.5). Typically, testing size is 20–25 feet wide to 30–40 feet high, (Fig. 10.6).

FIGURE 10.1 Testing frame prepared for building envelope mock-up. (*Courtesy of Architectural Testing, Inc.*)

FIGURE 10.2 Building envelope cladding being applied to mock-up. (*Courtesy of Architectural Testing, Inc.*)

FIGURE 10.3 Completed mock-up should include appropriate flashing and termination details. (*Courtesy of Architectural Testing, Inc.*)

FIGURE 10.4 Testing of completed mock-up. (*Courtesy of Architectural Testing, Inc.*)

FIGURE 10.5 Typical portion of envelope testing—masonry walls and punch windows. (*Courtesy of Architectural Testing, Inc.*)

FIGURE 10.6 Partial envelope—three stories in height—prepared for testing. (*Courtesy of Architectural Testing, Inc.*)

Particular attention should be given to transition areas from glass and metal to stone, masonry, or concrete areas; parapet areas; and horizontal-to-vertical and other changes in plane, including building corners, (Fig. 10.7). Testing of these transitional details ensures their effectiveness against water infiltration. Unfortunately, termination details such as above-grade areas to below-grade areas are not usually feasible for testing purposes. In addition, structural steel supports used in the testing mock-up (Figs. 10.8 and 10.9) often cause different test results if the envelope is to be applied to a more rigid frame such as a structural concrete framework.

Testing is also limited to air infiltration and water infiltration. Tests do not include weathering analysis that often deters envelope component effectiveness during movement cycles such as thermal expansion and contraction.

Thinner cladding materials used today are subject to stress by thermal movement and wind loading. Transition and termination details are different, based on the thickness of material and the movement stress that is expected with the in-place envelope. This makes testing mandatory. It is less costly to correct problems that appear in design and to construct a mock-up than to repair or replace an entire envelope after it is completed and tested by natural forces and weathering.

Testing will also reveal problems that might occur with coordinating the different trades involved in a single envelope design. For example, how, when, and who installs the through-wall flashing that runs continuously in a masonry, precast concrete, and window wall? Is the same flashing application detailing applicable in all these instances, and so forth?

FIGURE 10.7 Various termination and transition details are included for superior test results. (*Courtesy of Architectural Testing, Inc.*)

FIGURE 10.8 Structural supports for curtain wall facade. (*Courtesy of Architectural Testing, Inc.*)

Situations that arise during testing make it extremely important that mechanics installing a curtain wall at the job site be the same mechanics who install the mock-up construction for testing. Then, should problems arise in testing, they can be resolved with the knowledge carried to the job site. In the same manner it is important that a contractor supervisor be present and participate in mock-up construction and testing, to ensure continuity and quality of envelope job-site construction.

Mock-ups, in effect, become a partnering or teaming concept, with all partners—architect, owner, contractor, and subcontractor—involved. These partners work together to complete mock-up testing successfully and resolve any conflicts or problems before they occur at the job site.

One serious flaw that frequently occurs should not be allowed—using sealant materials to dam up leaks as they occur during testing. Often, discovered leakage is taken care of simply by applying sealant. This happens in areas such as perimeters of windows, joints in metal framing, and transition details. Allowing sealant application during testing goes directly against the actual purpose of testing.

FIGURE 10.9 Testing of curtain wall only, with no transition details. (*Courtesy of Architectural Testing, Inc.*)

If leakage occurs and the sealant was not part of original detailing, it should not be installed until determination of the leakage is resolved. Sealants are not long-life-cycle products compared to other envelope components, and they require much more maintenance than the typical envelope components of glass, metal masonry concrete, or stone.

Sealants can offer short-term solutions during testing but often do not function during weathering cycles such as thermal movement of the composite envelope. Even if sealants will perform in this detailing, if they are not part of the original design the owner's maintenance requirements are increased for envelopes that include sealants and their short life cycles. Such attempts at quick fixes should not be permitted, considering the effect and the costs committed for testing, as well as the possible long-term effects of using sealants as a stopgap measure.

ASTM provides evaluation criteria for the selection of independent laboratory testing agencies. These tests include ASTM E-669-79, which provides criteria for evaluation of laboratories involved in testing and evaluating components of a building envelope, and ASTM E-548-84, which includes generic criteria for evaluating testing and inspection firms.

Mock-up testing of the envelope involves three types of tests, including

- Air infiltration and exfiltration
- Static water pressure
- Dynamic water pressure

Additional envelope testing can include

- Thermal cycling
- Seismic movement

Air infiltration and exfiltration testing

The air infiltration and exfiltration test determines envelope areas that will allow air to pass into or out of a structure. Although this test is typically for control of environmental conditions, if air can pass through an envelope, water can also pass through.

Wind loading can force water into a structure, or unequal air pressures between exterior and interior areas can actually suck water into a building. Therefore a building must be completely weatherproof to be completely waterproof. A completely waterproof building is therefore completely weatherproof.

Air infiltration or exfiltration tests are typically conducted according to ASTM-283. This test is used for measuring and determining any airflow through exterior curtain walls. This is a positive pressure test, meaning that positive pressures are applied to an envelope face.

To conduct this test, a sealed chamber is constructed to enclose the back of a composite envelope portion being tested completely. Air pressures can be lowered in the chamber by removing air and creating a vacuum in the chamber. Figure 10.10 shows a tare bag in place for an air infiltration test in accordance with ASTM-283.

The lower air pressure then draws air through the envelope from higher-pressure exterior areas. Air is drawn through any envelope deficiencies. This air penetration can be determined by measuring pressure differentials within the chamber. However, if air penetration is occurring, it is difficult to locate specific failure areas during testing.

This testing can be reversed by forcing additional air into the chamber to create higher chamber air pressures. This type of test creates negative pressure on an envelope face. (The explanation of negative and positive air pressure testing is similar to the explanation of negative and positive waterproofing systems.)

With this testing, air that is forced into a chamber can be mixed with fabricated smoke or colored dyes. This allows leakage areas to be easily identified when the colored air begins escaping through envelope components to the exterior. Areas of leakage can then be marked and later inspected for causes. Proper repairs can be completed and areas retested until weathertight.

When testing masonry mock-up panels or curtain walls that contain masonry portions with weeps, it is expected that a certain amount of air will penetrate the envelope through these weeps. The amount of air infiltration that is within satisfactory limits must be determined, and testing must be done to check that infiltration does not exceed these limits.

FIGURE 10.10 Tare bag in place for testing. (*Courtesy of Architectural Testing, Inc.*)

Static pressure water testing

Static pressure water testing uses the same apparatus and methods used with air testing, while at the same time applying water to the envelope face at a uniform and constant rate. Test standards typically used are according to ASTM-E331. Water is applied using spray nozzles equally spaced so as to provide uniform water application to all envelope components (Fig. 10.11).

By creating positive air pressure on the envelope (withdrawing air from inside the test chamber), water is sucked through an envelope at any point of failure. Areas of leakage are then marked for review and cause determination after testing has been completed.

Certain areas, such as weeps built into masonry walls or curtain wall framing, are subject to some water infiltration. As weeps are an integral part of an envelope design, allowable percentages of water infiltration through these areas must be determined. Test results can then be measured to determine the acceptability of infiltration through these areas.

FIGURE 10.11 Static pressure water test. (*Courtesy of Architectural Testing, Inc.*)

Dynamic pressure water testing

Dynamic pressure water testing applies water and wind conditions directly to an envelope face. Airplane prop engines, (Fig. 10.12), are used to force water against the composite envelope from spray nozzles equally spaced and mounted to frames, (Fig. 10.13). This test usually simulates the most severe conditions, such as hurricane- and tornado-force winds and rain conditions.

Water applied in this manner is forced both vertically and laterally along the envelope face to recreate conditions encountered in high-rise construction. This method of wind loading, with the addition of water, can force water into envelopes not capable of withstanding designed or expected wind loads.

Glass can bend or flex away from mullions or glazing joints, allowing sufficient space for water to penetrate into interior areas. Structural conditions can also change during wind loading to create gaps or voids or even failure of envelope components, (Fig. 10.14), allowing direct water infiltration.

Amounts of water and wind introduced onto the envelope can be variably applied to simulate maximum conditions expected at a particular job site. A combination of weather

FIGURE 10.12 Eighteen-cylinder airplane engine used to create dynamic pressure water testing. (*Courtesy of Architectural Testing, Inc.*)

conditions as severe as hurricane forces of 70 mph, plus winds and water at a rate of 8–10 inches/hour, are achievable. This is often the ultimate test for any envelope.

Mock-up testing summary

All three mock-up tests offer excellent previewing of a proposed envelope design. Cost permitting, all three should be completed to accurately determine areas of potential problems. As previously discussed, areas of failure should not merely be sealed using sealants that are not part of the original design. This defeats the purpose of testing, and problems will recur in field construction when envelope watertightness is dependent on improperly designed and applied sealant material.

When envelopes containing masonry walls that combine dampproofing, flashing, and weep systems are tested, water will undoubtedly enter the envelope as designed. The water entering must not exceed the capability of the backup systems to redirect all entering water to the exterior.

Note that laboratory mock-ups can also be used for color and texture approval, limiting the number of mock-ups required at job sites and lowering overall costs of laboratory testing, (Fig. 10.15). (See Table 10.2.)

FIGURE 10.13 Spray nozzles equally spaced in preparation for dynamic pressure water test. (*Courtesy of Architectural Testing, Inc.*)

TABLE 10.2 **Mock-Up Testing Advantages and Disadvantages**

Advantages	Disadvantages
Allows testing of designs and review of problems before actual construction	May not incorporate sufficient termination and tr ansition details
Involves all project participants in reviewing design and offering suggestions for review	Often does not simulate exact job-site structural conditions
Can create conditions beyond the worst expected at actual project site	Does not account for testing after thermal movement and structural settlement changes in actual envelope construction

JOB-SITE TESTING

Tests done at a job site can be as simple or as scientific as required by immediate concerns. Simple water testing using a garden hose and water source is probably the most frequently used means of job-site testing on both new and existing envelopes. Static and dynamic pressure testing can also be accomplished at project sites by laboratories or consulting firms that can provide portable equipment to complete these types of tests, (Fig. 10.16).

FIGURE 10.14 Failure of components during test. (*Courtesy of Architectural Testing, Inc.*)

Test chambers are built at job sites directly over a sample wall portion including curtain and precast wall units, (Fig. 10.17). Portions being tested should include as many termination and transition details as possible, including any changes in plane. Such job-site tests are limited by actual areas tested, but offer the advantage of testing under actual conditions as compared to laboratory mock-ups.

Mock-up panels are often required at job sites to check for color and texture before acceptance by the architect. With only a few more construction requirements, a mock-up can often be made into fully testable units at the job site.

Mock-ups, besides allowing for watertightness testing, can be used at the site for instructional purposes. This provides an initial means of interaction for all subcontractors involved, to make them aware of their role in the overall effectiveness of a watertight envelope. This is especially useful in areas where many subcontractors are involved, such as a building parapet and coping.

The subcontractors are able to work together to develop the working schedules, patterns, and quality required to ensure envelope success before actual installation. Such a process can become an actual partnering exercise in any team building, total quality management (TQM), or partnering program undertaken by an owner and contractor.

There are too many positive benefits that can be derived from testing envelope components at a job site to justify not testing. At a minimum, testing should be done immediately after completion of the first building envelope portion during construction. This testing can often reveal deficiencies that can be corrected and eliminated in the remaining areas of construction. Testing can also reveal potential areas of cost savings, better materials, or details that can be incorporated into the remaining envelope portions. (See Table 10.3.)

FIGURE 10.15 Mock-up that can also be used for color and texture approval. (*Courtesy of Architectural Testing, Inc.*)

TABLE 10.3 Job-Site Testing Advantages and Disadvantages

Advantages	Disadvantages
Testing can be done under actual construction conditions at a project site	Extreme testing conditions of testing with mock-ups are not possible at job sites
More details of terminations and transitions can be included in actual testing	Larger envelope portions are difficult to test accurately at the job site
Costs are lower, since mock-ups are not necessary to construct	Tests are often completed after construction when problems occur

MASONRY ABSORPTION TESTING

All masonry and concrete claddings absorb water and moisture, requiring appropriate divertor systems such as dampproofing and flashings to be installed to prevent water infiltration to interior spaces. Since it is expected that these finishes will absorb moisture, they

FIGURE 10.16 Completion of field test by laboratory testing firm. (*Courtesy of Architectural Testing, Inc.*)

will not typically need to be tested unless the cladding is suspected of contributing to water infiltration due to excess water absorption.

ASTM E 514, "Standard Test Method for Water Penetration and Leakage through Masonry," is available to measure the amount of water penetrating or being absorbed by masonry claddings. While the test provides for measuring the water entering a mock-up masonry panel, ASTM provides no guide in determining what is excessive water penetration or exceeds the limits of typical divertor systems. Normal industry standards can be estimated to include excellent to good ratings for a masonry absorbing rate of no more than $^1/_2$ gallon per hour. Poor results are estimated at any rate above 1 gallon per hour.

Most absorption will occur at the masonry joints rather than through the masonry units. Therefore, laboratory testing of a specific masonry cladding will rarely duplicate the actual in-place conditions. When actual conditions are suspected of contributing to water infiltration, testing can be completed at the site to determine if the masonry and masonry joints are yielding too much water penetration for the divertor systems to manage properly.

This field test can be done simply by a water hose test as described in Chapter 11; or, if accurate measurements are necessitated for whatever reason, a "MAT (masonry absorption test) Tube" test can be completed. The MAT tube consists of a calibrated test tube that is attached directly over existing mortar joints and filled with water; the amount of water absorbed during specific time intervals is then recorded. Figure 10.18 details a typical MAT Tube.

The tubes which holds 5 milliliters of water, terminates in a circular flat bottom that is attached to the masonry using sot putty. The circular area provides a surface test area of

FIGURE 10.17 Job-site testing of curtain wall and precast units. (*Courtesy of Architectural Testing, Inc.*)

one square inch. This rather small test area requires the MAT test to be repeated in numerous locations on the masonry wall. Included in the testing should be specific areas of mortar joints, including the top, middle, and bottom of head joints and bed joints, as shown in Fig. 10.19. Test results should be plotted against time and joint location, and summarized for review.

Again, there are no standard results pinpointing excessive moisture absorption using a MAT test. However, if a joint absorbs 5 ml of water in less than 5 minutes, that is contributing to leakage that might exceed the divertor system's capability. Measurements exceeding absorption of 5 ml in 5 minutes necessitate a review of corrective measures to be taken to repair the mortar joints, including those described in Chapter 7.

SEALANT JOINT MOVEMENT

If an envelope is experiencing sealant joint failures, it may be necessary to measure the movement occurring at the joint to determine if it exceeds the capability of the sealant material. Sealant manufacturers can provide a relatively simple device to accurately measure joint movement, as shown in Fig. 10.20.

The joint movement indicator base (with an adjustable setscrew) is firmly attached to one side of the joint, and the opposite side receives a scribe plate. The setscrew will etch a record of joint movement onto the scribe plate, (Fig. 10.21). Typically the movement indicator is left in place for a sufficient period of time, an entire weather season, to record both expansion and contraction movement and thus provide a total amount of joint movement occurring.

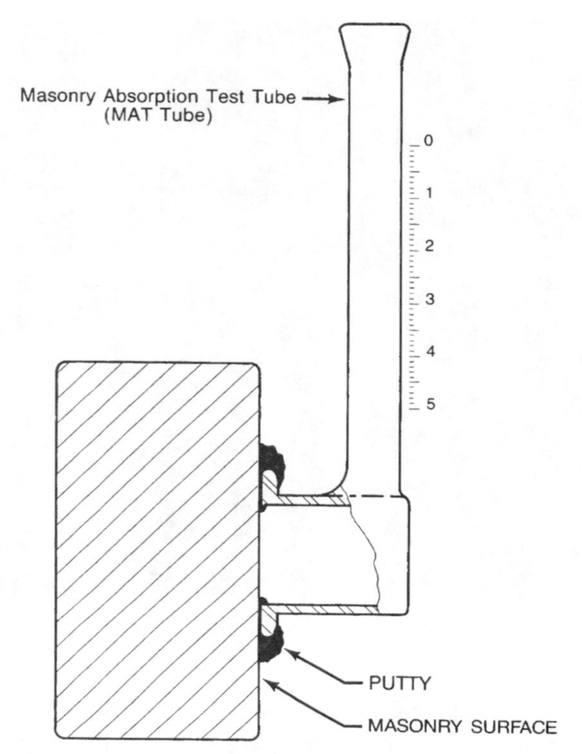

FIGURE 10.18 Detail of mat tube attached to a masonry surface. (*Courtesy of Saver Systems*)

To determine the total amount of joint movement, divide the measured amount of movement on the scribe line toward the joint (thermal expansion in the substrate makes the joint smaller) by the original joint width. For example, if a joint scribe line is $^3/_8$ inch and the joint was originally $^1/_2$ inch, movement was .375/.5 = .75 or 75% expansion.

Then measure the amount of scribe line movement away from the joint to find the contraction movement (joint becomes larger), and divide by the original joint width. For example, $^1/_4$-inch scribe and $^1/_2$-inch original joint width, movement was .25/.5 = .50 or 50% expansion.

The total joint movement in the above situation would be .75 + .50 = 1.25, or a total of 125% movement. So, if the joint was designed for 100% movement, it is likely that the actual movement at the joint exceeds the capability of the joint material.

MANUFACTURER TESTING

Often, manufacturers are depended upon to provide all the testing and information considered for inclusion of their product into a composite envelope. Without proper testing, this can lead to numerous problems.

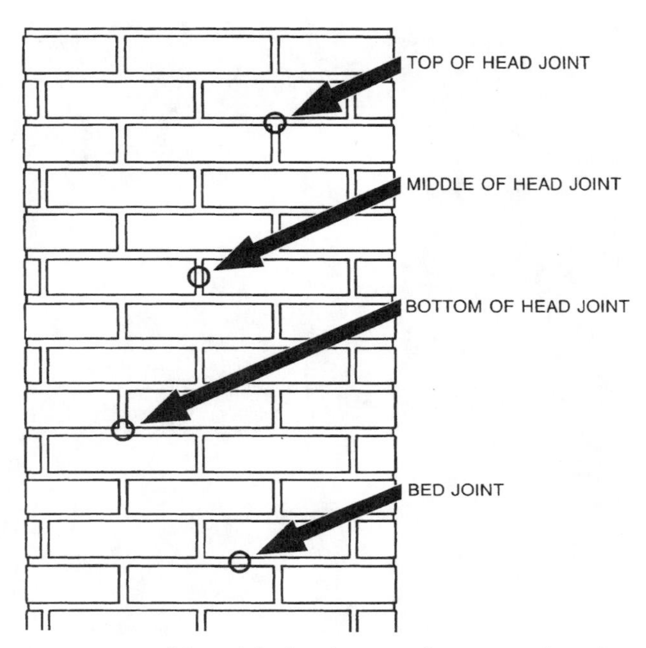

FIGURE 10.19 Mortar joint location types for water testing using mat tube. (*Courtesy of Saver Systems*)

FIGURE 10.20 Sealant joint movement testing. (*Courtesy of Dow Corning*)

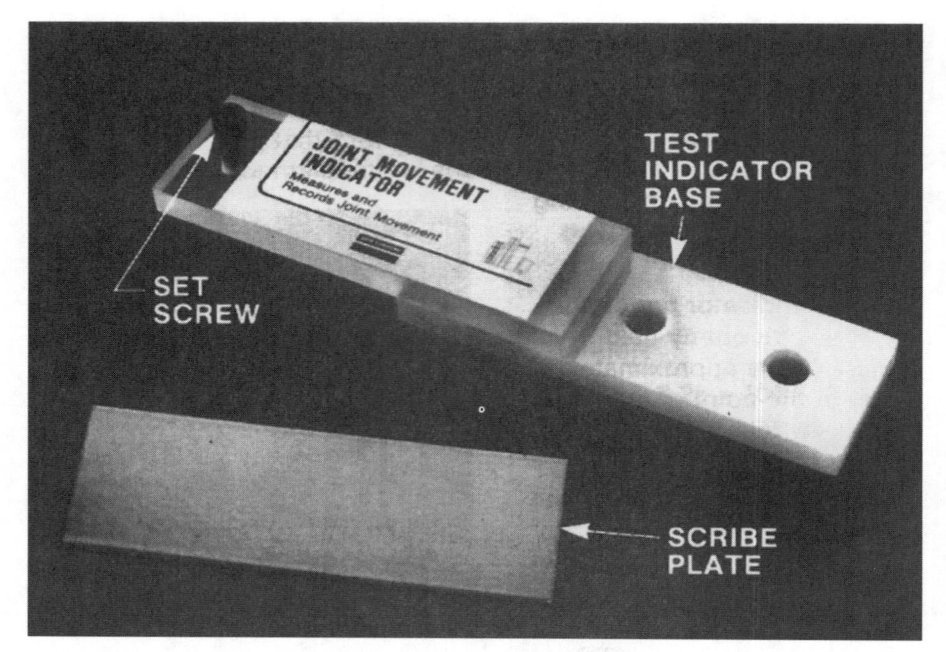

FIGURE 10.21 Components of sealant joint movement indicator. (*Courtesy of Dow Corning*)

Manufacturers are concerned solely with the materials or systems they manufacture. They do not provide the necessary information to properly evaluate their products' usefulness in a composite envelope. Their material is not checked or tested for compatibility with adjacent materials used in proposed envelopes.

The specified termination and transition details are often not those tested or typically used by a product manufacturer. A manufacturer often provides insufficient instructions for incorporating proper details for expected conditions and compatibility of their products with other envelope components.

Most manufacturers will, however, offer detailing suggestions and complete laboratory or site tests, if required to ensure the inclusion of their product in a project. Manufacturers have technical resources available to them that are not immediately available to a designer or building owner.

Manufacturers should become involved in the design and testing process that can bring a project to a successful completion. Their intricate knowledge of their materials or systems and suggestions for termination and transition detailing should be consulted as a basis for preliminary design requirements.

Manufacturers often are capable of providing laboratory analysis and testing of their products under the proposed project conditions. This provides a means to determine acceptability of present design requirements or to suggest alternate designs. It also determines the compatibility of their products with other envelope components.

Manufacturers representing the various envelope components can become involved with the final designing of a composite envelope, reviewing and suggesting revisions to ensure that the proposed detailing will work uniformly for all included products. When testing has been completed at job sites or with laboratory mock-ups, manufacturers should be invited to review the tests and results and to offer their opinions and suggestions.

TESTING DEFICIENCIES

Unfortunately, often envelopes still experience water infiltration after completion of testing. This can be caused by a variety of problems, including:

- Insufficient termination and transition detailing in the test parameters
- Repair of defects found during testing by insufficient methods, including sealing with low-performance sealants
- Substitution of products, materials, or systems after the completion of testing
- Testing that does not reveal the long-term incompatibility of products, resulting in short life cycles or water infiltration
- Long-term weathering cycles not included in the testing
- Expected detrimental elements such as acid rain, road salts, ultraviolet weathering, and freeze–thaw cycling not being included in the testing
- Actual field conditions not duplicated in the laboratory or mock-up testing; e.g., water at the actual job site containing chemicals detrimental to masonry admixtures, and dry weather preventing proper curing of mortar
- Mock-ups constructed under laboratory conditions not possible at the site, including expertise of mechanics working on the actual building envelope
- Performance requirements not as demanding as required by actual job-site conditions; e.g., wind loading (especially at upper building portions), thermal movement, and wear and durability required at locations such as loading docks
- Mock-ups not accounting for structural loading or settlement that will occur in actual building conditions

One major problem with envelope testing is a lack of standardized tests that allow review of typical transition and termination details. Further, there is a lack of testing designed specifically for waterproofing products. Tests supplied or used for waterproofing materials and systems are often those applied to roofing materials or other envelope systems.

The fact that water infiltration continues to plague building projects is evidence that either insufficient testing exists, or that tests and their results are often not considered seriously enough to warrant proper resolution of problems. All too often, mock-ups are eventually made to pass requirements once sufficient quantities of sealants have been applied.

To ensure that test results are properly used, the following procedures should be followed:

- Any infiltration that occurs during testing should be carefully documented.
- Determination of the leakage cause should be completed before attempts at repair are completed.

- Repairs or redesigns incorporated into an envelope should be reviewed for compatibility with other components, and their life-cycling must be adequate and equal to that of other components.
- Mechanics and supervisors should review proposed redesigns or repairs and be aware of their importance.
- After completion of repairs and redesign, envelopes should be retested to ensure their adequacy.
- Manufacturers should be consulted, to approve use of their materials and of any redesign or repairs.
- Warranties should be reviewed, to ensure that they are not affected by repairs or redesigns.

Proper pretesting and resolution of design or construction flaws can prevent most of the problems that occur after completion of a building envelope. Successful testing must include adequate representative portions of all terminations and transitions incorporated into an envelope design. It is also mandatory that any leakage be reviewed and properly repaired, to ensure the longevity and compatibility of the repair method.

CHAPTER 11
LEAK INVESTIGATION AND DETECTION

INTRODUCTION

Whenever an existing structure is experiencing water infiltration, there are standard measures that can be taken to determine the source of leakage and make appropriate repairs. Investigating and pinpointing building envelope breaches does not require scientific measuring or expensive equipment, and the steps required to complete an investigation are not technically difficult. By applying a few basic guidelines, determining the area(s) of intrusion can be addressed by most anyone.

Whenever leakage is occurring, it is imperative to recognize that it is very likely that the 90%/1% and 99% principles are contributing to create the problem. The 90%/1%, principle as described in Chapter 1, recognizes that the majority of leakage will occur at the terminations and transitions within the building envelope and not directly through the water barrier or divertor systems themselves. In addition, the 99% principle recognizes that in only 1% of the cases it is the material or system that will have failed or is causing the problem, as compared to a 99% chance that it is related to the original installation labor techniques.

If the leakage is attributable to causes that do follow these principles (e.g., the 1% chance that it is a material failure), the cause of leakage should be so obvious that no investigation is necessary. For example, a material failure should be easily observable in the form of material completely disbonded from the substrate, uncured material, brittle or cracked material, or other obvious signs of failure. If, however, the cause likely falls within the principle guidelines, then the material might have been applied too thin, transition detailing between different envelope systems been inadequate, transitions from a horizontal to vertical substrate improperly installed, divertor drainage means clogged, or a variety of other similar problems not directly associated with any specific envelope component or system.

LEAKAGE INVESTIGATIONS

When the leakage is being caused by one of these waterproofing principles, the process in determining the source of infiltration usually requires several important and progressive actions. These measures determine not only the area of leakage but also the cause and contributing factors that must be corrected to eliminate the infiltration completely. A leak detection process should include the following actions to adequately locate and address the problem:

- Reviewing any available records, documentation, or information on leakage problem
- Original construction document review
- Inspection
- Testing
- Investigation
- Remedial action plan
- Corrective measures implementation

These steps should be followed in the above order, to conduct an efficient and effective program that leads to the problem being corrected and the envelope properly functioning for an extended life cycle.

Reviewing leak documentation

In most situations leaks will have been documented in some manner, and this information should be reviewed first to determine if the situation could be narrowed to specific portions of the envelope. This information can be particularly advantageous if it is very specific as to where the leaks appear and under what conditions such as "heavy, wind-driven rains from the east."

This documentation often provides a specific area where the leaks are occurring but not necessarily related to where the infiltration begins. It is appropriate to ask building occupants to be as specific as they can about the leakage; for instance, instead of "leaks in suite 250," reporting such as "leaks running down from ceiling tiles above far NE window, starting immediately with any rainfall."

Water always seeks the path of least resistance into a structure, and this often results in water entering and traveling along structural elements of the building before entering occupied spaces. This could include water entering through cracked precast units, running along the structural supports and into the occupied areas several feet away from the actual point of entry.

The information provided should be used as the foundation for performing the additional tasks necessary to pinpoint the leakage and make necessary corrections. This preliminary information should narrow the required investigation and study by

- Providing a general location of infiltration
- Enabling one to determine an initial plan of action for further study
- Determining how serious the problem is by the quantity of water documented as intruding into the interior spaces
- Determining if outside support or consulting services are necessary to locate and solidus or correct the situation.

The initial survey of leakage documentation is often very basic; however, it is useful as an adequate starting point to develop the resources necessary to plan corrective measures without wasting unnecessary time or costs. For example, if the leaks are minimal and evidently from poor maintenance, such as clogged roof drains, then action can be taken without performing further testing, hiring consultants, or spending money on repairs that are not necessary. This preliminary documentation can provide sufficient evidence that suggests outside assistance is necessary and if emergency repairs are necessitated to prevent further damage and related escalating costs associated with this damage.

The initial leak reports are usually not sufficient to immediately determine the cause of leakage. It often takes further investigation, including inspections and testing, to provide sufficient knowledge for effective decisions on how to best make necessary repairs. The original construction documents also can provide important clues to the leakage cause(s).

Document review

Prior to the actual inspection, a through review of all available construction and as-built documents should be made, highlighting the general area of the leakage. Make note of any questionable termination and transition detailing of all envelope components. Determine the barrier line (described in Chapter 8), and specifically note if it was properly "closed" and hence ensuring that all individual components of the envelope are transitioned into the adjacent component or system with watertight connections. If the original documents do not clearly detail transition requirements, this should become an area for further visual field inspection.

A basic guide to the review of documents should include

1. Reviewing all individual envelope components. At the building elevation in question, the as-built drawings should be used to determine each of the individual components used in construction. Decide if any are likely contributors to the leakage that require visual inspection in the field. For example glass is not a likely contributor, but precast panels that have cracked can cause leakage. Highlight all individual waterproofing and roofing systems used in the envelope construction. Make note of each of these systems for field inspection. When reviewing these systems remember the 99% principle: that 99% of leakage is attributable to installation problems and not to actual waterproofing or roofing system failures.
2. Determine the barrier line. Determine if the documents required a weathertight envelope barrier or if there are areas of concern. Do the envelope transition barrier systems to divertor systems function properly? Document all areas that require further study in the field inspection.
3. Study termination and transition detailing. Keep in mind the 90%/1% principle, that 90% of all leakage can be attributed to 1% of the envelope area. Highlight areas that have no specific detailing provided for in this 1%; these areas should then be carefully inspected in the field. Also, make copies of all provided transitional detailing to ensure in the field inspection that it was installed as required.
4. Highlight all waterproofing secondary or backup systems specified in the original documents. These systems usually can not be inspected visually in the field without completing destructive testing that requires the removal of some or all of the envelope components to determine if these secondary systems were properly installed. Noting where these systems should have been installed can assist the inspector during the field inspection if they are related to the actual area of water infiltration. For example, in Fig. 11.1 note the multiple secondary systems, including the membrane flashing around the pipe penetration and the sealant joints under the vent cover. Neither of these systems would be evident in a visual inspection of the building envelope, but they play important roles in waterproofing at this 1% of the envelope area.

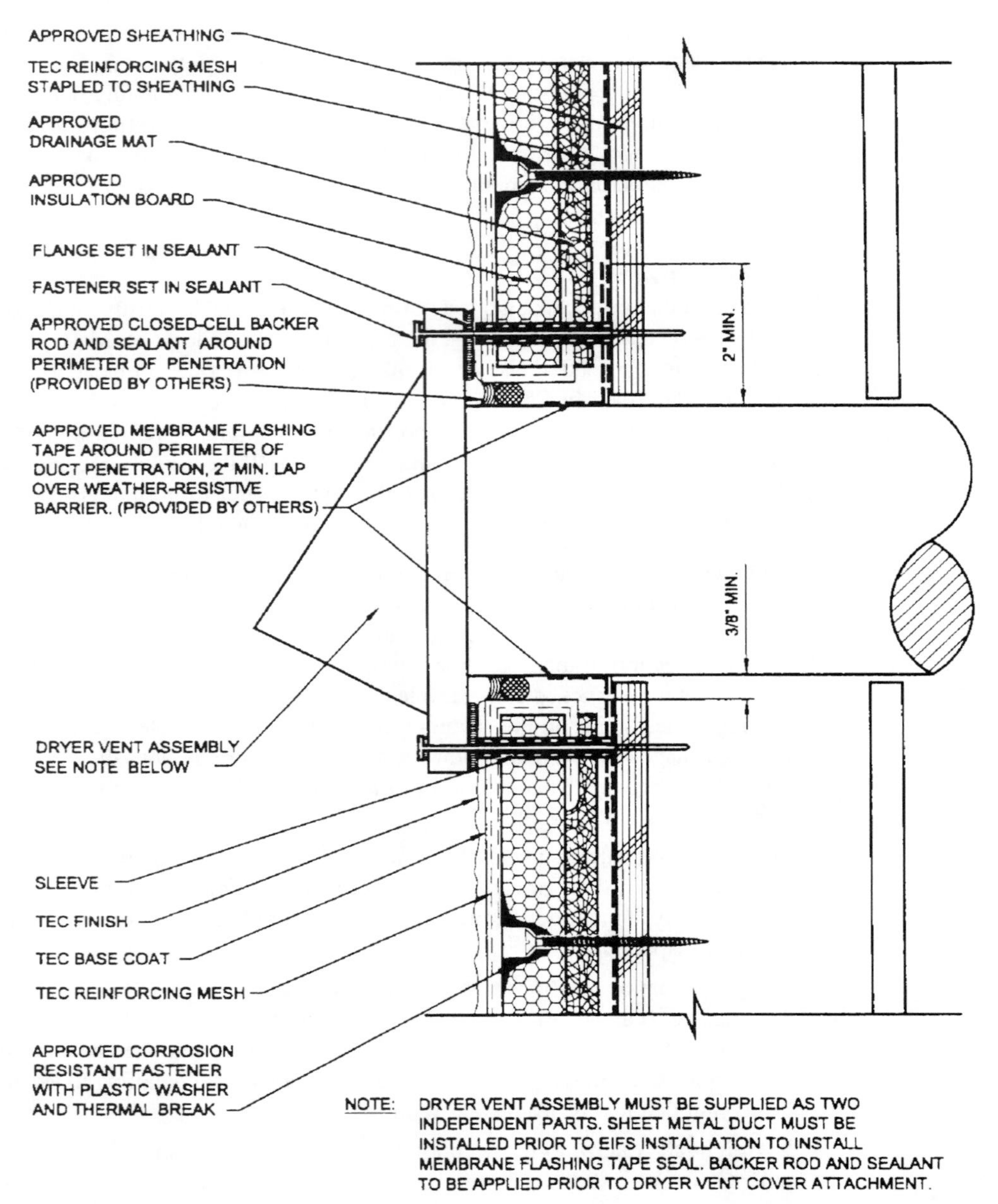

FIGURE 11.1 Construction detailing that might require destructive testing to determine the actual cause of water infiltration. (*Courtesy of TEC Specialty Products, Inc.*)

5. Make note of all unusual construction techniques. Often, designs require installations that local craftspeople cannot duplicate properly during construction. These areas might include highly decorative envelope finishes such as copper standing seam copings that should be carefully inspected in the field for contributing to the existing leakage.
6. As appropriate, review the structural, mechanical and landscape drawings. These documents might provide some insight into the causes of leakage, such as mechanical protrusions in a planter that have inadvertently damaged the waterproofing membrane.
7. If available, review shop-drawing submittals. These provide insight into the systems and products used in the original construction, including any warranties and product capabilities. The shop drawings also might be useful in determining if incompatible systems were installed, such as butyl sealants in contact with urethane membranes.

Once all available construction documents have been carefully reviewed and specific notes taken relating areas requiring further study, a field inspection can begin. The document notes should be taken on the field inspection, as well as any drawings or details that need to be compared to actual in-place conditions.

Inspection

After the records pertaining to reported leakage have been reviewed, a visual inspection is in order to determine what are the possible causes and where they might be located. A visual inspection can provide immediate evidence of possible leakage causes, but in many cases testing is required to either verify the cause or actually determine where the envelope has been breached.

Visual inspections will often provide evidence of the contributing factors of water infiltration, including failed sealant joints, faulty or cracked mortar joints, improperly functioning transition or termination detailing, and clogged drainage systems. When a visual inspection reveals these evident failures, corrective measures might be planned immediately and the area previously prone to leaks monitored to determine if the corrective measures have resolved the problem.

In many situations however, the visual inspection will not provide immediate evidence of obvious failure or breaches in the envelope. This is especially true with divertor systems, such as flashing systems, which allow water to enter the envelope but then divert the entering water back out to the exterior. In this type of construction, visual inspections are not able to investigate the actual components of the divertor systems since they are hidden behind the building envelope facade components.

Should leakage be occurring in such areas, either a water test has to be conducted to confirm that the leakage is occurring within the envelope components, or a destructive visual test must be made. The later is completed by removing portions of the envelope facade to expose the backup or divertor systems. This can be a costly method, and in most cases it is best to first complete a water test in the area to confirm that leakage is attributable to the systems in question before a destructive inspection is commenced.

In similar situations, access to the original construction documents can also provide insight into the causes without having to complete destructive testing. The documents, especially as-built drawings, should provide sufficient details as to the methods used to construct the divertor system without having to remove façconstruction methods used.

Testing can then be commenced to determine if the divertor system is contributing to water infiltration to interior areas.

The documents can also be useful when leakage is occurring at above-grade portions of the envelope that are not readily accessible for inspections. For instance, in multistory buildings, the envelope might not be accessible without scaffolding or swing-stage scaffolds. Documents and visual inspection from interior areas looking out through windows or curtain wall components might be the best initial means to conduct a leakage investigation. If this initial inspection is not sufficient for determining the cause, scaffolding or some means of accessing the area such as a hydraulic lift or available window-washing equipment would be required to complete the inspection. If this is the case, a water test might be scheduled at the same time to confirm any opinions formed from the visual inspection.

Below-grade areas also are difficult to inspect visually, since the exterior portions of the envelope are not accessible. Below-grade inspections are usually limited to visual reviews of the landscaping and site drainage on the exterior. On the interior side of below-grade area, finishes applied to the structure such as drywall also hamper visual inspections. More often than not, some destructive removal of finishes or landscaping is required to view and inspect below-grade leakage problems. Note, however, that destructive removal of surrounding areas can further damage the envelope, particularly on below-grade waterproofing systems that can be easily damaged during backfill removal.

The purpose of any visual inspection is to confirm the construction methods described in the building documents, document any obvious failures in the envelope components, reveal any poor or improper maintenance, and either provide recommendations for repairs that should prevent further leakage or outline the steps necessary for further testing of the envelope to determine the exact cause of leakage. Figure 11.2 provides a detailed guide for use in visual inspections of common building envelopes. It is useful to ensure that all possible contributing factors to the leakage are inspected. The inspection guide can be used to highlight areas that require maintenance or repairs before they also contribute to future weatherproofing problems.

Obvious problems such as deteriorated sealants, insufficient drainage, and cracks or holes in the substrate can be repaired immediately, and the area monitored to confirm that the problem has be resolved. If however, these corrective measures do not resolve the problem and no other obvious possible causes are evident, then testing is required to determine the reasons of water infiltration.

Testing

Testing for water leaks is not a science. Water tests do not require an engineering degree to be performed properly. Only if the situation might result or be involved in legal action would it be necessary for the testing to be monitored by an engineer or consultant who has the credentials to testify in court as to the cause(s) of leakage. Even in this situation, it is often not necessary to spend vast sums of money on elaborate equipment or destructive tests to document the leakage.

Mother Nature does not use calibrated funnels, measured amounts of water, and other elaborate equipment to create the leakage; therefore the same should not be required of the water test equipment. If a leak occurs, the exact amount of water infiltrating during a specific time frame is not useful information to anyone. A water test is necessary only to document and determine where and how the water is bypassing the envelope barrier and divertor systems.

ENVELOPE AREA	North Elevation	South Elevation	East Elevation	West Elevation
Walls:				
Substrate & Backing (e.g. EIFS, stone, masonry				
Flashings				
Divertor systems				
Transitions between substrates				
Waterproofing applications (clear repellents, coatings)				
Directional changes				
Horizontal abutments				
Materials delaminating				
Louvers/AC unit/other penetrations				
Deteriorated expansion/control joints				
Cracked mortar joints				
Efflorescence				
Clogged weep holes				
Deteriorated mortar joints				
Spalling of stone, concrete, or masonry				
Structural cracks in substrate				
Rusting of shelf angles, anchors				
Broken window or curtain wall panes				

FIGURE 11.2 Envelope inspection form.

Roofing:				
Surface alligatoring or cracking				
Disbonding from substrate				
Delamination of plys				
Seam splits				
Roof perimeters:				
Parapets (substrate or structural cracks, failed flashings)				
Coping (cracks, sealant failures, improper flashing)				
Wall and counterflashing				
Reglets				
Pitch pockets				
Mechancial and electrical penetrations, transition detailing				
Protrusion detailing and flashing at pipes, lighting, drains				
Overall drainage				
Scuppers, gutters, roof drains				
Horizontal expansion joints				
Valley, hip and ridge flashings				
Windows/Curtain Wall				
Perimeter joints				
Head, sill, and jamb flashing				

FIGURE 11.2 (*Continued*) Envelope inspection form.

Frame weeps				
Glazing, structural and non-structural				
Sealants				
Threshold				
Internal seals				
Gaskets				
Horizontal Areas				
Perimeter details				
Drainage				
Penetration detailing				
Railings and edge protection				
Column penetration detailing				
Substrate cracks				
Horizontal expansion joints and juncture to vertical				
Other Envelope Highlights				
Vertical expansion joints				
Site and below-grade drainage				
Waterstops				
Landscaping				

FIGURE 11.2 (*Continued*) Envelope inspection form.

Therefore, the best advice is for any water test to keep it as simple as possible. Duplicating rainfall only takes a water hose and sufficient water pressure. Anything else is usually not necessary, and likely used to impress the client rather than being useful in determining the reasons for water infiltration.

It is also important to remember that water tests can be overdone. Massive amounts of water applied on certain portions of any envelope, such as masonry facades using damp-proofing and flashing as a divertor system, can cause water infiltration even if the envelope is functioning properly. If the amount of water applied to the envelope exceeds anything considered as a normal weather cycle, the results are meaningless.

Using the information gathered from the leak documentation, construction documents, and inspection, testing parameters should be outlined, in particular the specific area to be tested. On above-grade portions of the envelope, most testing is done with a simple water hose and spray nozzle that simulates general rainfall. Conducting a water test requires a minimum of two people, one applying the water on the exterior and another inside to determine when water begins infiltrating. Radios should be available to enable the parties involved to talk to each other during the test, advising when water begins entering and when it is time to move testing to another location.

Vertical envelope testing

Testing should always begin at the lowest possible point of the area in question, and move upward only after determining that the lower areas are not contributing to the leakage. The test also should be limited to a controlled area of the surface and not allowed to overspray adjacent components. Figure 11.3 shows the progression of steps in a water test on a typical masonry wall with a punch window.

Testing the lower elevations first before moving up to the window reveals if the masonry wall is contributing to infiltration. Test the window sill area first and then the jambs and then the window head flashing and sealant. Each area of the test should be completed using a specific amount of time, providing time for the water to travel into the structure. This typically takes a minimum of ten to twenty minutes, unless infiltration becomes obvious sooner. If water infiltration does not appear, the testing should move to the next higher elevation.

Referring again to Fig. 11.1, note the importance of starting water tests at the lowest elevations. As an example, suppose that the membrane flashing was the cause of leakage occurring in the building, permitting water that penetrates the EIFS system above from entering into the building rather than being diverted to the drainage systems provided for this water to exit beneath the pipe. It might first appear that water is entering directly at the pipe penetration when in fact it is not. If water testing began above the pipe penetration, it might appear to confirm such an assumption.

If the test is properly conducted by applying water at lower elevations, then moving directly over the pipe, water should not infiltrate the envelope. However, once the water test is raised to an elevation above the pipe, water entering through the EIFS systems is properly diverted to the drainage mat down to the pipe where the membrane flashing that has failed permits the water to enter into the envelope. In this case, a review of the documents in conjunction with the test results should provide evidence that the leakage is probably occurring due to the failed pipe flashing.

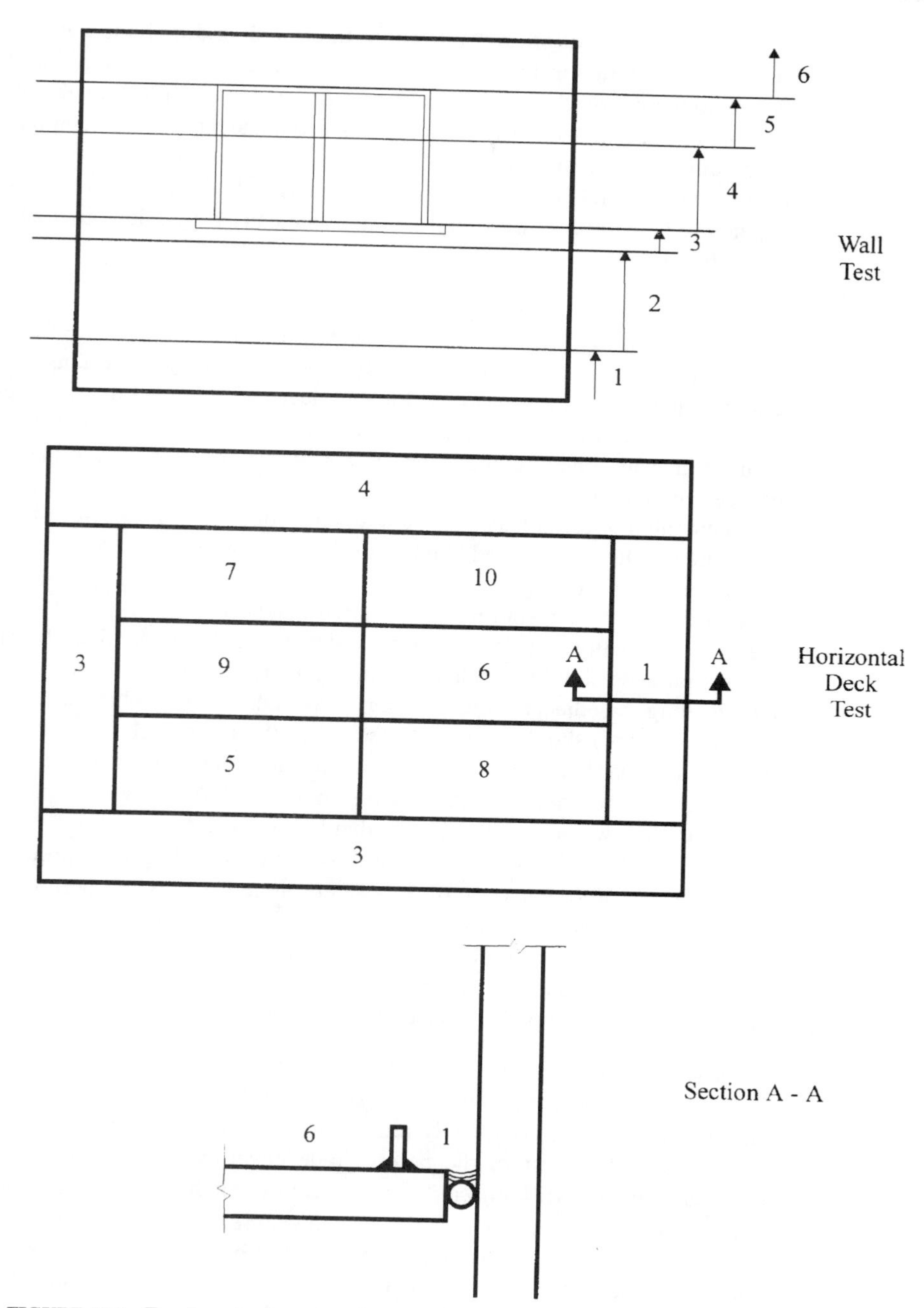

FIGURE 11.3 Envelope testing procedures.

If leakage starts at a lower elevation, the test should be halted until water penetration stops, then restarted at higher elevations to verify that there are not several contributing factors to the interior leakage. If leakage is not evident after testing all areas outlined in the original test parameters, testing should move to adjacent envelope components. Water travels a path of least resistance, and leakage might be entering the envelope through an area far removed from where it appears on the interior. For instance, a leak occurring on what appears to be a vertical wall might actually be caused by a leak in a balcony deck above the wall area.

Horizontal envelope testing

On horizontal surfaces, such a plaza decks, it is often necessary to flood-test the deck to determine the areas of leakage. To perform such testing, deck drains must be completely plugged, including the secondary drainage level of two level drains. Then a specific area must be closed off, usually be laying 2×4 lumber down that is sealed to the deck to prevent water from traveling under the lumber. Water is then added to the area until it reaches a certain height, usually 1 inch. The water should be left standing until water infiltration is documented or for at least 24 hours to verify that water is not traveling from this area to other areas. The areas tested should not be so large as to prevent pinpointing the actual areas causing the leakage.

Usually it is best to start testing of horizontal surfaces adjacent to any vertical envelope area, which usually represents an area subject to the 90%/1% principle. The area sealed off for this purpose should be relatively small, approximately one foot wide by a reasonable length along the adjacent wall. Similar areas in the deck or plaza areas (planters, mechanical penetrations) should then be tested. Then the main deck areas are tested, proceeding in a checkerboard pattern, not testing an area adjacent to a tested area until it has had sufficient time to be dried. Divertor boards should be left in place so that subsequent testing does not allow water to enter an area that has already been tested.

The most difficult areas to test are those with loose-laid waterproofing or roofing systems, or divertor systems. Both of these envelope systems can permit water to travel and infiltrate the interior spaces far from the actual point of breach in the envelope facade. In addition, there might actually be several breaches in the envelope that all contribute to the same interior leak, since water follows a path of least resistance that can be "fed" by water infiltrating the envelope from several sources.

Below-grade envelope testing

Below-grade areas are among the most difficult to test accurately. Often the surrounding landscaping must be completely removed for testing. In such situations, additional damage is likely to occur to the below-grade envelope or waterproofing systems by the removal of the backfill adjacent to the structure. Such damage prevents the accurate documentation of the original causes for leakage, eliminating any proper documentation for legal situations. After such complete removal of backfill, the entire waterproof membrane must be replaced due to the damage caused during excavation. This of course defeats the purpose of testing.

Therefore, it is always best to attempt to pinpoint leakage without having to resort to the destructive removal of any components or adjacent landscaping. Water applied at the surface is likely to penetrate the soil irregularly, and there is no way to control the test parameters.

Further exacerbating the problem is the recognition that water will enter interior spaces through the area of least resistance, this usually being the floor-wall juncture of below-grade structures, far from where the waterproof membrane has been breached. Repairs made from the negative side, interior, of the basement or below-grade areas, are only likely to cause the leak to move to the next weakest point.

Leakage reporting can often provide answers to problems associated with below-grade areas. If the leakage is documented as starting shortly after or immediately during rainfall, leaks may be caused directly by the rainwater and surface runoff compounded by poor or inadequate drainage. In this case the site conditions should be closely inspected during rainfall for evidence of ponding water, slow drainage away from the building, or clogged drains. Leakage in the envelope might be at higher elevations, possibly at the transition between below-grade and above-grade envelope components.

If the leakage begins after rainfall, leakage is probably being created by a rise in the groundwater level. In these situations, the leakage is likely to be at or near the lower portions of vertical areas, most probably the wall-to-floor intersection. This leakage is often evident by leakage through the interior near this juncture, often appearing behind interior baseboards. It is unlikely that the leakage can be pinpointed without substantial damage being done to the existing site conditions and positive-side waterproofing systems.

If sufficient information cannot be recognized from the leak reporting, construction documents, and whatever visual inspection is possible to determine a reasonable cause for the below-grade leakage, it is highly likely that a complete repair to the areas affected is required. This would eliminate any specific repair to only the area causing the leakage, but is likely to be as cost effective as destructive testing, which will likely require the complete replacement of the waterproofing system in any event. If sump pumps are present (frequently in residential construction) and no water is appearing in the sump, this is a likely indication that the below-grade drainage systems are clogged and prevented from directing the water to the sump area for removal. It is recommended that an attempt be made to clean these drains out first and monitor any improvement in the leakage after the sump pump is functioning properly.

In these situations, if testing provisions cannot be created to pinpoint the leak cause, rather than testing, it may be best to proceed to the investigation and remedial-action planning steps. Often in below-grade areas, this means recommending negative-side repairs or additional drainage applications to move water away from the structure before it can travel to the interior areas. These remedial systems are presented in Chapter 7.

Destructive testing

In certain situations however, destructive testing is the only applicable means to correctly determine the cause of water infiltration and permit the proper repair method to be selected. Destructive testing involves the removal of the outer layer of the envelope to expose interior components for inspection and testing.

Destructive testing is typically only required when divertor systems are involved. Divertor systems involve envelope components that permit water to enter that is later redirected back to the exterior by a combination of dampproofing and flashings. Masonry facades, EIFS water drainage systems, and curtain walls are examples of divertor systems. It is often difficult to pinpoint the exact cause of failure in these systems without removing the primary barrier to expose the divertor systems.

Figure 11.1 presents a situation where destructive testing might be necessary. Assume again that the pipe membrane flashing has failed, permitting water entering the EIFS drainage system to enter the envelope at this point. If the construction documents were not available it would be difficult, if not impossible, to determine the exact cause of leakage without removing the dryer vent cover, sealant, and possibly even a portion of the EIFS system at this area to inspect the membrane flashing.

Additional information on destructive testing is presented in Chapter 7. Obviously, it is recommended that you limit the amount of destructive testing completed on any envelope due to the difficulty in repairing the test area for watertightness and aesthetics. Testing equipment is available that can be used to assist in leak investigation and possibly prevent the need for destructive testing.

TESTING EQUIPMENT

There is a variety of testing equipment available for detecting leakage. Most of this equipment will verify the presence of leaks, but few can pinpoint specific areas of envelope failure. The equipment is often used in maintenance reporting, tracing any water infiltration into the envelope before it causes damage to interior areas.

Equipment includes moisture meters that can register the percent of moisture content in any substrate that can be probed. Moisture meters generally have two sets of metal, needle-like attachments that are inserted into a substrate to detect moisture. These meters are applicable to substrates such as wood, stucco, EIFS, paints, and other similar types of cladding. They do not work with stone, curtain wall, masonry, precast substrates.

While providing evidence of entrapped moisture and the relative amount of moisture, these meters cannot pinpoint the exact cause of leakage, only the areas affected by moisture. The meters can be useful to document the extent of leakage and limit the inspection for the problem to a small area by outlining where in the substrate moisture is present.

There are also a variety of thermographic infrared testers available. These tools can reveal the presence of entrapped moisture in envelopes and they are widely used in roof inspections. The equipment measures the amount of heat emitted from objects. The equipment is often used at night, when entrapped water retains heat accumulated during the daylight hours and releases it more slowly than surrounding areas. A warm zone on the measurements reflects the presence of water.

The equipment is used extensively on roofs, especially when single-ply or multi-ply asphalt materials have been used that permit water to travel beneath the roofing material and substrate. The infrared equipment can be used to locate the extent of water entrapped, and narrow the search for the actual breach in the membrane to a smaller area. The equipment can be used on all portions of the envelope, including vertical surfaces and plaza decks or balconies. While the equipment cannot pinpoint the cause of leakage, it can be useful to narrow the focus of the search to a limited area.

Nuclear testing equipment is also available, typically for use on roofs or plaza decks or other horizontal areas that might contain trapped moisture. Nuclear equipment works not by taking an X-ray of the substrate, but by sending a signal into the deck which is capable of measuring the hydrogen atoms that water contains. Since it is capable of measuring the amount of H atoms, the equipment is capable of providing an accurate reading of the amount of water in each specific grid of area tested. This can be useful in narrowing the search of the leakage problem.

All this equipment is useful in detailing the extent of moisture entrapped or present in the substrate, but cannot pinpoint the actual cause for water infiltration. The equipment must be used in conjunction with a visual inspection of the actual substrate conditions and appropriate water testing to verify the actual cause of water infiltration. There are techniques available and adopted from other industries that can assist in pinpointing leakage that is not capable of being visually inspected. This includes fiber-optic endoscopes.

Endoscopes have a tiny camera attached to the end of a flexible cable that permits the probe to be inserted directly into envelope components with minimal or no damage. The camera relays the picture back to a portable viewing station that can be adapted to tape the inspection on a video (VCR) format. The tape can then be viewed later, comparing it with the construction documents in addressing the leakage problem.

This equipment can be used to inspect the inter-wythe of a masonry wall without removing the outer masonry units to gain access for inspection.

Other envelope components that are multilayered can be accessed by the probe with minimal destructive damage, including curtain walls and their anchoring and natural stone facades applied with metal supports.

It is possible that the endoscope can be used to inspect below-grade exterior surfaces without necessitating the complete removal of backfill or other landscaping. Small bore holes adjacent to the envelope can be made, then lowering the probe into the hole. If necessary, the hole can be flushed with water to remove soil and debris from the substrate prior to the endoscope inspection.

TEST RESULTS

As the test progresses, all procedures and results should be documented for review. If the test is to be used as evidence in legal situations, videotaping can be used to document test procedures and results. Once an area begins to leak, the test should be stopped until water infiltration stops, then retested immediately to certify the leakage.

The documentation of the tests does not complete the leak detection process. All information gathered, from the initial leak reports to the test results, should be accumulated and compiled for further investigation. All written documentation, pictures, and recordings including video should be used together to verify the exact causes of the leakage and remedial planning.

INVESTIGATION

Testing does not necessarily confirm the cause of leakage, as it may only have isolated the specific area where the envelope has been breached. This area, though, may include several different envelope components, transition, or termination detailing, or involve divertor systems. To fully understand the cause of leakage and make appropriate remedial decisions, including steps to prevent the same situation from occurring at other similar areas of the envelope, an investigation and final determination of leakage and envelope breaches should be compiled.

For example, a precast facade might be first thought to be leaking only through failed sealant joints, but under investigation procedures it might be determined that a secondary seal or backup joint should prevent infiltration at these areas. The resulting investigation might lead to the conclusion that crackage in the precast substrate is permitting water infiltration that bypasses the joints into the structure. The investigation would then produce a remedial action plan that includes not only resealing the primary joint seals, but also the repair of precast cracking and application of a clear repellant to the entire precast facade. In another situation, the transition detailing might be found insufficient in certain envelope areas, and recommendations made to repair or replace all similar detailing throughout the envelope for long term life-cycle maintenance. This total plan, based on a thorough post-testing investigation, would not only resolve the current problems but also prevent the occurrence of similar leakage in other envelope elevations.

The personnel conducting the inspections and testing might not be comfortable in conducting the investigation alone. In this case the complete package of documentation can be provided to consultants, waterproofing manufacturers, product distributors, or engineers, to study and to provide their recommendations and outline of remedial actions.

It would be advisable to have the documentation reviewed by several different sources and review each of their recommendations. Manufacturers reviewing the situation might look for answers that involve selling their product; this might an acceptable solution, but it would be best to have a confirmed second opinion before expending money to implement the repair procedures.

The investigation can be useful to determine several alternative repair methods, including temporary fixes and long-term repairs and associated maintenance to improve the life cycle of the entire envelope. The building owner can then make appropriate management decisions on the actions that best fit their immediate and long-term requirements.

The investigation can also be used to determine if the problem has occurred in the past or is likely to repeat itself in other areas. Whatever the results, the complete reports created during the investigation should be carefully documented for future reference if necessary. The documentation collected should be reviewed in conjunction with Chapter 8, which reviews terminations and transitions, the likely cause of the majority of leakage problems in structures of all types.

REMEDIAL ACTION PLAN

From the investigation, a remedial action plan should be outlined. While in many cases only the current leakage problem is addressed, it is recommended that the entire envelope be reviewed to prevent similar problems from occurring elsewhere in the future. Certainly it is less costly to take a proactive remedial action approach than to have to continually spend the money necessary to correct not only the leakage but the resulting damage to building components and interior contents. The remedial plan should also address any routine maintenance steps that should be taken to extend the life of all existing as well as new envelope components.

The remedial action plan should outline the repairs necessary, listing all systems and materials necessary. Often the assistance gathered from the manufacturers, distributors, and representatives in the investigation stage is used to create a guide specification for the remedial repairs. The remedial plan should be very specific, outlining the actual envelope renovations and the required materials and systems necessary in these renovations.

The plan might include only written descriptions in the form of guide specifications that can be completed with "in-house staff" or used as a basis for receiving bids and proposals from waterproofing and roofing contractors. Scaled drawings might be necessary if the action plan is complicated, to ensure that the work is completed properly.

The firms offering assistance in the investigation stage are likely to offer to prepare these documents for their clients, especially if their products are included in the remedial action plan. Should a contractor not involved in the investigation stage desire to substitute methods, systems, or products, these should be carefully reviewed to ensure they meet the intent of the original action plan and can successfully provide the protection required.

Chapter 7, Remedial Waterproofing, can be used as a resource in planning a remedial action plan. Chapter 14 provides a list of manufacturers that also can provide assistance in both the investigation and remedial planning stages. Chapter 13 provides a series of guide specifications that can be consolidated into the specifications for materials or systems used in the repair procedures.

CORRECTIVE MEASURES IMPLEMENTATION

Once a remedial action plan has been completed, it should be implemented as soon as possible to prevent further damage to the envelope. Building owners should carefully review the proposal received, to correct the situation if outside assistance is required. Consultants and material distributors can be helpful in preparing the bid and proposal documents for a remedial action plan.

Selecting incompetent or inexperienced contractors might prevent the work from being completed properly or even create additional problems through damage to surrounding portions of the envelope. Chapter 10 provides a standard qualification format that can assist in selecting qualified contractors to complete the repairs. This form should be requested from every potential source for completing the required fieldwork.

Once potential bidders have been "short-listed" by reviewing the qualifications of potential offerers, a bid package must be prepared to ensure that all contractors are bidding on the identical scope of work. Otherwise, the building owner might find themselves comparing "apples to oranges," and not being able to accurately compare the proposals and prices. A general outline of a proposal package should include

- A general description of the scope of work that includes the results the owner expects after the work is completed.
- Insurance and bonding requirements for the contractors.
- Outline of who will supervise the contractor's work, such as in-house staff or outside consultants, including their rights to inspect and reject work and approve payment applications.

- A payment outline including retainage to be withheld.
- Detailed listing of the repairs to be made, using specifications, drawings, and details for each specific item.
- A listing of acceptable products that will be permitted for use on the work. Note specifically what if any substitutions will be approved.
- A list of procedures and standard references to clearly define the quality of the completed product.
- Requirements for any mockups and testing before work can commence.
- Specific schedule requirements, including any penalties for late delivery of the project.
- Work condition outlines including work times, tenant disruption avoidance, cleanup requirements.
- Warranties required, including any bondable guarantees.

Owners should carefully examine warranties before the work is released. The contracting or implementation stage should address the guarantees required for any work performed. These warranties should cover both labor and materials, and be bonded or insured by the material manufacturer as presented in Chapter 10.

Once the remedial work is completed, it should be carefully monitored to verify that the intended results have been created and to determine if any further action is necessary to properly maintain and protect the building envelope. All documents from these actions should be maintained, to facilitate any future investigations or envelope investigations required in the future.

CHAPTER 12
SAFETY

INTRODUCTION

Owners, contractors, and employees must abide by construction safety regulations that prevent unsanitary, unsafe, and hazardous working conditions that affect an individual's health and safety. Disregard of government regulations may result in punishments ranging from fines to imprisonment. What is more important, improper working conditions can cause death or severe injury to employees and bystanders (Fig.12.1).

Laws and regulations that govern the use and installation of a waterproofing system must be thoroughly researched. Such regulations are available from government offices that implement policies, including

- Occupational Safety and Health Administration (OSHA)
- Environmental Protection Agency (EPA)
- Department of Transportation (DOT)
- State and local regulations

These agencies have specific and detailed regulations for the use and installation of building envelope materials, general construction practices, and hazardous chemical use. Field inspectors ensure compliance with regulations and are empowered to levy fines, violation notices, and penalties that can lead to prison sentences. For complete protection, contractors completing work must have written and enforced safety policies and hazardous waste programs, ensuring compliance with government regulations.

OCCUPATIONAL SAFETY AND HEALTH ADMINISTRATION

OSHA has set extensive standards for occupational safety and health in all workplaces. Specific OSHA regulations and enforcement policies cover all construction worksites, including renovation of existing structures. No contractors are immune from OSHA guidelines. Regulations for the construction industry cover a broad scope, including:

- General safety and health
- Environmental provisions
- Material handling, storage, use, and disposal

FIGURE 12.1 Poor safety procedures can often result in accidents. (*Courtesy of Western Group*)

- Ladders and scaffolding
- Personal protective equipment
- Fire protection
- Signs, signals, and barricades
- Hand and power tool requirements and usage
- Welding and cutting
- Electrical wiring and equipment
- Floor and wall openings
- Cranes, elevators, and lift hoists
- Motorized vehicles and equipment
- Excavations, trenching, and shoring
- Demolition
- Explosives
- Power transmission
- Rollover and overhead protection

Whereas all these standards may relate to a specific installation, the first four are most important for waterproofing systems and building envelopes. The following data present a general review of these regulations but are not meant as a substitution for a complete review of current regulations.

General safety and health provisions

General safety and health regulations include specific requirements for first aid and safety training of personnel. Minimum standards for first aid equipment, including adequate fresh drinking water and sanitation facilities, must be at job sites and readily accessible. Emergency numbers, such as those for hospitals, must be posted in a conspicuous place. Illumination levels, sound levels, and requirements for protection of chemical gas vapors and dirt are specifically regulated. These provisions also include standards for handling and removal of asbestos.

Personal protection

The OSHA regulation section on personal protection contains specific requirements for personnel safety and storage of fire hazardous materials. Requirements include hard hats, eye and face protection, and respiratory protection. Fire protection and firefighting equipment must be provided at all projects where hazardous materials or systems are being used.

Specific regulations cover the types of fire extinguishers required and storage requirements for fire-rated materials. These materials are referred to as *red label materials*, because of the red warning labels attached. Common waterproofing systems requiring warning labels include sealants, solvents, and deck coatings.

Signs, signals, and barricades

Regulations for signs, signals, and barricades are for the safety of construction personnel and pedestrians near or on a construction site. Signs warning of specific dangers are regulated as to size, lettering, and colors. Barricades are required to deter the passage of unauthorized pedestrians or vehicular traffic into a dangerous area.

Material handling, storage, and disposal regulations

Material handling, storage, and disposal regulations cover requirements during material use and storage at job sites and the proper disposal. Also regulated are lifting and rigging equipment used to transport materials vertically, and the means by which materials are placed into appropriate refuse containers. Specific regulations for disposing of hazardous materials are governed by the Environmental Protection Agency, not OSHA.

Ladders and scaffolding

Ladders and scaffolding constitute an important part of OSHA standards, as an issue causing frequent accidents and accidental deaths. The size, type, construction, and placement of ladders are specifically detailed. No ladder should be permitted at or used on a construction site that is not clearly marked as OSHA-approved. Ladders used for access to roofs or other landings must extend 36 in above the landing and be tied securely to structures to prevent slippage or displacement during their use.

Suspended or swing-stage scaffoldings, which present a dangerous working condition, should be carefully inspected. Both scaffold construction and its rigging must conform to safety regulations set forth by OSHA. (See Figs. 12.2–12.4.)

Mechanics must wear a safety harness (safety belts are no longer acceptable) securely attached to a structural building component. Attachment must be independent of scaffolding rigging and be tied to building items such as structural columns. Plumbing stack pipes,

FIGURE 12.2 Proper low-rise rigging and safety procedures. (*Courtesy of Sto*)

FIGURE 12.3 High-rise rigging requiring use of building's structural components.

FIGURE 12.4 High-rise scaffolding supported by building's structural components.

exhaust fans, and similar items will not withstand the force of a falling person or scaffold, and should not be part of the rigging attachment under any circumstances.

This OSHA section also covers the composition and attachment of suspension scaffolding in detail. Actual stage deck construction, including use of toe boards, specification of size and height of back rails, and use of safety netting, is regulated. Scaffolds must be constructed to support a load of at least four times the intended load. Regulations also cover material composition and the size of cable used to support scaffolding, and requirements for attachment to a structure. Cable must carry at least six times the expected loading.

Stack or tubular scaffolding requirements include composition height limitations, planking, and anchoring. Scaffoldings are set on foundations rated to support the maximum loading to be encountered. Scaffolds must be secured at a minimum 30 feet horizontally and 26 feet vertically. Any scaffolding constructed used that is more than 125 feet high must be designed and approved by a registered engineer.

These are just a few of OSHA's regulations. Anyone endeavoring to complete a construction, no matter how minor in size, must abide by all OSHA regulations.

DEPARTMENT OF TRANSPORTATION

The federal Department of Transportation has greatly expanded its regulations affecting transportation and delivery of hazardous materials of more than 1000 pounds per shipment. This is roughly equivalent to two drums of materials, or 100 gallons Since many waterproofing materials are considered hazardous materials, DOT regulations affect their transportation and delivery. Federal auditors, state and city police, and weight-station personnel enforce these regulations. Drivers of transport vehicles, as well as company management personnel, can be found in nonconformance of regulations and can be penalized or imprisoned.

DOT regulations include requirements for both the vehicle in which the materials are transported and the driver transporting the materials. Vehicle requirements include vehicle registration and assigning of registration numbers, the latter being conspicuously posted on both sides of a vehicle.

The purpose of registration is to ensure that vehicles meet DOT guidelines and to provide contact with companies for monitoring by DOT. Vehicles transporting hazardous materials must display appropriate warning placards (e.g., flammable, combustible, and radioactive). These signs are posted only when a vehicle contains such materials; otherwise placards must be covered or removed.

Regulations also require that a vehicle file be maintained. This file includes identification numbers, maintenance records, and pre- and post-inspection records which indicate that any deficiencies found were corrected. Proper shipping records must be available to drivers while transporting materials. All materials must be shipped by DOT-approved methods and in DOT-approved packaging. Vehicles transporting hazardous materials must have adequate safety provisions such as fire extinguishers.

DOT also has strict regulations governing drivers of vehicles transporting hazardous materials. Drivers must have a commercial driver's license, a record of a physical examination every 24 months, and certification of having passed a written examination on the *Federal Motor Carrier Safety Regulations*. They must also pass an employment check, including past employers, and a motor vehicle records check for accident and traffic violations.

Most waterproofing materials are governed by DOT hazardous, flammable, or dangerous categories. This requires a thorough knowledge of applicable laws and required record keeping. Waterproofing contractors should contact the Department of Transportation to receive appropriate information relating to transportation of hazardous materials.

STATE AND LOCAL AGENCIES

In addition to federal regulations, most state and local governments have regulations and enforcement policies unique to their specific area. For instance, state Department of Transportation agencies may add to the federal requirements concerning transporting hazardous systems and other building envelope components.

Local codes enforce hazardous waste collection and related storage sites. Most states require registration for hazardous waste disposal. OSHA regulations are also enforced locally by state and local governments, and local agencies should be contacted for specific regional regulations.

Ignorance of the law is not an excuse for avoiding requirements affecting the use, installation, transportation, and disposal of hazardous, flammable, or dangerous materials. Efforts must be made to be acquainted completely with these laws and regulations, in order to avoid possible penalties and even imprisonment.

MATERIAL SAFETY DATA SHEETS

The right-to-know law protects employees from exposure to dangerous or hazardous chemicals. This law requires material safety data sheets to be published by every manufacturer of such materials and to be provided to all employers for distribution to employees. These data sheets must be readily available at every project, for review by employees to educate themselves on dangers of the material's use and risks to health. Specific safety and handling requirements to ensure the safe and proper material usage are also included.

These safety data sheets have a specific standard (OSHA 29 CFR 1910.1200) and must include the following information:

- Manufacturer's name and address
- Chemical name and family
- Any hazardous ingredients in the material
- Physical properties
- Fire and explosive data
- Health hazard data
- Reactivity data
- Spillage or leak procedures
- Protection information
- Special precautions

Figure 12.5 is a reproduced MSDS sheet using the standard Occupational Safety and Health Administration form 1218-0072. It specifically itemizes the requirements of 29 CFR 1910.1200 in a standard form that allows manufacturers to merely "fill in the blanks." Figure 12.6 details a customized MSDS sheet that still complies with 29 CFR 1910.1200.

All safety precautions listed on sheets must be complied with. OSHA inspectors will not only inspect usage compliance but also ensure that data sheets are available to all employees. Appropriate protection for employees handling hazardous materials must be provided (Fig. 12.7).

Should an accident occur involving a hazardous material, data sheets are given to doctors treating the patient. This allows review of the material's chemical composition and of appropriate treatment options. Additionally, should a material spill or leak at a job site, data sheets provide steps to be taken and means for proper disposal.

ENVIRONMENTAL PROTECTION AGENCY

Any organization that generates hazardous waste must register with the U.S. Environmental Protection Agency and receive an EPA identification number. The EPA issues regulations that monitor the use, storage, transfer, shipping, or disposal of hazardous waste. This agency conducts on-site inspections and is authorized to enforce and prosecute firms or employees that do not comply with such regulations. Recently OSHA has agreed

Material Safety Data Sheet
May be used to comply with
OSHA's Hazard Communication Standard,
29 CFR 1910.1200. Standard must be
consulted for specific requirements.

U.S. Department of Labor
Occupational Safety and Health Administration
(Non-Mandatory Form)
Form Approved
OMB No. 1218-0072

IDENTIFY *(As Used on Label and List)*
Professional® Water Sealant

Note: Blank spaces are not permitted. If any item is not applicable, or no information is available, the space must be marked to indicate that.

Section I

Manufacturer's Name Professional Products of Kansas, Inc.	Emergency Telephone Number Chemtrec 1-800-424-9300
Address *(Number, Street, City State, and ZIP Code)* 4456 S. Clifton	Telephone Number for Information 316-522-9300
Wichita, Kansas 67216	Date Prepared March 18, 1994
	Signature of Preparer *(optional)*

Section II – Hazardous Ingredients/Identity Information

Hazardous Components (Specific Chemical Identity; Common Name(s))	OSHA	ACGIH TLV	Other Limits Recommended	% *(optional)*
Mineral Spirits (Petroleum Distillate)	500 ppm	100 ppm	-----	-----
C.A.S. # 64742-88-7				

Section III – Physical/Chemical Characteristics (solvent)

Boiling Point (Degrees F)	320-369	Specific Gravity (H_2O = 1)	.792
Vapor Pressure (mm Hg.) @ 20° C	2.7	Melting Point	N.A.
Vapor Density (AIR = 1)	4.83	Evaporation Rate (BuA=1) less than 1	N.D.

Solubility in Water
<0.1%

Appearance and Odor
Clear, Colorless Liquid, Petroleum Odor.

Section IV – Fire and Explosion Hazard Data

Flash Point (Method Used)	Flammable Limits	LEL	UEL
ASTM D-93 105° F	N.A.	1.1	5.0

Extinguishing Media
Dry chemical - CO^2 or Regular foam, Water spray.

Special Fire Fighting Procedures
Use water to cool fire-exposed containers. If a leak or spill has not ignited, use water spray to disperse the vapors.

Unusual Fire and Explosion Hazards
NONE

(Reproduce Locally) OSHA 174, Sept. 1985

FIGURE 12.5 Standard MSDS information sheet. (*Courtesy Professional Products*)

Section V – Reactivity Data

Stability	Unstable		Conditions to Avoid
	Stable	X	Heat, sparks, open flame

Incompatibility *(Materials to Avoid)*
Oxidizing materials

Hazardous Decomposition or Byproducts
Carbon monoxide and carbon dioxide may be formed on burning in limited air supply.

Hazardous Polymerization	May Occur		Conditions to Avoid
	Will Not Occur	X	Strong Oxidizers

Section VI – Health Hazard Data

Route(s) of Entry:	Inhalation?	Skin?	Ingestion?
Eyes (splash)	Yes	Limited	Unlikely

Health Hazards *(Acute and Chronic)*
If inhaled in large quantities, may cause dizziness or eventually asphyxiation.

Carcinogenicity:	NTP?	IARC Monographs?	OSHA Regulated?
	Not listed	Not listed	Not listed

Signs and Symptoms of Exposure
Eye, Skin, Respiratory Tract Irritation. Redness and swelling of eyes and skin. Inhalation of vapors over a long period may cause headache, nausea, dizziness.

Medical Conditions
Generally Aggravated by Exposure
Not recognized

Emergency and First Aid Procedures
Eye - Flush with water. Skin - Wash with soap and water.
Ingestion - DO NOT INDUCE VOMITING. GET MEDICAL ATTENTION.

Section VII – Precautions for Safe Handling and Use

Steps to Be Taken in Case Material is Released or Spilled
Ventilate areas. Contain spill by using absorbing compound or rags and shovel up.

Waste Disposal Method
Dispose of in a facility approved under RCRA regulation for hazardous waste.

Precautions to Be Taken in Handling and Storing
Store away from heat and flame. Use in well-ventilated locations. Avoid breathing of mist or vapor over extended periods.

Other Precautions
WARNING! Causes irritation to eyes on contact. May cause irritation to skin with repeated contact. COMBUSTIBLE LIQUID.

Section VIII – Control Measure

Respiratory Protection *(Specify Type)*
Respirator approved by MSHA or NIOSH as appropriate.

Ventilations	Local Exhaust	Special
	Adequate to meet occupational exp. limits	Exp. limit for total product: (see below)
	Mechanical	Other
		Stoddard solvent, 100 ppm TWA (ACGIH) & OSHA.

Protective Gloves	Eye Protection
Resistant to Chemicals.	Avoid splashing; goggles or glasses should be worn.

Other Protective Clothing or Equipment
Protective covering as required to prevent extended skin contact.

Work/Hygienic Practices
Standard practices for working with petroleum distillate.

Section IX – Additional Regulatory Information

This product contains no substances listed in SARA TITLE III, SECTION 313 in more than deminimus concentrations.

Page 2

© U.S.G.P.O. 1986-491-529/45775

FIGURE 12.5 (*Continued*) Standard MSDS information sheet. (*Courtesy Professional Products*)

94 Grove St., Peterborough, NH 03458

Material Safety Data Sheet

Prepared According to the OSHA Hazard Communication Standard (29 CFR 1910.1200).

Product Names
CIM Premix for:
CIM 1000, CIM 1061
CIM 1000 Trowel Grade
CIM 800 Roofing Grade

Description
All CIM premixes are:
Asphalt/resin portion of 2-component urethane coatings.

Emergency Telephone
CHEMTREC (800) 424-9300
C.I.M. Industries, Inc. (603) 924-9481
Prepared by:
R. H. Stephens, CIM Industries, Inc.
March 25, 1997

CAUTION! ***Flammable Liquid — Keep out of reach of children.*** ***May cause eye and skin irritation. Prolonged or repeated contact with skin can be harmful.***

HAZARDOUS CONSTITUENTS

Component	CAS#	TLV	PEL	% Range	Primary Hazard
Petroleum asphalt	8052-42-4	5 mg/m^3 (Note1)	n/a	20 to 80%	n/a
Amine compounds		n/a	n/a	up to 10%	Irritant
Aliphatic hydrocarbon	8052-41-3	100 ppm	100 ppm	up to 30%	Flammable liquid
Aromatic Petroleum Distillates	64742-95-6	100 ppm	100 ppm	up to 2%	Flammable liquid

[1] applies to fumes from hot asphalt and is not likely to present a hazard when CIM Premix is used as directed.

	HEALTH EFFECTS	EMERGENCY & FIRST AID PROCEDURES	SPECIAL PROTECTION
Eyes	May cause eye irritation.	Flush eyes immediately with fresh water for at least 15 minutes while holding the eyelids open. If irritation persists, see a doctor.	Wear chemical safety goggles.
Skin	May cause skin irritation. Prolonged or repeated exposure may dry the skin. LD_{50} (rabbit)>5 g/kg.	Remove heavily contaminated clothing and wash skin thoroughly with soap and water. DO NOT use solvents or thinners to remove materials from skin. Asphalt can be removed with vegetable oil or mineral oil.	Skin contact can be minimized by wearing protective clothing and solvent resistant gloves.
Inhalation	Breathing solvent vapor can cause central nervous system effects including dizziness, weakness, fatigue, and headache and possible unconsciousness and even death. LC_{50}>2000 ppm.	Move the person to fresh air and apply oxygen if breathing is difficult. If breathing has stopped, apply artificial respiration. Call a doctor.	Use in well ventilated areas only. Wear an OSHA approved type C air supplied respirator if ventilation is inadequate to keep solvent inhalation vapors below the TLV.
Ingestion	This material contains solvents. An aspiration hazard may exist which could cause chemical pneumonitis which is sometimes fatal. Ingestion can cause gastrointestinal irritation, nausea, vomiting and diarrhea. LD_{50} (rat)>5 g/kg.	Give water or milk to drink and telephone for medical advice. Consult medical personnel before inducing vomiting. If medical advice cannot be obtained, then take the person and product container to the nearest medical emergency treatment center or hospital.	Avoid airborne mists which can be inhaled or swallowed. Use protective mask, if necessary.

All information is based on data of which we are aware and is believed to be correct as of the date hereof. Since the information contained herein may be applied under conditions beyond our control and with which we may be unfamiliar and since data made available subsequent to the date hereof may suggest modifications of the information, we do not assume any responsibility for the results of its use. This information is furnished upon the condition that the person receiving it shall make his own determination of the suitability of the material for his particular purpose.

FIGURE 12.6 Customized MSDS information sheet. (*Courtesy C.I.M. Industries*)

FIRE PROTECTION

Flammable Liquid: Solvents contained in this product evaporate and form vapor (fumes) which can catch fire and burn with explosive violence. Invisible vapor spreads easily and can be set on fire by many sources such as cigarettes, pilot lights, welding equipment, electrical motors and switches, and static discharge. Fire hazard is greater as liquid temperature rises.

Flash Point: 101°F

Autoignition Temp.: >500°F

Flammability Limits: 1% lower limit, 6% upper limit

Extinguishing Media: CO_2, Dry Chemical, Foam, Water Fog, Halon

Special Fire Fighting Procedures: For fires involving this material, do not enter any enclosed or confined fire space without proper protective equipment, including self-contained breathing apparatus. See Hazardous Decomposition Products. Read the entire MSDS.

NFPA Hazard Rating: Health 1; Flamability 2; Reactivity 0; Special 0

DOT Hazard: Coating Solution, n.o.s, Class 3, UN1139, PG III

PHYSICAL PROPERTIES

Solubility: Miscible in all proportions with most light halogenated hydrocarbon solvents; soluble to less than 300 ppm in water.

Appearance (Color, Odor, etc.): Black liquid with kerosene odor

Boiling Point: ca 310°F (155°C)

Melting Point: n/a

Specific Gravity: 0.9@ 20/20°C

Vapor Pressure: Approximately 3mm Hg @ 68°F (20°C)

Vapor Density (Air=1): Approximately 4.9

Percent Volatile (Volume): less than 15%

ENVIRONMENTAL PROTECTION

Environmental Impact: This material, if not activated with CIM Activator, may be toxic to aquatic organisms and should be kept out of sewage and drainage systems and all bodies of water.

Precautions if Material is Released or Spilled: Eliminate all open flame in vicinity of spill or released vapor. Clean up small spills using appropriate techniques such as absorbent materials or pumping. Where feasible and appropriate, remove contaminated soil. Follow prescribed procedures for reporting and responding to larger releases.

Waste Disposal Methods: Place contaminated materials in disposable containers and dispose of in a manner consistent with applicable regulations. Contact local environmental or health authorities for approved disposal of this material.

Regulatory Status: This product does not contain constituents known to be a carcinogen, mutagen, teratogen or reproductive toxin. This product contains certain aromatic solvents subject to the reporting requirements of section 313 of SARA Title III. Spills in excess of 10,000 lb. must be reported to the appropriate federal, state, and local authorities.

REACTIVITY DATA

Stability (Thermal, Light, etc.): Stable

Incompatibility (Materials to Avoid): May react with strong oxidizing materials.

Hazardous Decomposition Products: Incomplete combustion can produce carbon monoxide. Normal combustion forms carbon dioxide and water vapor and may produce oxides of nitrogen.

Hazardous Polymerization: Will not occur.

n/a = Not Applicable
NDA = No Data Available

ADDITIONAL HEALTH DATA

Cim Premix is used with CIM Activator to form an elastomeric coating for waterproofing and corrosion protection. Consult the MSDS for CIM Activator. Avoid inhalation of the activated CIM mixture which contains isocyanates and may result in sensitization and allergic response in some individuals.

No association has been established between industrial exposure to petroleum asphalt and cancer in humans. The International Agency for Research on Cancer has determined there is limited evidence of carcinogenicity for undiluted steam-refined asphalts in experimental animals and insufficient evidence of carcinogenicity for undiluted steam-refined asphalts in humans. These asphalt sources are not constituents of CIM Premix

SPECIAL PRECAUTIONS

READ AND OBSERVE ALL PRECAUTIONS ON PRODUCT LABEL.

DO NOT USE OR STORE near flame, sparks or hot surfaces.

USE ONLY IN WELL VENTILATED AREA. Keep container closed.

DO NOT weld, heat or drill container. Replace cap or bung. Emptied container still contains hazardous or explosive vapor or liquid.

CAUTION! Do not use pressure to empty drum or explosion may result.

3/97

FIGURE 12.6 (*Continued*) Customized MSDS information sheet. (*Courtesy C.I.M. Industries*)

FIGURE 12.7 Employee protective wear. (*Courtesy Coastal Construction Products*)

to joint project inspections of job sites with EPA inspectors. These regulations are part of the Resource Conservation and Recovery Act (RCRA). Specific regulations apply for waste generators based on the amount of waste the firm generates on a monthly basis.

Conditionally exempt are firms that generate less than 25 gallons of hazardous waste. Small-quantity firms produce 25–300 gallons per month; generator firms produce more than 300 gallons per month.

Since waterproofing systems often fall under the classification of hazardous materials, they are governed by the RCRA solvents, thinners, and primers used in conjunction with waterproofing, and are also regulated. Examples of hazardous waterproofing materials include coatings containing solvents, such as solvent-based urethanes and most epoxies. Materials become hazardous waste when any of the following situations occurs:

- Unused materials remain in their original container.
- Materials reach their shelf life (usable time).
- Spillage of the material occurs.
- Hazardous materials are introduced into nonhazardous systems.

A proper identification number is obtained from the federal agency by completing EPA form *Notification of Hazardous Waste Activity*. State and local government agency registrations must also be complied with.

Once all necessary licenses have been obtained, proper storage and periodic disposal methods are implemented. EPA, state, and local governments regulate the amount and means of hazardous waste storage. Only recognized authorized and licensed companies should be used to dispose of accumulated waste materials. EPA regional offices or local governments provide listings of such firms.

Building owners should be aware of these regulations and ensure that contractors who complete work on their structures are in compliance with them. Additionally, owners should not permit contractors to dump waste materials into building waste receptacles or trash dumpsters. This action may result in fines and penalties being levied against the building owner.

VOLATILE ORGANIC COMPOUNDS

The Environmental Protection Agency has imposed stricter standards relating to emissions from volatile organic compounds (VOC). The standards implemented address Section 183(e) of the Clean Air Act as amended in 1990. The new standards specifically regulate "architectural coatings," defined by the EPA as coatings applied to the surfaces of stationary structures. This includes all waterproofing and roofing products.

EPA notes that approximately 9% of VOC consumer and commercial emissions come from architectural coating products. VOCs are the main component in forming ground-level ozone. Exposure to ground-level ozone can damage lung tissue and cause serious respiratory damage. It is the intent of these new regulations concerning VOC to limit the creation of ground-level ozone in the immediate future.

The new regulations have directly affected most waterproofing products, with the exception of water-based coatings and sealers. Any product with VOCs, such as liquid-applied below-grade membranes, deck coatings, clear repellents, and adhesives used to apply materials are now regulated by EPA. Many waterproofing products are now being reformulated to meet the standards, and this may result in inferior products until manufacturers are able to perfect their new formulations and test the product to ensure it meets industry standards.

All waterproofing product containers must carry labels that clearly display the maximum VOC content in the coating. EPA regulations limit the weight of VOC in the waterproof coatings as listed in Section 183(e). This weight limit includes: 5.0 pounds per gallon of VOC for waterproofing products; 3.3 pounds per gallon for concrete protective coatings; and 7.1 pounds per gallon for clear repellents.

Contractors and building owners should carefully consider the implications of this new regulation, and ensure that products or systems selected meet the code while still meeting the service requirements for providing the necessary waterproofing protection.

CHAPTER 13

GUIDE SPECIFICATION FOR WATERPROOFING

The guide specifications in this chapter are included to assist in the preparation of project specifications for both new construction and remedial projects. The specifications are reproduced from manufacturers' recommendations and should form the basic specification for each waterproofing requirement on a building project. The specifications can be revised as appropriate, including the approved equivalent products for each category.

The specifications follow the Construction Specifications Institute numbering system. Chapter 14 can be used as a resource to contact the manufacturers for suggested provisions for any unique job-specific requirement that is necessary. Included in the specifications:

Figure	Specification Number	Specification type
13.1	2715	Prefabricated Drainage System
13.2	3252	Waterstop, Installation
13.3	3700	Concrete restoration—Epoxy and grout injection
13.4	7100	Fluid Liner for Civil Projects
13.5	7100	Fluid Applied Waterproofing for below-grade and above-grade
13.6	7111	Sheet Waterproofing membrane for below-grade and above-grade
13.7	7161	Cementitious Waterproofing Membrane
13.8	7162	Crystalline Waterproofing
13.9	7190	Water repellents
13.10	7540	Hot Applied Waterproofing
13.11	7570	Deck Coating Applications
13.12	7900	Sealants and Expansion Joints
13.13	9900	Paint/Elastomeric Coatings

This suggested guide specification has been developed using the current edition of the Construction Specifications Institute (CSI) "Manual of Practice," including the recommendations for the CSI 3 Part Section Format. Additionally, the development concept and organizational format of the American Institute of Architects (AIA) MASTERSPEC Program have been recognized in the preparation of this guide specification. Neither CSI nor AIA endorses the use of specific manufacturers and products.

SECTION 02715
FOUNDATION DRAINAGE
Prefabricated Drainage Composites)

Mop Notes (CSI) Manual of Practice Notes), Ed Notes (Editing Notes), and Coord Notes (Drawings and Specification Coordination Notes) are provided throughout this guide specification and are outlined in smaller typeface. Delete notes in final copy of project specification, coordinate specification with project drawings.

PART 1 GENERAL

1.01 SUMMARY

Ed Note: This Section specifies prefabricated drainage for both vertical and horizontal applications. Drainage course material requires selection based upon site-specific soil conditions and flow rates required.

Ed Note: Revise below to suit project conditions. Select drainage composites appropriate to project requirements. This section does not include sheet membrane waterproofing; refer to Division 7 Waterproofing Specification Section. Coordinate requirements of the two sections with each other to avoid conflicting requirements.

A. Section Includes: Prefabricated Drainage Composite.
   1. Types of Prefabricated Drainage Composites include:
      a. Prefabricated drainage composite for below grade, vertical wall applications.

Ed Note: Select appropriate system from above or below.

      b. Prefabricated drainage composite for below grade, vertical wall applications.

Ed Note: Revise below to suit project conditions and specification preparation requirements. Add section numbers per CSI "MASTERFORMAT" and Office Specification Practices.

B. Related Sections:

1. Earthwork: Refer to Division 2
2. Landscaping: Refer to Division 2
3. Pipe and fittings: Refer to Division 2
4. Storm sewage systems: Refer to Division 2
5. Cast-in-place concrete: Refer to Division 3
6. Structural pre-cast concrete: Refer to Division 3
7. Masonry: Refer to Division 4
8. Waterproofing: Refer to Division 7
9. Dampproofing: Refer to Division 7
10. Insulation: Refer to Division 7
11. Other appropriate project specific specifications sections

Mop Note: Below paragraph should describe products and work included in this section which are covered by alternates. Include alternate descriptions. Coordinate this section with Division 1 alternate section. Bid

FIGURE 13.1 Guide specification for section 2715, prefabricated drainage system. (*Courtesy TC MiraDRI*)

documents and bid forms (if any). Refer to specification coordination sheets included with this section for additional guidance on the use of alternates.

C. Alternates: Products and installation included in this section are specified by alternates. Refer to Division 1 Alternates Section for alternates description and alternate requirements.
1. Alternate Manufacturers: Refer to Part 2 Products herein.

Mop Note: Below article should list industry standards referenced in this section, along with acronym, alpha/numeric designations, titles, and dates. This article does not require compliance with standards, but is merely a listing to establish title and date of references.

1.02 REFERENCES (INDUSTRY STANDARDS)

A. General: Refer to Division 1 References Section.

1.03 SYSTEM DESCRIPTION

A. Performance Requirements: Provide prefabricated drainage composites which has been manufactured and installed to maintain drainage without defects, damage, or failure.

1.04 SUBMITTALS

A. General: Prepare, review, approve and submit specified submittals in accordance with "Conditions of the Contract" and Division 1 Submittals Sections. Product data, shop drawings, samples and similar submittals are defined in "Conditions of the Contract."

B. Product Data: Submit manufacturer's product data for drainage composites specified.

C. Shop Drawings: Submit shop drawings showing layout, profiles, and product components, including accessories for drainage composites.

Ed Note: Retain below as appropriate for drainage composites selected.

D. Samples: Submit verification samples for prefabricated drainage composites.

E. Quality Assurance/Control Submittals

1. The specified properties of drainage panels must be supported by test results from an independent laboratory, documenting the specified flow rate in the plane of the core and creep performance of the polymer core. The testing conditions shall comply with ASTM D-4716 as follows:

a. Hydraulic Gradient: 1.0 for vertical installations and 0.08 for horizontal installations.

b. Normal Pressure (pressure imposed perpendicular to the plane of the core): Equal to 3600 psf.

c. Creep: Model long term compression of the prefabricated drainage composite system and determine if the drain product flow channels become restricted with time. Long term creep/drainage performance shall be determined by measuring flow after 300 continuous hours under the above referenced normal pressure. The test method shall utilize a loading system which models the soil/drainage product interaction.

FIGURE 13.1 (*Continued*) Guide specification for section 2715, prefabricated drainage system. (*Courtesy TC MiraDRI*)

d. Flow Direction: Flow shall be measured on only one side of the core. Where the core geometry differs in principal directions, flow shall be measured in both directions, simulating water flowing vertically down a wall and horizontally across the face of the wall to accurately determine maximum flow rate in critical principal direction.

2. Certificates (Qualification Data): Submit product certificates signed by manufacturer certifying materials comply with specified performance characteristics and physical requirements, showing full time quality control, and that prefabricated drainage composites are being supplied by a single-source manufacturer.

F. Closeout Submittals:

1. Warranty: Submit warranty documents specified herein.
2. Project Record Documents: Submit project record documents for installed materials in accordance with Division 1 Project Closeout (Project Record Documents) Section.

1.05 QUALITY ASSURANCE

A. Qualifications:
1. Installer Qualifications: Installer experienced (as determined by contractor) to perform work of this section, who has specialized in the installation of work similar to that required for this project, who can comply with manufacturer's warranty requirements, and who is an approved applicator as determined by drainage manufacturer.
2. Manufacturer Qualifications: Manufacturer capable of providing field service representation during construction, approving acceptable installer, application method and conducting final inspection of the drainage composites assembly.

Ed Note: If mock-up is required for special project requirements or job conditions, add text. Typically mock-ups may not be required.

B. Pre-Installation Meetings: Conduct pre-installation meeting to verify project requirements, substrate conditions, manufacturer's installation instructions and manufacturer's warranty requirements.

C. Pre-Installation Testing: In accordance with manufacturer's recommendations and warranty requirements, conduct pre-installation testing of substrates to receive drainage composites.

FIGURE 13.1 (*Continued*) Guide specification for section 2715, prefabricated drainage system. (*Courtesy TC MiraDRI*)

1.06 DELIVERY, STORAGE, AND HANDLING

A. Delivery

1. Materials should be delivered in manufacturer's original, unopened packaging with labels attached.

B. Storage

1. Store materials in the original unopened packages.

C. Handling

1. All materials must be handled in a manner to prevent damage. Any material damaged must be removed from the project area and replaced with new material.

2. Miradrain materials must be handled in accordance with Mirafi guidelines.

1.07 PROJECT CONDITIONS/SITE CONDITIONS

A. When Miradrain is installed in conjunction with a waterproofing product, the Miradrain must be compatible with the waterproofing product and installed by methods acceptable to the waterproofing product manufacturer.

B. The outfall for any drainage pipe used with the drainage panels shall be coordinated with the site drainage.

> Ed Note: Below pertains to the American Institute of Architects (AIA) "General Conditions of the Contract for Construction," which places responsibility for means, methods and techniques of construction upon the contractor.

C. Project Conditions: Refer to "Conditions of the Contract" for control and responsibility of project site including means, methods and techniques of construction by the contractor.

D. Field Measurements: Verify actual measurements/openings by performing field measurements before fabrication; show recorded measurements on shop drawings. Coordinate field measurements, fabrication schedule with construction progress to avoid construction delays.

> Ed Note: Below warranty article requires coordination with the owner. Below assumes the use of the American Institute of Architects (AIA) "Conditions of the Contract for Construction."

FIGURE 13.1 (*Continued*) Guide specification for section 2715, prefabricated drainage system. (*Courtesy TC MiraDRI*)

1.08 WARRANTY

A. Project Warranty: Refer to "Conditions of the Contract" for project warranty provisions. Submit manufacturer's single-source warranty directly from drainage composites manufacturer for drainage system.

B. Manufacturer's Warranty Requirements: Submit, for Owner's acceptance, manufacturer's standard warranty document executed by authorized company official. Manufacturer's warranty is in addition to, and not a limitation of, other rights Owner may have under the Contract Documents.

1. Beneficiary: Issue warranty in the legal name of the project Owner.

Ed Note: Coordinate below with manufacturer's warranty requirements.

2. Warranty Period; ___ years commencing on Date of Substantial Completion

3. Warranty Acceptance: Owner is sole authority who will determine acceptability of manufacturer's warranty documents.

Ed Note: Select available Manufacturer's Warranty Program from below and revise below as appropriate. Consult with Mirafi for specific Warranty Terms and Conditions. For Manufacturer's Material Warranty Program select standard geotextile Mirafi 700XG Woven Fabric for horizontal applications or 140NC Nonwoven Fabric for vertical applications.

C. Manufacturer's Warranty Program:

1. Manufacturer's Material Warranty: Material only warranty, excludes labor.
   a. Standard Waterproofing and Drainage Duration: 5 year material warranty.
   b. Extended Waterproofing and Drainage Duration: 10 year material warranty.

Ed Note: For Manufacturer's Site Specific Warranty Program select Mirafi Filterweave 40/10, Filterweave HP500 Fabric, Filterweave 70/20 Woven or Mirafi 180N Nonwoven Fabrics. Below Warranty is available when selection of Site Specific fabric is in accordance with Mirafi project specific written recommendations, based upon review of project's soil report submitted by architect. Additionally, application must be performed by a Mirafi approved applicator. Warranty coverage is for drainage performance of Mirafi "Miradrain" Drainage Composite for the duration of the warranty.

2. Manufacturer's Site Specific Drainage Warranty: Manufacturer's site specific drainage warranty covers the Miradrain drainage system guaranteeing "no clogging drainage" performance of the Miradrain drainage composite for the duration of the warranty.
   a. Site Specific Drainage Duration: 5 year no clogging drainage warranty.

Contact a Mirafi Representative for further details on warranty provisions.

FIGURE 13.1 (*Continued*) Guide specification for section 2715, prefabricated drainage system. (*Courtesy TC MiraDRI*)

PART 2 PRODUCTS

Ed Note: Retain below article for proprietary specification. Do not use phrase "or equal"/"or approved equal"; use of such phrases is not considered good professional specification writing practice. Because of the differing interpretations of the phrase among the contracting parties, coordinate below article with Division 1 material and equipment (product options and substitutions) section, and bidding documents, if any.

2.01 MANUFACTURERS

A. Acceptable Manufacturer: Mirafi Moisture Protection Products, Nicolon/Mirafi Group, a Division of Nicolon Corp.
   1. Address: 3500 Parkway Lane, Suite 500, Norcross, GA 30092
      a. Telephone: (800) 234-0484; Fax (404) 729-1829.

Ed Note: Select appropriate system from below.

   2. Warranted Drainage System: Miradrain Prefabricated Drainage Composite. Provides system drainage components from a single source supply in accordance with manufacturer's warranty requirements.

Ed Note: Retain below as appropriate to drainage system selected and to project requirements. Consult with Mirafi for specific recommendations for project. Select fabric from manufacturers recommendations and warranty program requirements. Mirafi offers two groups of fabrics; one group is manufacturer's standard geotextile group including 700XG Woven Fabric for Miradrain 9000 and 140NC Nonwoven Fabric for Miradrain 2000, 5000, 6000, 6200 and 8000. The other group includes manufacturer's Site-Specific Geotextile Filtration fabrics: Filterweave 40/10, Filterweave 70/20, Filterweave HP500 Woven Fabric and Mirafi 180N Nonwoven Fabric.

      a. Miradrain Prefabricated Drainage Composites:

         1. Miradrain 6000 3-dimensional dimpled core and geotextile fabric for lagging and underslab applications.
         2. Miradrain 6200 3-dimensional dimpled core and geotextile fabric for vertical applications over waterproofing membranes.
         3. Miradrain 9000 3-dimensional dimpled core and geotextile fabric for horizontal plaza, roof deck and planter applications
         4. Miradrain 8000 3-dimensional dimpled core and geotextile fabric for Hydrocarbon Resistive applications.
         5. Miradrain 2000 3-dimensional dimpled core and Mirafi 140 NC geotextile fabric for shallow-depth foundation walls.
         6. Miradrain 5000 3-dimensional dimpled core and Mirafi 140 NC geotextile fabric for double sided drainage applications in landscaping and landfill applications.

FIGURE 13.1 (*Continued*) Guide specification for section 2715, prefabricated drainage system. (*Courtesy TC MiraDRI*)

Ed Note: Retain below with appropriate drainage composites above. Select fabric based upon site-specific soil conditions and Mirafi recommendations. Refer to Miradrian Site Specific Geotextile Fabric Selection Guide shown below. Consult with Mirafi for recommendations and site-specific requirements in accordance with project-specific soils report. Below listed performance requirements and terminology are standard for geotextile testing methodology.

MIRADRAIN, SITE-SPECIFIC GEOTEXTILE SELECTION GUIDE**

	Coarse Sand	Sand	Silty Sand	Clayey Sand	Sandy Silt	Silt	Clayey Silt	Sandy Clay	Silty Clay	Clay
Woven Fabrics										
HP 500	✔									
FW 40/10		✔	✔							
FW 70/20				✔		SPECIFIC* TESTING REQUIRED				
Nonwoven Fabric***										
180N								✔	✔	✔

*Due to instability of silty soils, Mirafi requires that an individual soil test and filtration design be performed by Mirafi engineering services. Contact a Mirafi representative for further details.

**Please note that this selection guide is merely a reference chart and is not meant to replace sound geotextile engineering design practices.

***Nonwoven fabrics are to be used in vertical applications only.

b. Miradrain Standard Fabrics:

	140NC (MD 2000, 5000, 6000, 6200, 8000)	700XG (MD 8000, 9000)
Flow Rate (ASTM D 4491)	120 gal./min./ft² (4903 l/min/m²)	100 gal./min./ft² (4086 l/min/m²)
Apparent Opening Size (ASTM D 4751)	70 (0.21) U.S. Sieve (mm)	40 (0.42) U.S. Sieve (mm)
Grab Tensile Strength (ASTM D 4632)	110 lb., Machine Direction (0.49 kN)	365 lb., Machine Direction (1.62 kN)
Mass Per Unit (ASTM D 5261)	4.0 oz/yd² (136 g/m²)	5.6 oz/yd² (190 g/m²)
Puncture (ASTM D 4833)	70 lb. (0.31 kN)	115 lb. (0.51 kN)

FIGURE 13.1 (*Continued*) Guide specification for section 2715, prefabricated drainage system. (*Courtesy TC MiraDRI*)

c. Miradrain Site Specific Fabrics:

	HP 500 (MD 6000, 6200, 8000, 9000)	FW 40/10 (MD 6000, 6200, 8000, 9000)	FW 70/20 (MD 6000, 6200, 8000, 9000)	180N (MD 6000, 6200, 8000)
Flow Rate (ASTM D 4491)	115 gal./min/ft² (4698 l/min/m²)	70 gal./min/ft² (2860 l/min/m²)	35 gal./min/ft² (1430 l/min/m²)	110 gal./min/ft² (4494 l/min/m²)
Apparent Opening Size (ASTM D 4751)	30 (0.595) U.S. Sieve (mm)	40 (0.42) U. S. Sieve (mm)	60 (0.25) U.S. Sieve (mm)	80 (0.177) U. S. Sieve (mm)
Grab Tensile (ASTM 4632)	400 lb. Machine Direction (1.78 kN)	265 lb. Machine Direction (1.18 kN)	275 lb. Machine Direction (1.22 kN)	200 lb. Machine Direction (0.89 kN)
Mass per unit (ASTM D 5261)	7.4 oz/yd² (250 g/m²)	5.0 oz/yd² (170 g/m²)	6.1 oz/yd² (207 g/m²)	8.0 oz/yd² (271 g/m²)
Puncture (ASTM D 4833)	145 lb. (0.81 kN)	125 lb. (0.56 kN)	145 lb. (0.65 kN)	130 lb. (0.58 kN)

Ed note: Retain below prefabricated drainage composite as appropriate to project requirements and site-specific soil conditions. Coordinate below composite with above drainage fabric.

3. Material(s)/System(s) Testing: Prefabricated drainage composites:

	Miradrain 6200	Miradrain 8000	Miradrain 9000
Compressive Strength - psf (kN/m²) (ASTM D 1621)	15,000 (719)	15,000 (719)	18,000 (862)
Water Flow Rate * - gpm/ft width (l/min/width) (ASTM D 4716)	15 (188)	15 (188)	18 (226)
Thickness - in (mm) (ASTM D 1777)	0.40 (11.3)	0.38 (10.8)	0.38 (10.8)

	Miradrain 6000	Miradrain 2000	Miradrain 5000
Compressive Strength - psf (kN/m²) (ASTM D 1621)	15,000 (719)	10,800 (518)	15,000 (719)
Water Flow Rate * - gpm/ft width (l/min/width) (ASTM D 4716)	15 (188)	9 (113)	15 (188)
Thickness - in (mm) (ASTM D 1777)	0.40 (11.3)	0.25 (7.1)	0.40 (11.3)

*Normal stress of 3,6000 psf
Hydraulic gradient of 1.0 for 300 + hours

FIGURE 13.1 (*Continued*) Guide specification for section 2715, prefabricated drainage system. (*Courtesy TC MiraDRI*)

Ed Note: Retain below for alternate method of specifying manufacturers and products. Add alternate number designation and coordinate with other specific project alternates.

2.02 ALTERNATES AND SUBSTITUTIONS

Ed Note: Presently there is only one manufacturer who markets a single-source drainage composite with Site Specific fabrics. Other manufacturers market a drainage composite system with standard geotextile fabrics. Below alternate should be for a drainage composite with either Standard or Site Specific fabrics.

A. Alternate (Manufacturers/Products): In lieu of providing below specified base bid/contract manufacturer, provide below specified alternate manufacturers. Refer to Division 1 Alternates Section.
   1. Base Bid/Contract Manufacturer/Product: Mirafi Moisture Protection Products, Nicolon/Mirafi Group.
      a. Product: Prefabricated Drainage Composite
         1. Drainage: Miradrain Prefabricated Drainage Composite

Ed Note: Number below alternates in accordance with the project requirements and office specification practices. Determine alternate type. Verify alternate manufacturer/product designations. Add product attributes/characteristics for product equivalency, such as drainage composite, including drainage core and fabric.

   2. Alternate #______ Manufacturer/Product: W. R. Grace
      a. Product
         1. Drainage: Hydroduct Drainage Composite

B. Substitutions: Refer to Division 1 General Requirements for substitution requirements.

2.03 RELATED MATERIALS (Specified in Other Sections)

A. Concrete: Refer to Division 3 Concrete Section.

B. Insulation: Refer to Division 7 Insulation Section.

C. Waterproofing: Refer to Division 7 Waterproofing Section.

Ed Note: Usually retain below in accordance with Mirafi warranty requirements.

2.04 SOURCE QUALITY CONTROL

A. Single Source Responsibility: Obtain drainage composite and accessories from a single source supplier for required warranty.

FIGURE 13.1 (*Continued*) Guide specification for section 2715, prefabricated drainage system. (*Courtesy TC MiraDRI*)

PART 3 EXECUTION

3.01 MANUFACTURER'S INSTRUCTIONS/RECOMMENDATIONS

A. Compliance: Comply with manufacturer's product data, including product technical bulletins, product catalog installation instructions and product packaging instructions.

3.02 EXAMINATION

A. Site Verification of Conditions: Verify substrate conditions (which have been previously installed under other sections) are acceptable for product installation in accordance with manufacturer's instructions. Do not proceed with drainage installation until substrate conditions are acceptable for compliance with manufacturer's warranty requirements.

3.03 PREPARATION

A. Adjacent Surfaces Protection: Protect adjacent work areas and finish surfaces from damage during installation operations.

B. Concrete Surface Preparation: Prepare concrete surfaces to receive drainage composite. Surfaces shall be smooth, free of depressions, voids, protrusions, clean and free of other surface contaminants which may impair the performance of drainage and manufacturer's warranty requirements.
   1. Cast-in-Place Concretes: Decks shall be monolithic, smooth, free of voids, spalled areas, laitenance, honeycombs, and sharp protrusions. Refer to Division 3 Concrete Section for concrete strength, density, finish, curing methods and other concrete requirements.
   2. Precast Concrete Decks: Decks shall be mechanically secured to minimize differential movement and each joint between precast units shall have an installed backer rod. Grout precast units as recommended by manufacturer.
   3. Shotcrete: Surface shall be monolithic and smooth with no indulations, irregularities or exposed wire mesh.

D. Substrate Cleaning: Clean substrate which is to receive drainage. Remove loose debris and other harmful contaminants which will affect performance of drainage composite.

Ed Note: Coordinate below article with Part 2 Products article to include each product requiring installation.

DWG Coord Note: Coordinate below article with Nicolon/Mirafi recommended drainage details number consecutively No. 1 thru No. 11.

FIGURE 13.1 (*Continued*) Guide specification for section 2715, prefabricated drainage system. (*Courtesy TC MiraDRI*)

3.04 DRAINAGE COURSE INSTALLATION (FILTER FABRIC/CORE)

A. Vertical Surfaces or Planters - MD-2000, MD-6000, MD-6200, MD-8000, or MD-9000

1. Completed Walls - Position the panel with the flat side against the wall and filter fabric toward the soil/drainage side. A washer headed concrete nail may be used to attach the panel against the concrete wall. Alternate methods of the attachments use concrete nails driven through wood strips placed over the Miradrain or Mirastick construction adhesive. Most methods used to secure protection board are acceptable.

2. Over Miradri Waterproofing Membrane - Place the Miradrain over the waterproofing membrane. Use two sided Mirabond tape or Mirastick construction adhesive to secure the Miradrain to the waterproofing membrane. Most methods used to secure protection board are acceptable.

3. Against Shoring Systems or Vertical Soil Excavations - Position the panel with the fabric facing the shoring or soil. Nails of sufficient length should be used to pin the panel directly against the shoring or soil. The fabric should lay flat against the shoring or soil to minimize voids. The flat side of the Miradrain may then receive the waterproofing material. Concrete or shotcrete may be placed directly against the waterproofing or back side of the Miradrain if waterproofing is not used.

4. Buttress Drainage and Landslide Repair - Miradrain should be rolled out fabric side down onto the properly prepared subgrade. A "chimney" drain type drain pattern should be formed.

   Spacing between Miradrain chimney drains will be dictated by the engineer. Chimney drains should be connected to the continuous horizontal collector panel by overlapping a minimum of 12 inches into the collector panel.

   Miradrain panels may be secured to the subgrade by ballasting with soil or nailing through the Miradrain panel into the underlying hillside. Fabric flaps must be folded over onto the core and secured with duct tape or soil ballast.

5. Edge Drain or Trench Drain -
   Miradrain 5000 should be laid out in 50 foot lengths adjacent to the previously cut/excavated trench.

   Panel end laps may be connected by overlapping the panels and applying locking clips or buttoning of the dimples. Flexible corrugated polyethylene or rigid PVC pipe, which has been wrapped with Mirafi filter fabric should be placed in the bottom of the trench. The Miradrain panel should be lowered into the trench beside the pipe and temporarily secured to the trench wall by nailing or propping. The Fabric side of the panel must face the direction from which the water is flowing. Fabric overlaps on the top of the Miradrain panels should be draped over the back of the core. Backfilling should be completed immediately.

6. Connecting Adjacent Panels - Connect adjacent panels at the longitudinal edge by pulling the filter fabric back to expose the flange. The flangeless panel edge should be placed on top of the flange of the adjacent panel and butted dimple to dimple. All connections should be completed in shingle fashion so that moisture will flow with the overlap and not against it. Overlap fabric in the direction of water flow. Cover all terminal edges with the filter fabric flap by tucking in behind the core.

B. Horizontal Surfaces - MD-9000 or MD-8000 (hydrocarbon-resistive applications)

1. Plaza Decks, Parking Decks, and other Split Slab Construction - MD-9000

FIGURE 13.1 (*Continued*) Guide specification for section 2715, prefabricated drainage system. (*Courtesy TC MiraDRI*)

Attach the Miradrain 9000 panels by either placing temporary ballast on top of the Miradrain or adhering the panels to the waterproofing membrane with Mirastick construction adhesive or Mirabond two sided tape.

Connect adjacent panels at the longitudinal edge by pulling the filter fabric back to expose the flange. Place a small amount of two sided tape or Mirastick on the flange. The flangeless edge of the panel should be placed on top of the flange of the adjacent panel and butted dimple to dimple. All connections should be completed in shingle fashion so that moisture will flow with the overlap and not against it. Overlap fabric in the direction of water flow. Cover all terminal edges with the filter fabric flap by tucking it behind the core.

Miradrain 9000 should be channeled into an internal drain or perimeter drain system.

Concrete, sand, grout, or pavers may be placed directly on the Miradrain 9000 fabric side. Caution should be taken not to place point loads on the Miradrain that might puncture the filter fabric on the Miradrain. When concrete is poured against Miradrain, use proper chuting techniques and avoid high drop heights.

2. Floor Slabs and Concrete Lined Channels - MD-6000 or MD-9000
The subgrade shall be graded to a 2% minimum slope and clear of rubble, rock, large soil clods, etc. The Miradrain should be placed fabric side down, directly on the subgrade. Installation should proceed from the higher elevation of the slope and work downward. Connection of panel ends shall be achieved with four row dimple overlapping and taping of terminal edges.

3. Planters - MD-9000
Miradrain 9000 should be placed fabric side to the inside (soil side) of the planter. The planter walls should be covered with Miradrain 9000. Allow a 3" (7.5 cm) fabric overlap at the bottom of vertical panels in order to cover the intersection of wall and bottom sections. Any exposed panel edges must be covered with supplemental pieces of fabric to prevent soil intrusion into the flow channels.

3.05 DISCHARGE CONNECTIONS

A. Drainage - Where drainpipe is indicated, place the drainpipe next to the core. Wrap the drainpipe or rock-pipe drain combination with an auxiliary piece of Mirafi filter fabric.

At weep holes, cut a hole in the core corresponding to the size and location of the weep hole. Avoid cutting a hole in the fabric by cutting the backside of the core between the dimples. A four dimple square area cut between the dimples (2 1/2 square inch) should be sufficient for most applications.

Plaza Drains - Create openings in the Miradrain core to correspond with all discharge holes in the drain at the structural deck level. Fabric must be placed over these holes to prevent intrusion of soil, grout, sand, or concrete into the drainage core.

FIGURE 13.1 (*Continued*) Guide specification for section 2715, prefabricated drainage system. (*Courtesy TC MiraDRI*)

B. Terminal Connections and Protrusions - Cover all terminal edges with the integral fabric flap by tucking it around the edge of the core and securing it. At protrusions, cut the core around the protrusion, cut an "X" in the fabric, and tape the fabric around the protrusion. Dirt and concrete must not infiltrate the core.

3.06 RELATED PRODUCTS INSTALLATION REQUIREMENTS

A. Earthwork Backfilling: Refer to Division 2 Earthwork Section for installation requirements.

B. Waterproofing: Refer to Division 7 Waterproofing Section for installation requirements.

3.07 FIELD QUALITY CONTROL

A. Manufacturer's Field Services: Upon Owner's request, provide manufacturer's field service consisting of product use recommendations and periodic site visits for inspection of product installation in accordance with manufacturer's instructions and warranty requirements.

3.08 CLEANING AND PROTECTION

A. Cleaning: Remove temporary coverings and protection of adjacent work areas. Repair or replace damaged installed products. Clean installed products in accordance with manufacturer's instructions prior to owner's acceptance. Remove construction debris from project site and legally dispose of debris.

B. Protection: Protection installed products finished surfaces from damage during construction.

END OF SECTION 02715

FIGURE 13.1 *(Continued)* Guide specification for section 2715, prefabricated drainage system. (*Courtesy TC MiraDRI*)

SECTION 3252
WATERSTOP STANDARD SPECIFICATION

PART 1 - GENERAL

1.01 Summary

A. All Polyvinyl Chloride (PVC) Waterstops embedded in structural concrete and spanning construction (non movement), expansion, and contraction (control) joints to create a flexible barrier which prevents the movement or seepage of water and certain other fluids from one side of the joint to the other in primary containment structures.

B. Chemical Resistant Waterstop, Thermoplastic Rubber (TPER) All TPER Waterstops embedded in structural concrete and spanning construction (non movement), expansion, and contraction (control) joints to create a flexible barrier which prevents the movement or seepage of certain hazardous chemicals from one side of the joint to the other, in secondary containment structures.

C. Preformed Asphalt Plastic Adhesive Waterstop (PAPA) An asphaltic compound used to provide a watertight barrier in concrete construction and cold joints in new pours or when joining existing concrete to a new pour in a primary containment structure for water or certain other fluids.

1.02 REFERENCES

A. US Army Corps of Engineers CRD-C-572-74, Specification for Polyvinyl Chloride Waterstop.

B. American Society for Testing Materials (ASTM)
1. D-638-91 Tensile and Elongation
2. D-746-79 (1987) Low Temperature Brittleness
3. D-747-90 Stiffness in Flexure
4. D-792-91 Specific Gravity (PVC)
5. D-2240-91 Durometer Hardness
6. D-570-81 (1988) Water Absorption
7. D-1203-89 Method A Volatile Loss
8. D-4-80 Hydrocarbon Plastic Content
9. D-6-80 Volatile Matter
10. D-71-84 Specific Gravity (Plasticized Asphalt)
11. D-113-85 Ductility
12. D-36-86 Softening Point
13. D-217-85 Penetration Cone (150 GTL)
14. D-92-85 Flash Point °F
15. D-471 (166 hour immersion) Change in Properties
16. D-412 Tensile Strength & Elongation of Rubber (TPER)

C. American Association of State Highway & Transportation Officials (AASHTO)
1. T-111 Inert Material

D. Federal Specifications SS-S
1. 210 A Section 4.5.3 Flow Resistance
2. 210 A Section 3.6 Chemical Resistance

Page 1

FIGURE 13.2 Guide specification for section 3252, waterstop. (*Courtesy Vinylex Corporation*)

1.03 SUBMITTALS

A. Product Data
   1. Submit data of complete physical properties for PVC, TPER, and PAPA Waterstop.

B. Samples
   1. Submit 6 inch samples of PVC, TPER, and PAPA Waterstop.

C. Laboratory Test Reports
   1. Submit independent test reports and/or reports certified by a professional engineer indicating average properties of PVC TPER, and PAPA Waterstop materials and showing conformance to the requirements specified in this section.

D. Quality Control Submittals
   1. Certificates of Compliance: Submit written certificates that PVC Waterstop supplied on this project meet or exceed the physical property requirements of current US ACOE CRD-C-572.

   2. Provide structural calculations showing that the Waterstop submitted (PVC, TPER) will withstand the stresses created by the following design conditions:

      Vertical Movement (Y axis) _______ inches
      Shear Movement (Z axis) _______ inches
      Expansion Movement (X axis)_______ inches
      Head of Water _______ feet
      Nominal Joint Opening _______ inches

   and not exceed 25% of the Ultimate Tensile Strength (after accelerated extraction).

   3. Manufacturer's Instructions: Submit instructions for materials specified in this section.

   4. Submit two samples of factory made fittings (ell's tee's or crosses) required on this project. No field made fittings are allowed. Butt splices only will be permitted in the field.

1.04 QUALITY ASSURANCE

A. Mock Ups:

   1. Welding Demonstration:

      a. Demonstrate ability to weld acceptable butt splices in PVC or TPER Waterstops before installing in form.

      b. Quality of welded joints, both factory and field, will be subject to acceptance by the engineer.

FIGURE 13.2 (*Continued*) Guide specification for section 3252, waterstop. (*Courtesy Vinylex Corporation*)

PART 2 - PRODUCTS

2.01 MATERIALS

A. Waterstops:

1. Material and Type
   a. PVC: Provide Waterstop compounded from virgin poly vinyl chloride. Ribbed Type center bulbs are the only type allowed, unless another style is specifically specified in this section.
   b. TPER: Thermoplastic Rubber certified as capable of withstanding the following hazardous chemicals:
      (1)____________________@_______°C
      (2)____________________@_______°C

   for 166 hours in a secondary containment application.
2. Manufacturer's of PVC: One of the following, or equal:
   a. VINYLEX CORPORATION.
   b.
3. Manufacturer's of TPER: One of the following, or equal:
   a. VINYLEX CORPORATION.
   b
4. Manufacturer's of PAPA: One of the following, or equal:
   a. VINYLEX CORPORATION.
   b.
5. Provide PVC Waterstop meeting the following requirements:

Physical Characteristics	Test Method	Required Results
Specific Gravity	ASTM D-792-91	1.4 maximum
Hardness	ASTM D-2240-91	76 ± 8 shore A
Tensile Strength	ASTM D638-91	2,000 psi minimum, after accelerated extraction
Ultimate Elongation	ASTM D638-91	350% minimum after accelerated extraction
Alkali Extraction	CRD-C-572-74	7 day weight change between minus 0.1% and plus 0.25%
Low Temperature Brittle Point	ASTM D-746 (1987)	10 of 10 must pass @ minus 35°F
Water Absorption	ASTM D-570-81 (1988)	not more than 0.15% after 24 hrs
Stiffness in Flexure	ASTM D-747-90	Not less than 750 psi
Tear Resistance	ASTM D-624-91	300 pounds per inch minimum

6. Provide TPER Waterstops meeting the following minimum requirements:

Physical Characteristics	Test Method	Required Results
Hardness	ASTM D-2240-91	70 to 80 shore A
Specific Gravity	ASTM D-792-91	not exceeding 1.0
Tensile Strength	ASTM D-412	1150 psi minimum
Ultimate Elongation	ASTM D-412	350% minimum
100% Modulus	ASTM D-412	450 psi minimum
Tear Strength	ASTM D-624-91	150 pounds per inch minimum
Brittle Point	ASTM D-749-79 (1987)	minus 80°F minimum

Page 3

FIGURE 13.2 (*Continued*) Guide specification for section 3252, waterstop. (*Courtesy Vinylex Corporation*)

Chemical Resistance	ASTM D-471 (166 hr immersion)	Ultimate Tensile Retention - 60% minimum 100% Modulus Retention - 75% minimum

7. Provide PAPA meeting the following minimum requirements:

Physical Characteristics	Test Method	Required Results
Hydrocarbon Plastic content % by weight	ASTM D-4-80	50% to 70%
Inert Mineral Ash Filler, % by weight	AASHTO T-111	30% to 50%
Volatile Matter, % by weigh	ASTM D-6-80	2.0 maximum
Specific Gravity @ 77°F	ASTM D 71-84	1.20 to 1.35
Ductility @ 77°F	ASTM D 113-85	5.0 minimum
Softening Point, Ring and Ball @ 77°F	ASTM D-36-86	320 minimum
Penetration Cone @ 77°F, 150 g 5 sec in 0.1 mm	ASTM D 217-85	50 to 120
Flash Point, Cleveland open cup °F	ASTM D-92-85	600°F minimum

PART 3 - EXECUTION

3.01 INSTALLATION

A. Waterstops - PVC and TPER General

1. Install Waterstops in concrete joints where indicated
2. Carry Waterstops in walls into lower slabs and join to Waterstops in slabs with appropriate fittings.
3. In water bearing structures: Provide all joints with Waterstops, whether indicated on the drawings or not.
4. Provide Waterstops that are continuous.
5. Set Waterstops accurately to position and line as indicated on the drawings.
6. Secure Waterstops using factory installed hog rings or factory pre-punched holes in the outermost rib with tie wire. Secure at intervals of not more that 15 inches so that the Waterstop does not move during the pour of concrete.
7. Do not drive nails, screws, or other fasteners through the Waterstop at any time or location.
8. Terminate Waterstops 3 inches from the top of finished surfaces of walls and slabs, unless otherwise specified on the drawings.

A. Waterstops - PAPA, General

1. Install Waterstops in concrete joints where indicated.
2. Carry Waterstops in walls into lower slabs and join to Waterstops in slabs with appropriate overlaps.
3. In water bearing structures: Provide all joints with Waterstops, whether indicated on the drawings or not.
4. Provide Waterstops that are continuous.
5. Set Waterstops accurately to position and line as indicated on the drawings.
6. Prime all surfaces to receive Waterstop with primer as recommended by the manufacturer. Allow primer to dry as recommended by the manufacturer.
7. Terminate Waterstops 3 inches from the top of finished surfaces of walls and slabs, unless otherwise specified on the drawings.
8. Do not install PAPA in expansion or contraction joints.
9. Do not attempt to splice PAPA with PVC or TPER Waterstop.

spec.ws

FIGURE 13.2 (*Continued*) Guide specification for section 3252, waterstop. (*Courtesy Vinylex Corporation*)

P.O. BOX 10615, COSTA MESA, CA 92627
PH (714) 662-4445 FAX (714) 662-4446
e-mail: webac@webac.com
internet: http://www.webac.com

WEBAC *America Corp.* **WEBAC stops WATER**

SPECIFICATION - INJECTION GROUTING

1 GENERAL REQUIREMENTS

1.1 Scope

a) Furnish all labor, materials, tools and equipment, and perform all operations necessary for water fix with injection grouting work.

b) Operation of drilling test and grout holes and the injection of grouting shall proceed as indicated in the sequence of work.

c) Injection of grout is for filling of voids and for the repair of water leaking cracks and joints.

d) Holes shall be drilled as required, mechanical packers shall be installed and grout shall be injected under such pressure so as not to damage the existing structure.

1.2 QUALITY ASSURANCE

a) Components shall be installed in accordance with manufacturers specifications.

b) The work shall be performed by a firm with the necessary experience installing this type of material.

2. MATERIALS

2.1 CHEMICAL GROUT

The chemical grout shall be a semi-flexible or flexible water activated hydrophobic polyurethane injection grout with accelerator. The grouting compound shall be a MDI based, 100% solvent free and non-flammable polymer. Grout shall be applied to defective joint or crack by use of a mechanical packer or injection port.

All materials shall be delivered to the site in undamaged, unopened containers bearing the manufacturer's original labels. Grouting shall be performed in accordance with the manufacturer's recommendations.

When the grout is mixed with 0.5% to 10% of accelerator, the mixture shall react upon contact with moisture to a polyurethane foam under expanding its volume.

3. CONSTRUCTION METHODS

3.1 DRILLING HOLES

The contractor shall, where required by inspection or ordered, drill holes in the existing structure for grout work.

PAGE 1 OF 3

FIGURE 13.3 Guide specification for section 3700, epoxy or chemical grout injection. (*Courtesy Webac Corporation*)

P.O. BOX 10615, COSTA MESA, CA 92627
PH (714) 662-4445 FAX (714) 662-4446
e-mail: webac@webac.com
internet: http://www.webac.com

WEBAC *America Corp.* **WEBAC stops WATER**

SPECIFICATION - INJECTION GROUTING

3.2 CHEMICAL GROUT

a) Storage

Immediately upon receipt at the work site, all chemical grouting materials shall be stored in secured, weather tight structure. A sufficient quantity of chemical grout and other components shall be stored at the work site to insure that grouting operations will not be delayed by shortages.

b) Mixing and Handling

Mixing and handling of the chemical grout shall be in accordance with the recommendations of the manufacturer and all applicable safety codes and shall be performed in such manner as to minimize hazard to personnel. It is the responsibility of the contractor to provide appropriate protective measures to insure that chemicals or foam produced by the chemicals are under control of the contractor at all times.

c) Pumps

Hand operated, air driven or electrical pumps can be used. One to five gallons per minute delivery is normally satisfactory. Pumps must be flushed with washing agent for at least two minutes before and after the grouting operation.

d) Equipment

All chemical grouting equipment shall be of a type, capacity and mechanical condition suitable for doing the work. The equipment shall be compatible with the chemicals to be handled and shall be maintained in proper operating condition at all times.

3.3 PROCEDURES

Remove loose, disintegrating and porous concrete at leaking cracks and joints. The size of the holes to be drilled depends on the size and kind of mechanical packer which is to be used and is of the greatest importance for a tight fit of the port. Drill holes in an angle to the wall alternately left and right near the crack to intersect leaking crack inside the structure. Insert packer and tighten properly. The injection of straight construction joints with seal in place may call for a different routine. In some cases it is beneficial to drill holes straight into the joint or seal to avoid unnecessary damage to the concrete. The distance of the holes to be drilled depends on the width of the crack and the dimensions of the structure. The distance of the ports to each other shall be appropriate to allow sufficient penetration of gout to seal the leak completely. If necessary, crack or joints shall be patched with quick setting cement or epoxy before injection. When packers are installed grout shall be injected under sufficient pressure to fill the crack or joint and to seal the leak. For dry cracks inject water first. If possible grout flow shall be monitored by using the 'port to port travel' method. To monitor 'port to port travel' of grout, zerk fittings of the ports not being injected shall be detached until grout starts to leak from the openings.

Grouting should begin at the lowest elevation and proceed upward on vertical cracks or shall begin at one end and proceed toward the other end on horizontal cracks. Injection of chemical grout shall continue at the first injection port until grout is observed at the next port. The valve shall then be closed and the supply line disconnected. Grout shall then be injected into the next port. This operation shall be repeated until the crack has been completely grouted.

PAGE 2 0F 3

FIGURE 13.3 (*Continued*) Guide specification for section 3700, epoxy or chemical grout injection. (*Courtesy Webac Corporation*)

P.O. BOX 10615, COSTA MESA, CA 92627
PH (714) 662-4445 FAX (714) 662-4446
e-mail: webac@webac.com
internet: http://www.webac.com

WEBAC *America Corp.* **WEBAC stops WATER**

SPECIFICATION - INJECTION GROUTING

When serious grout leaks are observed at the surface, they shall be plugged with rags dipped in chemical grout or by other approved methods. The injection pressure shall be adequate to the structure to be sealed. Foaming polyurethanes increase pressure while reacting!

On completion of grouting and after the grout has completely cured, fittings shall be removed and any remaining holes shall be filled with appropriate material.

COMMENT:

If possible, the chemical grout shall be injected into the crack when the space is at the widest aspect of the moving cycle. The material will be in compression or neutral and unnecessary tension will be avoided.

3.4 SAFETY PRECAUTIONS

To prevent any over-exposure to the chemicals used the following precautions shall be observed in strict compliance.

a) Technical data sheets and MSDS must be read and understood.

b) Each employee who works with the products must wear protective clothing, safety goggles, face shields, gloves and boots.

c) A portable eye wash unit must be provided near the work site for an immediate use in an emergency.

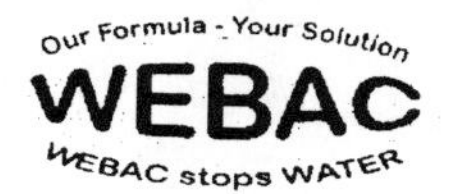

FIGURE 13.3 (*Continued*) Guide specification for section 3700, epoxy or chemical grout injection. (*Courtesy Webac Corporation*)

SECTION 07100
LIQUID BOOT®
FLUID APPLIED LINER FOR CANALS, PONDS, TANKS, RESERVOIRS
VERSION 4.0

(Note: These Specifications may have changed. Visit our website at www.liquidboot.com or call LBI Technologies, Inc. at (714) 575-9200 for most recent version)

PART 1 - GENERAL

1.01 **DESCRIPTION -**
General and Supplementary Conditions and Division 1 - General Requirements applies to this section. Provide fluid applied liner for canals, ponds, tanks, or reservoirs as indicated, specified and required.

A. Work in this section - principal items include:
   1. Fluid applied liners for canals, ponds, tanks, reservoirs, ditches, etc.

B. Related work not in this section:
   1. Excavation and backfilling.
   2. Parge coat on masonry to receive membrane.
   3. Mortar beds or concrete toppings over membranes.
   4. Latex waterproofing.
   5. Damp-proofing.
   6. Flashing and sheet metal.
   7. Joint sealers.
   8. Soil sterilant.
   9. Sand blasting.

1.02 QUALITY ASSURANCE -
A. Liner contractor/applicator shall be trained and approved by liner manufacturer, LBI Technologies, Inc. (LBI).

B. A pre-installation conference shall be held prior to application of liner to assure proper substrate and installation conditions, to include contractor, applicator, architect/engineer and special inspector (if any).

1.03 **SUBMITTALS** - (Refer to section 01300 for procedures)
A. Project Data - Submit manufacturer's product data and installation instructions for specific application.

B. Samples - Submit representative samples of the following for approval:
   1. Liner material.
   2. Protection Board and/or Protection Mat.
   3. Prefabricated Drainage Mat.
   4. Geotextiles.

1.04 **DELIVERY, STORAGE AND HANDLING -**
Deliver materials to site in original unbroken packages bearing manufacturer's label showing brand, weight, volume and batch number. Store materials at site in strict compliance with manufacturer's instructions. Do not allow materials to freeze in containers.

F:\data\specbook\reser40 1

FIGURE 13.4 Guide specification for section 7100, fluid membrane for civil projects. (*Courtesy LBI Technologies*)

1.05 JOB CONDITIONS -

A. Protect all adjacent areas not to be lined. Where necessary, apply masking to prevent staining of surfaces to remain exposed wherever membrane abuts to other finish surfaces.

B. Perform work only when existing and forecasted weather conditions are within manufacturer's recommendations for the material and product used.

C. Minimum clearance of 24 inches is required for application of product.

D. Ambient temperature shall be within manufacturer's specifications. (Greater than +32°F/+0°C).

E. All plumbing, electrical, mechanical and structural items to be under or passing through the liner shall be positively secured in their proper positions and appropriately protected prior to membrane application.

F. Liner shall be installed before placement of reinforcing steel. When not possible, all exposed reinforcing steel shall be masked by General Contractor prior to membrane application.

G. Expansion joints must be filled with a conventional waterproof expansion material.

1.06 PRODUCT WARRANTY -

LBI Technologies, Inc. (LBI) warrants its products to be free of defects. This warranty only applies when the product is applied by LBI Approved Applicators. As factors which affect the result obtained from this product -- including weather, equipment utilized, construction, workmanship and other variables -- are all beyond the manufacturer's control, LBI warrants only that the material conforms to its product specifications. Under this warranty LBI will replace at no charge any product not meeting these specifications within 12 months of manufacture, provided it has been applied in accordance with LBI written directions for use recommended as suitable for this product. Warranties are available for a longer period upon request and mutual written consent. This warranty is in lieu of any and all other warranties expressed or implied (including any implied warranty of merchantability or fitness for a particular use), and LBI shall have no further liability of any kind including liability for consequential or incidental damages resulting from any defects or delays caused by replacement or otherwise.

F:\data\specbook\reser40 2

FIGURE 13.4 *(Continued)* Guide specification for section 7100, fluid membrane for civil projects. *(Courtesy LBI Technologies)*

PART 2 - PRODUCTS

2.01 MATERIALS -

A. Fluid applied liner system - LIQUID BOOT®; a single course, high build, polymer modified asphaltic emulsion. Water borne and spray applied at ambient temperatures. A minimum thickness of 80 dry mils, unless specified otherwise. Non-toxic and odorless. LIQUID BOOT® Trowel Grade has similar properties with greater viscosity and is trowel applied. Manufactured by LBI Technologies, Inc., Anaheim, CA (714) 575-9200.

B. Fluid applied liner physical properties:

Elongation	ASTM D412	1,332% w/o reinforcement (90% Recovery)
Tensile Strength	ASTM D412	58 p.s.i. w/o reinforcement
Hydrostatic Head Resistance	ASTM D751	Tested to 138 feet (60 psi)
Tensile Bond Strength to Concrete	ASTM D413	2,556 lbs/ft² uplift force
Tensile Strength With 8 ounce Non-woven Geotextile both sides	ASTM D751	196 p.s.i. (Same as Geotextile tested separately)
Elongation With 8 ounce Non-woven Geotextile both sides	ASTM D751	100 % (Same as Geotextile tested separately)
Puncture Resistance With 8 ounce Non-woven Geotextile both sides	ASTM D4833	286 pounds (Travel of probe = 0.756 in.) (Same as Geotextile tested separately)
Hydrostatic Resistance (Burst) With 8 ounce Non-woven Geotextile both sides	ASTM D751	610 p.s.i. (Same as Geotextile tested separately)
Water Penetration Rate	ASTM D2434	$<7.75 \times 10^{-9}$ cm/sec
Water Vapor Permeability	ASTM E96	0.24 Perms
Water Vapor Transmission	ASTM E96	0.10 grains/h-ft²
Methane Gas Permeability	ASTM D1434	None Detected
Hydrogen Sulfide Gas Permeability	ASTM D1434	None Detected
Radon Permeability	Tested by the U.S. Dept. of Energy	Zero Permeability to Radon (222 Rn)
Cold Bend Test	ASTM D146	Passed - No cracking @ -25°F
Freeze-Thaw Resistance (100 Cycles)	ASTM A742	Meets criteria - No spalling or disbondment
Accelerated Weathering and Ultraviolet Exposure	ASTM D822	No adverse affect after 500 hours
Flame Spread	ASTM E108	Class A (Comparable to UL790)
Acid Exposure (10% H_2SO_4 for 90 Days)	ASTM D543	Less than 1% change of weight
Oil Exposure (30 wt. for 28 days)	ASTM D543	Less than 1% change of weight
Soil Burial	ASTM D4068	Passed - No change to elongation or gas permeability properties
Electrical Volume Resistivity	ASTM D257	1.91×10^{10} ohms-cm
Toxicity Test	22 CCR 66696	Passed - CCR Bioassay-Flathead Minnow
Potable Water Containment	ANSI/NSF 61	NSF Certified for tanks >300,000 gallons

C. Geotextile - Typar3401 non-woven geotextile, unless otherwise specified and approved by membrane manufacturer.The heat-rolled side shall be used as the applicator surface. Some projects may require a heavier geotextile (Typar 3631 or 3801).

D. Cold Joints, Cracks, Form Tie Holes: Hardcast CRT 1602 Type 3" wide.

F:\data\specbook\reser40 3

FIGURE 13.4 (*Continued*) Guide specification for section 7100, fluid membrane for civil projects. (*Courtesy LBI Technologies*)

PART 3 - EXECUTION

3.01 **EXAMINATION** -
All surfaces to be lined shall be inspected and approved by the applicator at least one day prior to commencing work.

3.02 **SURFACE PREPARATION** -
Provide 24 inch minimum clearance out from surfaces to receive the liner. The application surface shall be prepared and provided to the applicator in accordance with manufacturer's specifications listed below:

NOTE: Before lining tank, canal or reservoir, determine that it is structurally sound and able to support the weight and pressure of its contents when full. All cracks, joints, and holes must be grouted with appropriate materials sufficient to support the weight and pressures of the contents.

A. **Concrete/Shotcrete/Masonry/Asphalt**
Concrete surfaces shall be light broom finish or smoother, free of any dirt, algae, debris, loose material, release agents or curing compounds. Fill all voids more than ¼ inch deep and ¼ inch wide. Masonry joints shal be struck smooth. All penetrations shall be prepared in accordance with manufacturer's specifications. Provide a ¾ inch minimum cant of LIQUID BOOT® Trowel Grade, or other suitable material, at all horizontal to vertical transitions and other inside corners of 120° or less.

All form tie holes must be completely grouted from the inside to outside of wall with non-shrink grout as approved by engineer.

Install Hardcast reinforcing tape over all cold joints, cracks and form tie holes (after holes are grouted).

B. **Dirt**
The sub-grade shall be moisture conditioned and compacted to a minimum relative compaction of 90 percent or as specified by civil/geotechnical engineer. The finished surface shall be smooth, uniform, free of debris and standing water. Remove all stones or dirt clods greater than ¼ inch. (NOTE: Aggregate sub-bases shall be rolled flat.) Final sub-grade preparation shall not precede the membrane application by more than 72 hours. All penetrations shall be prepared in accordance with manufacturer's specifications. All form stakes that penetrate the membrane shall be of rebar which shall be bent over and left in the slab.

Trenches shall be cut oversize to accommodate liner and protection course with perpendicular to sloped sides and maximum obtainable compaction. Adjoining grade shall be finish graded and compacted. Excavated walls shall be vertical to sloped back, free of roots and protruding rocks. Specific sub-grade preparation shall be designed by a qualified civil or geotechnical engineer.

Soil sterilant must be sprayed onto dirt subbase for all canals, ponds, and reservoirs before geotextile is rolled out. Spray apply soil sterilant at the sterilant manufacturer's recommended rate.

C. Steel and Other Metal Surfaces
Remove all rust and loose material from the surface with a wire brush or brush blast. Repair all weak and/or perforated areas. Remove all factory oils from new substrates.

3.03 **INSTALLATION** -

3.03.10 **INSTALLATION ON DIRT SURFACES AND MUDSLABS**

A. Provide a continuous trench at least one foot (1') wide by one foot (1') deep at perimeter.

B. Roll out geotextile on sub-grade with the heat-rolled side facing up. (Note: For mudslabs, use 6 oz/yd^2 non-woven geotextile.) Overlap seams a minimum of six inches (6"). Lay geotextile tight at all inside corners. Spray LIQUID BOOT® within the seam overlap to a thickness of 80 mils minimum.

Terminate perimeter ends by making at least two complete turns around one inch (1"), schedule 40 plastic pipe. Place pipe in bottom of trench and backfill with clean, fine-grained soil.

F:\data\specbook\reser40 4

FIGURE 13.4 *(Continued)* Guide specification for section 7100, fluid membrane for civil projects. *(Courtesy LBI Technologies)*

C. Refer to section 3.03.40, "Sealing Around Penetrations", for procedures to seal around penetrations.

D. Spray apply LIQUID BOOT® onto geotextile to an 80 mils minimum dry thickness. Increase thickness to 100 dry mils if shotcrete is to be applied directly to membrane. If a second coat is required, remove any standing water from the membrane before proceeding with the second application.

E. Do not penetrate membrane. Keep membrane free of dirt, debris and traffic until a water or a protective cover is in place. It is the responsibility of the General Contractor to insure that the membrane and the protection system is not penetrated.

F. If foot traffic is anticipated on the membrane, create walking paths with protection board or geotextile.

G. Fill canal, pond, tank, or reservoir with water within 72 hours after membrane has completely cured.

3.03.20 INSTALLATION ON CONCRETE/SHOTCRETE/MASONRY/ASPHALT

Due to the numerous variables affecting concrete (ie: water content, mix specifications, cement source, "free-lime" percentage, calcium content, pumped vs. poured, environmental conditions at the time of concrete placement, admixtures, acidity, type of finish, curing conditions, etc., etc.) every job will require pre-testing of LIQUID BOOT® to determine the installation procedure. Follow the procedures below carefully.

A. Refer to section 3.03.40, "Sealing Around Penetrations", for procedures to seal around penetrations.

B. Provide a ¾ inch minimum cant of LIQUID BOOT® Trowel Grade, or other suitable material, at all horizontal to vertical transitions and other inside corners of 120° or less. **Allow to cure a minimum of 24 hours before the application of LIQUID BOOT®.**

C. Delineate a test area <u>on site</u> with a minimum dimension of 10 feet by 10 feet (3m by 3m). Apply LIQUID BOOT® to a thickness of 80 mils and let it cure for **24 hours**. Observe for blisters. If minor or no blistering occurs, proceed to the next step. (See note regarding blisters below). If blistering does occur, apply a thin (10 mil) tack coat of LIQUID BOOT® "A" side without catalyst to the concrete surface and allow to cure before proceeding. On horizontal concrete surfaces a 4 oz/yd² nonwoven geotextile may be applied onto the concrete in lieu of the tack coat. (See also information regarding blister repair on page 7.)

D. Spray apply LIQUID BOOT® to an 80 mils minimum dry thickness. Increase thickness to 100 dry mils if shotcrete is to be applied directly to membrane. If a second coat is required, remove any standing water from the membrane before proceeding with the second application.

E. Do not penetrate membrane. Keep membrane free of dirt and debris and traffic until water or a protective cover is in place. It is the responsibility of the General Contractor to insure that the membrane and the protection system is not penetrated.

F. If foot traffic is anticipated on the membrane, create walking paths with protection board or geotextile.

G. Fill canal, pond, tank, or reservoir with water within 72 hours after membrane has completely cured.

NON-HORIZONTAL SURFACES: Spray on non-horizontal surfaces should begin at the bottom and work towards the top. This method allows the product to adhere to the surface before hitting catalyst runoff.

NOTE: Due to the nature of concrete as a substrate, it is normal for some blistering to occur. This is caused by either concrete's tendency to off-gas when sealed, or water that is temporarily trapped between the concrete and the membrane. With time and the applied pressure of backfill or over-slab, blisters will absorb into the concrete without detriment to the membrane.

A small number of blister heads should be sampled and checked for proper membrane thickness. If the samples have the required membrane thickness (80 mils minimum), <u>then the remaining blisters should not be punctured or cut</u>. If the samples have less than the minimum 80 mils, then the area can either be resprayed to obtain the proper thickness, or the blisters can be cut out and the area resprayed or patched with LIQUID BOOT® Trowel Grade.

F:\data\specbook\reser40 5

FIGURE 13.4 (*Continued*) Guide specification for section 7100, fluid membrane for civil projects. (*Courtesy LBI Technologies*)

3.03.30 INSTALLATION ON STEEL (TANKS AND SIMILAR SURFACES)

A. Refer to Section 3.03.40, "Sealing Around Penetrations", for procedures to seal around penetrations.

B. Provide a ¾ inch minimum cant of LIQUID BOOT® Trowel Grade at all horizontal to vertical transitions and other inside corners of 90° or less. **Allow to cure a minimum of 24 hours before the application of LIQUID BOOT®.**

NOTE: WALLS AND CEILINGS SHOULD BE COMPLETED A MINIMUM OF 24 HOURS AND ALL RESIDUAL WATER REMOVED PRIOR TO COMMENCING APPLICATION ON FLOORS.

C. WALLS & CEILINGS: Spray apply a thin (10 mil) tack coat of LIQUID BOOT® **"A" side only (no catalyst).** Let it cure until tacky, then proceed to the next step.

D. WALLS & CEILINGS: Spray apply LIQUID BOOT® to an 80 mils minimum dry thickness. Increase thickness to 100 dry mils if shotcrete is to be applied directly to membrane. Spray on non-horizontal surfaces should begin at the bottom and work towards the top.

NOTE: WALLS AND CEILINGS SHOULD BE COMPLETED A MINIMUM OF 24 HOURS AND ALL RESIDUAL WATER REMOVED PRIOR TO COMMENCING APPLICATION ON FLOORS.

E. FLOORS: Spray apply a thin (10 mil) tack coat of LIQUID BOOT® **"A" side only (no catalyst).** Let it cure until tacky, then proceed to the next step.

F. FLOORS: Spray apply LIQUID BOOT® to an 80 mils minimum dry thickness. Increase thickness to 100 dry mils if shotcrete is to be applied directly to membrane.

G. Remove all residual moisture from the floor and continue to ventilate for a minimum of 72 hours following the application to allow the membrane to cure completely.

H. Take steps to insure that any initial blast of material which may hit the membrane during the fill process is adequately dissipated so that membrane is not damaged.

I. Do not penetrate membrane. Keep membrane free of dirt and debris and traffic until the tank is filled or a protective cover is in place.

3.03.40 SEALING AROUND PENETRATIONS

A. Clean and etch all penetrations. All metal penetrations shall be etched with a 10% muriatic acid solution.

B. Roll out geotextile on sub-grade with the heat-rolled side facing up. Overlapping seams a minimum of six inches (6"). Cut the geotextile around penetrations so that it lays flat on the sub-grade. Lay geotextile tight at all inside corners. Spray LIQUID BOOT® within the seam overlap to a thickness of 80 mils minimum. (**NOTE:** This step may be omitted with applications not requiring geotextile).

C. Apply 80 mils minimum dry thickness. LIQUID BOOT® Trowel Grade in a three inch (3") wide ring around the penetration and up the penetration a minimum of three inches (3").

D. Allow Trowel Grade to cure completely before proceeding to step "E".

E. Spray apply LIQUID BOOT® to an 80 mils minimum dry thickness around the penetration, completely encapsulating the collar assembly and to a height one and one half inches (1½") minimum above the Trowel Grade collar. Spray apply LIQUID BOOT® to surrounding areas as specified for the particular application.

F. Allow LIQUID BOOT® to cure completely before proceeding to step "G".

G. Wrap penetration with polypropylene cable tie at a point two inches (2") above the base of the penetration. Tighten the cable tie firmly so as to squeeze the cured membrane collar.

F:\data\specbook\reser40 6

FIGURE 13.4 *(Continued)* Guide specification for section 7100, fluid membrane for civil projects. *(Courtesy LBI Technologies)*

3.04 FIELD QUALITY CONTROL -

Field Quality Control is a very important part of all LIQUID BOOT® applications. Applicators should check their own work for coverage, thickness, and all around good workmanship before calling for inspections.

When thickness or integrity is in question the membrane should be tested in the proper manner as described below. However, over-sampling defeats the intent of inspections. Inspectors should always use visual and tactile measurement to guide them. Areas suspected of being too thin to the touch should be measured with the gauges to determine the exact thickness. With practice and by comparing tactile measurements with those of the gauges, fingers become very accurate tools.

3.04.10 ON DIRT AND OTHER SOFT SUBSTRATES

A. Samples to be inspected may be cut from the membrane and geotextile sandwich to a maximum area of 2 square inches, per 500 square feet. Measure the thickness with a mil-reading caliper. Deduct the plain geotextile thickness to determine the thickness of LIQUID BOOT membrane. Mark the test area for repair.

B. Voids left by sampling are to be patched with geotextile overlapping the void by a minimum of two inches (2"). Apply a thin tack coat of LIQUID BOOT® under the geotextile patch. Then spray or trowel apply LIQUID BOOT® to an 80 mils minimum dry thickness, extending at least three inches (3") beyond geotextile patch.

3.04.20 ON CONCRETE/SHOTCRETE/MASONRY & OTHER HARD SURFACES

A. Membrane may be checked for coverage with a blunt-nose depth gauge, taking four (4) readings over a one square inch area, every 500 square feet. Record the minimum reading. Mark the test area for repair.

B. Test areas are to be patched over with LIQUID BOOT® to an 80 mils minimum dry thickness, extending a minimum of one inch (1") beyond the test perimeter.

NOTE: Due to the nature of concrete as a substrate, it is normal for some blistering to occur. This is caused by either concrete's tendency to off-gas when sealed, or water that is temporarily trapped between the concrete and the membrane. With time and the applied pressure of backfill or over-slab, blisters will absorb into the concrete without detriment to the membrane.

A small number of blister heads should be sampled and checked for proper membrane thickness. If the samples have the required membrane thickness (80 mils minimum), then the remaining blisters should not be punctured or cut. If the samples have less than the minimum 80 mils, then the area can either be resprayed to obtain the proper thickness, or the blisters can be cut out and the area resprayed or patched with LIQUID BOOT® Trowel Grade.

F:\data\specbook\reser40 7

FIGURE 13.4 *(Continued)* Guide specification for section 7100, fluid membrane for civil projects. *(Courtesy LBI Technologies)*

SECTION 07100
LIQUID BOOT®
FLUID APPLIED WATERPROOFING
VERSION 2.0

(Note: These Specifications may have changed. Visit our website at www.liquidboot.com or call LBI Technologies, Inc. at (714) 575-9200 for most recent version)

PART 1 - GENERAL

1.01 DESCRIPTION -
General and Supplementary Conditions and Division 1 - General Requirements applies to this section. Provide fluid applied waterproofing as indicated, specified and required.

A. Work in this section - principal items include:
1. Fluid applied waterproofing on buildings, planter and site retaining walls.
2. Between slab waterproof membrane.

B. Related work not in this section:
1. Excavation and backfilling.
2. Parge coat on masonry to receive waterproof membrane.
3. Mortar beds or concrete toppings over waterproof membranes.
4. Latex waterproofing.
5. Damp-proofing.
6. Flashing and sheet metal.
7. Joint sealers.
8. Soil sterilant.

1.02 QUALITY ASSURANCE -
A. Waterproofing contractor/applicator shall be trained and approved by waterproof membrane manufacturer, LBI Technologies, Inc. (LBI).

B. A pre-installation conference shall be held prior to application of waterproof membrane to assure proper substrate and installation conditions, to include contractor, applicator, architect/engineer and special inspector (if any).

1.03 SUBMITTALS - (Refer to section 01300 for procedures)
A. Project Data - Submit manufacturer's product data and installation instructions for specific application.

B. Samples - Submit representative samples of the following for approval:
1. Waterproof membrane material.
2. Protection Board and/or Protection Mat.
3. Prefabricated Drainage Mat.
4. Geotextiles.

1.04 DELIVERY, STORAGE AND HANDLING -
Deliver materials to site in original unbroken packages bearing manufacturer's label showing brand, weight, volume and batch number. Store materials at site in strict compliance with manufacturer's instructions. Do not allow materials to freeze in containers.

F:\data\specbook\water20 1

FIGURE 13.5 Guide specification for section 7100, fluid-applied waterproofing. (*Courtesy LBI Technologies*)

1.05 **JOB CONDITIONS** -

A. Protect all adjacent areas not to be waterproofed. Where necessary, apply masking to prevent staining of surfaces to remain exposed wherever membrane abuts to other finish surfaces.

B. Perform work only when existing and forecasted weather conditions are within manufacturers recommendations for the material and product used.

C. Minimum clearance of 24 inches is required for application of product.

D. Ambient temperature shall be within manufacturer's specifications. (Greater than +32°F/+0°C).

E. All plumbing, electrical, mechanical and structural items to be under or passing through the waterproof membrane shall be positively secured in their proper positions and appropriately protected prior to membrane application.

F. Waterproof membrane shall be installed before placement of reinforcing steel. When not possible, all exposed reinforcing steel shall be masked by General Contractor prior to membrane application.

G. Expansion joints must be filled with a conventional waterproof expansion joint material.

1.06 **PRODUCT WARRANTY** -

LBI Technologies, Inc. (LBI) warrants its products to be free of defects. This warranty only applies when the product is applied by LBI Approved Applicators. As factors which affect the result obtained from this product -- including weather, equipment utilized, construction, workmanship and other variables -- are all beyond the manufacturer's control, LBI warrants only that the material conforms to its product specifications. Under this warranty LBI will replace at no charge any product not meeting these specifications within 12 months of manufacture, provided it has been applied in accordance with LBI written directions for use recommended as suitable for this product. Warranties are available for a longer period upon request and mutual written consent. This warranty is in lieu of any and all other warranties expressed or implied (including any implied warranty of merchantability or fitness for a particular use), and LBI shall have no further liability of any kind including liability for consequential or incidental damages resulting from any defects or delays caused by replacement or otherwise.

F:\data\specbook\water20 2

FIGURE 13.5 (*Continued*) Guide specification for section 7100, fluid-applied waterproofing. (*Courtesy LBI Technologies*)

PART 2 - PRODUCTS

2.01 MATERIALS -

A. Fluid applied waterproofing system - LIQUID BOOT®; a single course, high build, polymer modified asphaltic emulsion. Water borne and spray applied at ambient temperatures. A minimum thickness of 80 dry mils, unless specified otherwise. Non-toxic and odorless. LIQUID BOOT® Trowel Grade has similar properties with greater viscosity and is trowel applied. Manufactured by LBI Technologies, Inc., Anaheim, CA (714) 575-9200.

B. Fluid applied waterproofing physical properties:

Elongation	ASTM D412	1,332% w/o reinforcement (90% Recovery)
Tensile Strength	ASTM D412	58 p.s.i. w/o reinforcement
Hydrostatic Head Resistance	ASTM D751	Tested to 138 feet (60 psi)
Tensile Bond Strength to Concrete	ASTM D413	2,556 lbs/ft² uplift force
Tensile Strength With 8 ounce Non-woven Geotextile both sides	ASTM D751	196 p.s.i. (Same as Geotextile tested separately)
Elongation With 8 ounce Non-woven Geotextile both sides	ASTM D751	100 % (Same as Geotextile tested separately)
Puncture Resistance With 8 ounce Non-woven Geotextile both sides	ASTM D4833	286 pounds (Travel of probe = 0.756 in.) (Same as Geotextile tested separately)
Hydrostatic Resistance (Burst) With 8 ounce Non-woven Geotextile both sides	ASTM D751	610 p.s.i. (Same as Geotextile tested separately)
Water Penetration Rate	ASTM D2434	$<7.75 \times 10^{-9}$ cm/sec
Water Vapor Permeability	ASTM E96	0.24 Perms
Water Vapor Transmission	ASTM E96	0.10 grains/h-ft²
Methane Gas Permeability	ASTM D1434	None Detected
Hydrogen Sulfide Gas Permeability	ASTM D1434	None Detected
Radon Permeability	Tested by the U.S. Dept. of Energy	Zero Permeability to Radon (222 Rn)
Cold Bend Test	ASTM D146	Passed - No cracking @ -25°F
Freeze-Thaw Resistance (100 Cycles)	ASTM A742	Meets criteria - No spalling or disbondment
Accelerated Weathering and Ultraviolet Exposure	ASTM D822	No adverse affect after 500 hours
Flame Spread	ASTM E108	Class A (Comparable to UL790)
Acid Exposure (10% H_2SO_4 for 90 Days)	ASTM D543	Less than 1% change of weight
Oil Exposure (30 wt. for 28 days)	ASTM D543	Less than 1% change of weight
Soil Burial	ASTM D4068	Passed - No change to elongation or gas permeability properties
Electrical Volume Resistivity	ASTM D257	1.91×10^{10} ohms-cm
Toxicity Test	22 CCR 66696	Passed - CCR Bioassay-Flathead Minnow
Potable Water Containment	ANSI/NSF 61	NSF Certified for tanks >300,000 gallons

C. Protection - On vertical surfaces, use: MIRADRI 200V or other protection as approved by the manufacturer.

On horizontal surfaces, use: MIRADRI 300HV; ⅛ inch thick, asphaltic impregnated, felt and fiberglass protection mat; or other protection as approved by the manufacturer. Two inches (2") of clean sand may be used where there is light construction activity and is approved by project architect or engineer.

D. Prefabricated Drainage Mat - A formed plastic sheet, one side smooth with concave dimples, the other side bonded to a geotextile.

E. Geotextile - Typar 3401 non-woven geotextile, unless otherwise specified and approved by membrane manufacturer. The heat-rolled side shall be used as the application surface. Some projects may require a heavier geotextile (Typar 3631 or 3801).

F. Cold Joints, Cracks, Form Tie Holes: Hardcast CRT 1602 Type 3" wide.

F:\data\specbook\water20 3

FIGURE 13.5 (*Continued*) Guide specification for section 7100, fluid-applied waterproofing. (*Courtesy LBI Technologies*)

PART 3 - EXECUTION

3.01 **EXAMINATION** -
All surfaces to be waterproofed shall be inspected and approved by the applicator at least one day prior to commencing work.

3.02 **SURFACE PREPARATION** -
Provide 24 inch minimum clearance out from surfaces to receive the waterproof membrane. The application surface shall be prepared and provided to the applicator in accordance with manufacturer's specifications listed below:

A. **Concrete/Shotcrete/Masonry**
Concrete surfaces shall be light broom finish or smoother, free of any dirt, debris, loose material, release agents or curing compounds. Fill all voids more than ¼ inch deep and ¼ inch wide. Masonry joints shall be struck smooth. All penetrations shall be prepared in accordance with manufacturer's specifications. Provide a ¾ inch minimum cant of LIQUID BOOT® Trowel Grade, or other suitable material, at all horizontal to vertical transitions and other inside corners of 120° or less.

All form ties holes must be completely grouted from the inside to outside of wall with non-shrink grout as approved by engineer.

Install Hardcast reinforcing tape over all cold joints, cracks and form tie holes (after holes are grouted).

B. **Dirt & Gravel**
The sub-grade shall be moisture conditioned and compacted to a minimum relative compaction of 90 percent or as specified by civil/geotechnical engineer. The finished surface shall be smooth, uniform, free of debris and standing water. Remove all stones or dirt clods greater than ¼ inch. (NOTE: Aggregate sub-bases shall be rolled flat). Final sub-grade preparation shall not precede the membrane application by more than 72 hours. All penetrations shall be prepared in accordance with manufacturer's specifications. All form stakes that penetrate the membrane shall be of rebar which shall be bent over and left in the slab.

Trenches shall be cut oversize to accommodate waterproof membrane and protection course with perpendicular to sloped sides and maximum obtainable compaction. Adjoining grade shall be finish graded and compacted. Excavated walls shall be vertical or sloped back, free of roots and protruding rocks. Specific sub-grade preparation shall be designed by a qualified civil or geotechnical engineer.

If organic materials with potential for growth (ie: seeds or grasses) exist within the subbase, spray apply soil sterilant at the sterilant manufacturer's recommended rate.

C. **Lagging**
Lagging shall be held securely in place. All sharp edges and nails shall be removed or protected so as not to penetrate the membrane.

3.03 **INSTALLATION** -

3.03.10 **INSTALLATION ON DIRT SURFACES AND MUDSLABS**

A. Roll out geotextile on sub-grade with the heat-rolled side facing up. (Note: For mudslabs, use 6 oz/yd^2 non-woven geotextile.) Overlap seams a minimum of six inches (6"). Lay geotextile tight at all inside corners. Spray LIQUID BOOT® within the seam overlap to a thickness of 80 mils minimum.

Line trenches with geotextile extending at least six inches (6") onto adjoining sub-grade if slab and footings are to be sprayed separately. Overlap seams a minimum of six inches (6"). Lay geotextile tight at all inside corners. Spray LIQUID BOOT® within the seam overlap to a thickness of 80 mils minimum.

B. Refer to section 3.03.40, "Sealing Around Penetrations", for procedures to seal around penetrations.

F:\data\specbook\water20 4

FIGURE 13.5 (*Continued*) Guide specification for section 7100, fluid-applied waterproofing. (*Courtesy LBI Technologies*)

C. Spray apply LIQUID BOOT® onto geotextile to an 80 mil minimum dry thickness. Increase thickness to 100 dry mils if shotcrete is to be applied directly to membrane. If a second coat is required, remove any standing water from the membrane before proceeding with the second application.

D. Do not penetrate membrane. Keep membrane free of dirt, debris and traffic until a protective cover is in place. It is the responsibility of the General Contractor to insure that the membrane and the protection system is not penetrated.

E. If protection is required, lay down and butt join protection mat, roofing cap sheet, 6 oz/yd^2 geotextile or other protection as approved by the manufacturer on horizontal waterproof membrane after surface moisture has evaporated from membrane.

Two inches (2") of sand, which is free from debris, may be used in place of the protection options listed above. If sand option is used, the sand may be applied at any time after the membrane is applied.

On vertical surfaces apply 3/8 inch polystyrene protection board, 6 oz/yd^2 geotextile or other protection as approved by the manufacturer.

3.03.20 INSTALLATION ON CONCRETE/SHOTCRETE/MASONRY

Due to the numerous variables affecting concrete (ie: water content, mix specifications, cement source, "free-lime" percentage, calcium content, pumped vs. poured, environmental conditions at the time of concrete placement, admixtures, acidity, type of finish, curing conditions, etc., etc.) every job will require pre-testing of LIQUID BOOT® to determine the installation procedure. Follow the procedures below carefully.

A. Refer to section 3.03.40, "Sealing Around Penetrations", for procedures to seal around penetrations.

B. Provide a ¾ inch minimum cant of LIQUID BOOT® Trowel Grade, or other suitable material, at all horizontal to vertical transitions and other inside corners of 120° or less. **Allow to cure a minimum of 24 hours before the application of LIQUID BOOT®.**

C. Delineate a test area <u>**on site**</u> with a minimum dimension of 10 feet by 10 feet (3m by 3m). Apply LIQUID BOOT® to a thickness of 80 mils and let it cure for **24 hours**. Observe for blisters. If minor or no blistering occurs, proceed to the next step. (See note regarding blisters below). If significant blistering does occur, apply a thin (10 mil) tack coat of LIQUID BOOT® "A" side without catalyst to the entire concrete surface and allow to cure before proceeding. (See also information regarding blister repair on page 7).

D. Spray apply LIQUID BOOT® to an 80 mil minimum dry thickness. Increase thickness to 100 dry mils if shotcrete is to be applied directly to membrane. If a second coat is required, remove any standing water from the membrane before proceeding with the second application.

E. Do not penetrate membrane. Keep membrane free of dirt and debris and traffic until a protective cover is in place. It is the responsibility of the General Contractor to insure that the membrane and the protection system is not penetrated.

F. If protection is required, install pursuant to protection manufacturer's instructions, after surface moisture has evaporated from membrane.

NOTE: If water testing is to be performed, conduct before placing protection course

NON-HORIZONTAL SURFACES: Spray on non-horizontal surfaces should begin at the bottom and work towards the top. This method allows the product to adhere to the surface before hitting catalyst runoff.

F:\data\specbook\water20 5

FIGURE 13.5 *(Continued)* Guide specification for section 7100, fluid-applied waterproofing. *(Courtesy LBI Technologies)*

NOTE: Due to the nature of concrete as a substrate, it is normal for some blistering to occur. This is caused by either concrete's tendency to off-gas or water that is temporarily trapped between the concrete and the membrane. With time and the applied pressure of backfill or over-slab, blisters will absorb into the concrete without detriment to the membrane.

A small number of blister heads should be sampled and checked for proper membrane thickness. If the samples have the required membrane thickness (80 mils minimum), then the remaining blisters should not be punctured or cut. If the samples have less than the minimum 80 mils, then the area can either be resprayed to obtain the proper thickness, or the blisters can be cut out and the area resprayed or patched with LIQUID BOOT® Trowel Grade.

3.03.30 INSTALLATION ON LAGGING

A. Securely nail 8 oz. non-woven geotextile over lagging and soldier piles keeping geotextile tight to lagging wall. Overlap seams a minimum of six inches (6").

B. Roll out geotextile vertically with the heat-rolled side facing out and staple to lagging using ⅜ long staples 12" on center. Overlap seams a minimum of six inches (6"). Spray LIQUID BOOT® within the seam overlap to a thickness of 80 mils minimum. Do not staple top layer of geotextile at overlap.

C. Refer to section 3.03.40, "Sealing Around Penetrations", for procedures to seal around penetrations.

D. Provide a ¾ inch minimum cant of LIQUID BOOT® Trowel Grade, or other suitable material, at all horizontal to vertical transitions and other inside corners of 120° or less. **Allow to cure a minimum of 24 hours before the application of LIQUID BOOT®.**

E. Spray apply LIQUID BOOT® to a minimum thickness of 80 mils (100 mils if installing shotcrete walls). Remove any standing water.

F. Do not penetrate membrane. Keep membrane free of dirt and debris until a protective cover is in place. It is the responsibility of the General Contractor to insure that the membrane and the protection system is not penetrated.

3.03.40 SEALING AROUND PENETRATIONS

A. Clean and etch all penetrations. All metal penetrations shall be etched with a 10% muriatic acid solution.

B. For applications requiring geotextile, roll out geotextile on sub-grade with the heat-rolled side facing up, overlapping seams a minimum of six inches (6"). Cut the geotextile around penetrations so that it lays flat on the sub-grade. Lay geotextile tight at all inside corners. Spray LIQUID BOOT® within the seam overlap to a thickness of 80 mils minimum.

C. Apply 80 mils minimum dry thickness LIQUID BOOT® Trowel Grade in a three inch (3") wide ring around the base of penetration and up the penetration a minimum of three inches (3").

D. Allow Trowel Grade to cure completely before proceeding to step "E".

E. Spray apply LIQUID BOOT® to an 80 mils minimum dry thickness around the penetration, completely encapsulating the collar assembly and to a height of one and one half inches (1½") minimum above the Trowel Grade collar. Spray apply LIQUID BOOT® to surrounding areas as specified for the particular application.

F:\data\specbook\water20 6

FIGURE 13.5 (*Continued*) Guide specification for section 7100, fluid-applied waterproofing. (*Courtesy LBI Technologies*)

F. Allow LIQUID BOOT® to cure completely before proceeding to step "G".

G. Wrap penetration with polypropylene cable tie at a point two inches (2") above the base of the penetration. Tighten the cable tie firmly so as to squeeze the cured membrane collar.

3.04 FIELD QUALITY CONTROL -

Field Quality Control is a very important part of all LIQUID BOOT® applications. Applicators should check their own work for coverage, thickness, and all around good workmanship **before** calling for inspections.

When thickness or integrity is in question the membrane should be tested in the proper manner as described below. However, over-sampling defeats the intent of inspections. Inspectors should always use visual and tactile measurement to guide them. Areas suspected of being too thin to the touch should be measured with the gauges to determine the exact thickness. With practice and by comparing tactile measurements with those of the gauges, fingers become very accurate tools.

3.04.10 ON DIRT AND OTHER SOFT SUBSTRATES

A. Samples to be inspected may be cut from the membrane and geotextile sandwich to a maximum area of 2 square inches, per 500 square feet. Measure the thickness with a mil-reading caliper. Deduct the plain geotextile thickness to determine the thickness of LIQUID BOOT membrane. Mark the test area for repair.

B. Voids left by sampling are to be patched with geotextile overlapping the void by a minimum of two inches (2"). Apply a thin tack coat of LIQUID BOOT® under the geotextile patch. Then spray or trowel apply LIQUID BOOT® to an 80 mils minimum dry thickness, extending at least three inches (3") beyond geotextile patch.

3.04.20 ON CONCRETE/SHOTCRETE/MASONRY & OTHER HARD SURFACES

A. Membrane may be checked for coverage with a blunt-nose depth gauge, taking four (4) readings over a one square inch area, every 500 square feet. Record the minimum reading. Mark the test area for repair.

B. Test areas are to be patched over with LIQUID BOOT® to an 80 mils minimum dry thickness, extending a minimum of one inch (1") beyond the test perimeter.

NOTE: Due to the nature of concrete as a substrate, it is normal for some blistering to occur. This is caused by either concrete's tendency to off-gas or by water temporarily trapped between the concrete and the membrane. With time and the applied pressure of backfill or over-slab, blisters will absorb into the concrete without detriment to the membrane.

A small number of blister heads should be sampled and checked for proper membrane thickness. If the samples have the required membrane thickness (80 mils minimum), then the remaining blisters should not be punctured or cut. If the samples have less than the minimum 80 mils, then the area can either be resprayed to obtain the proper thickness, or the blisters can be cut out and the area resprayed or patched with LIQUID BOOT® Trowel Grade.

FIGURE 13.5 (*Continued*) Guide specification for section 7100, fluid-applied waterproofing. (*Courtesy LBI Technologies*)

This suggested guide specification has been developed using the current edition of the Construction Specifications Institute (CSI) "Manual of Practice," including the recommendations for the CSI 3 Part Section Format. Additionally, the development concept and organizational format of the American Institute of Architects (AIA) MASTERSPEC Program have been recognized in the preparation of this guide specification. Neither CSI nor AIA endorses the use of specific manufacturers and products.

SECTION 07111
SHEET MEMBRANE WATERPROOFING AND DRAINAGE SYSTEM
(Self-Adhering Rubberized Asphalt Sheet Membrane)

Mop Notes (CSI Manual of Practice Notes), Ed Notes (Editing Notes) and Coord Notes (Drawing and Specification Coordination Notes) are provided throughout this guide specification and are outlined in a smaller typeface. Delete notes in final copy of project specification. Coordinate specification with project drawings.

PART 1 GENERAL

1.01 SUMMARY

Ed Note: This section combines a Membrane Waterproofing System and a Prefabricated Drainage Composite System into a single-source waterproofing and drainage system. Miradri Self-Adhering Waterproofing Membrane is a single-ply sheet membrane consisting of rubberized asphalt laminated to a polyethylene film for both vertical and horizontal applications. Miradrain prefabricated drainage composite is a 3-dimensional dimpled core laminated to a Mirafi geotextile filtration fabric. Drainage course material requires selection based upon site-specific soil conditions and flow rates required.

Ed Note: Revise below to suit project conditions. Select Waterproofing and Drainage System appropriate to project requirements. This section includes Prefabricated Drainage Composites; if not required, delete drainage requirements from this section and specify in other specification section. Coordinate requirements of the two sections with each other to avoid conflicting requirements.

A. Section Includes: Cold applied, self-adhering, rubberized asphalt sheet membrane waterproofing system, including substrate primer, sheet membrane, penetration flashing materials, mastics and prefabricated drainage composite (if required) for a single-source warranted system.
   1. Types of Waterproofing and Drainage System include:
      a. Sheet Membrane Waterproofing and Drainage System for below grade, vertical wall applications.

Ed Note: Select appropriate system from above or below.

      b. Sheet Membrane Waterproofing and Drainage System for horizontal, plaza deck and planter applications.

Ed Note: Revise below to suit project conditions and specification preparation requirements. Add section numbers per CSI "Masterformat" and Office Specification Practices.

B. Related Sections:
   1. Earthwork: Refer to Division 2
   2. Landscaping (For Landscaped Waterproofing Assemblies): Refer to Division 2
   3. Cast-In-Place Concrete: Refer to Division 3
   4. Structural Precast Concrete: Refer to Division 3
   5. Masonry: Refer to Division 4
   6. Insulation: Refer to Division 7
   7. Joint Treatments / Sealants: Refer to Division 7
   8. Mechanical/Electrical: Refer to Division 15/16
   9. Other appropriate project-specific specification sections.

FIGURE 13.6 Guide specification for section 7111, sheet membrane waterproofing. (*Courtesy TC MiraDri*)

Mop Note: below paragraph should describe products and work included in this section which are covered by alternates. Include alternate descriptions. Coordinate this section with Division 1 Alternate Section, bid documents and bid forms (If any). Refer to Specification Coordination Sheets included with this section for additional guidance on the use of alternates.

C. Alternates: Products and installation included in this section are specified by alternates. Refer to Division 1 Alternates Section for alternates description and alternate requirements.
   1. Alternate Manufacturers: Refer to Part 2 Products herein.

Mop Note: Below article should list industry standards referenced in this section, along with acronym, alpha/numeric designations, titles and dates. This article does not require compliance with standards, but is merely a listing to establish title and date of references.

1.02 REFERENCES (INDUSTRY STANDARDS)

A. General: Refer to Division 1 References Section.

1.03 SYSTEM DESCRIPTION

A. Performance Requirements: Provide waterproofing and drainage system which has been manufactured and installed to maintain leak-proof waterproofing system without defects, damage or failure.

1.04 SUBMITTALS

A. General: Prepare, review, approve and submit specified submittals in accordance with "Conditions of the Contract" and Division 1 Submittals Sections. Product data, shop drawings, samples and similar submittals are defined in "Conditions of the Contract."

B. Product Data: Submit manufacturer's product data for waterproofing and drainage system specified.

C. Shop Drawings: Submit shop drawings showing layout, profiles and product components, including accessories for waterproofing and drainage system.

Ed Note: Retain below as appropriate for waterproofing system selected.

D. Samples: Submit verification samples for waterproofing membrane and prefabricated drainage composites.

E. Quality Assurance/Control Submittals:
   1. Test Reports Certification: Submit certified test reports from an acceptable independent testing laboratory indicating compliance with applicable ASTM for referenced performance characteristics and physical properties.
   2. Certificates (Qualification Data): Submit product certificates signed by manufacturer certifying materials comply with specified performance characteristics and physical requirements, showing full time quality control. Waterproofing system components are supplied by a single-source waterproofing and drainage manufacturer.

F. Closeout Submittals:
   1. Warranty: Submit warranty documents specified herein.
   2. Project Record Documents: Submit project record documents for installed materials in accordance with Division 1 Project Closeout (Project Record Documents) Section.

FIGURE 13.6 *(Continued)* Guide specification for section 7111, sheet membrane waterproofing. *(Courtesy TC MiraDri)*

1.05 QUALITY ASSURANCE

A. Qualifications:
   1. Installer Qualifications: Installer experienced (as determined by contractor) to perform work of this section, who has specialized in the installation of work similar to that required for this project, who can comply with manufacturer's warranty requirements, and who is an approved applicator as determined by waterproofing/drainage system manufacturer.
   2. Manufacturer Qualifications: Manufacturer capable of providing field service representation during construction, approving an acceptable installer, recommending appropriate application methods and conducting a final inspection of the waterproofing and drainage system/assembly.

Ed Note: If Mock-up is required for special project requirements or job conditions, add text. Typically mock-ups may not be required.

B. Pre-Installation Meetings: Conduct pre-installation meeting to verify project requirements, substrate conditions, manufacturer's installation instructions and manufacturer's warranty requirements.

C. Pre-Installation Testing: In accordance with manufacturer's recommendations and warranty requirements, conduct pre-installation testing of substrates to receive waterproofing.

1.06 DELIVERY, STORAGE, AND HANDLING

A. Ordering: Comply with manufacturer's ordering instructions and lead time requirements to avoid construction delays.

B. Packing, Shipping, Handling and Unloading: Deliver materials in manufacturer's original, unopened, undamaged containers with identification labels intact, including applicable UL labels.

C. Storage and Protection: Store materials protected from exposure to harmful weather conditions and at temperature conditions recommended by manufacturer.
   1. As recommended by manufacturer and to comply with warranty requirements, store waterproofing materials at a temperature between 60° F. (15.5° C.) and 80° F. (26.6° C.) If exposed to lower temperatures, restore materials to 60° F. (15.5° C.) minimum temperature before application.

1.07 PROJECT CONDITIONS/SITE CONDITIONS

A. Environmental Requirements/Conditions:
   1. Weather Conditions: Neither commence nor proceed with waterproofing application when surface temperature is below 25° F. (-3.9° C.).
   2. Inclement Weather: Neither commence nor proceed with waterproofing and drainage application when inclement weather conditions conflict with manufacturer's installation requirements and warranty provisions.

Ed Note: Below pertains to the American Institute of Architects (AIA) "General Conditions of the Contract for Construction", which places responsibility for means, methods and techniques of construction upon the contractor.

B. Project Conditions: Refer to "Conditions of the Contract" for control and responsibility of project site including means, methods and techniques of construction by the contractor.

C. Field Measurements: Verify actual measurements/openings by performing field measurements before fabrication; show recorded measurements on shop drawings. Coordinate field measurements, fabrication schedule with construction progress to avoid construction delays.

FIGURE 13.6 *(Continued)* Guide specification for section 7111, sheet membrane waterproofing. *(Courtesy TC MiraDri)*

Ed Note: Below warranty article requires coordination with the owner. Below assumes the use of the American Institute of Architects (AIA) "Conditions of the Contract for Construction".

1.08 WARRANTY

A. Project Warranty: Refer to "Conditions of the Contract" for project warranty provisions. Submit manufacturer's single-source warranty directly from waterproofing/drainage system manufacturer for waterproofing system, including prefabricated drainage composites.

B. Manufacturer's Warranty Requirements: Submit, for Owner's acceptance, manufacturer's standard warranty document executed by authorized company official. Manufacturer's warranty is in addition to, and not a limitation of, other rights Owner may have under the Contract Documents.
   1. Beneficiary: Issue warranty in the legal name of the project Owner.

Ed Note: Coordinate below with Manufacturer's Warranty Requirements.

   2. Warranty Period: ___ years commencing on Date of Substantial Completion
   3. Warranty Acceptance: Owner is sole authority who will determine acceptability of manufacturer's warranty documents.

Ed Note: Select available Manufacturer's Warranty Program from below and revise as appropriate. Consult with Mirafi for specific Warranty Terms and Conditions. For Manufacturer's Material Warranty Program select standard geotextile Mirafi 700XG Woven Fabric for horizontal applications or 140NC Nonwoven Fabric for vertical applications.

C. Manufacturer's Warranty Program: (available with Mirafi Standard fabrics 140NC and 700XG)
   1. Manufacturer's Material Warranty: Material only warranty, excludes labor.
      a. Standard Waterproofing and Drainage Duration: 5 year material warranty.
      b. Extended Waterproofing and Drainage Duration: 10 year material warranty.

Ed Note: For Manufacturer's Site Specific Warranty Program select Mirafi Filterweave 40/10, Filterweave HP500 Fabric, Filterweave 70/20 Woven or Mirafi 180N Nonwoven Fabrics. Below Warranty is available when selection of Site Specific Fabric is in accordance with Mirafi project-specific written recommendations, based upon review of project's soil report submitted by architect. Additionally, application must be performed by a Mirafi approved applicator. Warranty coverage is for Watertightness of "Miradri" Waterproofing Membrane and drainage performance of "Miradrain" Drainage Composite for the duration of the Warranty.

   2. Manufacturer's Site Specific Watertightness Warranty: Manufacturer's site specific watertightness warranty covers components of "below-grade" waterproofing and drainage system, including membrane and prefabricated drainage composites.
      a. Site Specific Watertightness Duration: 5 year watertightness and drainage warranty.

Contact a Mirafi representative for further details on warranty provisions.

PART 2 PRODUCTS

Ed Note: Retain below article for proprietary specification. Do not use phrase "or Equal" / "or Approved Equal"; use of such phrases is not considered good, professional specification writing practice, because of the differing interpretations of the phrase among the contracting parties. Coordinate below article with Division 1 material and equipment (Product Options and Substitutions) section and bidding documents, if any.

2.01 MANUFACTURERS

A. Acceptable Manufacturer: Mirafi Moisture Protection Products, Nicolon/Mirafi Group.
   1. Address: 3500 Parkway Lane, Suite 500, Norcross, GA 30092
      a. Telephone: (800) 234-0484; Fax: (404) 729-1829.

FIGURE 13.6 (*Continued*) Guide specification for section 7111, sheet membrane waterproofing. (*Courtesy TC MiraDri*)

Ed Note: Select appropriate system from below. Verify each component of waterproofing and drainage system. Determine if prefabricated drainage composites are to be used on the project.

2. Warranted Waterproofing And Drainage System: Miradri Rubberized Asphalt Sheet Membrane Waterproofing and Miradrain Drainage System, including primer, membrane, drainage composite (if required), penetration flashings and accessories. Provide system components from a single source supplier in accordance with manufacturer's warranty requirements.
   a. Miradri 860/861 Rubberized Asphalt Waterproofing System (60 mils total minimum thickness).
      1. Primer: Miradri All Weather Primer
      2. Rubberized Asphalt Sheet Membrane: Miradri 860/861.
      3. Mastic: Miradri M-800 Mastic.
      4. Liquid Membrane: Miradri LM-800 Liquid Membrane.
      5. Miradri 200V/300HV Protection Course (if drainage not required).

Ed Note: Retain below as appropriate to waterproofing and drainage system selected and to project requirements. Consult with Mirafi for specific recommendations for project. Select fabric from manufacturers recommendations and warranty program requirements. Mirafi offers two groups of fabrics: one group is manufacturer's standard geotextile group including 700XG Fabric for Miradrain 9000 and 140NC Fabric for Miradrain 2000, 5000, 6000, 6200 and 8000. The other group includes manufacturer's site-specific geotextile filtration fabrics: Filterweave 40/10, Filterweave 70/20, Filterweave HP500 Fabric and Mirafi 180N Fabric.

   b. Miradrain Prefabricated Drainage Composites:
      1. Miradrain 6200 3-dimensional dimpled core and geotextile fabric for vertical applications
      2. Miradrain 9000 3-dimensional dimpled core and geotextile fabric for horizontal applications and planters.
      3. Miradrain 8000 3-dimensional dimpled core and geotextile fabric for Hydrocarbon Resistive applications.

Ed Note: Retain below with appropriate drainage composites above. Select fabric based upon site-specific soil conditions and Mirafi recommendations. Refer to Miradrain Site Specific Geotextile Fabric Selection Guide shown below. Consult with Mirafi for recommendations and site-specific Requirements in accordance with project-specific soils report. Below listed performance requirements and terminology are standard for geotextile testing methodology.

MIRADRAIN, SITE-SPECIFIC GEOTEXTILE SELECTION GUIDE**

<table>
<tr><th>Woven Fabrics</th><th>Coarse Sand</th><th>Sand</th><th>Silty Sand</th><th>Clayey Sand</th><th>Sandy Silt</th><th>Silt</th><th>Clayey Silt</th><th>Sandy Clay</th><th>Silty Clay</th><th>Clay</th></tr>
<tr><td>HP 500</td><td>✔</td><td></td><td></td><td></td><td colspan="3" rowspan="6">SPECIFIC*
TESTING
REQUIRED</td><td></td><td></td><td></td></tr>
<tr><td>FW 40/10</td><td></td><td>✔</td><td>✔</td><td></td><td></td><td></td><td></td></tr>
<tr><td>FW 70/20</td><td></td><td></td><td></td><td>✔</td><td></td><td></td><td></td></tr>
<tr><td></td><td colspan="4"></td><td colspan="3"></td></tr>
<tr><td>Nonwoven Fabric***</td><td colspan="4"></td><td colspan="3"></td></tr>
<tr><td>180N</td><td></td><td></td><td></td><td></td><td>✔</td><td>✔</td><td>✔</td></tr>
</table>

*Due to instability of silty soils, Mirafi requires that an individual soil test and filtration design be performed by Mirafi engineering services. Contact a Mirafi representative for further details.

**Please note that this selection guide is merely a reference chart and is not meant to replace sound geotextile engineering design practice.

***Nonwoven fabrics are to be used in vertical applications only.

FIGURE 13.6 (*Continued*) Guide specification for section 7111, sheet membrane waterproofing. (*Courtesy TC MiraDri*)

c. Miradrain Standard Fabrics:

	140NC (MD 6200, 8000)	700XG (MD 9000, 8000)
Flow Rate (ASTM D 4491)	120 gal./min./ft² (4903 l/min/m²)	100 gal./min./ft² (4086 l/min/m²)
Apparent Opening Size (ASTM D 4751)	70 (0.21) U.S. Sieve (mm)	40 (0.42) U.S. Sieve (mm)
Grab Tensile Strength (ASTM D 4632)	110 lb., Machine Direction (0.49 kN)	365 lb., Machine Direction (1.62 kN)
Mass Per Unit (ASTM D 5261)	4.0 oz/yd² (136 g/m²)	5.6 oz/yd² (190 g/m²)
Puncture (ASTM D 4833)	70 lb. (0.31 kN)	115 lb. (0.51 kN)

d. Miradrain Site Specific Fabrics:

	HP 500 (MD 6200, 9000, 8000)	FW 40/10 (MD 6200, 9000, 8000)	FW 70/20 (MD 6200, 9000, 8000)	180N (MD 6200, 8000)
Flow Rate (ASTM D 4491)	115 gal./min/ft² (4698 l/min/m²)	70 gal./min/ft² (2860 l/min/m²)	35 gal./min/ft² (1430 l/min/m²)	110 gal./min/ft² (4494 l/min/m²)
Apparent Opening Size (ASTM D 4751)	30 (0.595) U.S. Sieve (mm)	40 (0.42) U. S. Sieve (mm)	60 (0.25) U. S. Sieve (mm)	80 (0.177) U. S. Sieve (mm)
Grab Tensile (ASTM 4632)	400 lb. Machine Direction (1.78 kN)	265 lb. Machine Direction (1.18 kN)	275 lb. Machine Direction (1.22 kN)	200 lb. Machine Direction (0.89 kN)
Mass per unit (ASTM 5261)	7.4 oz/yd² (250 g/m²)	5.0 oz/yd² (170 g/m²)	6.1 oz/yd² (207 g/m²)	8.0 oz/yd² (271 g/m²)
Puncture (ASTM D 4833)	145 lb. (0.81 kN)	125 lb. (0.56 kN)	145 lb. (0.65 kN)	130 lb. (0.58 kN)

Ed Note: Retain below prefabricated drainage composites as appropriate to project requirements and site specific conditions. Coordinate below composite with above drainage fabric.

3. Material(s)/System(s) Testing: Prefabricated drainage composites:

	Miradrain 6200	Miradrain 8000*	Miradrain 9000*
Compressive Strength - psf (kN/m²) (ASTM D 1621)	15,000 (719)	15,000 (719)	18,000 (862)
Water Flow Rate * - gpm/ft width (l/min/width) (ASTM D 4716)	15 (188)	15 (188)	18 (226)
Thickness - in (mm) (ASTM D 1777)	0.40 (11.3)	0.38 (10.8)	0.38 (10.8)

* Normal stress of 3600 psf
Hydraulic gradient of 1.0 for 300+ hours

FIGURE 13.6 (*Continued*) Guide specification for section 7111, sheet membrane waterproofing. (*Courtesy TC MiraDri*)

4. Materials(s)/System(s) Testing: Miradri 860/861 rubberized asphalt membrane meeting the following physical properties, test method and result:

Property	Test Method	Unit	Typical Value
Thickness		mils (mm)	60 min (1.5)
Pliability @ -35°F (-37.2°C)	ASTM D 146[1]		No Effect
Tensile Strength - Membrane	ASTM D 412	psi (kN/m²)	250 (1720) min
Tensile Strength - Film	ASTM D 412	psi (kN/m²)	5000 (34,500) min
Elongation	ASTM D 412[2]	%	300 min
Puncture Resistance - Membrane	ASTM E 154[3]	lb. (N)	40 (178) min
Peel adhesion, Concrete	ASTM D 903	lb./in. (N/m)	5 (876) min
Permeance	ASTM E 96, Method B	Perms (na/Pa *s*m²)	0.05 (2.5) max
Water Absorption	ASTM D 570	%	0.2 max
Resistance to hydrostatic head	ASTM D 751	lb (kg)	150 (45) min
Cycling Over Crack @ -15°F (-31°C), 100 cycles	ASTM C 836		No Effect
Exposure to fungi in soil	GSA-PBS 07115		No Effect
Soil Burial	GSA-PBS-4-07121	16 weeks	No Effect

[1]180° bend over a 1/4" (6.25 mm) mandrel at -35°F (-37.2°C)
[2]Ultimate failure of rubberized asphalt
[3]Stretched by a blunt object

Ed Note: Retain below for alternate method of specifying manufacturers and products. Add alternate number designation and coordinate with other specified project alternates.

2.02 ALTERNATES AND SUBSTITUTIONS

Ed Note: Presently there are only two manufacturers who market a single-source manufactured waterproofing and drainage system. Other manufacturers either market a waterproofing system or a drainage composite system as separate systems, but not as a single source manufactured system. Below alternate is for a combined waterproofing and drainage system from one manufacturer.

A. Alternate (Manufacturers/Products): In lieu of providing below specified base bid/contract manufacturer, provide below specified alternate manufacturers. Refer to Division 1 Alternates Section.
   1. Base Bid/Contract Manufacturer/Product: Mirafi Moisture Protection Products, Nicolon/Mirafi Group.
      a. Product: Sheet Membrane Waterproofing & Drainage System.
         1) Waterproofing: Miradri 860/861 Waterproofing Membrane
         2) Drainage: Miradrain Drainage Composite.

FIGURE 13.6 *(Continued)* Guide specification for section 7111, sheet membrane waterproofing. *(Courtesy TC MiraDri)*

Ed Note: Number below alternates in accordance with project requirements and office specification practices. Determine alternate types. Verify alternate manufacturer/product designations. Add product attributes/characteristics for product equivalency, such as membrane sheet and drainage composite, including drainage core and fabric.

2. Alternate #____ Manufacturer/Product: W. R. Grace
   a. Product:
      1) Waterproofing: Bituthene Waterproofing Membrane
      2) Drainage: Hydroduct Drainage Composite

B. Substitutions: Refer to Division 1 General Requirements for substitution requirements.

2.03 RELATED MATERIALS (Specified In Other Sections)

A. Concrete: Refer to Division 3 Concrete Section.

B. Insulation: Refer to Division 7 Insulation Section

C. Sealants/Backer Rods: Refer to Division 7 Joint Treatment Section.

Ed Note: Usually retain below in accordance with Mirafi warranty requirements.

2.04 SOURCE QUALITY CONTROL

A. Single Source Responsibility: Obtain rubberized asphalt waterproofing and drainage system including primer, rubberized asphalt, drainage composite, penetration flashing and accessories from a single source supplier for required warranty.

PART 3 EXECUTION

3.01 MANUFACTURER'S INSTRUCTIONS/RECOMMENDATIONS

A. Compliance: Comply with manufacturer's product data, including product technical bulletins, product catalog installation instructions and product packaging instructions.

3.02 EXAMINATION

A. Site Verification of Conditions: Verify substrate conditions (which have been previously installed under other sections) are acceptable for product installation in accordance with manufacturer's instructions. Do not proceed with waterproofing and drainage system installation until substrate conditions are acceptable for compliance with manufacturer's warranty requirements.

3.03 PREPARATION

A. Adjacent Surfaces Protection: Protect adjacent work areas and finish surfaces from damage during installation operations.

B. Surface Preparation: Surfaces to receive waterproofing shall be dry, clean, frost free, and free of laitenance, dust, dirt, oil, harmful curing compounds, or other foreign matter which may impair the bond or performance of the waterproofing and drainage system and which do not comply with manufacturer's warranty requirements.

C. Concrete Deck Surface Preparation: Prepare concrete deck surfaces to receive waterproofing and drainage system. Surfaces shall be smooth, free of depressions, voids, protrusions, clean and free

FIGURE 13.6 (*Continued*) Guide specification for section 7111, sheet membrane waterproofing. (*Courtesy TC MiraDri*)

of harmful curing compounds, form release agents and other surface contaminants which may impair the bond or performance of waterproofing and drainage system and manufacturer's warranty requirements.

1. Cast-In-Place Concrete/Composite Decks: Decks shall be monolithic, smooth, free of voids, spalled areas, laitenance, honeycombs and sharp protrusions. Refer to Division 3 Concrete Section for concrete strength, density, finish, curing methods and other concrete requirements.
2. Precast Concrete Decks: Decks shall be mechanically secured to minimize differential movement and each joint between precast units shall have an installed backer rod. Grout precast units as recommended by manufacturer.

D. Substrate Cleaning: Clean substrate which is to receive waterproofing and drainage system. Remove loose debris and other harmful contaminants which will affect performance of waterproofing and drainage system.

E. Adhesion Test: Determine substrate cleanability and acceptance by applying a test patch of membrane material to surface and check adhesion according to manufacturer's test procedures and warranty requirements.

Ed Note: Coordinate below article with Part 2 Products Article to include each product requiring installation.

DWG Coord Note: Coordinate below article with Mirafi recommended waterproofing and drainage system details numbered consecutively No. 1 thru No. 15.

3.04 MEMBRANE INSTALLATION

A. Primer (Surface Conditioning): Apply Miradri All Weather Primer to concrete substrate at approximate rate of 400 sq. ft. per gallon (7.3m²/L) depending on surface texture and condition and in accordance with manufacturer's recommendations and warranty requirements. Let primer dry for approximately 30 minutes or when dry to the touch, after which Miradri membrane may be applied. Reprime surface if primed area has been left exposed for 24 hours.

DWG Coord Note: Coordinate below with manufacturer recommended installation details.

B. Joints And Crack Treatment:

1. Cracks: Seal cracks (1/8" wide or less) in concrete with 60 mils minimum thick coat of Miradri LM-800 Liquid Membrane or prestrip joint 9" (22.86 cm) wide Miradri 860/861 membrane strip.

DWG Coord Note: Coordinate below with manufacturer recommended installation details.

2. Horizontal To Vertical Transitions and Footings: Treat transitions either with 9" (22.86 cm) strips of Miradri Membrane or a 6" (15.24 cm) wide, 90-mil thick strip of Miradri LM-800 Liquid Membrane. At inside horizontal corners apply a minimum 3/4" (1.9 cm) fillet of Miradri LM-800.
3. Expansion Joints: Treat expansion joints with Miradri 3-ply joint treatment as recommended by manufacturer's installation details. (Anticipated expansion joint movement: 1/2" (1.21 cm))

DWG Coord Note: Coordinate below with manufacturer recommended installation details.

4. Membrane Flashing At Drains: At drains, apply either a double layer of Miradri Membrane or place Miradri LM-800 Liquid Membrane from inside edge of drain extending out 12" (30.48 cm) beyond the drain. Install field sheet (36" (91.44 cm) by 36" (91.44 cm) minimum) centered in drain basin as recommended by manufacturer.

FIGURE 13.6 *(Continued)* Guide specification for section 7111, sheet membrane waterproofing. *(Courtesy TC MiraDri)*

5. Flashings and Protrusions: at posts and penetrations, apply either a double layer of Miradri Membrane extending out at least 6" (15.24 cm) in each direction and seal with Miradri M-800 Mastic or apply Miradri LM-800 Liquid Membrane around posts and protrusions, overlapping sheet membrane a minimum of 2" (5.08 cm). Seal cut edge terminations with Miradri M-800 Mastic.
6. Interior and Exterior Corners: Prestrip interior and exterior corners with 9" (22.86 cm) wide Miradri 860/861 membrane strip.

C. Membrane Installation, Vertical Surfaces:
1. Apply membrane to primed surfaces vertically, in sections of 8' (2.43 m) in length or less. On walls higher than 8' (2.43 m), apply two or more sections with upper section overlapping lower section with 5" (12.7 cm) minimum overlap. Side laps shall be a minimum of 2.5" (6.35 cm) and end laps shall be a minimum of 5" (12.7 cm). Use hand roller and firmly press membrane when placed on vertical surface.
2. Apply a bead of Miradri M-800 Mastic at each termination of the membrane. Trowel bead to a flat surface approximately 1/8" (.31 cm) thick by 3/4" (1.9 cm) wide. Work mastic into cut edge terminations.
3. Provide reglet or counter flashing where membrane terminates on a vertical surface above grade or on a parapet wall.
4. Seal laps occurring within 12" (30.48 cm) of a 90° change of direction with a trowelled bead of Miradri M-800 Mastic.

D. Membrane Installation, Horizontal Surfaces
1. Apply membrane to primed surfaces horizontally, starting at low point and working to high point in a shingling technique. Side laps shall be a minimum of 2.5" (6.35 cm) and end laps a minimum of 5" (12.7 cm). Use roller weighing a minimum of 75 lbs (34 kg) to firmly press membrane onto surface without air pockets.
2. Terminate turned up membrane on vertical walls into a reglet or under counter flashing or with a termination bar. Press membrane firmly to wall surfaces and seal with trowelled bead of Miradri M-800 Mastic.
3. Seal laps occurring within 12" (30.48 cm) of a 90° change of direction with a trowelled bead of Miradri M-800 Mastic.

3.05 DRAINAGE COURSE INSTALLATION (FILTER FABRIC/CORE)

A. Vertical Surfaces or Planters - MD-6200, MD-8000 or MD-9000

1. Completed Walls - Position the panel with the flat core side against the wall and filter fabric toward the soil side. A washer headed concrete nail may be used to attach the panel against the concrete wall. Alternate methods of attachments use concrete nails driven through wood strips placed over the Miradrain, or use Mirastick construction adhesive. Most methods used to secure protection board are acceptable.

2. Over Miradri Waterproofing Membrane - Place the Miradrain over the waterproofing membrane. Use two sided Mirabond tape or Mirastick construction adhesive to secure the Miradrain to the waterproofing membrane. Most methods used to secure protection board are acceptable.

3. Connecting Adjacent Panels - Connect adjacent panels at the longitudinal edge by pulling the filter fabric back to expose the flange. The flangeless panel edge should be placed on top of the flange of the adjacent panel and butted dimple to dimple. Panel end attachments can also be completed by pulling the filter fabric back to expose two rows of dimpled core. The end of the next panel may be placed over two dimples and interlocked. All connections should be completed in shingle fashion so that moisture will flow with the overlap and not against it. Overlap fabric in

FIGURE 13.6 (*Continued*) Guide specification for section 7111, sheet membrane waterproofing. (*Courtesy TC MiraDri*)

the direction of water flow. Cover all terminal edges with the filter fabric flap by tucking in behind the core.

B. Horizontal Surfaces - MD-9000, or MD-8000

1. Plaza Decks, Parking Decks, Planters and Split Slab Applications- MD-9000 - Attach the Miradrain 9000 panels by either placing temporary ballast on top of the Miradrain or adhering the panels to the waterproofing membrane with Mirastick construction adhesive or Mirabond two sided tape.

   Connect adjacent panels at the longitudinal edge by pulling the filter fabric back to expose the flange. The flangeless edge of the panel should be placed on top of the flange of the adjacent panel and butted dimple to dimple. Panel end attachments should be completed by pulling the filter fabric back to expose two rows of dimpled core. The end of the top panel may be placed over the two dimples of the bottom panel and interlocked. All connections should be completed in shingle fashion. Overlap fabric in the direction of water flow. Cover all terminations with the filter fabric flap by tucking it behind the core.

   Miradrain 9000 should be channeled into an internal drain or perimeter drain system.

   Concrete, sand, grout, or pavers may be placed directly on the Miradrain 9000 fabric side. Caution should be taken not to place point loads on the Miradrain that might puncture the filter fabric on the Miradrain. When concrete is poured against Miradrain, use proper chuting techniques and avoid high drop heights.

2. Floor Slabs and Concrete Lined Channels - MD-9000 The subgrade shall be graded to a 2% minimum slope and clear of rubble, rock, large soil clods, etc. The Miradrain should be placed fabric side down, directly on the subgrade. Installation should proceed from the higher elevation of the slope and work downward. Connection of panel ends shall be achieved with four row dimple overlapping and taping of terminal edges.

3. Planters - MD-9000 Miradrain 9000 should be placed fabric side to the inside (soil side) of the planter. The planter walls should be covered with Miradrain 9000. Allow a 3" (7.5 cm) fabric overlap at the bottom of vertical panels in order to cover the intersection of wall and bottom sections. Any exposed panel edges must be covered with supplemental pieces of fabric to prevent soil intrusion into the flow channels.

3.06 DISCHARGE CONNECTIONS

A. Drainage - Where drainpipe is indicated, place the drainpipe next to the core. Wrap the drainpipe or rock-pipe drain combination with an auxiliary piece of Mirafi filter fabric.

   At weep holes, cut a hole in the core corresponding to the size and location of the weep hole. Avoid cutting a hole in the fabric by cutting the backside of the core between the dimples. A four dimple square area cut between the dimples (2 1/2 square inch) should be sufficient for most applications.

   Plaza Drains - Create openings in the Miradrain core to correspond with all discharge holes in the drain at the structural deck level. Fabric must be placed over these holes to prevent intrusion of soil, grout, sand, or concrete into the drainage core.

B. Terminal Connections and Protrusions - Cover all terminal edges with the integral fabric flap by tucking it around the edge of the core and securing it. At protrusions, cut the core around the

FIGURE 13.6 *(Continued)* Guide specification for section 7111, sheet membrane waterproofing. *(Courtesy TC MiraDri)*

protrusion, cut an "X" in the fabric, and tape the fabric around the protrusion. Dirt and concrete must not infiltrate the core.

3.07 RELATED PRODUCTS INSTALLATION REQUIREMENTS

A. Earthwork Backfilling: Refer to Division 2 Earthwork Section for installation requirements.

B. Sealants and Backer Rods: Refer to Division 7 Joint Treatment Section for installation requirements.

3.08 FIELD QUALITY CONTROL

A. Membrane Water Test: In accordance with waterproofing and drainage manufacturer's warranty requirements, conduct a water test on installed horizontal waterproofing system. Water test waterproofing area by ponding water a minimum depth of 2" (50.8 mm) for a period of 48 hours to check integrity of membrane installation. Consult with structural engineer for deck loading limits before conducting a water test.
   1. If leaks should occur, drain water completely, repair membrane installation and retest waterproofing area.

B. Manufacturer's Field Services: Upon Owner's request, provide manufacturer's field service consisting of product use recommendations and periodic site visits for inspection of product installation in accordance with manufacturer's instructions and warranty requirements.

3.09 CLEANING AND PROTECTION

A. Cleaning: Remove temporary coverings and protection of adjacent work areas. Repair or replace damaged installed products. Clean installed products in accordance with manufacturer's instructions prior to owner's acceptance. Remove construction debris from project site and legally dispose of debris.

B. Protection: Protect installed product's finished surfaces from damage during construction.

END OF SECTION 07111

FIGURE 13.6 (*Continued*) Guide specification for section 7111, sheet membrane waterproofing. (*Courtesy TC MiraDri*)

SECTION 07161

CEMENTITIOUS MEMBRANE WATERPROOFING

PART 1 - GENERAL

1.1 SECTION INCLUDES

A. Cementitious Membrane Waterproofing to new or existing concrete, cmu-block or brick walls, above or below grade, on positive or negative water pressure side, for hydrostatic pressures tested up to 234 feet of water head (100 psi) as shown on drawings and as specified in this section.

B. Related Sections:
1. See section 03300 Concrete

1.2 REFERENCES

A. ASTM C 109 - Standard Test Method for Compressive Strength of Hydraulic Cement Mortars.

B. ASTM C 321 - Standard Test Method for Bond Strength of Chemical-Resistant Mortars.

C. ASTM C 348 - Standard Test Method for Flexural Strength of Hydraulic Cement Mortars.

D. ASTM C 596 - Standard Test Method for Drying Shrinkage of Mortar Containing Portland Cement.

E. COE CRD-C 48 - Method of Test for Water Permeability of Concrete; U.S. Army Corps of Engineers.

1.3 SUBMITTALS

A. General:
Submit manufacturer's certification that proposed materials, details and systems as indicated and specified fully comply with manufacturer's details and specifications. If any portion of Contract Documents do not conform to manufacturer's standard recommendations, submit notification of portions of design that are at variance with manufacturer's specifications.

B. Product Data:
1. Submit manufacturer's descriptive literature and product specifications for each product.

07161-1Cementitious Membrane Waterproofing

FIGURE 13.7 Guide specification for section 7161, cementitious waterproofing. (*Courtesy Vandex*)

2. Submit laboratory tests or data that validate product compliance with the performance criteria specified.

1.4 QUALITY ASSURANCE

A. Manufacturer Qualifications:
1. Company specializing in manufacturing Products specified in this Section with minimum 20 years documented experience.
2. ISO 9001 certified; submit copy of certificate.

B. Installer Qualifications:
1. Acceptable to manufacturer with documented experience on at least 5 projects of similar nature in past 5 years and/or training provided by the product manufacturer.

1.5 DELIVERY, STORAGE AND HANDLING

A. Deliver, store off the ground and covered, handle and protect products from moisture in accordance with manufacturer's instructions.

B. Deliver materials in manufacturer's unopened containers, fully identified with brand, type, grade, class and all other qualifying information. Provide Material Safety Data Sheets for each product.

C. Take necessary precautions to keep products clean, dry and free of damage.

1.6 WARRANTY

A. Warrant installed waterproofing to be free of leaks and defects for 1 year from date of acceptance, with the exception of structural cracks in the waterproofed concrete.

1.7 SYSTEM REQUIREMENTS

A. Coordinate waterproofing work with work of other trades.

B. Provide materials and accessories in timely manner so as not to delay Work.

1.8 PROJECT CONDITIONS

A. Maintain surfaces to be waterproofed and surrounding air temperature at not less than 40 degF (5 degC) for at least 48 hours before, during and after application

07161-2Cementitious Membrane Waterproofing

FIGURE 13.7 (*Continued*) Guide specification for section 7161, cementitious waterproofing. (*Courtesy Vandex*)

of waterproofing.

D. Do not apply materials to frozen or frost-filled surfaces.

E. Exercise caution when temperatures exceed 90 degF (32 degC). It may be necessary to apply waterproofing during times when the sun is not at its strongest (i.e. early morning, evening or night).

PART 2 - PRODUCTS

2.1 MANUFACTURERS

A. Provide products of Vandex Sales & Services, Inc., Columbia, MD. Phone (410)964-1410, Fax (410)964-1526.

2.2 MATERIALS

A. Waterproofing Material: Cementitious, non-polymer, ready-mixed, efflorescence-free surface waterproofer with hydrophobic properties, that requires just the addition of water, resistant to water and moisture but vapor permeable for all standard applications (vertical, overhead and horizontal surfaces not exposed to vehicular traffic).
1. Product: VANDEX BB 75.
2. Color: Concrete gray
3. Aggregate: Powder
4. Compressive Strength: (ASTM C-109) 6300 psi (43.4 MPa) at 28 days
5. Flexural Strength: (ASTM C-348) 1100 psi (7.6 MPa) at 28 days
6. Bond Strength: (ASTM C-321) 300 psi (2.1 MPa) at 14 days
7. Permeability:(CRD C-48) 0.00 cm/sec permeability at 100 psi (0.7 MPa) or 234 ft. (70 m) over 20 days testing period on negative side (1/8" (3 mm) thickness).
8. Microbiological growth: not supported
9. Toxic Metals: not detectable
10. Organoleptic/Physical test: no change

B. Water: Clean, clear, non alkaline and free of salts and other harmful elements; potable.

2.3 ACCESSORY MATERIALS

A. Patching Compound: Ready-mixed, non-polymer,

07161-3Cementitious Membrane Waterproofing

FIGURE 13.7 (*Continued*) Guide specification for section 7161, cementitious waterproofing. (*Courtesy Vandex*)

cementitious waterproofing and repair mortar recommended by waterproofing manufacturer for patching, honeycombs, seal strips (coves, reglets), etc.

1. Product: VANDEX UNI MORTAR 1 Z.
2. Compressive Strength: (ASTM C-109) 7600 psi (52 MPa) at 28 days
3. Flexural Strength: (ASTM C-348) 700 psi (4.8 MPa) at 28 days
4. Shrinkage: (ASTM C-596) -0.093% (28 days); +0.-073% (120 days)
5. Permeability: No water ingress with 100 psi (0.7 MPa) or 234 ft. (70 m) hydrostatic pressure. (Sample 3/8" (10 mm) thickness.)

B. Plugging Compound for Active Water Penetrations: Pulverized rapid-setting cement.
   1. Product: VANDEX PLUG.
   2. Compressive Strength: (ASTM C-109) 3400 psi (23.4 MPa) at 24 hrs
   3. Flexural Strength: (ASTM C-348) 450 psi (3.1 MPa) at 48 hrs

PART 3 - EXECUTION

3.1 EXAMINATION

A. Examine substrates and adjoining construction, and conditions under which Work is to be installed. Do not proceed with Work until unsatisfactory conditions are corrected.

B. Verify the following substrate conditions before application of cementitious membrane waterproofing:
   1. That substrate condition is satisfactory and in accordance with manufacturer's instructions.
   2. That concrete surfaces have open pores and wood float finish on horizontal surfaces.
   3. That concrete surfaces are free of voids, spalled areas, loose aggregate and sharp protrusions, and with no coarse aggregate visible.
   4. That curing compounds or surface hardeners incompatible with waterproofing have not been used on concrete.

3.2 PREPARATION

A. Protect adjacent surfaces not designated to receive waterproofing.

FIGURE 13.7 (*Continued*) Guide specification for section 7161, cementitious waterproofing. (*Courtesy Vandex*)

B. Substrate preparation:

¶***

SELECT AND SPECIFY ARTICLE BELOW DEPENDING ON SUBSTRATE.

***§

1. Concrete surfaces:
   a. Remove remaining concrete fins and projections, and general surface dirt.
   b. Concrete to receive cementitious membrane waterproofing treatments must have a clean surface and an open pore system to ensure maximum mechanical bonding. Remove all cement laitance, grease, oil and other contaminants. Use steam cleaning, high-pressure water blasting, wet or dry sand blasting, wire brush or other methods recommended by waterproofing manufacturer.
   c. Rout out faulty construction joints and visible cracks not subject to movement, exceeding 0.012" (0.3 mm) in width to approx. 3/4" (20 mm) width and minimum 3/4" (20 mm) depth. Remove all protrusions, work back to sound concrete and chisel out any spalled or honeycombed areas. Roughen formtie holes.

2. CMU-Block and brick walls:
   a. Remove existing coatings, paints, etc..
   b. Use high-pressure water or other suitable means to remove existing efflorescence and general dirt, to produce surfaces suitable for application of waterproofing.
   c. Patch all holes, cracks, weak mortar joints with patching compound.

C. Stop active water leakages as per manufacturer's plugging specifications.

D. Rinse surfaces to be waterproofed several times so that the substrate is thoroughly saturated. Surfaces shall be moist but not wet when waterproofing system is applied. Remove any surface water on horizontal surfaces.

3.3 INSTALLATION

A. Mix waterproofing material in proportions recommended by manufacturer.

B. Apply waterproofing material in quantities as per manufacturer's specifications to:

1. Cavity fill:
   a. At cleaned and prepared faulty construction joints, cracks, formtie holes, prime areas with

FIGURE 13.7 (*Continued*) Guide specification for section 7161, cementitious waterproofing. (*Courtesy Vandex*)

patching compound in slurry consistency, and while "green" (tacky) fill voids with patching compound in mortar consistency flush to surface. Leave "scratched" surface finish for subsequent waterproofing material to bond.

b. Laminate patching compound in 2 to 3 layers as per manufacturer's instructions for larger spalled or honeycombed areas.

¶***

SELECT AND SPECIFY SEAL STRIPS/REGLETS IN ARTICLE BELOW IN CASE OF NO WATERSTOPS PRESENT, OR IN ADDITION TO WATERSTOPS. DELETE IF NOT REQUIRED.

****

§

2. Horizontal and vertical construction joints:

a. Install seal strips/reglets in pre-formed 1" x 1" (25 mm x 25 mm) joints. Prime joint area with patching compound in slurry consistency, and while "green" (tacky) fill cavity with patching compound in mortar consistency flush to surface. Leave "scratched" surface finish for subsequent water-proofing material to bond.

¶***

DELETE ANY OF THE FOLLOWING ARTICLE THAT ARE NOT REQUIRED ON THE PROJECT.

****

§

3. Horizontal surfaces and vertical surfaces - not subject to vehicular traffic:

a. Damp proofing:
Apply waterproofing material in one coat at 1/16" (1.5 mm) thickness (5.5 lb/sq.yd., approx. 90 sq.ft. per 55 lb bag).

b. Pressureless surface and seepage:
Apply waterproofing material in one coat at 1/12" (2.0 mm) thickness (7.3 lb/sq.yd., approx. 70 sq.ft. per 55 lb bag).

c. Surfaces under pressure + cmu-block and brick walls:
Apply waterproofing material in 2 (two) coats of 1/16" (1.5 mm) each, total 1/8" (3.0 mm) thickness (11 lb/sq.yd., approx. 45 sq.ft. per 55 lb bag, including 2 coats). Apply 2nd coat after 1st coat has sufficiently hardened. Leave parallel, horizontal brush stroke finish on 1st coat if 2nd coat is applied next day. Pre-water before applying 2nd coat.

d. Surface finish shall be - brushed - troweled smooth - finish.

07161-6Cementitious Membrane Waterproofing

FIGURE 13.7 (*Continued*) Guide specification for section 7161, cementitious waterproofing. (*Courtesy Vandex*)

3.4 CURING

A. Protect exposed waterproofed surfaces from rain, frost and drying out.

B. Moisture cure exposed waterproofed surfaces as per manufacturer's instructions.

¶***
*

SELECT AND SPECIFY ARTICLE BELOW IF ACRYLIC MODIFIED CEMENT WATERPROOFING IS TO BE OVER COATED WITH PAINT.

****
§

3.5 DECORATION

A. Follow manufacturer's general instructions for surfaces treated with Cementitious Membrane Waterproofing which are to be painted, coated or tiled.

3.6 ACCEPTANCE

A. Remove materials left over and any foreign material resulting from the work from the site.

B. Clean adjacent surfaces and materials.

END OF SECTION

Project:.......(9/24/98)

07161-7Cementitious Membrane Waterproofing

FIGURE 13.7 (*Continued*) Guide specification for section 7161, cementitious waterproofing. (*Courtesy Vandex*)

SECTION 07162

CEMENTITIOUS CAPILLARY/CRYSTALLINE WATERPROOFING

PART 1 - GENERAL

1.1 SECTION INCLUDES

A. Cementitious Capillary/Crystalline Waterproofing on concrete structures and surfaces as shown on drawings and as specified in this section.

B. Related Sections:
1. See section 03300 - Cast-in-Place Concrete

1.2 REFERENCES

A. ASTM C 109 - Standard Test Method for Compressive Strength of Hydraulic Cement Mortars.

B. ASTM C 321 - Standard Test Method for Bond Strength of Chemical-Resistant Mortars.

C. ASTM C 348 - Standard Test Method for Flexural Strength of Hydraulic Cement Mortars.

D. ASTM C 596 - Standard Test Method for Drying Shrinkage of Mortar Containing Portland Cement.

E. ASTM C 944 - Standard Test Method for Abrasion Resistance of Concrete or Mortar Surfaces by the Rotating-Cutter Method.

F. COE CRD-C 48 - Method of Test for Water Permeability of Concrete; U.S. Army Corps of Engineers.

G. NSF 61 - Drinking Water Systems - Health Effects; NSF International, Inc.

1.3 SUBMITTALS

A. General:
Submit manufacturer's certification that proposed materials, details and systems as indicated and specified fully comply with manufacturer's details and specifications. If any portion of Contract Documents do not conform to manufacturer's standard recommendations, submit notification of portions of design that are at variance with manufacturer's specifications.

B. Product Data:

07162-1 Crystalline Waterproofing

FIGURE 13.8 Guide specification for section 7162, crystalline waterproofing. (*Courtesy Vandex*)

1. Submit manufacturer's descriptive literature and product specifications for each product.
2. Submit laboratory tests or data that validate product compliance with the performance criteria specified.
3. Submit copy of test report with magnified photos demonstrating crystalline growth within the concrete.
4. Submit manufacturer's literature showing product's capability to post-seal cracks up to 0.012" (0.3 mm) which appear after the application.
5. Submit copy of NSF certification for applications in connection with potable water.

1.4 QUALITY ASSURANCE

A. Manufacturer Qualifications:
   1. Company specializing in manufacturing Products specified in this Section with minimum 20 years documented experience.
   2. ISO 9001 certified. Submit copy of certificate.

B. Installer Qualifications:
   1. Acceptable to manufacturer with documented experience on at least 5 projects of similar nature in past 5 years and/or training provided by the product manufacturer.

1.5 DELIVERY, STORAGE AND HANDLING

A. Deliver, store off the ground and covered, handle and protect products from moisture in accordance with manufacturer's instructions.

B. Deliver materials in manufacturer's unopened containers, fully identified with brand, type, grade, class and all other qualifying information. Provide Material Safety Data Sheets for each product.

C. Take necessary precautions to keep products clean, dry and free of damage.

1.6 WARRANTY

A. Warrant installed waterproofing to be free of leaks and defects for ______ year(s) from date of acceptance, with the exception of structural cracks in the waterproofed concrete which are 0.012" (0.3 mm) or wider.

1.7 SYSTEM REQUIREMENTS

A. Coordinate waterproofing work with work of other

07162-2 Crystalline Waterproofing

FIGURE 13.8 (*Continued*) Guide specification for section 7162, crystalline waterproofing. (*Courtesy Vandex*)

trades.

B. Provide materials and accessories in timely manner so as not to delay Work.

1.8 PROJECT CONDITIONS

A. Maintain surfaces to be waterproofed and surrounding air temperature at not less than 40 degF (5 degC) for at least 48 hours before, during and after application of waterproofing.

B. Do not apply materials to frozen or frost-filled surfaces.

C. Exercise caution when temperatures exceed 90 degF (32 degC). It may be necessary to apply waterproofing during times when the sun is not at its strongest (i.e. early morning, evening or night).

PART 2 - PRODUCTS

2.1 MANUFACTURERS

A. Provide products of Vandex Sales & Services, Inc., Columbia, MD. Phone (410)964-1410, Fax (410)964-1526.

****
SELECT AND SPECIFY/EDIT COLORS IN ARTICLE BELOW TO SUIT PROJECT REQUIREMENTS. DELETE INAPPLICABLE ITEMS.
****

2.2 MATERIALS

A. Capillary/Crystalline Waterproofing: Blend of rapid-hardening portland cement, specially treated quartz sand and a compound of active chemicals with the following characteristics:

1. Product: VANDEX SUPER
2. Color: Cement ________ (gray / white)
3. Aggregate: Powder
4. Potable Water Certification: NSF Standard 61
5. Compressive Strength: (ASTM C-109) 10200 psi (70 MPa), 28 days
6. Flexural Strength: (ASTM C-348) 730 psi (5.0 MPa) at 28 days
7. Bond Strength: (ASTM C-321) 690 psi (4.8 MPa) at 14 days
8. Abrasion: (ASTM C-944) 1.28 g (10 kg/ft^2 on 4000 psi conc., 28 days)
9. Sulfate Resist.: (ASTM C-452) 0.0012 percent (28 days)
10. Bond of Reinf.: (ASTM C-321) No loss of bond due

07162-3 Crystalline Waterproofing

FIGURE 13.8 (*Continued*) Guide specification for section 7162, crystalline waterproofing. (*Courtesy Vandex*)

	to waterproofing material.
11. Permeability: (CRD C-48)	0.00 cm/sec permeability at 210 psi (1.5 MPa) or 484 ft. (148 m) over 20 days testing period on negative side.

B. Water: Clean, clear, non alkaline and free of salts and other harmful elements; potable.

2.3 ACCESSORY MATERIALS

A. Patching Compound: Ready-mixed cementitious waterproofing and repair mortar recommended by waterproofing manufacturer for honeycombs, tie holes, seal strips (fillets/coves, reglets), etc.
   1. Product: VANDEX UNI MORTAR 1 Z.
   2. Compressive Strength: (ASTM C-109) 7600 psi (52 MPa) at 28 days
   3. Flexural Strength: (ASTM C-348) 700 psi (4.8 MPa) at 28 days
   4. Shrinkage: (ASTM C-596) -0.093% (28 days); +0.073% (120 days)

B. Plugging Compound for Active Water Penetrations: Accelerating agent for capillary/crystalline waterproofing products or pulverized rapid-setting cement.
   1. Product: VANDEX QUICKBINDER or VANDEX PLUG.
   2. Compressive Strength: (ASTM C-109) 3400 psi (23.4 MPa) at 24 hrs
   3. Flexural Strength: (ASTM C-348) 450 psi (3.1 MPa) at 48 hrs
   4. Potable water certification: NSF Standard 61

PART 3 - EXECUTION

3.1 EXAMINATION

A. Examine substrates and adjoining construction, and conditions under which Work is to be installed. Do not proceed with Work until unsatisfactory conditions are corrected.

B. Verify the following substrate conditions before application of capillary/crystalline waterproofing:
   1. That substrate condition is satisfactory and in accordance with manufacturer's instructions.
   2. That concrete surfaces have open pores and wood float finish on horizontal surfaces.
   3. That concrete surfaces are free of voids, spalled

07162-4 Crystalline Waterproofing

FIGURE 13.8 *(Continued)* Guide specification for section 7162, crystalline waterproofing. (*Courtesy Vandex*)

areas, loose aggregate and sharp protrusions, and with no coarse aggregate visible.
4. That curing compounds or surface hardeners incompatible with waterproofing have not been used on concrete.

3.2 PREPARATION

A. Protect adjacent surfaces not designated to receive waterproofing.

B. Substrate preparation:
1. Remove remaining concrete fins and projections, and general surface dirt.
2. Remove grease, oil and other contaminants. Use steam cleaning, high-pressure water blasting, wet or dry sand blasting, wire brush or other methods recommended by waterproofing manufacturer to produce surfaces suitable for application of waterproofing.
3. Follow manufacturer's instructions to clean and prepare surfaces and seal cracks and joints.
4. Rout out faulty construction joints and visible cracks not subject to movement, exceeding 0.012" (0.3 mm) in width to approx. 3/4" (19 mm) width and minimum 3/4" (19 mm) depth.
5. Remove all protrusions, work back to sound concrete and chisel out any spalled or honeycombed areas.
6. Roughen formtie holes.
7. Stop active water leakages as per manufacturer's plugging specifications.

C. Rinse surfaces to be waterproofed several times so that the concrete is thoroughly saturated. Surfaces shall be moist but not wet when waterproofing system is applied. Remove any surface water on horizontal surfaces.

3.3 INSTALLATION

A. Mix waterproofing material in proportions recommended by manufacturer.

B. Apply waterproofing material in quantities as per manufacturer's specifications and recommendations.

C. Cavity Fill:
1. Prime cavities at cleaned and prepared faulty construction joints, cracks, formtie holes, etc. with waterproofing material and fill flush to surface with patching compound in mortar consistency.
2. Laminate patching compound in 2 to 3 layers as per

07162-5 Crystalline Waterproofing

FIGURE 13.8 (*Continued*) Guide specification for section 7162, crystalline waterproofing. (*Courtesy Vandex*)

manufacturer's instructions for larger spalled or honeycombed areas.

****

SELECT AND SPECIFY SEAL STRIPS/REGLETS IN ARTICLE BELOW IN CASE OF NO WATERSTOPS PRESENT, OR IN ADDITION TO WATERSTOPS. DELETE IF NOT REQUIRED.

****

D. Horizontal and Vertical Construction Joints:
   1. Prime seal strips/reglets in pre-formed 1" x 1" (25 x 25 mm) cavities with waterproofing material and fill flush to surface with patching compound in mortar consistency.

****

INCLUDE ANY OF THE FOLLOWING METHODS THAT ARE APPLICABLE TO THE PROJECT.

USE DRY-SPRINKLE METHOD FOR SLABS EXPOSED TO MECHANICAL WEAR OR ABRASION, SUCH AS IN PARKING GARAGES, WASTE WATER TANKS, ETC.)

****

E. Horizontal surfaces:
   1. Dry-sprinkle waterproofing material to freshly poured slabs at 2.0 lb/sq.yd. (1.0 kg/sq.m) and power or hand trowel.

   2. Dry distribute to prewatered mud slab or existing slab at 2.25 lb/sq.yd. (1.2 kg/sq.m) immediately prior to casting the structural slab or new topping slab.

   3. Brush or spray apply waterproofing material in slurry consistency, in one coat on existing slabs.
      a. For standard applications, apply at rate of 2.0 lb/sq.yd. (1.0 kg/sq.m).
      b. For applications in contact with salt or waste water, apply at rate of 2.5 - 2.8 lb/sq.yd. (1.4 to 1.5 kg/sq.m).
      c. Spread material evenly and work it well into the surface.

F. Vertical Surfaces:
   1. Apply base coat of waterproofing material in slurry consistency at uniform rate of 1.25 - 1.4 lb/sq.yd. (0.7 to 0.75 kg/sq.m). Apply using appropriate compressed-air spray equipment, stiff masonry brush or stiff broom.
   2. After base coat has reached initial set but is still "green" (tacky), apply finish slurry coat of waterproofing material at 1.25 - 1.4 lb/sq.yd. (0.7 to 0.75 kg/sq.m). Apply so that final brush or broom strokes leave parallel, uniform texture.

07162-6 Crystalline Waterproofing

FIGURE 13.8 *(Continued)* Guide specification for section 7162, crystalline waterproofing. *(Courtesy Vandex)*

3.4 CURING

A. Follow manufacturer's general instructions for curing and hardening of waterproofing material.

B. Protect surfaces from rain, frost and drying out.

****
INCLUDE THE PARAGRAPH BELOW IF WATERPROOFING IS TO BE OVERCOATED WITH PAINT, EPOXY OR TILE MORTAR.
****

3.5 PREPARATION FOR DECORATION, COATING AND TILING

A. All surfaces treated with Capillary/Crystalline Waterproofing which are to be coated, painted or tiled shall be left to cure for 4 weeks. At the end of the curing period, the surfaces shall be saturated with water and neutralized with a 1:8 solution of muriatic acid. Rinse waterproofed areas thoroughly with clean water.

****
THE FOLLOWING ARTICLE APPLIES ONLY TO WATER TANKS, RESERVOIRS, MARINE AQUARIUMS, ETC.
****

3.6 ADJUSTING

A. Following application and completion of related work, as required, but well prior to completion of entire project, fill tanks to capacity and allow to stand not less than 2 weeks. Should leakage occur after this period, drain tanks to perform repairs. Notify Owner prior to draining tanks.

B. Stop leakage due to curing and shrinkage cracks in concrete which can develop during this period by installing plugs, seal-strips and additional surface treatment at no additional cost to the Owner. Following all required repairs, re-test by refilling tank and allow to stand not less than 1 week. Follow this procedure until all leakage is eliminated.

C. Thoroughly rinse all tanks and reservoirs with water and with 100 ppm chlorine water solution if this is included in the waterproofing contract.

3.7 ACCEPTANCE

A. Remove left over materials and any foreign material resulting from the work from the site.

B. Clean adjacent surfaces and materials.

07162-7 Crystalline Waterproofing

FIGURE 13.8 *(Continued)* Guide specification for section 7162, crystalline waterproofing. *(Courtesy Vandex)*

END OF SECTION

Project: .... (9/24/98)

07162-8 Crystalline Waterproofing

FIGURE 13.8 (*Continued*) Guide specification for section 7162, crystalline waterproofing. (*Courtesy Vandex*)

PRODUCT GUIDE SPECIFICATION

SECTION 07190
formally 07180

WATER REPELLENTS

****

This section is based on products manufactured by:

CHEMPROBE TECHNOLOGIES, INC.
2805 Industrial lane
Garland, TX 75041

PH : 800/760-6776 PH : 972/271-5551
FX : 972/271-5553

Specifier Notes : This specification covers the selection, preparation of surface, and application of Chemprobe's Water Repellent products to exterior vertical and horizontal concrete, brick, stucco, concrete masonry, etc.

Chemprobe Technologies, Inc. manufactures a full line of water repellent products for exterior surfaces.

****

PART 1 - GENERAL

1.01 SECTION INCLUDES

A This section covers the preparation, materials, services, and equipment required in conjunction with the application of a clear water repellent on all above grade, vertical and horizontal, exterior exposed surfaces of masonry, concrete, brick, stucco and [ ] .

1.02 RELATED SECTIONS

****

Specifier Notes : Edit this section as necessary to include other specific sections.

****

A Section 03300 - Cast In Place Concrete.

B Section 03400 - Precast Concrete.

C Section 03900 - Concrete Restoration & Cleaning.

D Section 04400 - Stone.

E Section 04800 - Masonry Assemblies.

F Section 04900 - Masonry Restoration & Cleaning.

G Section 07180 - Traffic Coatings.

H Section 09630 - Masonry Flooring.

I Section 09910 - Paints.

J Section 09960 - High Performance Coatings.

Section 07190 - 01

FIGURE 13.9 Guide specification for section 7190, water repellents. (*Courtesy Chemprobe Technologies*)

1.03 SUBMITTALS

A Submit under provisions of Section 01300 - Submittals.

B Product Specification Data: Submit manufacturer's technical literature, specifications, and application instructions for the specified clear water repellent material.

C Samples: Obtain samples of water repellent for sample application. Sample application is covered in section 1.04 QUALITY ASSURANCE.

D Applicator Qualifications: Submit certification stating applicator has a minimum of three (3) years experience using the specified product. Provide a list of several most recently completed projects where the specified material was used. Include the project name, location, architect and method of application.

****

Specifier Notes : Delete the following section if there are no VOC restrictions.

****

E Environmental Regulations: Submit certification stating the water repellent material to be applied is in compliance with federal, state and local environmental Volatile Organic Compounds (VOC) regulations.

1.04 QUALITY ASSURANCE

A Manufacturer: A firm with no less than ten (10) years experience in manufacturing the products specified in this section.

B Applicator Qualification: A firm with no less than three (3) years experience in the application of the products specified in this section. In addition, applicator must state the intended use of the proper application equipment and that it has been well maintained.

C Mock-Up:

1 Apply water repellent per manufacturer's application instructions as directed by the Architect to substrate material which matches actual job conditions. Determine the acceptability of appearance and optimum coverage rate required for application.

2 After sample treatment has cured in accordance with manufacturers recommendations, water test to verify that substrate is coated with sufficient water repellent to effectively repel liquid water from the surface.

3 Obtain Architect and/or Project Owner approval prior to full scale application of water repellents.

****

Specifier Notes : Coordinate pre-application meeting with section 01040 - Coordination and 01200 - Project Meetings

****

D Pre-Application Meeting: Convene a pre-application meeting prior to the start of application of the specified material. Attendance by a representative of each of the following organizations is requested; the application firm, the architectural firm, and the water repellent manufacturer. Notify each of the attendees at least three (3) days prior to the meeting time.

Section 07190 - 02

FIGURE 13.9 *(Continued)* Guide specification for section 7190, water repellents. (*Courtesy Chemprobe Technologies*)

1.05 **PRODUCT DELIVERY**

A Material Delivery: Deliver materials to the job site in original sealed containers, clearly marked with manufacturer's name, brand name, and type of material. Verify the product matches that of the original sample applied on the mock - up wall.

B Storage & Protection: Store materials inside if possible, away from sparks and open flame. Store in a secure area to avoid tampering and contamination. Water based materials must be kept from freezing. Store and handle in accordance with manufacture's written instructions.

1.06 **PROJECT CONDITIONS**

A Surface Preparation: Surface must be free of cracks, dirt, oils, paint or other contaminants which may effect the appearance or performance of the water repellent material.

B Environmental Requirements:
1 Air and substrate temperature must be above 40° F (5° C) or below 95° F (35° C) unless otherwise specified by manufacturer.
2 Do not proceed with application if the substrate is wet or contains frozen water.
3 Do not apply material when rain is predicted within 48 hours; or earlier that five (5) days after the substrate became wet.
4 Do not apply materials in high or gusty winds.

C Protection:
1 Special precautions should be taken to avoid vapor transmission (fumes) from entering the building being treated. Ventilation systems and fresh air intakes should be turned off and closed.
2 Protect shrubs, metal, wood trim, glass, asphalt and other building hardware during application from overspray.
3 Do not permit spray mist or liquid to drift onto surrounding properties.

1.07 **SCHEDULING**

A Architect shall be notified not less that 48 hours before each application of water repellent is scheduled.

****

Specifier Notes : Delete the following section if no warranty is required or if the water repellent work is covered under the terms of a general project warranty specified elsewhere. Edit as necessary to obtain a warranty from the contractor only, or to suit other project specific criteria.

****

1.08 **WARRANTY**

A The contractor and applicator shall jointly and severally warrant water repellent material against failure in material and workmanship for a period of five (5) years from the date of application.

B Submit completed manufacturer's written "Request For Warranty Form" to manufacturer ten (10) days prior to application.

C After substantial completion of the project, submit manufacturer's written "5 Year Warranty Application" to manufacturer for processing. Upon receiving validated warranty, submit copies to Architect and building owner.

Section 07190 - 03

FIGURE 13.9 (*Continued*) Guide specification for section 7190, water repellents. (*Courtesy Chemprobe Technologies*)

PART 2 - PRODUCTS

2.01 MANUFACTURER

Chemprobe Technologies, Inc.
2805 Industrial Lane
Garland, TX 75041
PH: 800/760-6776
PH: 972/271-5551 FX: 972/271-5553

2.02 WATER REPELLENT

A GENERAL: All products shall contain Siloxane and or Silane. No fillers, sterates, or paraffins. Products containing Siloxane only, shall have a range of three (3) to seven (7) percent solids. Products containing Silane only, shall be either twenty (20) or fourty (40) percent solids. Silane-siloxane blend products shall contain a minimum of fifteen (10) percent solids.

****

Specifier Notes : Refer to Chemprobe's product sheets or contact a Chemprobe Representative to determine the appropriate water repellent to be specified. Edit this list of water repellents to contain only those products and substrates to be included in the work.

****

1 PRIME A PELL 200 : Solvent based siloxane.

2 PRIME A PELL Plus : Solvent based silane - siloxane blend.

3 PRIME A PELL H_2O Block Formula : Water based siloxane.

4 PRIME A PELL H_2O Brick & Precast Formula : Water based siloxane.

5 DECK A PELL 15 : Solvent based silane - siloxane blend.

6 DECK A PELL 20 : Solvent based silane - siloxane blend.

7 DUR A PELL 10 : Water based silane - siloxane blend.

8 DUR A PELL 20 : Water based silane - siloxane blend.

9 DUR A PELL 20"S" : Alcohol Base Silane.

10 DUR A PELL 40"S" : Alcohol Base Silane.

11 DUR A PELL 100"S" : Alcohol Base Silane.

PART 3 - EXECUTION

3.01 EXAMINATION

A Verify the following:
1 The required joint sealants have been installed.
2 New masonry and mortar has cured a minimum of twenty one (21) days.
3 Surface to be treated is clean, dry and contains no frozen water.
4 Environmental conditions are appropriate for application.

3.02 PREPARATION

A Protection:
1 Special precautions should be taken to avoid vapor transmission (fumes) from entering the building being treated. Ventilation systems and fresh air intakes should be turned off and closed.
2 Protect shrubs, metal, wood trim, glass, asphalt and other building hardware during application from overspray.
3 Do not permit spray mist or liquid to drift onto surrounding properties or parking lots.

Section 07190 - 04

FIGURE 13.9 (*Continued*) Guide specification for section 7190, water repellents. (*Courtesy Chemprobe Technologies*)

3.03 APPLICATION

A Apply water repellents in accordance with manufacturer's written application instructions.

B Material must be applied using low pressure application equipment designed for water repellent application.

C Apply material as shipped by the manufacturer. Do not dilute.

D Apply treatment evenly until surface is totally saturated. Coverage rates are dependent on substrate material. Only one saturation coat is required.

3.04 FIELD QUALITY CONTROL

A The architect shall be contacted 48 hours prior to application so as to provide supervision as required. The architect or the architect's representative shall inspect the progress as the work proceeds. Do not apply any water repellent that is not specified by the architect.

B After water repellent has cured for 24 hours at low humidity and temperature between 70°-90° F or 48 hours at high humidity and low temperature between 50°-69° F, all surfaces shall be tested with a water spray. Recoat any area that indicates water absorption.

3.05 CLEANING

A At completion, remove from the job site, all excess material, debris, and waste resulting from this work. Dispose of water repellent containers according to state and local environmental regulations.

END OF SECTION

PAPSPEC1 02/13/98 Section 07190 - 05

FIGURE 13.9 (*Continued*) Guide specification for section 7190, water repellents. (*Courtesy Chemprobe Technologies*)

This suggested guide specification has been developed using the current edition of the Construction Specifications Institute (CSI) "Manual of Practice," including the recommendations for the CSI 3 Part Section Format. Additionally, the development concept and organizational format of the American Institute of Architects (AIA) MASTERSPEC Program have been recognized in the preparation of this guide specification. Neither CSI nor AIA endorses the use of specific manufacturers and products.

SECTION 07540
HOT FLUID-APPLIED RUBBERIZED ASPHALT WATERPROOFING SYSTEM

MOP NOTES (CSI MANUAL OF PRACTICE NOTES), ED NOTES (EDITING NOTES), AND COORD NOTES (DRAWING AND SPECIFICATIONS COORDINATION NOTES) ARE PROVIDED THROUGHOUT THIS GUIDE SPECIFICATION; ARE INDENTED TO THE FLUSH RIGHT; AND ARE CAPITALIZED IN A SMALLER TYPEFACE. DELETE NOTES IN FINAL COPY OF PROJECT SPECIFICATION. COORDINATE SPECIFICATION WITH PROJECT DRAWINGS.

PART 1 GENERAL

1.01 SUMMARY

ED NOTE: MIRASEAL 9100 HOT FLUID-APPLIED RUBBERIZED ASPHALT WATERPROOFING SYSTEM IS A SEAMLESS, SINGLE-COMPONENT, AVAILABLE EITHER AS A SINGLE-PLY (NON-REINFORCED) OR A FABRIC REINFORCED, MONOLITHIC WATERPROOFING SYSTEM FOR BOTH VERTICAL AND HORIZONTAL APPLICATIONS.

ED NOTE: REVISE BELOW TO SUIT PROJECT CONDITIONS. SELECT WATERPROOFING SYSTEM APPROPRIATE TO PROJECT REQUIREMENTS. COORDINATE SPECIFICATION OF INSULATION AND BALLAST/PAVERS BELOW WITH THIS SECTION OR WITH OTHER SPECIFICATION SECTIONS.

A. Section Includes: Hot fluid applied rubberized asphalt waterproofing system, including substrate primer, a monolithic rubberized asphalt membrane, separation sheet/protection course, drainage course (if required), extruded polystyrene insulation, filter fabric, asphaltic flashings, and ballast/pavers.
   1. Types of Hot Fluid Applied Rubberized Asphalt Waterproofing include:
      a. Hot Fluid Applied Rubberized Asphalt (Non-Reinforced) Waterproofing System.

ED NOTE: SELECT APPROPRIATE SYSTEM FROM ABOVE OR BELOW.

      b. Hot Fluid Applied Rubberized Asphalt (Reinforced) Waterproofing System.

ED NOTE: REVISE BELOW TO SUITE PROJECT CONDITIONS AND SPECIFICATION PREPARATION REQUIREMENTS. ADD SECTION NUMBERS PER CSI "MASTERFORMAT" AND OFFICE SPECIFICATION PRACTICES.

B. Related Sections:
   1. Cast-In-Place Concrete
   2. Structural Precast Concrete
   3. Lightweight Structural Concrete
   4. Masonry
   5. Rough Carpentry
   6. Building Insulation
   7. Accessories/Metal Flashing
   8. Joint Treatment
   9. Mechanical/Electrical

MOP NOTE: BELOW PARAGRAPH SHOULD DESCRIBE PRODUCTS AND WORK INCLUDED IN THIS SECTION WHICH ARE COVERED BY ALTERNATES. INCLUDE ALTERNATE DESCRIPTIONS. COORDINATE THIS SECTION WITH DIVISION 1 ALTERNATE SECTION, BID DOCUMENTS AND BID FORMS (IF ANY). REFER TO SPECIFICATION COORDINATION SHEETS INCLUDED WITH THIS SECTION FOR ADDITIONAL GUIDANCE ON THE USE OF ALTERNATES.

FIGURE 13.10 Guide specification for section 7540, hot-applied waterproofing. (*Courtesy TC MiraDRI*)

C. Alternates: Products and installation included in this section are specified by alternates. Refer to Division 1 Alternates Section for alternates description and alternate requirements.
   1. Alternate Manufacturers: Refer to Part 2 Products herein.

MOP NOTE: BELOW ARTICLE SHOULD LIST INDUSTRY STANDARDS REFERENCED IN THIS SECTION, ALONG WITH ACRONYM, ALPHA/NUMERIC DESIGNATIONS, TITLES, AND DATES. THIS ARTICLE DOES NOT REQUIRE COMPLIANCE WITH STANDARDS, BUT IS MERELY A LISTING TO ESTABLISH TITLE AND DATE OF REFERENCES.

1.02 REFERENCES (INDUSTRY STANDARDS)

A. General: Refer to Division 1 References Section.

ED NOTE: BELOW STANDARD IS CURRENT STANDARD FOR FLUID APPLIED WATERPROOFING PROPERTIES AND TEST METHODS, INCLUDING RELEVANT ASTM STANDARDS. COPY OF STANDARD AVAILABLE FROM NICOLON MIRAFI.

B. Canadian Government Specification Board (CGSB):
   1. CGSB 37-GP-50 Hot Applied Rubber Test, 1989

1.03 SYSTEM DESCRIPTION

A. Performance Requirements: Provide waterproofing system which has been manufactured and installed to maintain leak-proof waterproofing system without defects, damage, or failure.

1.04 SUBMITTALS

A. General: Prepare, review, approve, and submit specified submittals in accordance with "Conditions of the Contract" and Division 1 Submittals Sections. Product data, shop drawings, samples, and similar submittals are defined in "Conditions of the Contract."

B. Product Data: Submit manufacturer's product data for waterproofing system specified.

C. Shop Drawings: Submit shop drawings showing layout, profiles, and product components, including accessories for waterproofing system.

ED NOTE: RETAIN BELOW AS APPROPRIATE FOR WATERPROOFING SYSTEM SELECTED.

D. Samples: Submit verification samples for ballast/pavers.

E. Quality Assurance/Control Submittals:
   1. Test Reports Certification: Submit certified test reports from an acceptable independent testing laboratory indicating compliance with CGSB 37-GP-50 standard for rubberized asphalt membranes, including applicable ASTM procedures for referenced performance characteristics and physical properties.
   2. Certificates (Qualification Data): Submit product certificates signed by manufacturer certifying materials comply with specified performance characteristics and physical requirements, showing full time quality control, and waterproofing system components are being supplied by a single-source waterproofing manufacturer.
   3. Manufacturer's Field Reports: Submit manufacturer's field reports, including final inspection of installed waterproofing system.

FIGURE 13.10 (*Continued*) Guide specification for section 7540, hot-applied waterproofing. (*Courtesy TC MiraDRI*)

F. Closeout Submittals:
   1. Warranty: Submit warranty documents specified herein.
   2. Project Record Documents: Submit project record documents for installed materials in accordance with Division 1 Project Closeout (Project Record Documents) Section.

1.05 QUALITY ASSURANCE

A. Qualifications:
   1. Installer Qualifications: Installer experienced (as determined by contractor) to perform work of this section who has specialized in the installation of work similar to that required for this project who can comply with manufacturer's warranty requirements, and who is an approved applicator as determined by waterproofing system manufacturer.
   2. Manufacturer Qualifications: Manufacturer capable of providing field service representation during construction, approving acceptable installer, application method and conducting final inspection of the waterproofing system/assembly.

ED NOTE: USUALLY RETAIN BELOW FOR SPECIAL PROJECT REQUIREMENTS OR JOB CONDITIONS. OTHERWISE DELETE BELOW.

B. Mock-Ups (Field Constructed): Install at project site a job mock-up using acceptable products and manufacturer approved installation methods. Obtain Owner's and Architect's acceptance for manufacturer's warranty requirements including workmanship standard.

ED NOTE: SPECIFY BELOW MOCK-UP SIZE AND SUBSTRATE.

   1. Mock-Up Size:______________________________
   2. Mock-Up Substrate:______________________________
   3. Maintenance: Maintain mock-up during construction for workmanship comparison; remove and legal dispose of mock-up when no longer required.
   4. Incorporation: Mock-up may be incorporated into final construction upon Owner's approval.

C. Pre-Installation Meetings: Conduct pre-installation meeting to verify project requirements, substrate conditions, manufacturer's installation instructions and manufacturer's warranty requirements.

D. Pre-Installation Testing: In accordance with manufacturer's recommendations and warranty requirements, conduct pre-installation testing of substrates to receive waterproofing.

1.06 DELIVERY, STORAGE, AND HANDLING

A. Ordering: Comply with manufacturer's ordering instructions and lead time requirements to avoid construction delays.

B. Packing, Shipping, Handling, and Unloading: Deliver materials in manufacturer's original, unopened, undamaged containers with identification labels intact, including applicable UL labels.

C. Storage and Protection: Store materials protected from exposure to harmful weather conditions and at temperature conditions recommended by manufacturer.
   1. As recommended by manufacturer and to comply with warranty requirements, store fluid materials at a temperature between 60°F. (15.5°C.) and 80° F. (26.6° C.) If exposed to lower temperatures, restore materials to 60° F. (15.5° C.) minimum temperature before application.

FIGURE 13.10 *(Continued)* Guide specification for section 7540, hot-applied waterproofing. *(Courtesy TC MiraDRI)*

1.07 PROJECT CONDITIONS/SITE CONDITIONS

A. Environmental Requirements/Conditions:
   1. Weather Conditions: Neither commence nor proceed with waterproofing application when ambient temperature is below 0° F. (-17.7° C.). At temperatures below 32° F. (0° C.) contact manufacturer for approved primers and application recommendations for warranty compliance.

ED NOTE: BELOW PERTAINS TO THE AMERICAN INSTITUTE OF ARCHITECTS (AIA) "GENERAL CONDITIONS OF THE CONTRACT FOR CONSTRUCTION," WHICH PLACES RESPONSIBILITY FOR MEANS, METHODS AND TECHNIQUES OF CONSTRUCTION UPON THE CONTRACTOR.

B. Project Conditions: Refer to "Conditions of the Contract" for control and responsibility of project site including means, methods and techniques of construction by the contractor.

C. Field Measurements: Verify actual measurements/openings by performing field measurements before fabrication; show recorded measurements on shop drawings. Coordinate field measurements, fabrication schedule with construction progress to avoid construction delays.

ED NOTE: BELOW WARRANTY ARTICLE REQUIRES COORDINATION WITH THE OWNER. BELOW ASSUMES THE USE OF THE AMERICAN INSTITUTE OF ARCHITECTS (AIA) "CONDITIONS OF THE CONTRACT FOR CONSTRUCTION."

1.08 WARRANTY

A. Project Warranty: Refer to "Conditions of the Contract" for project warranty provisions. Submit manufacturer's single-source warranty directly from waterproofing system manufacturer for waterproofing system, including insulation and ballast/pavers.

B. Manufacturer's Warranty Requirements: Submit, for Owner's acceptance, manufacturer's standard warranty document executed by authorized company official. Manufacturer's warranty is in addition to, and not a limitation of, other rights Owner may have under the Contract Documents.
   1. Beneficiary: Issue warranty in the legal name of the project Owner.

      ED NOTE: COORDINATE BELOW WITH MANUFACTURER'S WARRANTY REQUIREMENTS.

   2. Warranty Period: ___ years commencing on Date of Substantial Completion
   3. Warranty Acceptance: Owner is sole authority who will determine acceptability of manufacturer's warranty documents.

ED NOTE: SELECT AVAILABLE MANUFACTURER'S WARRANTY PROGRAM FROM BELOW. REVISE BELOW AS APPROPRIATE. CONSULT WITH NICOLON MIRAFI FOR EXACT WARRANTY TERMS AND CONDITIONS.

C. Manufacturer's Warranty Program:
   1. Manufacturer's Material Warranty: Material only warranty, excludes labor.
      a. Duration:
         1) Standard Assembly: 2, 5 years.
         2) Fabric Reinforced Assembly: 2, 5, 10 years.
   2. Manufacturer's Watertightness Warranty: Watertightness warranty, includes labor and materials.
      a. Duration:
         1) Standard Assembly: 5, 10 years.
         2) Fabric Reinforced Assembly: 5, 10, 15 years.

FIGURE 13.10 (*Continued*) Guide specification for section 7540, hot-applied waterproofing. (*Courtesy TC MiraDRI*)

3. Manufacturer's Thermal Warranty: Thermal warranty, includes 90% retention of insulation original thermal value and remain on the deck at 70 mph wind gust.
   a. Duration:
      1) Standard Assembly: 5, 10 years.
      2) Fabric Reinforced Assembly: 5, 10, 15 years.
4. Manufacturer's Total System Warranty: Manufacturer's total system warranty (Miraseal Pinnacle System) covers components of waterproofing assembly, including membrane, flashing, insulation and pavers. Includes removal and replacement of pavers and overburden when installed in accordance with Mirafi requirements.
   a. Duration of Membrane/Flashing (Watertight Condition):
      1) Standard Assembly: 5, 10 years.
      2) Fabric Reinforced Assembly: 5, 10, 15 years.
   b. Duration of Insulation: (90% of insulation original thermal value; remain on the deck withstanding winds speeds not to exceed 70 mph).
      1) Standard Assembly: 5, 10 years.
      2) Fabric Reinforced Assembly: 5, 10, 15 years.
   c. Duration of Pavers: (Crack, Split or Disintegrate due to freeze-thaw).
      1) Standard Assembly: 5, 10 years.
      1) Fabric Reinforced Assembly: 5, 10 years.

PART 2 PRODUCTS

ED NOTE: RETAIN BELOW ARTICLE FOR PROPRIETARY SPECIFICATION. DO NOT USE PHRASE "OR EQUAL" / "OR APPROVED EQUAL"; USE OF SUCH PHRASES IS NOT CONSIDERED GOOD, PROFESSIONAL SPECIFICATION WRITING PRACTICE, BECAUSE OF THE DIFFERING INTERPRETATIONS OF THE PHRASE AMONG THE CONTRACTING PARTIES. COORDINATE BELOW ARTICLE WITH DIVISION 1 MATERIAL AND EQUIPMENT (PRODUCT OPTIONS AND SUBSTITUTIONS) SECTION, AND BIDDING DOCUMENTS, IF ANY.

2.01 MANUFACTURERS

A. Acceptable Manufacturer: Mirafi Moisture Protection Products, Nicolon Mirafi Group, a Division of Nicolon Corp.
   1. Address: 3500 Parkway Lane, Suite 500, Norcross, GA 30092
      a.Telephone: (404) 447 6272; Fax: (404) 448 5124

ED NOTE: SELECT APPROPRIATE SYSTEM FROM BELOW. VERIFY EACH COMPONENT OF WATERPROOFING SYSTEM. DETERMINE IF BALLAST OR PAVERS OR BOTH ARE TO BE USED ON THE PROJECT.

   2. Warranted Waterproofing System: Miraseal 9100 Hot Fluid Applied Rubberized Asphalt Waterproofing System, including primer, membrane, separator/protection course, drainage composite (if required), insulation, filter fabric, asphaltic flashings, and ballast/pavers. Provide system components from a single source supply in accordance with manufacturer's warranty requirements.
      a. Miraseal 9100 Hot Fluid Applied Rubberized Asphalt (Non-Reinforced) Waterproofing System (180 mils total minimum thickness).
      b. Miraseal 9100 Hot Fluid Applied Rubberized Asphalt (Fabric Reinforced) Waterproofing System, (90 mils minimum first coat thickness with fabric reinforcing sheet and 125 mils minimum second coat thickness).

ED NOTE: RETAIN BELOW FOR PROPRIETARY PRODUCT DESCRIPTIONS.

FIGURE 13.10 (*Continued*) Guide specification for section 7540, hot-applied waterproofing. (*Courtesy TC MiraDRI*)

3. Proprietary Product(s): Miraseal 9100 Hot Fluid Applied Rubberized Asphalt Waterproofing System.
   a. Primer: Mirafi 951 Primer. (Conforming to CGSB 37-GP-9 or ASTM D 41).
   b. Hot Fluid Applied Rubberized Asphalt: Miraseal 9100. (Conforming to the Property, Test Method and Results in CGSB 37-GP-50 and appropriate ASTM contained therein).
   c. Reinforcements: Miraseal Uncured Neoprene Sheet or Miraseal Polyester Reinforcement Fabric as recommended by waterproofing system manufacturer.

ED NOTE: BELOW LISTED PRODUCT RECOMMENDED BY NICOLON MIRAFI. OTHER PROPRIETARY PRODUCTS, SUCH AS MIRASEAL SBS PROTECTION COURSE AND MIRADRI PROTECTION COURSE 300 HV ARE AVAILABLE FROM NICOLON MIRAFI. REVISE BELOW TO SUIT PROJECT CONDITIONS AND JOB REQUIREMENTS. CONSULT WITH NICOLON MIRAFI FOR SPECIFIC RECOMMENDATIONS.

   d. Separation Sheet/Protection Course: 4 mil polyethylene sheet.
   e. Filter Fabric: Nicolon Mirafi PRM Fabric.
   f. Flashing Material: Miraseal Roof-Flash or Miraseal Uncured Neoprene, as recommended by manufacturer.

ED NOTE: RETAIN BELOW AS APPROPRIATE TO WATERPROOFING SYSTEM SELECTED AND TO PROJECT REQUIREMENTS. CONSULT WITH NICOLON MIRAFI FOR SPECIFIC RECOMMENDATIONS.

   g. Drainage Course: Miradrain 9000.

4. Waterproofing Manufacturer Endorsed Proprietary Products: In accordance with waterproofing manufacturer's warranty requirements, provide the following proprietary products:

ED NOTE: SELECTED APPROPRIATED ENDORSED PRODUCTS AS REQUIRED FOR PROJECT.

   a. Insulation: Extruded Polystyrene Insulation Board, UC Industries, FOAMULAR Insulation (Conforming to ASTM C 578-91, r-value of 5.0 per inch, Foamular 400 (40 psi) VI , or Foamular 600 (60 psi) VII (ASTM D-1621).
   b. Pavers: 2 feet by 2 feet (609.6 mm) maximum, 22 pounds/linear ft (32.7 kg/m) minimum, Hanover Pavers; color, texture, and pattern as selected by Architect.
   c. Ballast: Well screened and washed stone or gravel ballast (ASTM D 448-80) complying with Nicolon Mirafi warranty requirements for ballast.

ED NOTE: BELOW MAY BE DELETED OR RETAINED IN ACCORDANCE WITH ARCHITECT'S PRACTICE FOR PROPRIETARY TYPE SPECIFICATION METHOD OR FOR PERFORMANCE TYPE SPECIFICATION METHOD. REVIEW BELOW FOR USE WITH PUBLIC AND PRIVATE TYPE OF CONSTRUCTION PROJECTS.

3. Materials(s)/System(s) Testing: Hot fluid applied rubberized asphalt membrane meeting the following physical properties, test method, and result:

Property	Test Method	Result
• Flash Point	ASTM D-92 CGSB 37-GP-50	500 degrees F. (260 degrees C.)*
• Low Temperature Crack Bridging Capability	CGSB 37-GP-50	No Cracking, Adhesion Loss, or Splitting
• Water Vapor Permeability	ASTM E-96, Procedure E CGSB 37-GP-50	1.7 ng/Pa (s) M^2 max. (0.027 perm)

FIGURE 13.10 (*Continued*) Guide specification for section 7540, hot-applied waterproofing. (*Courtesy TC MiraDRI*)

• Water Resistance (5 days/50 degrees C.)	CGSB 37-GP-50	No Delamination, Blistering, Emulsification, or Deterioration
• Water Absorption	CGSB 37-GP-50	Gain in Weight: 0.35 g. max. Loss in Weight: 0.18 g. max.
• Elasticity/Ratio of Toughness to Peak Load	CGSB 37-GP-50	Min. toughness of 5.5 joules (48.67 in. pounds)/.04.
• Viscosity	CGSB 37-GP-50	2-15 seconds
• Heat Stability	CGSB 37-GP-50	No change in viscosity, penetration, flow or low temperature flexibility.
• Low Temperature (-25 degrees C.)	CGSB 37-GP-50	No delamination, adhesion loss, or cracking.
• Penetration	ASTM D-1191 CGSB 37-GP-50	At 77° F. (25° C.), max. 110. At 122°F. (50° C.), max. 200.
• Flow	ASTM D-1191 CGSB 37-GP-50	At 140° F. (60° C.), max. 3.0 mm, max.
• Softening Point	ASTM D-36	180° F. (82° C.)
• Elongation	ASTM D-1191	1000% min.
• Resiliency	ASTM D-3408	40% min.
• Bond To Concrete	ASTM D-3408	Pass

* Or alternatively not less than 77 degrees F (25 degrees C) above the manufacturers maximum recommended application temperature.

ED NOTE: RETAIN BELOW FOR ALTERNATE METHOD OF SPECIFYING MANUFACTURERS AND PRODUCTS. ADD ALTERNATE NUMBER DESIGNATION AND COORDINATE WITH OTHER SPECIFIED PROJECT ALTERNATES.

2.02 ALTERNATES AND SUBSTITUTIONS

A. Alternate (Manufacturers/Products): In lieu of providing below specified base bid/contract manufacturer, provide below specified alternate manufacturers. Refer to Division 1 Alternates Section.
   1. Base Bid/Contract Manufacturer/Product: Mirafi Moisture Protection Products, Nicolon Mirafi Group, a Division of Nicolon Corp.
      a. Product: Hot Fluid Applied Rubberized Asphalt (Non-Reinforced) Waterproofing System (180 mils total minimum thickness).

ED NOTE: RETAIN EITHER ABOVE OR BELOW WATERPROOFING SYSTEM.

FIGURE 13.10 *(Continued)* Guide specification for section 7540, hot-applied waterproofing. *(Courtesy TC MiraDRI)*

b. Product: Hot Fluid Applied Rubberized Asphalt (Fabric Reinforced) Waterproofing System (90 mils minimum first coat thickness with fabric reinforcing sheet and 125 mils minimum second coat thickness).

ED NOTE: NUMBER BELOW ALTERNATES IN ACCORDANCE WITH PROJECT REQUIREMENTS AND OFFICE SPECIFICATION PRACTICES. DETERMINE ALTERNATE TYPES. VERIFY ALTERNATE MANUFACTURER/PRODUCT DESIGNATIONS, ADD PRODUCT ATTRIBUTES/CHARACTERISTICS FOR PRODUCT EQUIVALENCY.

2. Alternate #____ Manufacturer/Product: American Hydrotech, Inc.
    a. Product: Monolithic Membrane 6125 Fluid Applied Waterproofing System.
3. Alternate # ___ Manufacturer/Product: Bakor, Inc.
    a. Product: Bakor 790.11 Fluid Applied Waterproofing System.

B. Substitutions: Refer to Division 1 General Requirements for substitution requirements.

2.03 RELATED MATERIALS (Specified In Other Sections)

A. Accessories: Refer to Division 7 Roof Accessories/Flashings for metal roof accessories and counter flashings.

B. Sealants/Backer Rods: Refer to Division 7 Joint Treatment Section for sealants and backer rods.

ED NOTE: USUALLY RETAIN BELOW, INCLUDING REQUIREMENTS FOR INSULATION AND BALLAST/PAVERS IN ACCORDANCE WITH NICOLON MIRAFI WARRANTY REQUIREMENTS.

2.04 SOURCE QUALITY CONTROL

A. Single Source Responsibility: Obtain hot fluid applied rubberized asphalt waterproofing system including primer, rubberized asphalt, separator/protection course, drainage composite, insulation, filter fabric, asphaltic flashing and ballast/pavers from a single source supplier for required warranty.

PART 3 EXECUTION

3.01 MANUFACTURER'S INSTRUCTIONS/RECOMMENDATIONS

A. Compliance: Comply with manufacturer's product data, including product technical bulletins, product catalog installation instructions and product packaging instructions.

3.02 EXAMINATION

A. Site Verification of Conditions: Verify substrate conditions (which have been previously installed under other sections) are acceptable for product installation in accordance with manufacturer's instructions. Do not proceed with waterproofing installation until substrate conditions are acceptable for compliance with manufacturer's warranty requirements.

3.03 PREPARATION

A. Adjacent Surfaces Protection: Protect adjacent work areas and finish surfaces from damage during installation operations.

FIGURE 13.10 *(Continued)* Guide specification for section 7540, hot-applied waterproofing. *(Courtesy TC MiraDRI)*

B. Surface Preparation: Surfaces to receive waterproofing shall be dry, clean, frost free, and free of laitenance, dust, dirt, oil, harmful curing compounds, or other foreign matter which may impair the bond or performance of the waterproofing system and which do not comply with manufacturer's warranty requirements.

C. Concrete Deck Surface Preparation: Prepare concrete deck surfaces to receive hot fluid applied rubberized asphalt waterproofing system. Surfaces shall be smooth, free of depressions, voids, protrusions, clean and free of harmful curing compounds, form release agents, and other surface contaminants which may impair the bond or performance of waterproofing system and manufacturer's warranty requirements.
1. Cast-In-Place Concrete/Composite Decks: Decks shall be monolithic, smooth, free of voids, spalled areas, laitenance, honeycombs, and sharp protrusions. Refer to Division 3 Concrete Section for concrete strength/density, finish, curing methods and other concrete requirements.
2. Precast Concrete Decks: Decks shall be mechanically secured to minimize differential movement and each joint between precast units shall have an installed backer rod.

D. Substrate Cleaning: Sweep substrate which is to receive waterproofing system. Substrate shall be blown clean using an air compressor to remove any remaining loose debris.

E. Adhesion Test: Determine substrate cleanability and acceptance by applying a test patch of membrane material to surface and check adhesion according to manufacturer's test procedures and warranty requirements.

ED NOTE: COORDINATE BELOW ARTICLE WITH PART 2 PRODUCTS ARTICLE TO INCLUDE EACH PRODUCT REQUIRING INSTALLATION.

DWG COORD NOTE: COORDINATE BELOW ARTICLE WITH NICOLON MIRAFI RECOMMENDED WATERPROOFING DETAILS NUMBERED CONSECUTIVELY NO. W-1 THRU NO. W-11.

3.04 MEMBRANE INSTALLATION

A. Primer (Surface Conditioner): Apply Mirafi 951 Primer to concrete substrate at approximate rate of 400-600 sq. ft./gallon (9.8 - 17.7 m²/L) depending on surface texture and condition.

DWG COORD NOTE: COORDINATE BELOW WITH MANUFACTURER RECOMMENDED INSTALLATION DETAILS.

B. Joints And Crack Treatment:
1. Precast Concrete Deck Joints:
    a. Joint Widths: Reinforce joint widths using Miraseal Reinforcement Fabric embedded in 125 mils thick of Miraseal 9100 Membrane. Extend reinforcement fabric 6 inches on each side of precast concrete joint widths.
    b. Joint Lengths: Reinforce joint lengths using 125 mils thick Miraseal 9100 Membrane and a 12 inch wide strip of Miraseal 9100 Reinforcement Fabric. Extend reinforcement 6 inches on each side of precast concrete joint lengths.
2. Cracks: Seal cracks (1/8 inch wide or less) in concrete with 125 mils minimum thick coat of Miraseal 9100 Membrane and a 12 inch wide strip of Miraseal 9100 Reinforcement Fabric. Extend membrane and reinforcement along length of crack.

DWG COORD NOTE: COORDINATE BELOW WITH MANUFACTURER RECOMMENDED INSTALLATION DETAILS.

FIGURE 13.10 *(Continued)* Guide specification for section 7540, hot-applied waterproofing. *(Courtesy TC MiraDRI)*

3. Decks To Vertical Transitions (Flashings):
   a. Apply Miraseal 9100 Membrane to provide a 125 mils thickness to vertical wall, a minimum of 8 inches up vertical surface and a minimum of 6 inches on the horizontal surface.
   b. Embed Miraseal 9100 Roof-Flash in hot membrane. Avoid wrinkling or fishmouth, extending roof-flash 6 inches on horizontal surface. Mechanically attach Miraseal 9100 Roof-Flash 12 inches on center. Lap roof-flash material a minimum of 3 inches. Apply 125 mils thickness of hot Miraseal 9100 Membrane, extending membrane 3 inches over vertical flashing.
4. Expansion Joints: At expansion joints, loop Miraseal 9100 Uncured Neoprene sheet a least 2 inches deep. Embed in a 125 mils thick hot membrane and extend Miraseal 9100 Uncured Neoprene sheet at least 6 inches on both sides of joint. Seal joint with at least 6 inch wide coat of hot Miraseal 9100 Membrane, 125 mils thick. Seal neoprene loop with hot Miraseal 9100 Membrane.

C. Membrane Preparation:
   1. Preparation: Heat Miraseal 9100 Membrane in double jacketed, oil bath melter with mechanical agitation, specifically designed for preparation of rubberized asphalt membrane. Heat membrane until membrane can be drawn-free flowing at a temperature range of between 350° F. (177° C.) and 410° F. (210° C.).

DWG COORD NOTE: COORDINATE BELOW WITH MANUFACTURER RECOMMENDED INSTALLATION DETAILS.

   2. Membrane Flashing At Drains:
      a. Coat area around drain with 125 mils thick hot Miraseal 9100 membrane.
      b. Place Miraseal 9100 Uncured Neoprene sheet or Miraseal SBS Protection Course over primed drain flange extending a minimum of 12 inches around flange.
      c. Apply 125 mils thick second coat of hot Miraseal 9100 Membrane over the Miraseal 9100 Uncured Neoprene sheet or Miraseal 9100 SBS Protection Course.
      d. Apply clamping ring to ensure a leak-proof seal between membrane and clamping ring. Avoid blocking drains from debris/material which may cause clogging.
   3. Flashings And Protrusions:
      a. At mechanical vent or pipe protrusions, hand/finger gusset Miraseal 9100 Uncured Neoprene sheet flashing into 125 mils thick hot Miraseal 9100 Membrane, extending membrane 6 inches on the horizontal and the required height. (If more than 8 inches in height, mechanically fasten sheet flashing.) Apply second layer of hot membrane 6 inches on horizontal surface and 3 inches on vertical surface. Install mechanical fastening at top of flashing.
      b. Pitch Pockets: Place pan on hot Miraseal 9100 membrane and attach to waterproofing. Set Mirafi flashing into 125 mils thick hot membrane over top of flange. Fill pitch pocket with hot membrane and/or sealer.

D. Membrane Application: Apply Miraseal 9100 membrane in single or multiple layers. Apply at a rate to provide a continuous, monolithic coat for both vertical and horizontal surfaces, free of air pockets, wrinkles or tears. Ensure bond to substrate.
   1. Non-Reinforced Membrane: Apply first layer of Miraseal 9100 at a minimum thickness of 90 mils, to form a continuous, seamless, and monolithic membrane (including application over previously reinforced areas).

ED NOTE: SELECT EITHER ABOVE OR BELOW, OR BOTH IF APPLICABLE TO PROJECT CONDITIONS.

FIGURE 13.10 (*Continued*) Guide specification for section 7540, hot-applied waterproofing. (*Courtesy TC MiraDRI*)

2. Reinforced Membrane: Apply Miraseal 9100 Reinforcement Fabric sheet and press into first layer of hot membrane. Overlap reinforcement fabric sheet 1/4 inch ensuring complete coverage. Apply second layer of membrane at a minimum thickness of 125 mils, for a total membrane thickness of 215 mils minimum.

E. Separation Sheet/Protection Course:
1. Embed 4 mil polyethylene sheet into hot membrane to insure a bond.
2. Overlap adjoining sheet edges (dry) a minimum of 2 inches - 3 inches (50.8 mm - 76.2 mm) to insure complete coverage.
3. Cover separation sheet/protection course by drainage course, insulation fabric, ballast/pavers as soon as possible, but within ten days of membrane installation.

3.05 DRAINAGE COURSE INSTALLATION (INSULATION, FILTER FABRIC SHEET, BALLAST/PAVER PLACEMENT)

A. General: In accordance with waterproofing manufacturer's warranty requirements, install drainage course, insulation, fabric sheet, and ballast/pavers as each waterproofing section area is completed.

B. Prefabricated Drainage Course Placement:
1. Install drainage course on horizontal and vertical surfaces in accordance with waterproofing manufacturer's recommendations.
2. Layout and position Miradrain 9000 Drainage Course and allow to lay flat. Cut and fit drainage course to perimeter edges and to penetrations.
3. Bond geotextiles overlap edges to adjacent drainage course geotextile with Mirastick or Mirabond Tape to insure geotextile integrity.
4. Place subsequent topping materials.

C. Insulation Placement:
1. Loose lay in a staggered manner and tightly butt together each of the insulation boards. Maximum acceptable opening between insulation boards is 3/8 inch (9.5 mm). Install insulation within 1/4 inch (6.25 mm) of each of the projections and penetrations.
2. For multi-layer insulation applications, the bottom insulation layer shall be the thickest layer and must be a minimum of 2 inches thick (50.8 mm). Stagger each insulation layer.

D. Filter Fabric Sheet Placement:
1. Position Mirafi PRM Waterproofing Fabric over insulation as follows:
   a. Overlap edges a minimum of 18 inches (457.2 mm).
   b. Extend filter fabric 2 inches to 3 inches (50.8 mm to 76.2 mm) above ballast/paver at perimeter and penetrations.
   c. Extend filter fabric to drain bases and bonnets, but do not cover drains or restrict water flow to drain.

ED NOTE: COORDINATE BELOW WITH PART 2 PRODUCTS AND WITH PROJECT DRAWINGS.

E. Ballast/Pavers Placement
1. Install ballast/pavers immediately following the placement of filter fabric. Install a minimum of 10 psf (48.8 kg/m^2) of ballast coverage over field of waterproofing.
2. Place a 4 foot (1219.2 mm) wide area of ballast, or a continuous row of pavers. Pavers shall provide a minimum of 22 pounds/linear ft. (32.7 kg/m) and be installed at required areas designated by insurance underwriting agency, the perimeter, penetrations, drains, and other locations where insulation boards terminate.

FIGURE 13.10 (*Continued*) Guide specification for section 7540, hot-applied waterproofing. (*Courtesy TC MiraDRI*)

3. Pavers may be used as walkways or to prevent stone scouring due to wind on ballastered projects. Paver blocks shall not exceed 2 feet by 2 feet (609.6 mm) in size.

3.06 RELATED PRODUCTS INSTALLATION REQUIREMENTS

A. Accessories/Flashings: Refer to Division 7 Roof Accessories/Flashing Section of installation requirements.

B. Sealants and Backer Rods: Refer to Division 9 Joint Treatment Section for installation requirements.

3.07 FIELD QUALITY CONTROL

A. Membrane Water Test: In accordance with waterproofing manufacturer's warranty requirements, conduct a water test on installed waterproofing system. Water test waterproofing area by ponding water a minimum depth of 2 inches (50.8 mm) for a period of 48 hours to check integrity of membrane installation.
   1. If leaks should occur, drain water completely, repair membrane installation, and retest waterproofing area.

B. Manufacturer's Field Services: Upon Owner's request, provide manufacturer's field service consisting of product use recommendations and periodic site visits for inspection of product installation in accordance with manufacturer's instructions and warranty requirements.

3.08 CLEANING AND PROTECTION

A. Cleaning: Remove temporary coverings and protection of adjacent work areas. Repair or replace damaged installed products. Clean installed products in accordance with manufacturer's instructions prior to owner's acceptance. Remove construction debris from project site and legally dispose of debris.

B. Protection: Protect installed product's finished surfaces from damage during construction.

END OF SECTION 07120

FIGURE 13.10 *(Continued)* Guide specification for section 7540, hot-applied waterproofing. *(Courtesy TC MiraDRI)*

Page 1 Section 07572

This Master Specification is prepared by Pacific Polymers in order to assist you in the preparation of your Project Specification. Pacific Polymers does not attest to the completeness or accuracy of this document as an architectural or engineering specification and advises you to be very careful in editing this document to assure that no errors or omissions are committed. The performance and procedural information regarding Elasto-Deck Products is accurate at the time of printing the document and we will use our best efforts to keep it up to date. If you have questions, please call us at (714) 898-0025 or (800) 888-8340 for clarifications.

Edit titles below to agree with titles remaining within the body of completed Project Specification.

TABLE OF CONTENTS

FIGURE 13.11 Guide specification for section 7570, deck coating. (*Courtesy Pacific Polymers*)

This Section specifies fluid applied, seamless, waterproofing systems for application to surfaces exposed to vehicular or pedestrian traffic. The systems are suitable for applications over structurally sound substrates such as concrete, masonry, metal and plywood.

The Elasto-Deck 5000 Systems are described. These systems utilize varying coating thicknesses and aggregates to compensate for varying wear conditions and aesthetic requirements, as follows:

1. **Elasto-Deck 5000 TC** specifies 3 separate areas of varying wear conditions, featuring separate and distinct treatment of base coats, top coats and non-skid aggregates:

   A. Parking & Pedestrian Traffic Areas
   B. Driving Aisles

2. **Elasto-Deck 5000 PDS**: Section 07571, specifies a variation of 5000 TC which utilizes a different top coating technique which employs an aliphatic top coat followed by an aliphatic spatter coat to provide for high wear protection.

3. **Elasto-Deck 5000 X2**: Section 07573, specifies maximum protection for pedestrian traffic decks which are located over occupied space or which will be subjected to extremely heavy traffic.

4. **Elasto-Deck 500 WS**: Section 07574, specifies maximum protection for pedestrian decks where walnut shells are used as the non-skid media, providing a very even appearance and ensuring the protection of the areas below.

5. **Elasto-Deck 5000 WDA**: Section 07575, specifies protection over areas which will see light traffic exposure. Incorporates an integral mineral non-skid media.

6. **Elasto-Deck 5000 MR**: Section 07576, specifies traffic bearing protection for mechanical room floors which are subject to petroleum, qualified chemical and water spillage.

We have used the technique of combining these traffic bearing waterproofing systems in one, comprehensive master specification in order to save a great deal of duplication and "paper".

Please edit carefully to include only those Articles which apply to your project. 7 finish colors standard.

FIGURE 13.11 *(Continued)* Guide specification for section 7570, deck coating. *(Courtesy Pacific Polymers)*

PART 1 GENERAL

1.1 REFERENCE

> If the following paragraphs are used, edit the optional phrases and, if necessary, the complete paragraphs, to agree with wording used in similar paragraphs of other Sections of the Project Specifications.

A. Requirements established within the portions of this Project Manual titled ["Conditions of the Contract"] [and] [Division 1, General Requirements,] are collectively applicable to work required of this Section.

1.2 WORK SUMMARY

A. Provide labor, materials, equipment and supervision necessary to complete the fluid applied [vehicular] [pedestrian] [mechanical room] deck coating work.

> Use the following sub-paragraphs if more than one deck coating system is to be used and edit so that application areas are properly described. If the following is to be used, move the line fragment "including the following" up to complete the last line preceding the paragraph.

, including the following:

1. Elasto-Deck 5000 TC System shall be installed [as indicated on the Drawings.] [at ______________________________ ______________________________.]

Install the 3 parts of this system as follows:

1a. Parking and pedestrian areas [as indicated on Drawings.] [at ______________________________ ______________________________.]

1b. Driving aisles and ramps and all other areas [as indicated on Drawings.] [at ______________________________ ______________________________.]

FIGURE 13.11 *(Continued)* Guide specification for section 7570, deck coating. *(Courtesy Pacific Polymers)*

Page 4 Section 07572

2. Elasto-Deck 5000 PDS System with "Spatter Coat" finish shall be installed [as indicated on Drawings.] [at____.]

   2a. Parking and pedestrian areas [as indicated on Drawings.] [at____.]

   2b. Driving aisles (except high radius turns) [as indicated on Drawings.] [at____]

   2c. Ramps, entrances, exits, high radius turns, ticketing machines, pay booths [as indicated on Drawings.] [at____.]

3. Elasto-Deck 5000 X2 System shall be installed [as indicated on Drawings.] [at____.]

4. Elasto-Deck 5000 WS System shall be installed [as indicated on Drawings.] [at____.]

5. Elasto-Deck 5000 WDA System shall be installed [as indicated on Drawings.] [at____.]

6. Elasto-Deck 5000 MR System shall be installed [as indicated on Drawings.] [at____.]

> Edit from and add to the following list cautiously. Wording in any Section whose work influences, or is influenced by, this Section should be carefully evaluated to assure complimentary conditions exist. Refer also to Appendix A, at the end of this Section, for additional suggestions.

1. Expansion Joints — Section 03250
2. Cast-in-Place Concrete — Section 03300
3. Concrete curing — Section 03370
4. Metal Decking — Section 05310
5. Nails, Screws, Fasteners and Adhesives — Section 06050
6. Plywood Decking — Section 06115
7. Parking & Traffic Control Markings — Section 02600
8. Metal Flashings — Section 07600

FIGURE 13.11 (*Continued*) Guide specification for section 7570, deck coating. (*Courtesy Pacific Polymers*)

Use "By Others" only to identify pertinent 'not-in-contract' activities.

1. *
2. *
3. *

1.3 QUALITY ASSURANCE

A. Qualifications: Manufacturer of the coating system[s] shall have a minimum of 5 years experience in the manufacture of fluid applied deck coatings. The System Applicator shall be licensed by the Manufacturer and shall have a minimum of 5 years experience in application of fluid applied deck coatings. See Certifications requirements specified under "Submittals" article of this Section.

1.4 SUBMITTALS

A. [No submittals required.]

***** (or) *****

Edit the following paragraphs to properly describe submittals and procedures required.

B. In compliance with [Section 01300] [General and Supplementary General Conditions] provide submittals identified in the following paragraphs.

C. Samples: Submit samples of coating system[s] applied to 1/4" plywood or similar rigid base. [Submit one sample of each color coating to be used on Project.]

D. Manufacturer's Literature: Submit complete Manufacturer's literature and technical data for the deck coating system[s] proposed.

E. Certification: Submit, via Contractor's transmittal, properly identified with Project Name, location and date, certification of Manufacturer's and Applicator's compliance with requirements of "Quality Assurance" article of this Section. Evidence of

FIGURE 13.11 *(Continued)* Guide specification for section 7570, deck coating. *(Courtesy Pacific Polymers)*

compliance for both Manufacturer and Applicator shall include a list of at least 5 projects of a similar nature which have been installed during the last 5 years.

F. Applicator's Certificate: Submit copy of Licensed Applicator Certificate issued by Manufacturer.

G. Maintenance Manual: Upon completion of the work required by this Section, submit one Maintenance Manual, identified with Project Name, location and date; type of coating material applied; and, surfaces to which system was applied, including sketches where necessary.

Include recommendations for periodic inspections, care and maintenance. Identify common causes of damage with instructions for temporary patching until permanent repairs can be made.

See also specimen guarantee attached to this document.

H. Guarantee: Upon completion and acceptance of the work required by this Section, submit an executed copy of the attached guarantee.

1.5 PRODUCT DELIVERY AND STORAGE

A. Deliver materials to jobsite in sealed, undamaged containers. Each container shall be identified with material name, date of manufacture and/or lot number.

1.6 JOB CONDITIONS

A. Install coating materials under conditions where all of the following conditions are met:

1. Concrete has not been treated with any substance which will adversely affect adhesion or performance.

2. Rain is not anticipated within 8 hours of application.

3. Substrate surface temperatures are above 40 deg. F. (5 deg. C.) and lower than 110 deg. F. (44 deg. C.).

4. Positive ventilation for interior applications can be continuously supplied throughout application period and 8 hours after.

FIGURE 13.11 (*Continued*) Guide specification for section 7570, deck coating. (*Courtesy Pacific Polymers*)

5. Open fires and spark producing equipment are not, and will not be, in application area until vapors have dissipated.

B. Post 'No Smoking' signs in area during and for at least 8 hours following application period.

C. Strictly adhere to special requirements of Manufacturer as modified by applicable rules and regulations of local, state and federal authorities having jurisdiction.

1.7 GUARANTEE

Edit the following paragraph carefully to effect proper coordination of the Contract Documents. See also specimen guarantee attached immediately following this Section.

A. Completed installation shall be guaranteed against defects of materials and workmanship, as defined on the guarantee [attached], for a period of [2] [3] [5] years, beginning with date of substantial completion of the deck coating system[s].

PART 2 PRODUCTS

2.1 GENERAL

A. Components shall be products of a single approved Manufacturer, or shall be certified by the approved Manufacturer as compatible with components produced by him.

2.2 DECK COATING

A. Coating material shall be polyurethane elastomer based, meeting or exceeding minimum physical properties listed in Table 1, and capable of producing a seamless, waterproof, traffic bearing deck coating resistant to fuels, oils, solvents and cleaning compounds. Color of top coating shall be [concrete grey] [aluminum grey] [limestone] [tan] [black] [beige] [off-white] [and installed as indicated on Drawings.].

FIGURE 13.11 *(Continued)* Guide specification for section 7570, deck coating. (*Courtesy Pacific Polymers*)

Before adding other Manufacturers to following list, carefully pre-qualify them in terms of 1) physical properties of coating material, 2) Availability of Manufacturer Licensed Applicators, and 3) Successful past history of installations extending over at least a 5 year period. Edit Table 1 to reflect performance requirements appropriate for acceptable materials.

B. If in compliance with this Specification, the following systems will be acceptable:

Select Manufacturers and Systems carefully from the following list. Delete those Systems which are not appropriate to this Project.

1. Manufacturer: Pacific Polymers, Inc., 12271 Monarch Street, Garden Grove, CA 92641 (714) 898-0025.

   1a. Elasto-Deck 5000 TC System (Traffic Compensated)
   1b. Elasto-Deck 5000 PDS System (Spatter Coat Finish)
   1c. Elasto-Deck 5000 X2 System (Pedestrian)
   1d. Elasto-Deck 5000 WS System (Pedestrian)
   1e. Elasto-Deck 5000 WDA System (Pedestrian)
   1f. Elasto-Deck 5000 MR System (Mechanical Room)

2. *

3. *

FIGURE 13.11 *(Continued)* Guide specification for section 7570, deck coating. *(Courtesy Pacific Polymers)*

TABLE 1 PROPERTIES OF CURED COATINGS

PROPERTY	MEASURING STANDARDS AND CONDITIONS	RESULTS, BASE COAT	RESULTS, TOP COAT
Product Type		Aromatic moisture curing polyurethane	Aliphatic moisture curing polyurethane
Tensile Strength See Note 1	ASTM D 412 Die "C" Pulled at 20 ipm	975 psi min. (.684 kgf/sq.mm)	2,600 psi min. (1.83 kgf/sq.mm)
Elongation See Note 1	ASTM D 412 Die "C" Pulled at 20 ipm	825% min.	190% min.
Moisture Vapor Transmission 15 dry mil	ASTM E 96 Procedure "B" See Note 2	3.7 perms +/- .6	0.8 perms +/- .13
Moisture Vapor Transmission 30 dry mil	ASTM E 96 Procedure "B" See Note	2.4 perms +/- .4	NA
Abrasion	ASTM C 501 30 mil dry film on 4" x 4" metal CS-17 wheel 1000 revs. with 1000 gram weight	No change of weight	No change of weight
Adhesive peel strength on primed concrete	ASTM D 903	90 pli	NA
Fire resistance	ASTM E 108 (U.L. 790)	System rated Class "A" on non-combustible substrate	System rated Class "A" on non-combustible substrate

Elasto-Deck 5000 Systems

Note 1 Base coat tests conducted on deaerated 40 mil (1mm) dry film, top coat tests conducted on deaerated 10 mil (0.25 mm) dry film, cured for 7 days at 77 deg. F. (25 deg. C.) at 50% relative humidity.

Note 2 Cured for 7 days at 77 deg. F. (25 Deg. C.) at 50% relative humidity.

FIGURE 13.11 *(Continued)* Guide specification for section 7570, deck coating. *(Courtesy Pacific Polymers)*

2.3 RELATED MATERIALS

A. Sealant: Elasto-Thane 230 or Elasto-Thane 227.

B. Backer Rod: Expanded polyethylene rod equal to "Ethafoam" by Dow Chemical.

Do not specify silicone sealants

C. Flashing Tape: Woven glass cloth tape, commercial grade.

D. Sheet Flashing: .050" thick, precured, commercial grade neoprene.

E. Aggregate: Cleaned and graded silicon carbide, 20 mesh; 24 mesh silica sand, commercially prepared ground walnut shells, as recommended by approved coating Manufacturer.

PART 3 EXECUTION

3.1 CONDITION OF SURFACES

> Edit following paragraphs to indicate the types of substrate conditions. Verify that requirements of other Sections do not conflict with following paragraphs.
>
> Also see Appendix "A" for finishing and other substrate requirements.

A. Before coating work is commenced, surface shall be reinspected and treated as necessary to remove laitance, loose material on the surface, grease, oil and other contaminants which will affect bond of the coating. Surfaces shall be left free of contaminants, broom and vacuum clean.

> The following paragraph requires a 'mat test' which is accomplished by placing a 2' x 2' non-breathing rubber or vinyl mat directly onto the substrate. The edges of the mat are to be taped to the surface. The mat is removed after a minimum of 4 daylight hours. If there is no visible condensate, the Applicator may begin coating operations.
>
> We also include descriptions for plywood surfaces which should be deleted if there are no plywood substrate surfaces on this Project.

FIGURE 13.11 *(Continued)* Guide specification for section 7570, deck coating. (*Courtesy Pacific Polymers*)

B. Concrete surfaces shall be visibly dry and pass a 4-hour rubber mat test (no condensate) prior to application of coating system. Mat shall be taped to deck on all sides.

C. Verify that only sodium silicate based curing compounds are used to cure concrete in coating areas.

D. Metal surfaces shall be dry, clean, free of grease, oil, dirt, rust and corrosion, other coatings and contaminates which could affect bond of coating system, and without sharp edges or offsets at joints.

E. Plywood decks shall be dry, clean, sound and well nailed and/or glued, free of voids and without offsets at joints. Ensure that all nail heads are driven flush with surface. Plywood edges shall be tongue and groove or supported on solid lumber framing or blocking to prevent differential deflection of the panels.

F. Commencement of coating installation implies acceptance of that surface area, as it regards the suitability of the surface to accept the coating systems.

3.2 PREPARATION

A. Thoroughly clean all surfaces to receive coating materials in strict compliance with Manufacturer's written instructions and recommendations. Remove oil and grease with a commercial grade alkaline cleaner; thoroughly rinse and dry. Prepare all concrete surfaces by sandblasting, blast tracking. or by etching with a 10-15% solution of muriatic acid. Flush all acid with clean water and allow to dry.

B. Rout or sawcut all cracks exceeding 1/16" in width and caulk with sealant, in accordance with approved coating Manufacturer's recommendations.

See Appendix "A" for reference to special condition on precast deck with topping or leveling slab.

FIGURE 13.11 (*Continued*) Guide specification for section 7570, deck coating. (*Courtesy Pacific Polymers*)

C. Caulk all expansion, control and construction joints to be overcoated by deck coating with coating Manufacturer's sealant.

D. Protect adjacent surfaces with drop cloths or masking as required.

3.3 FLASHINGS

A. Provide fluid applied flashings at all locations where a horizontal surface abutts a vertical surface and at all deck penetration as specified.

B. At locations of potential high movement such as wall/slab intersections which are not structurally and rigidly connected, provide 10" min. width of precured sheet flashing or reinforce coating with one layer of uncoated, woven fiberglass cloth. Where sheet flashings are used, they shall be free or unbonded to the substrate within 2" vertically and horizontally from meeting angle but shall be fully bonded for not less than 2" on vertical surface and 4" on horizontal surface. Do not use precured sheet flashings over expansion joints in horizontal surfaces.

C. At projections through deck coatings such as posts, vents, pipes, stanchions, railings and similar locations of potential slight movement, provide a 1/4" bead of coating Manufacturer's recommended sealant. Tool sealant to form a cove and allow to cure before overcoating.

3.4 PRIMER & DETAIL WORK

A. Primer: Prime all concrete, masonry and metal surfaces. Apply primers at coating Manufacturer's recommended rate. Concrete prime coat shall be allowed to completely dry but shall not be applied more than 8 hours preceding application of deck coating. Metal prime coat may be applied up to 1 hour prior to application of deck coating.

1A. On concrete which has not cured for 28 days, but which must be coated, use Elasto-Poxy Concrete Primer applied at the rate of 300 sq. ft. per gallon. Mix only for use over a 2 hour period (max.). Install deck coating base coat on the same day and within 4 hours following application of the primer. Allow primer to become slightly tacky to the touch before installing base coat.

B. Apply 25 mil dry film thickness of base coat material over all flashings (sheet flashings, sealant coves and rigid corners). Extend coating 2" beyond flashing out onto adjacent deck surface. Unless otherwise indicated on Drawings or where limited by height of base, extend coating a minimum of 1" above the top of the flashing and terminate in a neat straight line. Use masking tape for such purposes.

C. Apply 25 mil dry film thickness of base coat material over and for a distance of 1-1/2" on each side of all cracks.

FIGURE 13.11 (*Continued*) Guide specification for section 7570, deck coating. (*Courtesy Pacific Polymers*)

Do not specify coating to extend over any joints larger than 1" nominal width and/or any joints which may move in excess of 25% of nominal dimension. This requirement shall apply to detail coatings as well as deck coatings. In order to provide for aesthetic continuity it is allowable to specify a 'paint' coat of the top coating material to be applied over the cured joint.

D. Apply 25 mil dry film thickness of base coat material over and for a distance of 2" on each side of all expansion joints, control joints and construction joints to be coated.

The following table shows the number of coats, mil thicknesses of each coat and aggregates for the different types of deck coating systems. It is included to provide the Specifier with a comprehensive presentation of the various types of deck coating systems included in this document. Review the table with respect to the system or systems to be specified and edit the following articles as to coating type and thickness.

FUNCTION	Base #1	Base #2	Top #1	Top #2
	Dry Film Thickness in mils			
TC System				
Park & Ped	25	NA	15	NA
Aisles	30	NA	10	10
PDS System	25	25	10	10 avg.
X2 System	25	25	10	10
WS System	25	25	15	NA
WDA System	30	NA	10	10
MR System	30	NA	10	NA

FIGURE 13.11 *(Continued)* Guide specification for section 7570, deck coating. *(Courtesy Pacific Polymers)*

Grit	=	Aluminum Oxide or Silicon Carbide: #20-24 mesh System TC Park & Ped; System PDS.
	=	Various sandblasting aggregates as approved by Manufacturer; System TC, traffic aisles and high wear areas; System X2; System WDA; and System MR (if needed)
	=	Walnut shells: System WS (premixed into top coat in accordance with Manufacturer's instructions.

3.5 APPLICATION - ELASTO-DECK 5000 TC

A. Parking and Pedestrian Areas:

1a. Base Coat: Apply Elasto-Deck 5001 at the rate of 60 sq. ft. per gallon to produce a 25 mil dry film thickness. Coating may be installed with squeegee and backroll or spray.

a. Allow to cure 16 hours min. At temperatures less than 75 deg. F. (24 deg. C.) and/or relative humidity less than 50%, extend curing time.

2a. Top Coat: Apply Elasto-Glaze 6001 at the rate of 66 sq. ft. per gallon to produce a 15 mil dry film thickness. Coating may be applied by squeegee or spray. While coating is in a fluid condition, broadcast 20-24 mesh silicon carbide grit at the rate of 5-7 lbs./100 sq. ft. and backroll to ensure even distribution of the grit.

a. Allow to cure 16 hours min. At temperatures less than 75 deg. F. (24 deg. C.) and/or humidity less than 50%, extend curing time before allowing any traffic. At least 72 hours total cure time should elapse before admitting vehicular traffic.

B. Traffic Aisles:

1b. Base Coat: Apply Elasto-Deck 5001 at the rate of 45 sq. ft. per gallon to produce a 30 mil dry film thickness. Coating may be applied by squeegee and backroll or spray.

a. Allow to cure 16 hours min. At temperatures less than 75 deg. F. (24 deg. C.) and/or humidity less than 50%, extend curing time.

FIGURE 13.11 (*Continued*) Guide specification for section 7570, deck coating. (*Courtesy Pacific Polymers*)

Page 15 Section 07572

2b. Top Coat #1: Apply Elasto-Glaze 6001 AL at the rate of 100 sq. ft. per gallon to produce a 10 mil dry film thickness. Coating may be applied by squeegee and backroll or spray. While coating is in a fluid condition broadcast approved aggregate at the rate of refusal (completely bury top coat in grit). Allow to cure overnight and remove excess grit.

3b. Top Coat #2: Apply Elasto-Glaze 6001 AL at the rate of 100 sq. ft. per gallon to produce a 10 mil dry film thickness. Coating may be applied by spray or roller. Thinning of the coating material is not permitted for application purposes. Regardless of how applied, coating shall be backrolled to provide an even and consistently thick finish coat.

a. Allow to cure 16 hours min. At temperatures less than 75 deg. F. (24 deg. C.) and/or humidity less than 50%, extend curing time before allowing any traffic. At least 72 hours total cure time should elapse before admitting vehicular traffic.

In the turning radius, ramps, entrances and exits and other very high wear traffic areas, it is advised to apply the Elasto-Glaze 6001 AL Top Coat #1 (Broadcast Coat) at a rate of 66 ft.^{2}/gallon for a 15 dry mil thickness.

FIGURE 13.11 *(Continued)* Guide specification for section 7570, deck coating. *(Courtesy Pacific Polymers)*

3.5 APPLICATION - ELASTO-DECK 5000 PDS

A. All Areas:

1a. Base Coat #1: Apply Elasto-Deck 5001 at the rate of 60 sq. ft. per gallon to produce a 25 mil dry film thickness. Coating may be installed with squeegee and backroll or spray.

a. Allow to cure 16 hours min. At temperatures less than 75 deg. F. (24 deg. C.) and/or relative humidity less than 50%, extend curing time.

2a. Base coat #2: Apply Elasto-Deck 5001 at the rate of 60 sq. ft. per gallon to produce a 25 mil dry film thickness. Coating may be installed with squeegee and backroll or spray.

a. Allow to cure 16 hours min. At temperatures less than 75 deg. F. (24 deg. C.) and/or relative humidity less than 50%, extend curing time.

3a. Top Coat #1: Apply Elasto-Glaze 6001 AL at the rate of 100 sq. ft. per gallon to produce a 10 mil dry film thickness. Coating may be applied by squeegee and backroll or spray and backroll.

a. Allow to cure 16 hours min. At temperatures less than 75 deg. F. (24 deg. C.) and/or humidity less than 50%, extend curing time before allowing any traffic.

4a. Top Coat #2 (Spatter Coat): Apply Elasto-Glaze 6001 S at the rate of 100 sq. ft per gallon to produce a 10 mil average dry film thickness. While coating is in a fluid condition, broadcast 20-24 mesh carbide at the rate of 7 lbs./100 sq. ft.

a. Allow to cure 16 hours min. At temperatures less than 75 deg. F. (24 deg. C.) and/or humidity less than 50%, extend curing time before allowing any traffic. At least 72 hours total cure time should elapse before admitting vehicular traffic.

3.5 APPLICATION - ELASTO-DECK 5000 X2

A. Base Coat #1: Apply Elasto-Deck 5001 at the rate of 60 sq. ft per gallon to produce a 25 mil dry film thickness. Coating may be installed with notched trowel or squeegee and backroll or spray.

1a. Allow to cure 16 hours min. At temperatures less than 75 deg. F. (24 deg. C.) and/or humidity less than 50%, extend curing time before allowing any traffic.

FIGURE 13.11 (*Continued*) Guide specification for section 7570, deck coating. (*Courtesy Pacific Polymers*)

B. Base Coat #2: Apply Elasto-Deck 5001 at the rate of 60 sq. ft. per gallon to produce a 25 mil dry film thickness. Coating may be installed with notched trowel or squeegee and backroll or spray.

1b. Allow to cure 16 hours min. At temperatures less than 75 deg. F. (24 deg. C) and/or humidity less than 50%, extend curing time before allowing any traffic.

C. Top Coat #1: Apply Elasto-Glaze 6001 AL at the rate of 100 sq. ft. per gallon to produce a 10 mil dry film thickness. Coating may be installed with notched trowel or squeegee and backroll or spray. While coating is in a fluid condition, broadcast aggregate at the rate of refusal (completely bury the top coat in grit). Allow to cure overnight and remove excess grit by sweeping, air blast or vacuum.

D. Top Coat #2: Apply Elasto-Glaze 6001 AL at the rate of 100 sq. ft. per gallon to produce a 10 mil dry film thickness.

1d. Allow to cure 16 hours min. At temperatures less than 75 deg./. F. (24 deg. C.) and/or humidity less than 50%, extend curing time before allowing any traffic. At least 48 hours total cure time should elapse before admitting heavy foot traffic.

3.5 APPLICATION - ELASTO-DECK 5000 WS

A. Base coat #1: Apply Elasto-Deck 5001 at the rate of 60 sq. ft. per gallon to produce a 25 mil dry film thickness. Coating may be applied by notched trowel or squeegee and backroll or spray.

1a. Allow to cure 16 hours min. At temperatures less than 75 deg. F. (24 Deg. C.) and/or humidity less than 50%, extend curing time.

B. Base Coat #2: Apply Elasto-Deck 5001 at the rate of 60 sq. ft. per gallon to produce a 25 mil dry film thickness. Coating may be applied by notcned trowel or squeegee and backroll or spray.

1b. Allow to cure 24 hours min. At temperature less than 75 deg. F. (24 deg. C.) and/or humidity less than 50%, extend curing time.

C. Top Coat: Before beginning coating application, mix 1 gallon of ground walnut shell per 5 gallon pail of Elasto-Glaze 6001 Al (pigmented). Provide continuous agitation for the combined material. Apply mixed coating at the rate of 66 sq. ft. per gallon to produce a 15 mil dry film thickness. Mixed coating shall be applied by notched trowel or squeegee and backrolled.

FIGURE 13.11 (*Continued*) Guide specification for section 7570, deck coating. (*Courtesy Pacific Polymers*)

1c. Allow to cure 24 hours min. At temperatures less than 75 deg. F. (24 deg. C.) and/or humidity less than 50%, extend curing time before admitting any traffic. At least 72 hours total cure time should elapse before admitting heavy foot traffic.

3.5 APPLICATION - ELASTO-DECK 5000 WDA

A. Base coat: Apply Elasto-Deck 5001 at the rate of 45 sq. ft per gallon to produce a 30 mil dry film thickness. Coating may be applied by notched trowel or squeegee and backroll or spray.

1a. Allow to cure 16 hours min. At temperatures less than 75 deg. F. (24 deg. C.) extend curing time.

B. Top Coat #1: Apply Elasto-Glaze 6001 AL at the rate of 100 sq. ft per gallon to produce a 10 mil dry film thickness. Coating may be installed with notched trowel or squeegee and backroll or spray. While coating is in a fluid condition, broadcast approved aggregate at the rate of refusal (completely bury the top coat in grit). Allow to cure overnight and remove excess grit by sweeping, air blast or vacuum.

C. Top Coat #2: Apply Elasto-Glaze 6001 AL [pigmented] [clear] at the rate of 100 sq. ft per gallon to produce a 10 mil dry film thickness.

1c. Allow to cure 48 hours min. At temperatures less than 75 deg. F. (24 Deg. C.) and/or humidity less than 50%, extend curing time before admitting any traffic.

3.5 APPLICATION - ELASTO-DECK 5000 MR

A. Base Coat: Apply Elasto Deck 5001 at the rate of 60 sq. ft. per gallon to produce a 25 mil dry film thickness. Coating may be installed with notched trowel or squeegee and backroll or spray.

1a. Allow to cure 16 hours min. At temperatures less than 75 deg. F. (24 deg. C.) extend curing time.

The use of non-skid aggregate in the System 5000 MR is optional. If the floor will generally be dry and the traffic very occasional, you may wish to eliminate the aggregate. Edit the following paragraph to meet job requirements.

FIGURE 13.11 *(Continued)* Guide specification for section 7570, deck coating. *(Courtesy Pacific Polymers)*

B. Top Coat: Apply Elasto-Glaze 6001 AR at the rate of 100 sq. ft per gallon to produce a 10 mil dry film thickness. [While the coating is in a fluid condition, broadcast approved aggregate at the rate of 7 lbs. per 100 sq. ft. Backroll the coating immediately following the application of grit to ensure even dispersal and encapsulation of the grit.]

1b. Allow to cure 24 hours min. before admitting any pedestrian traffic. At temperatures less than 75 deg. F. (24 deg. C.) and/or humidity less than 50%, extend curing time. Allow 4 days cure before placing mechanical room in service.

3.6 CLEANING

> Use only cleaning solvents and detergents which are approved by coating Manufacturer and in compliance with regulations of local, state and federal agencies having jurisdiction.

A. Clean stains from adjacent surfaces with approved cleaning agent.

B. Remove all construction barricades, debris and other items of work, including empty containers, from the Project site.

C. Remove foreign matter from finished coating surfaces.

END OF SECTION 07572

FIGURE 13.11 *(Continued)* Guide specification for section 7570, deck coating. *(Courtesy Pacific Polymers)*

Page 20

Section 07572

NOTES TO ASSIST COORDINATION WITH WORK OF OTHER SECTIONS

1. GENERAL

   A. Employ caution and good engineering practice in design of substrates. Excessive movement of substrate can create shear failures in any coating/membrane system. Joints must be supported with dowels or other devices to eliminate differential deflection between panels.

   B. Finishing: In accordance with requirements established within Division 3, concrete surfaces to which coating is applied shall receive a finish equivalent to a trowel finish complying with ACI 301-72, para. 11.7.3, except that hand trowelling is not required, and followed by a fine hair broom to very lightly score the surface. Concrete Sections shall require removal of fins, ridges and other protrusions and patching of voids.

   C. Curing Time: All concrete to receive deck coatings must be in place for 28 days (min.). If a deck must be coated prior to full 28 day cure, consult with Licensed Applicator or Manufacturer regarding the use of Elasto-Poxy Primer. There will be a slight premium charge for the use of this procedure. Note that Elasto-Poxy Primer must also be used on decks cast in permanent metal pans or on split slabs with sheet or fluid applied membranes.

   D. Curing Methods: To assure adequate bonding of coating material, water curing is recommended. Sodium silicate based curing compounds are generally acceptable in lieu of water cure. Oil, wax or silicone based compounds are not acceptable. Specifier should verify the specific requirements of each approved Coating Manufacturer and assure that curing requirements specified are acceptable to all.

   E. Precast Concrete: Tooled joints shall be provided in leveling slab directly over joints between precast members (including ends), to act as crack control joints. Sawcutting of 'green' concrete may be substituted for tooling of joints. Such sawcuts must be 1" minimum in depth.

4. CONSTRUCTION OF PLYWOOD DECKS

   A. All plywood shall be identified as conforming to U.S. Product Standard PS 1-74 by the grade trademarks of the American Plywood Association. All plywood to receive deck coatings shall meet or exceed requirements for A-C plugged, Exterior Grade insofar as such grades govern face veneer inner ply adjacent to face veneer and sanding.

FIGURE 13.11 *(Continued)* Guide specification for section 7570, deck coating. *(Courtesy Pacific Polymers)*

B. Installation: Unless more stringent requirements are indicated on the Drawings or required by governing building code, installation of plywood shall be in accordance with recommendations of the American Plywood Association. Panel edges shall be tongue and groove or supported on solid lumber framing or blocking to prevent differential deflection. Space panels 1/16" at both ends and edges. Where wet or humid conditions are expected, these spacings may be increased (as much as doubled). Where adhesive is used to bond panel edges and ends, 'squeeze-out' shall be struck smooth and slightly concave at all of the joints. It is suggested that a polyurethane adhesive/sealant be specified for such bonding purposes.

C. Nailing: Use annular ring or spiral shank common nails, 6d for 5/8" and 3/4" plywood and 8d for 7/8" , 1", 1-1/8" and 1-1/4" plywood. Space nails 6" on center along panel edges and 10" on center along intermediate supports.

D. Protection: Protection of the deck during construction and preparation of joints between panels shall be in accordance with the recommendations of the coating system manufacturer.

FIGURE 13.11 (*Continued*) Guide specification for section 7570, deck coating. (*Courtesy Pacific Polymers*)

MANUFACTURER'S GUIDE SPECIFICATION:
MIGUTAN FP... Series WATERTIGHT SPLIT-SLAB/PLAZA-DECK EXPANSION JOINT SYSTEM

SECTION 07900—WATERTIGHT EXPANSION JOINT SEALANT SYSTEMS

NOTE: It is critical in ensuring the success of the project that this expansion joint be specified in the same section as the deck waterproofing system and is installed by the contractor responsible for installing the deck waterproofing system.

SUBMITTALS
General: Submit the following in accordance with *(Specifier: Insert appropriate reference to related sections here, ex. Conditions of Contract, Division 1, etc.)*

A. Product data in the form of manufacturer's product specifications, specific, detailed installation instructions that include specifics on integration of side flashing sheets with deck waterproofing system, and general recommendations for each type of expansion joint cover assembly indicated.
B. Shop drawings showing full extent of expansion joint cover assemblies; include details indicating profiles of each type of expansion joint cover assembly, splice joints between sections, joinery with other types, special end conditions, fasteners, and relationship to adjoining work and finishes with specific reference to tie-in with deck waterproofing system through integration with expansion joint system side flashing sheets.
C. Samples for each joint cover assembly of same materials to be used in work.

QUALITY ASSURANCE
A. Manufacturer's Instructions: In addition to requirements of these specifications, comply with manufacturer's instructions and recommendations for all phases of work, including preparation of substrate, applying materials, and protection of installed units.
B. Use only qualified, authorized applicator, thoroughly skilled and specially trained in the techniques of expansion joint installation, which has demonstrated the ability to install the materials to the manufacturer's satisfaction. The applicator must be the same contractor as will be installing the deck waterproofing system.
C. Single-Source Responsibility: Obtain expansion joint cover assemblies from one source from a single manufacturer.

PRODUCTS
ACCEPTABLE MANUFACTURERS
Manufacturers known to be able to supply materials as specified:

MIGUTAN FP... Series SYSTEM from EMSEAL JOINT SYSTEMS, LTD., 108 Milk Street, Suite 3, Westborough, MA, 01581-1228, PH: 508-836-0280, FX: 508-836-0281

MATERIALS: WATERTIGHT SPLIT-SLAB/PLAZA-DECK EXPANSION JOINT
General: The expansion joint assemblies installed on this project must meet all of the requirements of the specification and must have a minimum 10-year proven track record of leak-free installations under similar conditions to this project. Expansion joint system must consist of all of the following components:

A. Articulated aluminum retainer legs and pre-drilled mounting flanges (separate interlocking extrusions)
B. Extruded, heat-weldable, thermoplastic-rubber sealing insert (gland)
C. Heat-weldable PVC flashing membranes locked into reglet in top surface of retainer leg on each side of joint
D. Pre-drilled, striated stainless steel protective cappings with stainless steel machine screws and nylon counter-sink seals
E. Epoxy-mortar setting/leveling bed to act as dialetric separator as well as to ensure that the system is fully supported and at the appropriate elevation throughout its length.
F. Carbon Steel Grade II zinc dichromate yellow finish, UNC 16, anchors and nuts and hi-mod anchor epoxy.

Aluminum Extrusions
ASTM B 221, alloy 6063-T5. The retainer leg/mounting flange assembly shall consist of two or more (but not less than two) interlocked components which are free to slide relative to each other to provide positive interlocking of successive joint lengths and which also serve to accommodate relative longitudinal movements.

(Exceptions to articulated leg requirement: 1) Where the solid leg system (FP .../25) is used for transitions to vertical surfaces, the retainer leg/mounting flanges will be of a single-piece solid extrusion with mounting flanges no less than 4mm thick. 2) Where high-volume continuous heavy-weight vehicular traffic is expected the solid leg FP.../6000S must be used. The vertical solid-leg required in continuous-traffic applications must have a thickness of not less than 3/8-inch (6mm). Additionally the extrusion must incorporate a bracing "toe" that protrudes beyond the plane of the vertical leg, opposite to the mounting flange, to resist torque forces. Where solid leg models are required to be used, extrusion is to contain a channel into which a solid stainless steel pin must be inserted to positively align consecutive lengths.

All mounting flanges shall be extruded such that the edge of the flange over which the deck waterproofing will be laid will be sloped to eliminate any abrupt angles. All deck-mounted flanges shall contain a recessed channel to receive the anchors and nuts and which will be subsequently filled and will serve to hold the nuts below the plane of the waterproofing materials above. Where the solid leg system (FP.../25) is used on vertical surfaces, the anchor holes in the mounting flanges must be countersunk in the field to receive a countersunk anchor that will not protrude into the plane of the waterproofing materials above. The bottom surfaces of

FIGURE 13.12 Guide specification for section 7900, sealants. (*Courtesy Emseal*)

the flanges which will contact the setting-bed epoxy and the top surfaces of the mounting flanges which will be contacted by the waterproofing materials will be striated to provide a mechanical grip with the epoxy and waterproofing materials respectively.

The top surface of each retainer leg shall incorporate two reglets for receiving the sealing insert and side flashing sheets' sealing darts from above. Top surface of the extrusion shall be pre-drilled and tapped to receive stainless steel anchoring machine screws for the stainless steel protective capping/retaining strips.

Sealing Insert (Gland)
Sealing insert shall be thermoplastic rubber to enable heat welding of butt joins, directional changes (tees, ells) and transitions in plane (horizontal to vertical and vice versa). All directional changes and transitions shall be factory-fabricated to drawing dimensions on centerline provided by the contractor. Underside of flanges of gland shall have sealing darts that mate with the top of the supporting aluminum rail extrusion. Gland shall have flanges which extend beyond location of sealing darts and which overlap the PVC sealing darts also installed into reglet in top of supporting aluminum rail extrusion. Gland to be secured in compression to supporting aluminum rail extrusion with pre-drilled, striated stainless steel protective cappings with stainless steel machine screws and nylon counter-sink seals.

Side Flashing Sheets
The flashing side sheets shall be 300mm wide x 3mm thick thermoplastic PVC to allow heat welding at all joints for continuous waterproofing at all transitions, corners, upturns, etc. Double-width flashing sheets shall be supplied for use in models with leg heights over 6" (150mm). Flashing sheets shall have .5mm amplitude ribs at 2mm on-center spacing on both top and bottom surfaces to create labyrinth seal in integration with deck waterproofing materials. Flashing sheets shall have ribbed sealing darts inserted into channels on the top of the supporting aluminum rail extrusion.

Flashing sheets shall be "sandwiched" between two layers of the deck waterproofing system. Flashing materials must be PVC-based material to ensure adhesion with deck waterproofing materials. Flashing sheets shall have a short-term temperature resistance suitable for integration with hot-applied deck waterproofing systems.

Gland and Flashing Sheet Retainer Caps
Stainless steel, ASTM A167, Type 304 with 2B finish. Secure to extrusion with stainless steel machine screws, 600 millimeters on center, seated on countersink nylon washers. Retainer caps shall be removable to allow for expansion joint gland replacement. At locations where cover plates are required over top of the gland, add over one retainer-cap a specially fabricated stainless-steel cover plate to extend across the sealing gland and rest on the top of the opposite retainer cap. Stainless steel coverplate to be fabricated from minimum 11 gauge (3mm) thick, type 304 stainless steel.

Fabrication, Terminations and Transitions
Provide expansion joint system needed to accommodate joint size and structural movement specified. Expansion joint supporting aluminum rail extrusions to be factory set at mid-point of movement range and held at this dimension by spacers to be removed after attachment of the rails to the deck. Expansion joint lengths to be factory numbered in sequence. Furnish in lengths to minimize number of end joins. Provide heat-welded corners where joint changes directions or abuts other materials. All terminations & transition pieces, tee-joints, corners, curbs and cross-connections in insert portion as well as all aluminum components of joint assembly must be factory-fabricated and welded to provide continuity.

EXECUTION

SURFACE CONDITION:
Joint surfaces to receive seal shall be sound, smooth, straight, parallel, clean, dry and free of all visible contaminants. Applications of non-visible coatings or contaminants to surfaces of joint interface area prior to installation of seal shall be controlled by the Architect/Engineer in consultation with the expansion joint manufacturer.

INSTALLATION
Inspection: Manufacturer's technician shall be on site at commencement of installation for inspection of substrate preparation and demonstration of installation procedures. Bids must include a specific line item for manufacturer's technical service and will be considered incomplete and subject to disqualification if excluded. Technical service is defined as the paid, contracted service of a manufacturer's technician..

The following is a general summary of installation requirements. In all cases the manufacturer's standard written instructions or specific instructions of the manufacturer's technician are to be followed.

Fastening: Use epoxy anchoring devices and fasteners for securing expansion joint cover assemblies to in-place construction. Provide fasteners as supplied by expansion joint manufacturer as follows: 3/8-inch diameter x 4-inch long anchor, carbon-steel grade II, zinc-dichromate yellow finish, UNC 16, threaded end- to-end, with nut of the same material. *(Note to specifier: if conditions dictate the need for longer this must be disclosed and determined during design.)*

Use 2 part, hi-Mod epoxy, as supplied by expansion joint manufacturer, for securing of anchors into predrilled holes.

System to be leveled into and embedded in 2-part hi-mod epoxy-gel setting-bed as supplied by expansion joint manufacturer.

FIGURE 13.12 *(Continued)* Guide specification for section 7900, sealants. *(Courtesy Emseal)*

Curing, Fitting and Placement: Perform all cutting, and fitting required for installation of expansion joint covers. Horizontal flanges shall be set in an epoxy-mortar leveling bed with a minimum 3/8-inch thickness. Install joint cover assemblies in true alignment and proper relationship to expansion joints and adjoining finished surfaces measured from established lines and levels. Allow adequate space between rail lenghts to allow free movement for thermal expansion and contraction of metal to avoid buckling. Securely attach in place with all required accessories. Locate anchors at recommended intervals, and not less than 3 inches from each end.

Joinery and Continuity: Maintain continuity of expansion joint cover assemblies with end joints held to a minimum and metal members aligned according to sequential numbering on retainers, ensuring that joint is positively interlocked by staggering the sliding retainer leg/flange components or through the use of stainless steel joining pins in solid-leg models.

Installation of Extruded Preformed Seals: Integrate flashing sheets with deck waterproofing system materials according to manufacturer's detailed installation instructions. Install seals in continuous lengths to comply with manufacturer's instructions. All transitions and terminations shall be factory-welded wherever possible according to field-measurements and drawings on centerlir provided by the contractor. Site welding, where required, to be carried out only after suitable instruction by the expansion joint manufacturer.

CLEANING AND PROTECTION

Remove all waste materials from the site. Do not use waste materials for patching unless so directed by the manufacturer. Seal shall be cleaned of all foreign matter as recommended by the seal manufacturer. Leave work in a condition satisfactory to the Architect/Engineer.

FIGURE 13.12 (*Continued*) Guide specification for section 7900, sealants. (*Courtesy Emseal*)

GUIDE SPECIFICATION:
KONFLEX Series PARKING & HEAVY DUTY DECK EXPANSION JOINT SYSTEM by EMSEAL

SECTION 05800 or 07900–EXPANSION JOINT SEALANT SYSTEMS

SUBMITTALS
General: Submit the following in accordance with *(Designer: Insert appropriate section reference for submittals).*

A. Product data in the form of manufacturer's product specifications, installation instructions, and general recommendations for each type of expansion joint sealant system indicated.
B. Shop drawings showing full extent of expansion joint sealant system; include details indicating profiles of each type of expansion joint sealant system, splice joints between sections, joinery with other types, special end conditions, fasteners, and relationship to adjoining work and finishes. Include description of materials and finishes.
C. Samples for each expansion joint sealant system of same materials to be used in work.

QUALITY ASSURANCE
Manufacturer's Instructions: In addition to requirements of these specifications, comply with manufacturer's instructions and recommendations for all phases of work, including preparation of substrate, applying materials, and protection of installed units.

Use only qualified workers, thoroughly skilled and specially trained in the techniques of seal installation, which can demonstrate ability to install joint sealer to Architect/Engineer and Manufacturer's satisfaction.

Single-Source Responsibility: Obtain expansion joint systems from one source from a single manufacturer.

PRODUCTS
ACCEPTABLE MANUFACTURERS
Subject to compliance requirements, acceptable manufacturers would be those listed below who can provide evidence that systems offered meet or exceed the full scope of the following performance specification.

KONFLEX KB or KCR SERIES by EMSEAL JOINT SYSTEMS, LTD, 108 Milk Street, Suite 3, Westborough, MA 01581-1228, PH: 508-836-0280, FX: 508-836-0281

MATERIALS
Expansion joint system shall consist of the following components:
A. Heat-weldable, continuous Santoprene sealing insert
B. Pre-formed Santoprene nosing retainer pads with co-extruded high-density polyethylene anchoring plate and leading-edge deflectors
C. Chemically adhered Type 410 stainless steel anchoring bolts, washers and nuts
D. Hydrophilic Neoprene setting bed and bolt-hole sealant.

Santoprene Sealing Insert
Physical properties of the sealing insert shall be as set out in table below. Sealing insert shall be manufactured from Santoprene thermoplastic rubber to enable heat welding of butt joins, directional changes (tees, ells), transitions in plane (horizontal to vertical and vice versa), and terminations. The flanges shall have integral waterstops that mate to the nosing retainer pad.

Properties	ASTM Test Method	Sealing Inserts
Material	N/A	Santoprene Thermoplastic Rubber
Hardness	ASTM D-2240	73A ± 3 (KB Series) 67A ± 3 (KCR Series)
Specific Gravity	ASTM D-792	0.98
Tensile Strength	ASTM D-412	1200 psi
Ultimate Elongation	ASTM D-412	440 (KB Series) 420 (KCR Series)
100% Modulus	ASTM D-412	470
Tear Strength	ASTM D-624	159
Tension Set	ASTM D-412	14%
Compression Set 168 Hrs.	ASTM D-395	26% @ 73ºF 39% @ 212ºF
Brittle Point	ASTM D-746	-81ºF

Nosing Retainer Pads
Physical properties of nosing retainer pads as set out in table below. Nosing pad shall be heat-weldable, shock-absorbing Santoprene with a co-extruded high-density polyethylene (HDPE) anchor plate and nosing deflector. The nosing retainer pads shall be removable for long-term maintenance of the system. The nosing retainer pad shall have a channel to accept the sealing-insert waterstops.

FIGURE 13.12 *(Continued)* Guide specification for section 7900, sealants. *(Courtesy Emseal)*

C. Spray apply LIQUID BOOT® onto geotextile to an 80 mil minimum dry thickness. Increase thickness to 100 dry mils if shotcrete is to be applied directly to membrane. If a second coat is required, remove any standing water from the membrane before proceeding with the second application.

D. Do not penetrate membrane. Keep membrane free of dirt, debris and traffic until a protective cover is in place. It is the responsibility of the General Contractor to insure that the membrane and the protection system is not penetrated.

E. If protection is required, lay down and butt join protection mat, roofing cap sheet, 6 oz/yd² geotextile or other protection as approved by the manufacturer on horizontal waterproof membrane after surface moisture has evaporated from membrane.

Two inches (2") of sand, which is free from debris, may be used in place of the protection options listed above. If sand option is used, the sand may be applied at any time after the membrane is applied.

On vertical surfaces apply 3/8 inch polystyrene protection board, 6 oz/yd² geotextile or other protection as approved by the manufacturer.

3.03.20 INSTALLATION ON CONCRETE/SHOTCRETE/MASONRY

Due to the numerous variables affecting concrete (ie: water content, mix specifications, cement source, "free-lime" percentage, calcium content, pumped vs. poured, environmental conditions at the time of concrete placement, admixtures, acidity, type of finish, curing conditions, etc., etc.) every job will require pre-testing of LIQUID BOOT® to determine the installation procedure. Follow the procedures below carefully.

A. Refer to section 3.03.40, "Sealing Around Penetrations", for procedures to seal around penetrations.

B. Provide a ¾ inch minimum cant of LIQUID BOOT® Trowel Grade, or other suitable material, at all horizontal to vertical transitions and other inside corners of 120° or less. **Allow to cure a minimum of 24 hours before the application of LIQUID BOOT®.**

C. Delineate a test area <u>on site</u> with a minimum dimension of 10 feet by 10 feet (3m by 3m). Apply LIQUID BOOT® to a thickness of 80 mils and let it cure for **24 hours**. Observe for blisters. If minor or no blistering occurs, proceed to the next step. (See note regarding blisters below). If significant blistering does occur, apply a thin (10 mil) tack coat of LIQUID BOOT® "A" side without catalyst to the entire concrete surface and allow to cure before proceeding. (See also information regarding blister repair on page 7).

D. Spray apply LIQUID BOOT® to an 80 mil minimum dry thickness. Increase thickness to 100 dry mils if shotcrete is to be applied directly to membrane. If a second coat is required, remove any standing water from the membrane before proceeding with the second application.

E. Do not penetrate membrane. Keep membrane free of dirt and debris and traffic until a protective cover is in place. It is the responsibility of the General Contractor to insure that the membrane and the protection system is not penetrated.

F. If protection is required, install pursuant to protection manufacturer's instructions, after surface moisture has evaporated from membrane.

NOTE: If water testing is to be performed, conduct before placing protection course

NON-HORIZONTAL SURFACES: Spray on non-horizontal surfaces should begin at the bottom and work towards the top. This method allows the product to adhere to the surface before hitting catalyst runoff.

F:\data\specbook\water20 5

FIGURE 13.12 (*Continued*) Guide specification for section 7900, sealants. (*Courtesy Emseal*)

GUIDE SPECIFICATION:
THERMAFLEX TM Series MEMBRANE/NOSING PARKING-DECK EXPANSION JOINT SYSTEM by EMSEAL

SECTION 07900–EXPANSION JOINT SEALANT SYSTEMS

SUBMITTALS
General: Submit the following in accordance with *(Designer: Insert appropriate section reference for submittals).*

A. Product data in the form of manufacturer's product specifications, installation instructions, and general recommendations for each type of expansion joint sealant system indicated.
B. Shop drawings showing full extent of expansion joint sealant system; include details indicating profiles of each type of expansion joint sealant system, splice joints between sections, joinery with other types, special end conditions, fasteners, and relationship to adjoining work and finishes. Include description of materials and finishes.
C. Samples for each expansion joint sealant system of same materials to be used in work.

QUALITY ASSURANCE
Manufacturer's Instructions: In addition to requirements of these specifications, comply with manufacturer's instructions and recommendations for all phases of work, including preparation of substrate, applying materials, and protection of installed units.

Use only qualified workers, thoroughly skilled and specially trained in the techniques of seal installation, which can demonstrate ability to install joint sealer to Architect/Engineer's and Manufacturer's satisfaction.

Single-Source Responsibility: Obtain expansion joint systems from one source from a single manufacturer.

PRODUCTS
ACCEPTABLE MANUFACTURERS
Subject to compliance requirements, acceptable manufacturers would be those listed below who can provide evidence that systems offered meet or exceed the full scope of the following performance specification.

THERMAFLEX TM SERIES by EMSEAL JOINT SYSTEMS, LTD, 108 Milk Street, Suite 3, Westborough, MA 01581-1228, PH: 508-836-0280, FX: 508-836-0281

MATERIALS
Membrane/Nosing System: Expansion joint system shall be a heat-weldable, Santoprene, thermoplastic-rubber, double-celled extrusions with perforated flanges embedded in elastomeric concrete nosing.

System Components:
A. Sealing Insert:
Heat-weldable, double-celled Santoprene sealing insert.
B. Nosing Material:
Cold-applied and cured, 2-part polyurethane nosing with characteristics and physical properties as described below.

Santoprene Sealing Insert
Physical properties of the extruded sealing insert shall be as set out in following table. Sealing insert shall be manufactured from Santoprene thermoplastic rubber to enable heat welding of butt joins, directional changes (tees, ells), transitions in plane (horizontal to vertical and vice versa), and terminations.

Properties	ASTM Test Method	DOUBLE-BARRIER Seals
Material	N/A	Santoprene Thermoplastic Rubber
Hardness	ASTM D-2240	73A ± 3
Specific Gravity	ASTM D-792	0.98
Tensile Strength	ASTM D-412	1200 psi
Ultimate Elongation	ASTM D-412	410
100% Modulus	ASTM D-412	470
Tear Strength	ASTM D-624	159
Tension Set	ASTM D-412	14%
Compression Set 168 Hrs.	ASTM D-395	26% @ 73ºF 39% @ 212ºF
Brittle Point	ASTM D-746	-81ºF

Cold-Applied Nosing Material
Elastomeric nosing shall be a two-part, polyurethane resin (epoxies not permitted) mixed with aggregate with sand particles not to exceed 30-mesh. Nosing material shall be pourable grade material capable of encapsulating the perforated flanges in a single pour without the laying of a neat resin setting-bed or tack coat. The material shall flow so as to fill voids or irregularities in concrete blockout and beneath the perforated flanges of the gland. Material shall cure to flexible (non-rigid) state. So as to ensure proper mixing, flowability, and consolidation, mixed nosing material shall have an aggregate loading of ratio of liquid to aggregate not to exceed those set forth in Table 2 below.

FIGURE 13.12 *(Continued)* Guide specification for section 7900, sealants. *(Courtesy Emseal)*

Evaluation of physical properties of nosing materials will be based on elastomeric concrete (resins mixed with aggregate) and must meet or exceed test results set forth in the tables below. Physical properties of nosing materials based on resin (binder) only are not relevant to the application of the system and will not be considered.

TABLE 2: Physical Property of Nosing Mixed With Aggregate	ASTM Test Method	Test Results
Tensile Properties (after conditioning at 100-degF (37 degC) for 7 days):		
Tensile Strength, psi		600 minimum
Elongation, %		25 min.
Hardness	D-2240	50 D, 96 A
Compressive Properties		
Compressive Stress, psi (Mpa) 5% Deflection	ASTM D	800 (5.52) min.
Resilience, %, 5% Deflection	695	95 min.
Adhesion Properties		
Bond strength to concrete, min, psi (Mpa)		400 (70.05) Min.
Wet bond strength to concrete, min, psi (Mpa)		250 (43.78) Min.
Impact Resistance		
Ball drop, ft-lb. (Joule) @ -20ºF (-29ºC)		>10 (13.58) No Cracks
Ratio by Weight of Liquid Resin to Aggregate		Not to Exceed 1:1.6
Ratio by Volume of Liquid Resin to Aggregate		Not to Exceed 1: 1.3
Max. sand aggregate size		Not to exceed 30 mesh

Dimensions:
Nominal size of the sealing insert should be agreed between specifier and manufacturer. Joint-gap size at installation will vary according to temperature and shall be determined from a Temperature Adjustment Table produced in cooperation between manufacturer and designer.

Blockout size on each side of the joint-gap shall measure 3/4" deep x 3" wide (20mm x 75mm). If blockout is wider than standard requirement then additional nosing material will be required and must be included for in bids.

FABRICATION

General: Provide expansion joint sealant system needed to accommodate joint size, variation in adjacent surfaces, and structural movement. Furnish units in longest practicable lengths to minimize number of end joints. Provide factory-welded, mitered corners where joint changes direction, plane, abuts other materials or terminates. Santoprene sealing insert must be factory-fabricated for all terminations, transition pieces, tee-joints, corners, curbs and cross-connections to provide continuity of seal.

[NOTE TO SPECIFIER: INCLUDE THIS SECTION AS APPROPRIATE FOR THE CONDITIONS AT HAND WHERE SIGNIFICANT SHEAR MOVEMENTS ARE ANTICIPATED: for transitions experiencing significant horizontal-shear movement perpendicular to the direction of movement which the seal is primarily designed to accept, supply manufacturer's standard factory-fabricated shear transition. Shear-transition shall consist of standard seal with factory-modification to incorporate a welded shear pocket capable of accepting horizontal-shear movement up to the full, published movement rating of the specified seal.]

EXECUTION

SURFACE CONDITION:
Joint surfaces to receive seal shall be sound, smooth, straight, parallel, clean, dry and free of all contaminants. Applications of non-visible coatings or contaminants to surfaces of joint interface area prior to installation of seal shall be controlled by the Architect/Engineer in consultation with the seal manufacturer.

PREPARATION
Coordinate and furnish sealing insert, nosing materials, and instructions for installation of expansion joint sealant system including preparation of blockouts.

INSTALLATION

Inspection: Manufacturer's technician shall be on site at commencement of installation for inspection of substrate preparation and demonstration of installation procedures. Bids must include a specific line item for manufacturer's technical service and will be considered incomplete and subject to disqualification if excluded. Technical service is defined as the paid, contracted service of a manufacturer's technician.

Fitting, Placement and Curing,: Perform all cutting, and fitting required for installation of expansion joint sealing system in strict accordance with manufacturer's standard written and job-specific instructions. Install expansion joint sealant system in true alignment and proper relationship to expansion joints and adjoining finishes. Do not install expansion joint sealing system at a substrate temperature lower than 45°F (7°C).

CLEANING AND PROTECTION
Remove all waste materials from the site. Do not use waste materials for patching. Seal shall be cleaned of all foreign matter as recommended by the seal manufacturer. Leave work in a condition satisfactory to the Architect/Engineer.

FIGURE 13.12 (*Continued*) Guide specification for section 7900, sealants. (*Courtesy Emseal*)

GUIDE SPECIFICATION:
THERMAFLEX TCR Series MEMBRANE/NOSING PARKING-DECK EXPANSION JOINT SYSTEM by EMSEAL

SECTION 07900–EXPANSION JOINT SEALANT SYSTEMS

SUBMITTALS
General: Submit the following in accordance with *(Designer: Insert appropriate section reference for submittals).*

A. Product data in the form of manufacturer's product specifications, installation instructions, and general recommendations for each type of expansion joint sealant system indicated.
B. Shop drawings showing full extent of expansion joint sealant system; include details indicating profiles of each type of expansion joint sealant system, splice joints between sections, joinery with other types, special end conditions, fasteners, and relationship to adjoining work and finishes. Include description of materials and finishes.
C. Samples for each expansion joint sealant system of same materials to be used in work.

QUALITY ASSURANCE
Manufacturer's Instructions: In addition to requirements of these specifications, comply with manufacturer's instructions and recommendations for all phases of work, including preparation of substrate, applying materials, and protection of installed units.
Use only qualified workers, thoroughly skilled and specially trained in the techniques of seal installation, which can demonstrate ability to install joint sealer to Architect/Engineer and Manufacturer's satisfaction.

Single-Source Responsibility: Obtain expansion joint systems from one source from a single manufacturer.

PRODUCTS
ACCEPTABLE MANUFACTURERS
Subject to compliance requirements, acceptable manufacturers would be those listed below who can provide evidence that systems offered meet or exceed the full scope of the following performance specification.

THERMAFLEX TCR SERIES by EMSEAL JOINT SYSTEMS, LTD, 108 Milk Street, Suite 3, Westborough, MA 01581-1228, PH: 508-836-0280, FX: 508-836-0281

MATERIALS
Membrane/Nosing System: Expansion joint system shall be a heat-weldable, Santoprene, thermoplastic-rubber, Multi-celled extrusions with perforated flanges embedded in elastomeric concrete nosing.

System Components:

A. Sealing Insert:
Heat-weldable, multi-celled Santoprene sealing insert.
B. Nosing Material:
Cold-applied and cured, 2-part polyurethane nosing with characteristics and physical properties as described below.

Santoprene Sealing Insert
Physical properties of the multi-cell sealing insert shall be as set out in following table. Sealing insert shall be manufactured from Santoprene thermoplastic rubber to enable heat welding of butt joins, directional changes (tees, ells), transitions in plane (horizontal to vertical and vice versa), and terminations.

Properties	**ASTM Test Method**	**DOUBLE-BARRIER Seals**
Material	N/A	Santoprene Thermoplastic Rubber
Hardness	ASTM D-2240	67A ± 3
Tensile Strength	ASTM D-412	1000 psi
Ultimate Elongation	ASTM D-412	420
Tear Strength	ASTM D-624	159
Tension Set	ASTM D-412	14%
Compression Set 168 Hrs.	ASTM D-395	26% @ 73°F 39% @ 212°F
Brittle Point	ASTM D-746	-81°F

Cold-Applied Nosing Material
Elastomeric nosing shall be a two-part, polyurethane material (epoxies not permitted) mixed with aggregate with sand particles not to exceed 30-mesh. Nosing material shall be pourable grade material capable of encapsulating the perforated flanges in a single pour without the laying of a setting-bed or tack coat. The material shall flow so as to fill voids or irregularities in concrete blockout and beneath the perforated flanges of the gland. Material shall cure to flexible (non-rigid) state. So as to ensure proper mixing, flowability, and consolidation, mixed nosing material shall have an aggregate loading of ratio of liquid to aggregate not to exceed those set forth in Table 2 below.

FIGURE 13.12 *(Continued)* Guide specification for section 7900, sealants. *(Courtesy Emseal)*

Evaluation of physical properties of nosing materials will be based on elastomeric concrete (resins mixed with aggregate) and must meet or exceed test results set forth in the tables below. Physical properties of nosing materials based on resin (binder) only are not relevant to the application of the system and will not be considered.

TABLE 2: Physical Property of Nosing Mixed With Aggregate	ASTM Test Method	Test Results
Tensile Properties (after conditioning at 100-degF (37 degC) for 7 days): Tensile Strength, psi Elongation, % Hardness	D-2240	600 minimum 25 min. 50 D, 96 A
Compressive Properties Compressive Stress, psi (Mpa) 5% Deflection Resilience, %, 5% Deflection	ASTM D 695	800 (5.52) min. 95 min.
Adhesion Properties Bond strength to concrete, min, psi (Mpa) Wet bond strength to concrete, min, psi (Mpa)		400 (70.05) Min. 250 (43.78) Min.
Impact Resistance Ball drop, ft-lb. (Joule) @ -20ºF (-29ºC)		>10 (13.58) No Cracks
Ratio by Weight of Liquid Resin to Aggregate		Not to Exceed 1:1.6
Ratio by Volume of Liquid Resin to Aggregate		Not to Exceed 1: 1.3
Max. sand aggregate size		Not to exceed 30 mesh

Dimensions:
Nominal size of the sealing insert should be agreed between specifier and manufacturer. Joint-gap size at installation will vary according to temperature and shall be determined from a Temperature Adjustment Table produced in cooperation between manufacturer and designer.

Blockout size on each side of the joint-gap shall measure: for multi-cell gland systems ¾-inches deep x 3 ½-inches wide (20mm x 90mm).

FABRICATION
General: Provide expansion joint sealant system needed to accommodate joint size, variation in adjacent surfaces, and structural movement. Furnish units in longest practicable lengths to minimize number of end joins. Provide factory-welded, mitered corners where joint changes direction, plane, abuts other materials or terminates. Santoprene sealing insert must be factory-fabricated for all terminations, transition pieces, tee-joints, corners, curbs and cross-connections to provide continuity of seal.

[NOTE TO SPECIFIER: INCLUDE THIS SECTION AS APPROPRIATE FOR THE CONDITIONS AT HAND WHERE SIGNIFICANT SHEAR MOVEMENTS ARE ANTICIPATED: for transitions experiencing significant horizontal-shear movement perpendicular to the direction of movement which the seal is primarily designed to accept, supply manufacturer's standard factory-fabricated shear transition. Shear-transition shall consist of standard seal with factory-modified to incorporate a welded shear pocket capable of accepting horizontal-shear movement up to the full, published movement rating of the specified seal.]

EXECUTION
SURFACE CONDITION:
Joint surfaces to receive seal shall be sound, smooth, straight, parallel, clean, dry and free of all contaminants. Applications of non-visible coatings or contaminants to surfaces of joint interface area prior to installation of seal shall be controlled by the Architect/Engineer in consultation with the seal manufacturer.

PREPARATION
Coordinate and furnish sealing insert, nosing materials, and instructions for installation of expansion joint sealant system including preparation of blockouts.

INSTALLATION
Inspection: Manufacturer's technician shall be on site at commencement of installation for inspection of substrate preparation and demonstration of installation procedures. Bids must include a specific line item for manufacturer's technical service and will be considered incomplete and subject to disqualification if excluded. Technical service is defined as the paid, contracted service of a manufacturer's technician. The following is a general summary of installation requirements. In all cases the manufacturer's standard written instructions or specific instructions of the manufacturer's technician are to be followed.

Fitting, Placement and Curing: Perform all cutting, and fitting required for installation of expansion joint sealing system in strict accordance with manufacturer's standard written and job-specific instructions. Install expansion joint sealant system in true alignment and proper relationship to expansion joints and adjoining finished. Do not install expansion joint sealing system at a substrate temperature lower than 45°F (7°C).

CLEANING AND PROTECTION
Remove all waste materials from the site. Do not use waste materials for patching unless so directed by the manufacturer. Seal shall be cleaned of all foreign matter as recommended by the seal manufacturer. Leave work in a condition satisfactory to the Architect/Engineer.

FIGURE 13.12 *(Continued)* Guide specification for section 7900, sealants. *(Courtesy Emseal)*

MANUFACTURER'S GUIDE SPECIFICATION: *20H SYSTEM by EMSEAL*

SECTION 07915-- SEALANTS, CAULKING, & SEALS

PRODUCTS
20H SYSTEM by EMSEAL JOINT SYSTEMS, LTD, 108 Milk Street, Suite 3, Westborough, MA 01581-1228, PH: 508-836-0280, FX: 508-836-0281.

SUBMITTALS
General: Submit the following in accordance with *(Designer: Insert appropriate section reference for submittals).*

A. Product data in the form of manufacturer's product specifications, installation instructions, and general recommendations for each type of expansion joint sealant system indicated.
B. Samples of same materials to be used in work mounted in simulated substrates to show complete system installation.

QUALITY ASSURANCE
Manufacturer's Instructions: In addition to requirements of these specifications, comply with manufacturer's instructions and recommendations for all phases of work, including preparation of substrate, applying materials, and protection of installed units.

MATERIALS
For expansion joint and isolation joints in decks and below-grade walls and tunnels, sealant system shall be comprised of three components: 1) Impregnated foam compressed 5-times; 2) Field-applied epoxy adhesive primer; 3) Topcoat material.

Foam shall be open cell polyurethane foam impregnated with a neoprene and acrylic polymer modified water-based asphalt emulsion. Foam to have at least 90% open cell structure. Impregnation agent to have proven non-migratory characteristics. Impregnation density 9-10 lb/cuft (144-160 kg/cuM) and must not exceed 10 lb/cuft (160 kg/cuM). Compression when expanded in joint shall be at approximately 20% of its uncompressed dimension (5x compression).

Depth of seal as recommended by manufacturer but shall for all product sizes exceed the width of the joint. 20H foam seal to be installed into manufacturer's standard field-applied epoxy adhesive. Exposed surface to be sealed with Topcoat.

In below-grade applications the system is to be installed flush to the positive wall surface and the below-grade waterproofing materials are to be installed over the expansion joint sealant in the waterproofing material manufacturer's standard multiple-layer configuration.

Submittals must include samples for each expansion joint sealant system of same materials to be used in work. Manufacturer to certify that the expanding foam material has been tested and meets the values in the table below:

Property	Value	Test Method
Base Material	Open-cell, high density, polyurethane foam	N/A
Impregnation	Acrylic-Polymer Modified Asphalt	N/A
Color	Black	N/A
Density (uncompressed)	Not to exceed 9-10 lb/ft^3 (144-160 kg/m^3)	
Tensile Strength	21 psi min (145 kPa)	ASTM D3574
Elongation-Ultimate	150% min	ASTM D3574
Compression Set (70°C 50% RH after 72 hrs.)	3% max	ASTM D3574
Thermal Conductivity	0.34 BTU in/hr ft^2 °F (0.05 W/m °C)	ASTM C518
Water Vapor Transmission	0.011 perms	ASTM C355-64
Low Temperature Flexibility 32°F to −10°F (0°C to −23°C)	No cracking or splitting	ASTM C711

EXECUTION
The joint configuration and the joint surfaces shall be as detailed in the drawings and in accordance with the contract specifications and in compliance with requirements in the current material Tech Data available from the Manufacturer. All known detrimental conditions shall be reported immediately in writing. Field measurements of the depth and width of the joint shall be supplied to manufacturer before material is ordered.

Do not proceed with the installation of joint sealer if the joint is other than designed, until written notification of these conditions is submitted to the manufacturer and design professional, and a written acknowledgement with an order to proceed is provided.

Do not proceed with the installation of joint sealer under adverse weather conditions when joint to be sealed is damp, wet or frozen, or when temperatures are below or above the manufacturer's recommended limitations for installation. Consult manufacturer for specific instructions before proceeding.

Joint sealer/expansion joint material to be installed in strict accordance with the manufacturer's instructions and the advice of their official representative.

Remove all waste materials from site.

FIGURE 13.12 *(Continued)* Guide specification for section 7900, sealants. *(Courtesy Emseal)*

PRODUCT GUIDE SPECIFICATION

SECTION 09900

PAINTS

****

This section is based on products manufactured by:

CHEMPROBE TECHNOLOGIES, INC.
2805 Industrial lane
Garland, TX 75041
PH : 800/760-6776 PH : 972/271-5551
FX : 972/271-5553

Specifier Notes : This specification covers the selection, preparation of surface, and application of Chemprobe's concrete and masonry acrylic coatings to the exterior, above grade, vertical, concrete, brick, stucco and concrete masonry structures.

****

PART 1 - GENERAL

1.01 SECTION INCLUDES

A This section covers the preparation, materials, services, and equipment required in conjunction with the application of a concrete & masonry stain on all types of above grade, vertical, exterior exposed surfaces of masonry, concrete, brick, stucco and [ ] .

1.02 RELATED SECTIONS

****

Specifier Notes : Edit this section as necessary to include other specific sections.

****

A Section 03300 - Cast In Place Concrete.

B Section 03400 - Precast Concrete.

C Section 03900 - Concrete Restoration & Cleaning.

D Section 04400 - Stone.

E Section 04800 - Masonry Assemblies.

F Section 04900 - Masonry Restoration & Cleaning.

G Section 07190 - Water Repellents.

H Section 09630 - Masonry Flooring.

I Section 09960 - High Performance Coatings.

FIGURE 13.13 Guide specification for section 9900, elastomeric coatings. (*Courtesy Chemprobe Technologies*)

1.03 SUBMITTALS

A Submit under provisions of Section 01300 - Submittals.

B Product Specification Data: Submit manufacturer's technical literature, specifications, and application instructions for the specified concrete & masonry stain material for architect's approval.

C Samples: Obtain either liquid or draw down sample(s) of the specified concrete & masonry stain for sample application and color approval. Sample application is covered in section 1.04 QUALITY ASSURANCE.

D Applicator Qualifications: Submit certification stating applicator has a minimum of three (3) years experience using the specified product. Provide a list of several most recently completed projects where the specified material was used. Include the project name, location, architect and method of application.

****

Specifier Notes : Delete the following section if there are no VOC restrictions.

****

E Environmental Regulations: Submit certification stating the concrete & masonry stain material to be applied is in compliance with federal, state and local environmental Volatile Organic Compounds (VOC) regulations.

1.04 QUALITY ASSURANCE

A Manufacturer: A firm with no less than ten (10) years experience in manufacturing the products specified in this section.

B Applicator Qualification: A firm with no less than three (3) years experience in the application of the products specified in this section. In addition, applicator must state the intended use of the proper application equipment and that it has been well maintained.

C Mock-Up:

1 Apply concrete & masonry stain per manufacturer's application instructions as directed by the Architect to substrate material which matches actual job conditions. Determine the best method of application, optimum coverage rate, and number of coats required to produce the desired appearance.

2 After sample treatment has cured in accordance with manufacturer's recommendations, verify the substrate is coated with sufficient stain material to produce the desired appearance and color.

3 Obtain architect and/or project owner approval of coverage appearance and color prior to full scale application.

****

Specifier Notes : Coordinate pre-application meeting with section 01040 - Coordination and 01200 - Project Meetings

****

D Pre-Application Meeting: Convene a pre-application meeting prior to the start of application of the specified material. Attendance by a representative of each of the following organizations is requested; the application firm, the architectural firm, and the water repellent manufacturer. Notify each of the attendees at leas three (3) days prior to the meeting time.

Section 09900 - 02

FIGURE 13.13 (*Continued*) Guide specification for section 9900, elastomeric coatings. (*Courtesy Chemprobe Technologies*)

1.05 PRODUCT DELIVERY

A Material Delivery: Deliver materials to the job site in original sealed containers, clearly marked with manufacturer's name, brand name, and type of material and color or color formula. Verify the product matches that of the original sample applied on the mock - up wall.

B Storage & Protection: Store materials inside if possible, away from sparks and open flame. Store in a secure area to avoid tampering and contamination. Water based materials must be kept from freezing. Store and handle according to manufacture's written instructions.

1.06 PROJECT CONDITIONS

A Surface Preparation: Surface must be free of cracks, dirt, oils, paint or other contaminants which may effect the appearance or performance of the concrete & masonry stain material.

B Environmental Requirements:

1. Air and substrate temperature must be above 50° F (10° C) or below 95° F (35° C) unless otherwise specified by manufacturer.
2. Do not proceed with application if the substrate is wet or contains frozen water.
3. Do not apply material when rain is predicted within 48 hours; or earlier that five days after the substrate became wet.
4. Do not apply materials in high or gusty winds.

C Protection:

1. Special precautions should be taken to avoid vapor transmission (fumes) from entering the building being treated. Ventilation systems and fresh air intakes should be turned off and closed.
2. Protect shrubs, metal, wood trim, glass, asphalt and other building hardware during application from overspray.
3. Do not permit spray mist or liquid to drift onto surrounding properties.

1.07 SCHEDULING

A Architect shall be notified not less that 48 hours before each application of concrete & masonry stain is scheduled.

****

Specifier Notes : Delete the following section if no warranty is required or if the water repellent work is covered under the terms of a general project warranty specified elsewhere. Edit as necessary to obtain a warranty from the contractor only, or to suit other project specific criteria.

****

1.08 WARRANTY

A The contractor and applicator shall jointly and severally warrant masonry stain material against failure in material and workmanship for a period of five (5) years from the date of application.

B Submit completed manufacturer's written "Request For Warranty Form" to manufacturer ten (10) days prior to application.

C After completion of the specified material, submit manufacturer's written "5 Year Warranty Application" to manufacturer for processing. Upon receiving validated warranty, submit copies to Architect and building owner.

Section 09900 - 03

FIGURE 13.13 (*Continued*) Guide specification for section 9900, elastomeric coatings. (*Courtesy Chemprobe Technologies*)

PART 2 - PRODUCTS

2.01 MANUFACTURER

Chemprobe Technologies, Inc.
2805 Industrial Lane
Garland, TX 75041
PH: 800/760-6776
PH: 972/271-5551 FX: 972/271-5553

2.02 PAINT MATERIAL

A GENERAL: All products shall contain Methyl-Methacrylate. No fillers, extenders, or paraffins. All products shall have a range of twenty two (22) to thirty five (35) percent solids with a minimum of eighteen (18) percent acrylic resin.

****

Specifier Notes : Refer to Chemprobe's product sheets or contact a Chemprobe Representative to determine the appropriate concrete masonry stain and color to be specified. Edit this list to contain only those products, colors and specifications to be included in the work.

****

1 CONFORMAL STAIN - CUSTOM COLOR: CHEMPROBE's CONFORMAL Stain must be custom color matched to ***[include color specification here]*** and approved on mock-up panel prior to committing to large scale application. Unless otherwise required by custom color or fitness for particular purpose, stain material must be a minimum of twenty two (22) percent solids with a minimum of eighteen (18) percent acrylic resin.

2 CONFORMAL STAIN - STANDARD COLOR: Stain must be CHEMPROBE's ***[include a standard color from CHEMPROBE's Standard or Accent Color Cards here]*** and approved on a mock-up panel prior to committing to large scale application. Unless otherwise required by fitness for particular purpose, stain material must be a minimum of twenty two (22) percent solids with a minimum of eighteen (18) percent acrylic resin.

3 CONFORMAL CLEAR: CHEMPROBE's CONFORMAL CLEAR Unless otherwise required by fitness for particular purpose, must be a minimum of twenty two (22) percent solids with a minimum of eighteen (18) percent acrylic resin.

4 PHYLON 1422: Product must be a clear acrylic, containing twenty five (25) percent acrylic resin, no fillers or stearates. Product will provide a clear glossy wet look to surface.

PART 3 - EXECUTION

3.01 EXAMINATION

A Verify the following:

1 The required joint sealants have been installed.
2 New masonry and mortar has cured a minimum of twenty one (21) days.
3 Surface to be treated is clean, dry and contains no frozen water.
4 Environmental conditions are appropriate for application.
5 The need for a PRIME A PELL water repellent product. Refer to section 07190 - Water Repellents.

Section 09900 - 04

FIGURE 13.13 *(Continued)* Guide specification for section 9900, elastomeric coatings. *(Courtesy Chemprobe Technologies)*

3.02 PREPARATION

A Protection:

1 Special precautions should be taken to avoid vapor transmission (fumes) from entering the building being treated. Ventilation systems and fresh air intakes should be turned off and closed.

2 Protect shrubs, metal, wood trim, glass, asphalt and other building hardware during application from overspray.

3 Do not permit spray mist or liquid to drift onto surrounding properties or parking lots.

3.03 APPLICATION

A Apply concrete & masonry stain in accordance with manufacturer's written application instructions.

B Apply concrete & masonry stain as supplied by the manufacturer. Do not dilute or thin.

C Mix material well just prior to application using a power mixer to assure color uniformity.

D Material must be applied using the proper application equipment, and the same technique used on the mock-up sample panel.

E Apply treatment evenly until a uniform color and appearance is achieved.

3.04 FIELD QUALITY CONTROL

A The architect shall be contacted 48 hours prior to application so as to provide supervision as required. The architect or the architect's representative shall inspect the progress as the work proceeds. Do not apply any material that is not specified by the architect.

B After concrete & masonry stain has cured (minimum of 12 hours), verify color uniformity. Recoat any area that is unacceptable.

3.05 CLEANING

A At completion, remove from the job site, all excess material, debris, and waste resulting from this work. Dispose of containers according to state and local environmental regulations.

END OF SECTION

CONFSPE1 10/96 Section 09900 - 05

FIGURE 13.13 (*Continued*) Guide specification for section 9900, elastomeric coatings. (*Courtesy Chemprobe Technologies*)

CHAPTER 14
INFORMATION RESOURCES

The following listing of manufacturers, government agencies, associations, and so on is a resource for acquiring information on many of the waterproofing systems, materials, and envelope components and processes provided in this book. The manufacturers are listed for information purposes only, and the list is not intended as recommendations of the companies or materials.

Most manufacturers will provide information on their products as well as generic or similar products. They are also excellent sources for resources for reviewing termination and transition detailing and compatibility of adjacent products in an envelope.

Below-Grade Waterproofing

Waterstop

Cetco Building Materials Group
1500 W. Shure Drive
Arlington Heights, IL 60004
847/392-5800 Fax: 847/577-5571
Web site: *www.cetco.com*
E-mail: *sbyrd@cetco.com*

EarthShield
551 Birch Street
Lake Elsinore, CA 92530
888/836-5778
Fax: 909/674-1315
Web Site: *www.earthshield.com*
E-mail; *davidp@earthshie1d.com*

Greenstreak, Inc.
3400 Tree Court Industrial Boulevard
St. Louis, MO 63122
314/225-9400
Fax: 314/225-9854
Web site: *www.greenstreak.com*
E-mail: *info@greenstreak.com*

J P Specialties, Inc.
551 Birch Street
Lake Elsinore, CA 92530
800/821-3859
Fax: 909-674-1315
Web site: *www.jpspecialties.com*

Tamms Industries
3835 State Route 72
Kirkland, IL 60146
815/522-3394
Fax: 815/522-3257
Web site; *www.tamms.com*
E-mail: *sales@~tamms.com*

Vandex
PO Box 1440
Columbia, MD 20144
410/964-1410
Web site: *www.vandexus.com*
E-mail: *info@vandexus.com*

Vinylex Corporation
PO Box 7187
2636 Byington-Solway Road
Knoxville, TN 37921-0087
423/690-2211
Fax: 423/691-6273
Web site: *www.vinylex.com*
E-mail: *centr@vinylex.com*

Webtec, Inc.
PO Box 241166
Charlotte, NC 28224
704/398-0954
Fax: 704/394-7946
E-mail: *info@webtecgeos.com*

W.R. Meadows, Inc.
PO box 338
Hampshire, IL 60140-0338
847/683-4500
Fax: 847/683-4544
Web site: *www.wrmeadows.com*
E-mail: *wrmil@wrmeadows.com*

Drainage Protection Courses

American Hydrotech, Inc.
303 East. Ohio Street
Chicago, IL 60611-3387
312/337-4998
Fax: 312/661-0731
Web site: *www.hydrotechusa.com*

American Wick Drain Corporation
1209 Airport Road
Monroe, NC 28110
704/238-9200, 800/242-9425
Fax; 704/296-0690
Web site: *www.americanwick.com*
E-mail: *mobermeyer@~americanwick.com*

Carlisle Coatings & Waterproofing
8810 West 100th Street South
Sapulpa, OK 74060
800/338-8701
Fax: 918/227-0603
Web site: *www.carlisle-ccw.com*
E-mail: *info@carsile-ccw.com*

Cetco Building Materials Group
1500 W. Shure Drive
Arlington, Heights, IL 60004
847/392-58O
Fax: 847/577-5571
Web site: *www.cetco.com*
E-mail: *sbyrd@cetco.com*

Cosella-Dorken Products, Inc.
4655 Delta Way
Beamsville, ON, Canada L0R 1B4
905/563-3255
Fax: 905/563-5582
Web site: *www.deltams.com*
E-mail: *cdsales@niagara.com*

Schluter Systems Inc.
28 Hymus
Pointe Claire, QC H9R 1C9
800/667-8746
Fax: 514/630-0983
Web site: *www.schluter.com*
E-mail: *info@schluter.com*

TC MiraDRI
2170 Satellite Blvd., Ste. 350
DuLuth, GA 30097-4074
770/689-2627, 888/464-7234
Fax: 770/689-2628
E-mail: *W.Harvie@miradri.com*
Web site: *www.miradri.com*

W R Grace
Construction Products Division
62 Whittemore Ave.
Cambridge, MA 02140-1692
617/876-1400
Web site: *www.gcp-grace.com*
E-mail: *info@gcp-grace.com*

Cementitious positive and negative system

Anti Hydro International, Inc
45 River Road
Flemington, New Jersey 08822
908/284-9000, 800/777/1773
Fax: 908/284-9464
Web site: *www.anti-hydro.com*
E-mail: *sales@anti-hydro.com*

Bonsal
PO Box 241148
Charlotte, NC 28224-1148
704/525-7621
Fax: 704/529-5261
Web site: *www.bonsal.com*
E-mail:
commercialproducts@bonsal.com

Tamms Industries
3835 State Route 72
Kirkland, IL 60146
815/522-3394
Fax: 815/522-3257
Web site; *www.tamms.com*
E-mail: *sales@~tamms.com*

Vandex
PO Box 1440
Columbia, MD 20144
410/964-1410
Web site: *www.vandexus.com*
E-mail: *info@vandexus.com*

Fluid-applied systems

American Hydrotech, Inc.
303 East. Ohio Street
Chicago, IL 60611-3387
312/337-4998
Fax: 312/661-0731
Web site: *www.hydrotechusa.com*

Anti Hydro International, Inc
45 River Road
Flemington, New Jersey 08822
908/284-9000, 800/777-1773
Fax: 908/284-9464
Web site: *www.anti-hydro.com*
E-mail: *sales@anti-hydro.com*

Basement Systems
60 Silvermine Road
Seymour, CT 06483
800/638-7048
Fax: 203/881-5095
Web site: *www.basementsystems.com*
E-mail: *mailbox@basementsystems.com*

Carlisle Coatings & Waterproofing
8810 West 100th Street South
Sapulpa, OK 74060
800/338-8701
Fax: 918/227-0603
Web site: *www.carlisle-ccw.com*
E-mail: *info@carsile-ccw.com*

Karnak
330 Central Avenue
Clark, NJ 07066
732/388-0300
Fax: 732/388-9422
Web site: *www.karnakcorp.com*
E-mail: *info@karnakcorp.com*

LBI Technologies, Inc.
3873 East Eagle Drive
Anaheim, CA 92807-1722
714/575-9200
Fax: 714/575-9229
Web site: *www.liquidboot.com*
E-mail: *lbi@liquidboot.com*

Mer-Kote Products, Inc.
501 South Van Ness Avenue
Torrance, California 90501
310/775-2461
Fax: 310/320-4938
Web site: *www.merkote.com*
E-mail: *info@~merkote.com*

Mar-Flex Systems, Inc.
6866 Chrisman Lane
Middletown, Ohio 45042
513/422-7285
Fax: 513/422-7282
Web site: *www.mar-flex.com*
E-mail: *keepdry@mar-flex.com*

Nox-crete/Kinsman Corporation
PO box 8102
Omaha. NE 68108
402/341-9752
Fax: 402/341-9752
Web site: *www.nox-crete.com*
E-mail: *kinsman@~nox-crete.com*

Pacific Polymers
12271 monarch street
Garden Grove, CA 92841
714/898-0025
Fax: 714/898-5687
Web site: *www.pacpoly.com*
E-mail: *pacpoly@~aol.com*

Polymer Plastics Corp.
65 Davids Drive
Hauppauge, NY 11788
516/231-1300
Fax: 516/231-1329

Rubber Polymer Corporation
1135 West Portage Trail Ext.
Akron, OH 44313
330/945-7721
Fax; 330/945-9416
Web site: *www.rpclink.com*
E-mail: *rpclink@)worldnet.att.net*

Terry Industries
8600 Berck Boulevard
Hamilton, Ohio 45015
513/874-6859
Fax: *513/874-6870*
Web site: *www.houseguard.com*
E-mail: *kim@fuse.net*

TC MiraDRI
2170 Satellite Blvd., Ste. 350
DuLuth, GA 30097-4074
770/689-2627, 888/464-7234
Fax: 770/689-2628
E-mail: *W.Harvie@~miradri.com*
Web Site: *www.miradri.com*

W R Grace
Construction Products Division
62 Whittemore Ave.
Cambridge, MA 02140-1692
617/876-1400
Web site: *www.gcp-grace.com*
E-mail: *info@grp-grace.com*

W.R. Meadows, Inc.
PO box 338
Hampshire, IL 60140-0338
847/683-4500
Fax: 847/683-4544
Web site: *www.wrmeadows.com*
E-mail: *wrmil@wrmeadows.com*

Hot-applied systems

Carlisle Coating & Waterproofing
8810 West 100th Street South
Sapulpa, OK 74060
800/338-8701
Fax: 918/227-0603
Web site: *www.carlisle-ccw.com*
E-mail: *info@carsile-ccw.com*

TC MiraDRI
2170 Satellite Blvd., Ste. 350
DuLuth, GA 30097-4074
770/689-2627, 888/464-7234
Fax: 770/689-2628
E-mail: *W.Harvie@~miradri.com*
Web site: *www.miradri.com*

Texas Refinery Corp.
One Refinery Place
PO Box 711
Fort Worth, TX 76101
817/332-1161

Sheet systems

Carlisle Coating & Waterproofing
8810 West 100th Street South
Sapulpa, OK 74060
800/338-8701
Fax: 918/227-0603
Web site: *www.carlisle-ccw.com*
E-mail: *info@carsile-ccw.com*

Compotite Corporation
355 Glendale Boulevard
Los Angeles, CA 90026
213/483-4444, 800/221-1056
Fax: 213/483-4445
E-mail: *Anna@compotite.com*

NEI Advanced Composite Technology
50 Pine Road
Brentwood, NH 03833
603/778-8899
603/778-7455
Web site: *www.nei-act.com*
E-mail: *nei@nei-act.com*

The Noble Company
614 Monroe Street
Grand Haven, Michigan 49417
618/842-7844
Fax: 800/272-1519
Web site: *www.noblecompany.com*
E-mail: *sales@noblecompany.com*

Poly-Wall International, Inc.
Minneapolis, MN
800/846-3020
Fax: 612/780-0170
Web site: *www.poly-wall.com*
E-mail: *info@poly-wall.com*

Protecto Wrap Company
2255 South Delaware Street
Denver, CO 80223
303/777-3001
Fax: 303/777-9273
Web site: *www.protectowrap.com*
E-mail: *info@protectowrap.com*

Tamko Waterproofing
220 West 4th Street
Joplin. MO 64801
800/841-1923
Fax: 417/624-8935
Web site: *www.tamko.com*
E-mail: *info@tamko.com*

TC MiraDRI
2170 Satellite Blvd., Ste. 350
DuLuth, GA 30097-4074
770/689-2627, 888/464-7234
Fax: 770/689-2628
Web site: *www.miradri.com*
E-mail: *W.Harvie@~miradri.com*

W R Grace
Construction Products Division
62 Whittemore Ave.
Cambridge, MA 02140-1692
617/876-1400
Web site: *www.gcp-grace.com*
E-mail: *info@gcp-grace.com*

W.R. Meadows, Inc.
PO Box 338
Hampshire, IL 60140-0338
847/683-4500
Fax: 847/683-4544
Web site: *www.wrmeadows.com*
E-mail: *wrmil@~wrmeadows.com*

Clay systems

Cetco Building Materials Group
1500 W. Shure Drive
Arlington Heights, IL 60004
847/392-5800
Fax: 847/577-5571
Web site: *www.cetco.com*
E-mail: *sbyrd@cetco.com*

TC MiraDRI
2170 Satellite Blvd., Ste. 350
DuLuth, GA 30097-4074
770/689-2627, 888/464-7234
Fax: 770/689-2628
Web site: *www.miradri.com*
E-mail: *W.Harvie@~miradri.com*

Above-Grade Waterproofing

Clear Repellents

American Building Products, Inc
9720 South 60th Street
Franklin, WI 53132
414/421-4125.
Fax: 414/421-8696
Web site: *www.abrp.com*

Anti Hydro International, Inc
45 River Road
Flemington, New Jersey 08822
908/284-9000
Fax: 908/284-9464
Web site: *www.anti-hydro.com*
E-mail: *sales@anti-hydro.com*

Aquron Corporation
PO Box 758
Rockwall, Texas 75087
972/412-4649
Fax: 972/412-1543

Chemprobe Technologies, Inc.
2805 Industrial Lane
Garland, TX 75041
972/271-5551
Fax: 972/271-5553
Web site: *www.chemprobe.com*
E-mail: *info@chemprobe.com*

Chemical Products Industries, Inc.
7649 Southwest 34th Street
Oklahoma City, OK 73179
405/745-2070
Fax: 405/745-2276

Conklin Company, Inc.
PO Box 155
Shakopee, MN 55379
612/445-6010
Fax: 612/496-4285

Cresset Products
One Cresset Center, Box 367
Weston, OH 43569
419/669-2041
Fax: 419/669-2200
Web site: *www.cresset.com*
E-mail: *cresset@cresset.com*

Mer-Kote Products, Inc.
501 South Van Ness Avenue
Torrance, California 90501
(323) 775-2461
FAX (310) 320-4938
Web site: *www.merkote.com*
E-mail: *info@merkote.com*

Nox-crete/Kinsman Corporation
PO Box 8102
Omaha. NE 68108
402/341-9752
Fax: 402/341-9752
E-mail: *kinsman@nox-crete.com*

PK Professional Products of Kansas, Inc.
4456 South Clifton
Wichita, KS 67216
316/522-9300
Fax: 316/522-9346
Web site: *www.watersealant.com*
E-mail: *ppk@watersealant.com*

Rainguard Products Company
3471 Via Lido, Suite 213
Newport Beach, CA 92663
888/765-7070
Fax: 949-3450
Web site: *www.rainguard.com*
E-mail: *rainguardinfo@rainguard.com*

Saver Systems
1751 Sheridan Street
Richmond, Indiana 47374
765/966-5084
Fax: 765/935-4999
Web site: *www.saversystems.com*
E-mail: *sales@saversystems.com*

Sivento
PO Box 6821
220 Davidson Avenue
Somerset, NJ 08873-0919
800/828-0919
Fax: 732/560-6970
Web site: *siventoinc.com*

Tamms Industries
3835 State Route 72
Kirkland, IL 60146
815/522-3394
Fax: 815/522-3257
Web site; *www.tamms.com*
E-mail: *sales@tamms.com*

U. S. Epoxy, Inc.
1633-1 Sycamore Ave.
Bohemia, NY 11716
516/218-9426
Fax: 516/218-9429
Web site: *www.pwlnet.com/*epoxy
E-mail: *useepoxy@v~illagenet.com*

Waterlox Coatings Corporation
9808 Meech Avenue
Cleveland, OH 44105-4191
216/641-4877
Fax: 216/641-7213
Web site: *www.waterlox.com*
E-mail:*jhawkins@waterlox.com*

Cementitious Coatings

Bonsal
PO Box 241148
Charlotte, NC 28224-1148
704/525-7621
Fax: 704/529-5261
Web site: *www.bonsal.com*
E-mail:
commercialproducts@~bonsal.com

Laticrete International
1 Laticrete Park North
Bethany, CT 06524
203/393-0010
Fax: 203/393-1684
Web Site: *www.laticrete.com*
E-mail: *info@laticrete.com*

United Gilsonite Laboratories
PO Box 70
Scranton, PA 18501-0070
717/344-1202
Fax: 717/969-7634
Web site: *www.uglproducts.com*
E-mail: *uglpr@ugl.com*

Vandex
PO Box 1440
Columbia, MD 20144
410/964-1410
Web site: *www.vandexus.com*
E-mail: *info@vandexus.com*

W.R. Meadows, Inc.
PO Box 338
Hampshire, IL 60140-0338
847/683-4500
Fax: 847/683-4544
Web site: *www.wrmeadows.com*
E-mail: *wrmil@wrmeadows.com*

Elastomeric Coatings

Bonsal
PO Box 241148
Charlotte, NC 28224-1148
704/525-7621
Fax: 704/529-5261
Web site: *www.bonsal.com*
E-mail:
commercialproducts@bonsal.com

Edison Coatings
Phone: (203) 597-9727 or (800) 697-8055
Fax: (203) 597-8044 or (800) 697-8044
Web site: *www.edisoncoatins.com*
Email: *edison@edisoncoatings.com*

Poly-Wall International, Inc.
Minneapolis, MN
800/846-3020
Fax: 612/780-0170
Web site: *www.poly-wall.com*
E-mail: *info@poly~-wall.com*

Rainguard Products Company
3471 Via Lido, Suite 213
Newport Beach, CA 92663
888/765-7070
Fax: 949-3450
Web site: *www.rainguard.com*
E-mail: *rainguardinfo@~rainguard.com*

Tamms Industries
3835 State Route 72
Kirkland, IL 60146
815/522-3394
Fax: 815/522-3257
Web site; *www.tamms.com*
E-mail: *sales@tamms.com*

United Gilsonite Laboratories
PO Box 70
Scranton, PA 18501-0070
717/344-1202
Fax: 717/969-7634
Web site: *www.uglproducts.com*
E-mail: *uglpr@ugl.com*

United Coatings
19011 E. Cataldo
Greenacres, WA 99016
509/926-7143
Fax: 509/928-1116

VIP/Lighthouse Products
PO Box 1253
New Smyra Beach, FL 32170
904/423-7477
Fax: 904/423-3002
E-mail: *lighthousevip@att.net*

Deck Coatings

Carlisle Coating & Waterproofing
8810 West 100th Street South
Sapulpa, OK 74060
800/338-8701
Fax: 918/227-0603
Web site: *www.carlisle-ccw.com*
E-mail: *info@carsile-ccw.com*

Ensurco Duradek
N. Kansas City, MO
800/338-3568
Fax: 816/421-2924
Web site: *www.durdek.com*
E-mail: *duradek@kcnet.com*

Environmental Coating Systems, Inc.
3321 South Susan St.
Santa Ana, CA 92704
714/641-1340
Fax: 714/641-7006
Web site: *www.alldeck.com*
E-mail: *info@alldeck.com*

General Polymers
12355 Gladstone Avenue
Sylmar, CA 91342
818/365-9261
Fax: 818/361-8023
Web site: *www.generalpolymers.com*
E-mail: *cpopham@generalpolymers.com*

Golden Look International
3401 House Anderson Road
Euless, TX 76040
817/540-1900
Fax: 817/545-5600
Web site; *www.goldenlook.com*
E-mail: *dekfxr@aol.com*

Hill Brothers Chemical Company
1675 North Main Street
Ornage, CA 92867-3499
714/998-8800
Fax: 714/998-6310
Web site: *www.hillbrothers.com*
E-mail: *hillchem@hillbrothers.com*

Karnak
330 Central Avenue
Clark, NJ 07066
732/388-0300
Fax: 732/388-9422
Web site: *www.karnakcom.com*
E-mail: *info@karnakcorp.com*

Mer-Kote Products, Inc.
501 South Van Ness Avenue
Torrance, California 90501
(323) 775-2461
Fax: (310) 320-4938
Web site: *www.merkote.com*
E-mail: *info@merkote.com*

Neogard
6900 Maple Avenue
PO Box 35288
Dallas, TX 75235
214/353-1689
Fax: 214/357-7532
Web site: *www.neogard.org*
E-mail: *neogard@neogard.org*

Nox-crete/Kinsman Corporation
PO Box 8102
Omaha. NE 68108
402/341-9752
Fax: 402/341-9752
E-mail: *kinsman@nox-crete.com*

Pacific Polymers
12271 Monarch street
Garden Grove, CA 92841
714/898-0025
Fax: 714/898-5687
Web site: *www.pacpoly.com*
E-mail: *pacpoly@aol.com*

Polymer Plastics Corp.
65 Davids Drive
Hauppauge, NY 11788
516/231-1300
Fax: 516/231-1329

Tamms Industries
3835 State Route 72
Kirkland, IL 60146
815/522-3394
Fax: 815/522-3257
Web site; *www.tamms.com*
E-mail: *sales@tamms.com*

Clear Deck Sealers

Anti Hydro International, Inc
45 River Road
Flemington, New Jersey 08822
908/284-9000
Fax: 908/284-9464
Web site: *www.anti-hydro.com*
E-mail: *sales@~anti-hydro.com*

Chemprobe Technologies, Inc.
2805 Industrial Lane
Garland, TX 75041
972/271-5551
Fax: 972/271-5553
Web site: *www.chemprobe.com*
E-mail: *info@chemprobe.com*

Saver Systems
1751 Sheridan Street
Richmond, Indiana 47374
765/966-5084
Fax: 765/935-4999
Web site: *www.saversystems.com*
E-mail: *sales@saversystems.com*

Tamms Industries
3835 State Route 72
Kirkland, IL 60146
815/522-3394
Fax: 815/522-3257
Web site; *www.tamms.com*
E-mail: *sales@tamms.com*

XYPEX Chemical Corporation
13731 Mayfield Place
Richmond, British Columbia, Canada
V6V 2G604/273-5265
Fax: 604/270-0451
Web site: *www.xypex.com*
E-mail: *info@~xypex.com*

Protected membranes

American Hydrotech, Inc.
303 East Ohio Street
Chicago, IL 60611-3387
312/337-4998
Fax: 312/661-0731
Web site: *www.hydrotechusa.com*

Carlisle Coating & Waterproofing
8810 West 100th Street South
Sapulpa, OK 74060
800/338-8701
Fax: 918/227-0603
Web site: *www.carlisle-ccw.com*
E-mail: *info@carsile-ccw.com*

Compotite Corporation
355 Glendale Boulevard
Los Angeles, CA 90026
213/483-4444, 800/221-1056
Fax: 213/483-4445
E-mail: *anna@compotite.com*

NEI Advanced Composite Technology
50 Pine Road
Brentwood, NH 03833
603/778-8899
603/778-7455
Web site: *www.nei-act.com*
E-mail: *nei@nei-act.com*

The Noble Company
614 Monroe Street
Grand Haven, Michigan 49417
618/842-7844
Fax: 800/272-1519
Web Site: *www.noblecompany.com*
E-mail: *sales@noblecompany.com*

Protecto Wrap Company
2255 South Delaware Street
Denver, CO 80223
303/777-3001
Fax: 303/777-9273
Web site: *www.protectowrap.com*
E-mail: *info@protectowrap.com*

TEC Speciality Products, Inc
315 South Hicks Road
Palatine, IL 60067-6972
800/323-7407
Fax: 800/952-2368
Web site: *www.hbfuller.com*/products/tec

TC MiraDRI
2170 Satellite Blvd., Ste. 350
DuLuth, GA 30097-4074
770/689-2627, 888/464-7234
Fax: 770/689-2628
Web site: *www.miradri.com*
E-mail: *W.Harvie@~miradri.com*

W R Grace
Construction Products Division
62 Whittemore Ave.
Cambridge, MA 02140-1692
617/876-1400
Web site: *www.gcp-grace.com*
E-mail: *info@~gcp-grace.com*

W.R. Meadows, Inc.
PO box 338
Hampshire, IL 60140-0338
847/683-4500
Fax: 847/683-4544
Web site: *www.wrmeadows.com*
E-mail: *wrmil@wrmeadows.com*

Civil Waterproofing

Anti Hydro International, Inc
45 River Road
Flemington, New Jersey 08822
908/284-9000
Fax: 908/284-9464
Web site: *www.anti-hydro.com*
E-mail: *sales@anti-hydro.com*

Bonsal
PO Box 241148
Charlotte, NC 28224-1148
704/525-7621
Fax: 704/529-5261
Web site: *www.bonsal.com*
E-mail: *commercialproducts@~bonsal.com*

C.I.M. Industries, Inc.
94 Grove Street
Peterborough, NH 03458
603/924-9481
Fax: 603/924-9482
Web site: *www.cimind.com*
E-mail: *cimind@top.monad.net*

Rubber Polymer Corporation
1135 West Portage Trail Ext.
Akron, OH 44313
330/945-7721
Fax; 330/945-9416
Web site: *www.@rpclink.com*

Tamms Industries
3835 State Route 72
Kirkland, IL 60146
815/522-3394
Fax: 815/522-3257
Web site: *www.tamms.com*
E-mail: *sales@tamms.com*

Vandex
PO Box 1440
Columbia, MD 20144
410/964-1410
Web site: *www.vandexus.com*
E-mail: *info@vandexus.com*

EIFS

Bonsal
PO Box 241148
Charlotte, NC 28224-1148
704/525-7621
Fax: 704/529-5261
Web site: *www.bonsal.com*
E-mail: *commercialproducts@bonsal.com*

Parex, Inc.
PO Box 189
1870 Stone Mt.–Lithonia Rd.
Redan, GA 30074
770/482-7872
Fax: 770/482-6878
Web site: *www.parex.com*

TEC Speciality Products, Inc
315 South Hicks Road
Palatine, IL 60067-6972
847/776-4378, 800/323-7407
Fax: 847/776-4340
Web site: *www.hbfuller.com*/products/tec

Sealants

Backer Rod

Industrial Thermo Polymers Limited
153 Van Kirk Drive
Brampton, Ont. Canada L7A 1A4
905/846-3666
Fax: 905/846-0360

Nomaco
501 NMC Drive
Zebulon, NC 27597
919/269-6500
Fax: 919/269-7936
E-mail: *mschroeder@nomaco.com*

Tools

Albion Engineering Company
2080A Wheatsheaf Lane
Philadelphia, PA 19124-5091
215/535-3476
Fax: 215/744-4179
Web site: *www.ablioneng.com*
E-mail: *service@albioneng.com*

Premanufactured Protection Caps

Weathercap, Inc.
PO Box 1776
Slidell, LA 70459-1776
504/640-4000

Acrylics

Conklin Company, Inc.
PO Box 155
Shakopee,NM 55379
612/445-6010
Fax: 612/496-4285

General Sealants
300 South Turnbull Canyon Road
City of Industry, CA 91745
626/961-0211
Fax: 626/968-5140
Web site: *www.generalsealants.com*
E-mail: *dennis.Karinger@~gener-alsealants.com*

Polymer Plastics Corp.
65 Davids Drive
Hauppauge, NY 11788
516/231-1300
Fax: 516/231-1329

VIP/Lighthouse Products
PO Box 1253
New Smyra Beach, FL 321,70
904/423-7477
Fax: 904/423-3002
E-mail: *lighthousevip@att.net*

Latex

General Sealants
300 South Turnbull Canyon Road
City of Industry, CA 91745
626/961-0211
Fax: 626/968-5140
Web site: *www.generalsealants.com*
E-mail: *dennis.Karinger@gener-alsealants.com*

Urethane

Carlisle Coating & Waterproofing
8810 West 100th Street South
Sapulpa, OK 74060
800/338-8701
Fax: 918/227-0603
Web site: *www.carlisle-ccw.com*
E-mail: *info@carsile-ccw.com*

General Sealants
300 South Turnbull Canyon Road
City Of Industry, CA 91745
626/961-0211
Fax: 626/968-5140
Web site: *www.generalsealants.com*
E-mail: *dennis.Karinger@gener-alsealants.com*

Pacific Polymers
12271 Monarch street
Garden Grove, CA 92841
714/898-0025
Fax: 714/898-5687
Web site: *www.pacpoly.com*
E-mail: *pacpoly@aol.com*

W.R. Meadows, Inc.
PO Box 338
Hampshire, IL 60140-0338
847/683-4500
Fax: 847/683-4544
Web site: *www.wrmeadows.com*
E-mail: *wrmil@wrmeadows.com*

Polysulfides

W.R. Meadows, Inc.
PO box 338
Hampshire, IL 60140-0338
847/683-4500
Fax: 847/683-4544
Web site: *www.wrmeadows.com*
E-mail: *wrmil@wrmeadows.com*

Silicones

Dow Corning Corporation
Midland, MI 48686-0094
517/496-6000
Fax: 517/496-4586
Web site: www.dowcorning.com

GE Silicones
260 Hudson River Road
Waterford, New York 12188
Web site: www.ge.com/silicones
E-mail: *rick.morgan@gepex.ge.com*

NPC Colored Sealants
1208 S. 8th Ave.
Maywood, IL 60163
708/681-1040
Fax: 708/681-1424

Precompressed Foam

Emseal Joint Systems Ltd.
108 Milk Street, Suite 3
Westborough, MA 01581-1228
508/836-0280
Fax: 508/836-0281
Web site: *www.emseal.com*

Expansion Joints

Sealant systems

Emseal Joint Systems Ltd.
108 Milk Street, Suite 3
Westborough, MA 01581-1228
508/836-0280
Fax: 508/836-0281
Web site: *www.emseal.com*

Expanding foam

Emseal Joint Systems Ltd.
108 Milk Street, Suite 3
Westborough, MA 01581-1228
508/836-0280
Fax: 508/836-0281
Web site: *www.emseal.com*

Polytite Manufacturing Corporation
324 Rindge Road
Cambridge, MA 02140
617/864-0930
Fax: 617/864-0930
Web site: *www.polytite.com*
E-mail: *info@polytite.com*

W.R. Meadows, Inc.
PO Box 338
Hampshire, IL 60140-0338
847/683-4500
Fax: 847/683-4544
Web site: *www.wrmeadows.com*
E-mail: *wrmil@wrmeadows.com*

Hydrophilic expansion seals

Vandex
PO Box 1440
Columbia, MD 20144
410/964-1410
Web site: *www.vandexus.com*
E-mail: *info@~vandexus.com*

Bellows systems

Emseal Joint Systems Ltd.
108 Milk Street, Suite 3
Westborough, MA 01581-1228
508/836-0280
Fax: 508/836-0281
Web site: *www.emseal.com*

Preformed rubber systems

Conspec Systems, Inc.
6696 Route 405 Highway
Muncy, PA 17756
570/546-5941
Fax: 570/546-8022
Web site: *www.c-sgroup.com*

Emseal Joint Systems Ltd.
108 Milk Street, Suite 3
Westborough, MA 01581-1228
508/836-0280
Fax: 508/836-0281
Web site: *www.emseal.com*

Combination rubber and metal systems

Conspec Systems, Inc.
6696 Route 405 Highway
Muncy, PA 17756
570/546-5941
Fax: 570/546-8022
Web site: *www.c-sgroup.com*

Emseal Joint Systems Ltd.
108 Milk Street, Suite 3
Westborough, MA 01581-1228
508/836-0280
Fax: 508/836-0281
Web site: *www.emseal.com*

Metal systems

Emseal Joint Systems Ltd.
108 Milk Street, Suite 3
Westborough, MA 01581-1228
508/836-0280
Fax: 508/836-0281
Web site: *www.emseal.com*

Vertical metal and stucco systems

Metalex
PO box 399
1530 Artaius Parkway
Libertyville, IL 60048
847/362-8300
Fax: 847/362-7939
Web site; *www.metlx.com*
E-mail: *www.metlx.com*

Admixtures

Liquid and powder admixtures

Anti Hydro International, Inc
45 River Road
Flemington, New Jersey 08822
908/284-9000
Fax: 908/284-9464
Web site: *www.anti-hydro.com*
E-mail: *sales@anti-hydro.com*

Aquron Corporation
5240 Gordon Smith Drive
Rowlett, TX 75088
972/412-4649
Fax: 972/412-1543

Cresset Products
One Cresset Center, Box 367
Weston, OH 43569
419/669-2041
Fax: 419/669-2200
Web site: *www.cresset.com*
E-mail: *cresse@cresset.com*

Master Builders, Inc.
23700 Chagrin Blvd.
Cleveland, OH 44122-5554
216/831-5500
Web site: *www.masterbuilders.com*

Tamms Industries
3835 State Route 72
Kirkland, IL 60146
815/522-3394
Fax: 815/522-3257
Web site: *www.tamms.com*
E-mail: *sales@tamms.com*

W R Grace
Construction Products Division
62 Whittemore Ave.
Cambridge, MA 02140-1692
617/876-1400
Web site: *www.gcp@-grace.com*
E-mail: *info@~gcp-grace.com*

W.R. Meadows, Inc.
PO Box 338
Hampshire, IL 60140-0338
847/683-4500
Fax: 847/683-4544
Web site: *www.wrmeadows.com*
E-mail: *wrmil@wrmeadows.com*

XYPEX Chemical Corporation
13731 Mayfield Place
Richmond, British Columbia, Canada
V6V 2G604/273-5265
Fax: 604/270-0451
Web site: *www.xypex.com*
E-mail: *info@xypex.com*

Polymer concrete overlays

Bonsal
PO Box 241148
Charlotte, NC 28224-1148
704/525-7621
Fax: 704/529-5261
Web site: *www.bonsal.com*
E-mail:
commercialproducts@~bonsal.com

ITW Devcon
30 Endicott Street
Massachusetts 01923
800/933-8266

Master Builders, Inc.
23700 Chagrin Blvd.
Cleveland, OH 44122-5554
216/831-5500
Web site: *www.masterbuilders.com*

Mer-Kote Products, Inc.
501 South Van Ness Avenue
Torrance, California 90501
310/775-2461
Fax: 310/320-4938
Web site: *www.merkote.com*
E-mail: *info@merkote.com*

Remedial Waterproofing

Cleaning

American Building Products, Inc
9720 South 60th Street
Franklin, WI 53132
414/421-4125
Fax: 414/421-8696
Web site: *www.abrp.com*

Tamms Industries
3835 State Route 72
Kirkland, IL 60146
815/522-3394
Fax: 815/522-3257
Web site: *www.tamms.com*
E-mail: *sales@tamms.com*

Cementitious coatings and tuck-pointing

Bonsal
PO Box 241148
Charlotte, NC 28224-1148
704/525-7621
Fax: 704/529-5261
Web site: *www.bonsal.com*
E-mail:
commercialproducts@bonsal.com

Master Builders, Inc.
23700 Chagrin Blvd.
Cleveland, OH 44122-5554
216/831-5500
Web site: *www.masterbuilders.com*

Vandex
PO Box 1440
Columbia, MD 20144
410/964-1410
Web site: *www.vandexus.com*
E-mail: *info@vandexus.com*

Epoxy injection

Abatron, Inc.
5501-95th Avenue
Kenosha, WI 53144
414/653-200
Fax: 414/653-2019
Web site: *www.abatron.com*
E-mail: *info@abatron.com*

Advanced Epoxy Systems
5103 East Third St.
Katy, TX 77493
281/391-0454
Fax: 281/391-6699
Web site: *www.advancedeposy.com*
E-mail: *AESsales@fbtc.net*

ITW Devcon
30 Endicott Street
Massachusetts 01923
800/933-8266
Fax: 800/451-4563

U. S. Epoxy, Inc.
1633-1 Sycamore Ave.
Bohemia, NY 11716
516/218-9426
Fax: 516/218-9429
Web site: *www.pwlnet.com/*epoxy
E-mail: *useepoxy@villagenet.com*

Webac
PO Box 10615
Costa Messa, CA 92627
714/662-4445
Fax: 714/662-4446
Web site: *www/webac.com*
E-mail: *webac@webac.netcom.com*

Chemical grouts

Abatron, Inc.
5501-95th Avenue
Kenosha, WI 53144
414/653-200
Fax: 414/653-2019
Web site: *www.abatron.com*
E-mail: *info@abatron.com*

Cetco Building Materials Group
1500 W. Shure Drive
Arlington Heights, IL 60004
847/392-5800
Fax: 847/577-5571
Web site: *www.cetco.com*
E-mail: *sbyrd@cetco.com*

De Neef Construction Chemicals
PO Box 1219
Waller Texas 77484-1219
409/372-9185
Web-site: *www.deneef.com*
E-mail: *iske@deneef.com*

Vandex
PO Box 1440
Columbia, MD 20144
410/964-1410
Web site: *www.vandexus.com*
E-mail: *info@vandexus.com*

Webac
PO Box 10615
Costa Messa, CA 92627
714/662-4445
Fax: 714/662-4446
Web site: *www/webac.com*
E-mail: *webac@webac.netcom.com*

Electro-osmosis

Drytronic, Inc
1601 Caledonia St., Suite B
LaCrosse, WI 54603-3605
800/497-0579, 608/779-4199
Fax: 608/779-4198
E-mail: *drytronic@~pressenter.com*

Testing Facilities

American Test Lab
6795 NW 17th Avenue
Fort Lauderdale, FL 33309
954/973-0808
Fax: 954/973-0823
Web site: *www.atlsf.com*

Architectural Testing Inc.
130 Derry Court
York, PA 17402-9405
717/764-7700
Fax: 717/764-4129
Web site: *www.testati.com*
E-mail: *ati.york@worldnet.att.net*

Construction Consulting Laboratory
1601 Luna Road
Carrollton, Texas 75006
972/242-0556
Fax: 972/242-0345
Web site: *www.inch.net/conconlab.com*

Industry Associations

National Association of Waterproofing Contractors
25550 Chagrin Boulevard, Suite 403
Cleveland, OH 44122
800/245-6292
Fax: 216/595-8230
Web site: *www.apk.net/nawc*
E-mail: *nawc@apk.net*

Sealant Engineering & Associated Lines
2841 Main
Kansas City, Missouri 64108
816/472-8870
Fax: 816/472-7765

Sealant Waterproofing and Restoration Institute (SWRI)
816/472-7974
Fax: 816/472-7765
Web site: *www.swironline.org*
E-mail: *info@swrionline.org*

Special thanks to Coastal Construction Products for photographs used throughout the book:

Coastal Construction Products, Inc.
1901 Service Street
Jacksonville, FL 32207
904/398-7171
Fax: 904/396-5495

GLOSSARY

Above-grade waterproofing The prevention of water intrusion into exposed structure elements through a combination of materials or systems. These materials are not subject to hydrostatic pressure but are exposed to weathering and pollutant attack.

Abrasive cleaning A cleaning method that incorporates an abrasive material such as sand to remove dirt, stains, and paint from existing substrates.

Absorption An accumulation of water or moisture vapor directly into a material's structure of cells or fibers.

Accelerated weathering Controlled conditions applied in laboratory testing to condense greatly the weathering a waterproofing material would experience over a long life cycle. Test results are used to compare materials of different generic types or manufacturers.

Acrylic sealants Factory-mixed, one-component materials polymerized from acrylic acid. They are not used on joints subject to high movement due to their relatively low-movement capability.

Adhesion The ability of a waterproof material to bond to a substrate or other material during movement or stress.

Adhesive failure An after-installation failure of the bond between a waterproofing material, such as sealant, to the substrate surface.

Adhesive strength The ability of sealants to bond to a particular substrate, including adhesion during substrate movement.

Admixtures Materials added to masonry or concrete envelope components to enhance and improve in-place product performance.

Adsorption The surface absorption of water allowed by a waterproofing system. Testing for adsorption is carefully controlled under laboratory conditions to ensure uniform test results between different waterproofing materials.

Aliphatic Of or pertaining to materials such as urethane in which the molecular structure is arranged in open or straight chains of carbon atoms.

Alligatoring The cracking that occurs in a waterproofing material because of movement the material is not capable of withstanding. Alligatoring also occurs when substrate movement begins before final waterproofing material curing.

ASTM American Society for Testing and Materials, a nationally recognized and impartial society for the testing of building materials. Test results are used for comparisons among various types and manufacturers of materials.

As-built drawings Drawings and specifications completed after or during construction that represent the actual details used in construction. The as-built drawings might vary considerably from the original plans and specifications.

Backing materials Backer rods and backing tape that prevent three-sided adhesion in joint design. When joints have insufficient depth for backer-rod installation, tape is used at backs of joints, providing there is a firm substrate against which to install sealant. Backer rod is installed in joints where there is no backing substrate. Backing material also provides a surface against which to tool material and helps to maintain proper depth ratios.

Bag grouting Application of a cementitious waterproofing material to the entire face of a masonry envelope. The cementitious material is removed before it is completely set and cured by using burlap bags or stiff brushes. This is also referred to as face grouting because the entire face of the masonry facade is covered.

Barrier system A building envelope component or waterproofing system that completely repels water rather than diverting it. *See* divertor.

Base flashing Flashing that prevents water from wicking upward in capillary action in a masonry wall. *See* Flashing.

Bellows expansion systems Systems manufactured from vulcanized rubber into preformed joint sections. They are installed by pressurizing the joint cross section during adhesive curing, which promotes complete bonding to joint sides.

Below-grade waterproofing Use of materials that prevent water under hydrostatic pressure from entering a structure or its components. These systems are not exposed or subjected to weathering such as by ultraviolet rays.

Bentonite waterproofing systems Waterproofing materials composed primarily of montmorillonite clay, a natural material. Typically, bentonite waterproofing systems contain 85–90 percent of montmorillonite clay and a maximum of 15 percent natural sediments such as volcanic ash.

Blister A portion of a waterproofing material raised from the substrate because of negative vapor pressure or application over wet substrates.

Bond-breaker Specially manufactured tapes used to prevent three-side adhesion in sealant joints. *See* backer-rod.

Building envelope The combination of roofing, waterproofing, dampproofing, and flashing systems that act cohesively as a barrier, protecting interior areas from water and weather intrusion. These systems envelop a building from top to bottom, from below grade to the roof.

Butyl sealants Sealants produced by copolymerization of isobutylene and isoprene rubbers. Butyls are some of the oldest derivatives to be used for sealant materials.

Cant The angled or beveled transition between a horizontal and vertical substrate that provides for a smooth transition between this change-in-plane detail. Also refers to the application of waterproofing material, such as cementitious coatings or sealants, applied at this intersection.

Capillary admixtures Admixtures that react with the free lime and alkaline in a concrete or masonry substrate to form microscopic crystalline growth in the capillaries left by hydration. This crystalline growth fills the capillaries, resulting in a substrate impervious to further capillary action.

Capillary action A wick-like migration of water created by surface tension or molecular attraction.

Caulking Joint sealing material appropriate for interior joints that exhibit little or no movement.

Chemical cleaning A cleaning method using a variety of chemical formulations to remove a number of substrate stains including paint, rust, and pollutants.

Chemical grouts Similar in application to epoxy injection repairs; however, these materials are manufactured from hydrophobic liquid polymer resins. Chemical grouts are used for waterproofing cracks in a substrate and not for structural repairs.

Cladding A material or system, usually nonloadbearing, attached to the exterior surface of a structure that becomes part of the building envelope.

Closed-cell foams Backer-rod materials that are manufactured to prevent the movement of air, vapor, or moisture through their structure.

Cohesive strength The ability of a material's molecular structure to stay together internally during movement. Cohesive strength has a direct bearing on elongation ability.

Cold joint Another name for construction joint, typically nonmoving in nature.

Construction joint A joint formed at the intersection of two separate concrete placements.

Contaminants Foreign material found on a substrate such as oils, curing agents, dust, dirt, or rust.

Control joint A joint in building or envelope materials that allows for substrate movement.

Copings A protective, waterproof covering placed on the top of a structure such as a parapet wall.

Counterflashing Flashing that is surface-mounted or placed directly into walls with a portion exposed to flash various building elements, including roof flashings, waterproofing materials, building protrusions, and mechanical equipment, into the envelope. *See* Flashing.

Cure A process whereby, through evaporation, heat, or chemical reactions, a waterproof material attains its final performance properties.

Curing agent A separate material applied immediately after application to waterproofing materials or substrates. Curing agents enhance curing time and properties.

Dampproofing A system that is resistant to water vapor or minor amounts of moisture and that acts as a backup system to primary waterproofing materials. Dampproofing materials are not subject to weathering or water pressure.

Delamination Separation of envelope materials from the applied substrate due to movement or improperly applied materials.

Detailing joints Joints required as a component or part of complete waterproofing systems. They are used for watertightness at building details such as pipe penetrations and changes in plane before application of primary waterproofing materials.

Differential movement A phenomenon that occurs because materials have the individual characteristics of coefficients of movement that differ from surrounding envelope materials or systems. These differences will cause the materials to move at different rates during substrate movement.

Divertor system A building envelope component that repels water by redirecting it back to the exterior by the use of flashings and weeps. *See* a barrier system.

Drainage systems Prefabricated materials that facilitate the drainage of water away from the building envelope.

DOT Department of Transportation federal and state agencies that monitor and regulate the transportation and disposal of hazardous waste.

Dual-cell foams A backer rod that is comprised of both open- and closed-cell structures.

Efflorescence The staining and discoloring of masonry walls created by water-carried acids and chemicals.

Elasticity The measure of a sealant's ability to return to its original shape and size after being compressed or elongated. As with elongation, elasticity is measured as a percentage of its original length.

Elastomeric An adjective describing the ability of a waterproof material to return to its original shape and size after substrate movement during expansion or contraction.

Elongation The ability of a sealant to increase in length, then return to its original size. Limits of elongation are expressed as a percentage of original size. A material with a 200 percent elongation, for example, is capable of stretching to double its original size without splitting or tearing. Also, the increase in length of an applied waterproofing material or system during expansion of the substrate.

EPA Environmental Protection Agency, a federal agency created to enforce and monitor regulations set by Congress relating to the environment, particularly those dealing with hazardous waste materials.

Epoxy injection The injection of low-viscosity epoxy materials into substrate cracks to restore the monolithic nature of the substrate. These can be used on wood, concrete, masonry, natural stone, or metal substrates. If additional substrate movement occurs, the epoxy may crack.

Expanding-foam sealants These are composed of open-cell polyurethane foam, fully impregnated with a manufacturer's proprietary product formulation. These products include neoprene rubbers, modified asphalts, or acrylic materials.

Expansion joint A break or joint in structural elements of a building that will continue to experience movement by thermal expansion and contraction.

Exposed flashings Flashings used in a variety of methods and locations. They can be an integral part of a system, such as skylight construction, or applied to provide protection between two dissimilar materials, including cap flashings, coping flashings, gravel stops, and edge flashings.

Exterior Insulated Finish System (EIFS) A cladding and envelope system consisting of drainage, insulation board, base coat, and finish coat that resembles a stucco finish.

Face grouting *See* Bag grouting.

Flashing A material or system installed to redirect water entering through the building skin to the exterior. These are made from a variety of materials including noncorrosive metals and synthetic rubber sheet goods. Flashings are installed as backup systems for waterproofing or dampproofing systems. They are also used for waterproofing material transitions or terminations.

Floor flashing Flashing used in conjunction with shelf angles supporting brick or other facade materials. *See* Flashing.

Freeze–thaw cycle The cycle of water freezing, and subsequently thawing, related to building-related components.

Glazing The installation of glass panes to cladding components using structural sealants.

Gunite Pneumatically applied, small aggregate concrete or sand-cement mixtures, which are also referred to as *shotcrete*.

Head flashing Flashing installed above window head detail, just below adjacent facing material that the window abuts. *See* Flashing.

Hydration The process of adding water to cement, sand, and aggregate, to form a paste that cures, hardens, and shrinks to create the finished concrete or masonry product. During curing, water leaves this paste through a process called *dehydration*, which causes formation of microscopic voids and cracks in concrete. Once formed, these voids allow water absorption through the material.

Hydraulic cement Frequently referred to as "hot patch" materials, because of the heat generated during their extremely fast cure cycle. These materials are used to patch substrate cracks and small areas experiencing water leaking under hydrostatic pressure.

Hydrophilic The chemical properties that attract or absorb water.

Hydrophobic The chemical properties for repelling or diverting water away from a material or substrate.

Hydrophobic expansion systems Systems combining hydrophobic resins with synthetic rubber to produce hydrophobic expansion seals. The term *hydrophobic* refers to materials that swell in the presence of water. Thus, these materials require active water pressure to become effective water barriers.

Hydrostatic pressure Pressure applied to envelop materials by various heights of water at rest.

Infiltration The inadvertent leakage of water, moisture, vapor or air into a building envelope.

Isolation joints Joints that allow for any differential movement that will occur between two materials at junctures of these materials. For example, window frame perimeters require isolation joints when abutting other facade materials. These joints allow for differential movement at such locations as changes in structural components (e.g., spandrel beam meeting brick facing material).

Joint grouting Application of cementitious grout to all surfaces of existing mortar joints, to repair the structure and waterproof effectively.

Laitance A thin layer of unbonded cement paste on concrete substrate surfaces that must be removed before waterproofing material application.

Latex sealants Typically, acrylic emulsions or polyvinyl acetate derivatives. Latex materials have limited usage for exterior applications. They are typically used for interior applications when a fast cure time is desired for painting.

Mechanic A person trained in the proper and safe application of a particular waterproofing system.

Membrane A flexible or elastomeric waterproofing material applied to a building envelope.

Millage A unit for measuring the thickness of waterproofing products. One mil is equal to one-thousandth of an inch.

Modulus A measure of stress to strain; measured as tensile strength, expressed as a given percentage of elongation in pounds per square inch (lb/in^2). Modulus has a direct effect on elongation or movement capability.

Moisture content movement Movement caused by certain materials, particularly masonry substrates, swelling when subjected to wetting and subsequent drying. When this movement is calculated as an aggregate total of the entire facade area, it can be considerable.

Negative waterproofing systems Below-grade waterproofing systems applied to the interior or negative side of a structure, away from direct exposure to groundwater.

Organic A product comprised or manufactured of hydrocarbons, plants, animal, or their derivatives.

OSHA Occupational Safety and Health Administration, a federal agency that enforces workplace safety laws and regulations created by Congress.

Outgassing The deterioration of a closed-cell backer material that causes the sealant material to improperly cure and often leads to failures.

Overlays Cementitious materials used for restoring deteriorated horizontal concrete substrates.

Parapet flashing Flashing installed at the base of a parapet, usually at ceiling level. It is also used on the roof side of parapets as part of roof or counterflashing. *See* Flashing.

Parging The application of a cementitious material to a masonry or concrete substrate.

Perm The rate of water vapor transmission through a substrate or material of one grain per hour per square foot per inch of mercury pressure difference.

Permeability The ability of a waterproofing material or substrate to allow the passage of water vapor through itself without blistering.

pH The chemical measurement of a substrate's alkalinity or acidity.

Polymer concrete A modified concrete mixture formulated by adding natural and synthetic chemical compounds known as polymers. Although the priority chemical compounds (polymers) vary, the purpose of these admixtures is the same. They provide a dense, high-strength, low-shrinkage, and chemically and water-resistant concrete substrate.

Polysulfides Materials produced from synthetic polymers of polysulfide rubbers. Polysulfides make excellent performing sealants for most joint uses.

Polyurethane Any of various polymers that are produced by chemical reactions formed by mixing di-isocynate with a hydroxyl, and are used in making flexible and rigid foams, elastomers, and resins. Many polyurethanes are moisture-cured materials reacting to moisture in atmospheric conditions to promote curing. Other polyurethanes are chemically curing mixtures.

Positive waterproofing systems Waterproofing systems applied to substrate side with direct exposure to water or a hydrostatic head of water.

Pot life The length of time a waterproof material or system is workable or applicable after having been activated.

Poultice An absorbent material such as talc or fuller's earth that is applied to the envelope substrate to remove dirt and staining. The poultice absorbs the staining and dirt into itself, and then is removed by water-pressure cleaning.

Primer A separate material, usually in liquid form, applied to a substrate before actual waterproofing material application. Primers enhance adhesion properties of the waterproofing system.

Protected membrane A membrane applied between a structural slab and topping slab or other top layer protection such as tile. The topping slab or protection slab protects the membrane from weathering and traffic wear. This is also referred to as a *sandwich membrane*.

Protection layer A layer of rigid or semi-rigid boards or sheets installed directly over a waterproofing system application to protect the waterproofing material from damage during back-fill or concrete placement operations.

Reglet A formed or sawn groove in substrate (usually concrete) providing a transition point for two adjoining waterproofing systems.

Remedial flashings Flashings typically surface-mounted and applied directly to exposed substrate faces. These can include a surface-mounted reglet for attachment. They do not provide for redirecting entering water. Only by dismantling a wall or portion thereof can remedial through-wall flashings be installed.

Roofing That portion of a building that prevents water intrusion in horizontal or slightly inclined elevations. Although typically applied to the surface and exposed to the elements, roofing may also be internal, or "sandwiched" between other building components.

Sandwich membrane *See* Protective membrane.

Sealant A material applicable to exterior building envelope joints. Sealants are capable of withstanding continuous joint movement during weathering conditions without failing.

Shear movement Lateral movement in a substrate.

Sheet expansion systems Systems manufactured from neoprene or hypalon rubber sheets. Joint expansion and contraction is made watertight by installing these materials in a bellows or loop fashion.

Shelf angle Steel angle extrusion used over envelope openings to support masonry and precast and other cladding materials.

Shelf life The maximum time that packaged and unopened waterproofing materials can remain usable.

Shop drawings A drawing or written specifications prepared by the manufacturer or subcontractor that details the installation requirements for a particular building component. Shop drawings are submitted for review by the architect or engineer and then used to monitor quality of the finished product.

Shore hardness A measure of resistance to impact using a durometer gage. This property becomes important in choosing sealants subject to punctures or traffic, such as horizontal paver joints. A shore hardness of 25 is similar to a soft eraser; a hardness of 90 is equivalent to a rubber mallet.

Shotcrete *See* Gunite.

Silanes Water repellents that contain the smallest molecular structures of all silicone-based materials. The small molecular structure of the silanes allows the deepest penetration into substrates. Silanes must have silica present in substrates for the chemical action providing water repellency to take place. These materials therefore are inappropriate for substrates such as wood, metal, and limestone.

Silicone rubber A water repellent manufactured from silicone elastomers that has a millage usually greater than other silicone-based sealers.

Silicone water repellents Water repellents manufactured by mixing silicone solids (resins) into a solvent carrier. Most manufacturers base their formulations on a 5 percent solids mixture, in conformance with the requirements of federal specification SS-W-110C.

Silicones Silicone sealants are derivatives of silicone polymers produced by combining silicon, oxygen, and organic materials. Silicones have extremely high thermal stability and are used as abrasives, lubricants, paints, coatings, and synthetic rubbers. *See* Sealant.

Sill flashing Flashing installed beneath window or curtain wall sills. *See* Flashing.

Siloxanes Silicone masonry water repellents produced from the CL-silane material. Siloxanes are manufactured in two types—oligomerous (meaning short chain of molecular structure) alkylalkoxysiloxanes and polymeric (long chain of molecular structure) alkylalkoxysiloxanes. Most siloxanes produced now are oligomerous, due to a tendency for polymeric products to remain wet or tacky on the surface, attracting dirt and pollutants.

Sodium silicates Materials that react with the free salts in concrete, such as calcium or free lime, making the concrete surface more dense. Usually these materials are sold as floor hardeners.

Spall/spalling The crumbling or breaking off in small pieces of a concrete or masonry substrate, usually owing to freeze–thaw cycles or deterioration (rusting) of embedded structural elements such as reinforcing steel.

Structural movement Substrate movement caused by the curing process in concrete during settlement and/or the structural loading of a building.

Substrate Structure or envelope components to which waterproofing materials or systems are applied.

T-joint system A sealant system reinforced with metal or plastic plates and polymer concrete nosing on each side of the sealant. This system derives its name from a cross section of the joint, which is in the shape of a T.

Tackiness Stickiness of a waterproofing material's exposed surface after installation or during its final curing stage.

Tensile strength The ability of a waterproofing material to resist being pulled or stretched apart to a point of failure.

Thermal movement Movement, either expansion or contraction, caused by temperature changes.

Thiokol Trademark of the first commercial synthetic elastomeric produced by Thiokol Chemical Company.

Tooling The means of finishing mortar or sealants that have been applied into envelope joints.

Tuck-pointing The repair of existing mortar joints in masonry walls by removing the old material and filling with new mortar.

Ultraviolet A form of light energy positioned in the spectrum of sunlight beyond violet, the limit of visible light, that is an especially detrimental weathering condition to above-grade waterproofing materials and envelope components.

Viscosity The chemical property of liquid waterproofing materials to resist a change in shape.

Watercleaning A building cleaning method incorporating water by pressure, soaking, or steam to remove dirt and pollutants from a substrate.

Water penetration/infiltration The circumstance in which water accesses the interior areas of a building envelope.

Waterstops Preformed materials placed between construction joints in separate concrete placements to prevent passage of water between the joints that forms at this intersection.

Weep hole Opening in a masonry wall or sealant that allows the escape of water entering envelopes to the exterior. Exiting water is collected by a combination of dampproofing and flashing materials.

INDEX